# Orthogonal Series

TRANSLATIONS OF MATHEMATICAL MONOGRAPHS

VOLUME **75**

# Orthogonal Series

## B. S. KASHIN
## A. A. SAAKYAN

American Mathematical Society · Providence · Rhode Island

БОРИС СЕРГЕЕВИЧ КАШИН, АРТУР АРТУШОВИЧ СААКЯН

# ОРТОГОНАЛЬНЫЕ РЯДЫ

«НАУКА», МОСКВА, 1984

Translated from the Russian by Ralph P. Boas
Translation edited by Ben Silver

1980 *Mathematics Subject Classification* (1985 *Revision*). Primary 42-02, 42C15, 42C20, 42C30, 30B50, 30D55; Secondary 42A50, 42A55, 44A15, 42A61.

ABSTRACT. In this book we present the fundamental methods of the theory of orthogonal series. We study general orthonormal systems as well as specific systems (for example, the Haar and Franklin systems). We present both classical results and those obtained more recently. All propositions that fall outside the scope of university courses are provided with full proofs.

The book is intended for specialists in mathematical analysis, graduate students, and students in advanced courses in related fields.

Bibliography: 202 titles.

**Library of Congress Cataloging-in-Publication Data**

Kashin, B. S.
    [Ortogonal'nye riady. English]
    Ortogonal series/by B. S. Kashin, A. A. Saakyan; [translated from the Russian by Ralph P. Boas; translation edited by Ben Silver].
    p.   cm. – (Translations of mathematical monographs; v. 75)
    Translation of: Ortogonal'nye riady.
    Bibliography: p.
    Includes index.
    ISBN 0-8218-4527-6
    1. Series, Orthogonal.   I. Saakian, A. A. (Artur Artushovich)   II. Title.   III. Series.
QA404.5.K3413   1989                                                            89-333
515'.2433–dc19                                                                      CIP

Information on Copying and Reprinting can be found at the back of this volume.
The paper used in this book is acid-free and falls within the guidelines
established to ensure permanence and durability. ♾

This publication was typeset using $\mathcal{A}_{\mathcal{M}}\mathcal{S}$-TEX,
the American Mathematical Society's TEX macro system.

# Contents

# Preface

This book is concerned with the theory of general orthogonal series. The theory originated at the beginning of the present century as a natural generalization, based on Lebesgue integration, of the theory of trigonometric series, but has been developed most actively in the past twenty-five years. By now it has become clear that:

many propositions about properties of the trigonometric system are general in nature and remain valid for a broad class of orthonormal systems:

the study of systems of functions more general than orthonormal systems frequently reduces to the study of the latter:

in many problems, nonclassical orthonormal systems turn out to be "better" than classical systems:

the results and methods of the theory of general orthogonal systems have a variety of applications outside this theory.

These observations, which speak to the importance of the systematic study of the properties of a variety of orthonormal systems, are confirmed to a substantial extent by the contents of this book. We note in this connection that the book does not claim to be a complete treatment of the subject, and does not even touch on some important topics. We have also not attempted to present results in their most general form. Our principal aim is to present the main ideas and methods of the theory of orthogonal series. Our choice of material has been much influenced by Men'shov and Ul'yanov's seminar on the theory of real functions, which has long been active at Moscow University as a "continuous extension" of Luzin's twenty-five year seminar. We have also used the lectures on the theory of orthogonal series that the first author gave at Moscow University during 1979–1981. A significant number of the theorems proved in the book have not appeared in the monographic literature, and we hope that even specialists will find something new here. Nevertheless, we are largely oriented toward beginning mathematicians, and have therefore adhered to the rule of proving all propositions that fall outside the scope of university courses. We assume that the reader is familiar with the contents of Kolmogorov and Fomin's *Elements of the Theory of Functions and Functional Analysis*, and also with the basic theory of functions of a complex variable (for example, with the content of [107]). The essential additional material from the theory of functions and functional analysis is presented in two appendices. Information on the source of the results presented in the book is given in the Notes, which also contain commentaries and the proofs of some additional facts.

In addition, we thank our colleagues K. I. Oskolkov, A. A. Talalyan, K. Tandori, Z. Ciesielski, and P. Oswald for advice and assistance.

*B. S. Kashin*
*A. A. Saakyan*

# Summary of Notation

We use, without explanation, a number of standard notations (used, in particular, in [89]), for example $R^n$, $L^p(0, 1)$, $l^p$, and $C(0, 1)$. In addition:

$L^0(0, 1)$ and $L^0(R^1)$ are the spaces of functions that are measurable and finite almost everywhere on $[0, 1]$ or $R^1$.

$C(-\pi, \pi)$ and $L^p(-\pi, \pi)$ are the spaces of functions of period $2\pi$ on $R^1$, continuous or of summable $p$th power on $[-\pi, \pi]$, respectively.

$\mathscr{D}_N$, $N = 1, 2, \ldots$, is the $N$-dimensional space of piecewise constant functions defined on $[0, 1]$:

$$\mathscr{D}_N = \left\{ f(x) : f(x) = \text{const} = c_i \quad \text{if } x \in \left( \frac{i-1}{N}, \frac{i}{N} \right), \quad i = 1, \ldots, N; \right.$$
$$\left. f\left( \frac{i}{N} \right) = \frac{c_i + c_{i+1}}{2}, \quad \text{if } 1 \le i < N, \ f(0) = c_1, \ f(1) = c_N \right\}.$$

Similarly, for any interval $[a, b]$

$$\mathscr{D}_N(a, b) = \{ f(x), x \in [a, b] : g(t) = f(a + t(b - a)) \in \mathscr{D}_N \}.$$

The norm of the matrix $H = \{h_{ij}\}$, $i = 1, \ldots, m$, $j = 1, \ldots, n$ is the number

$$\|H\| = \sup_{\{x_i\}, \{y_j\} : \ \sum_1^m x_i^2 = \sum_1^n y_j^2 = 1} \sum_{i=1}^m \sum_{j=1}^n h_{ij} x_i y_j.$$

$m(E)$ (or $|E|$ if $E$ is an interval $(a, b)$) is the Lebesgue measure of the set $E \subset R^1$.

$\text{card } E$ is the number of elements of the finite set $E$.

If $x$ is an element of the Banach space $X$ and $y$ is a bounded linear functional on $X$, then $\langle x, y \rangle$ or $\langle y, x \rangle$ denotes the functional $y$ at the element $x$. When $X$ is a Hilbert space, $(x, y)$ is sometimes used instead of $\langle x, y \rangle$.

If $f(x)$ is a real or complex function, $\text{supp } f(x) = \{ x : f(x) \ne 0 \}$.

If $\{a_n\}_{n=1}^{\infty}$ and $\{b_n\}_{n=1}^{\infty}$ are two sequences of positive numbers, the notation $a_n \asymp b_n$ means that there are constants $C_1$, $C_2 > 0$ such that $C_1 \leq a_n/b_n \leq C_2$, $n = 1, 2, \ldots$.

We use the following abbreviations:

An O.N.S. is an orthonormal system—recall that a system of functions $\{\varphi_n(x)\}_{n=1}^{\infty} \subset L^2(a, b)$ is *orthonormal* if

$$\int_a^b \varphi_n(x)\varphi_m(x)\, dx = \begin{cases} 1 & \text{if } n = m, \\ 0 & \text{if } n \neq m, \end{cases} \quad n, m = 1, 2, \ldots;$$

A C.O.N.S. is a complete orthonormal system;

a.e.="almost everywhere."

A phrase of the form "see Theorem 5.1" or "see §3.2" is a reference to Theorem 1 of Chapter 5 or to §2 of Chapter 3. The chapter number is omitted for references within a chapter.

*Translator's note.* Definitions, theorems, corollaries, and propositions are numbered consecutively (independently of each other) in each chapter. Lemmas are numbered consecutively for each theorem.

# CHAPTER I

# Introductory Concepts and Some General Results

In this chapter we introduce concepts that will be used throughout the book. We also present a number of propositions of a general nature, whose proofs are given in Appendix 1.

## §1. Types of convergence

The majority of the theorems established in the book are propositions on the convergence or divergence of series. Hence we begin by formulating the kinds of convergence that will be used.

I. Convergence of series with respect to a norm [in particular, convergence of series of functions in the spaces $L^p(0, 1)$, $1 \leq p \leq \infty$, and in $C(0, 1)$].

II. Convergence in measure.

If we introduce the metric

$$\rho(f, g) = \int_0^1 \frac{|f(x) - g(x)|}{1 + |f(x) - g(x)|} \, dx \tag{1}$$

in the space $L^0(0, 1)$, then $L^0(0, 1)$ becomes a complete linear metric space. Convergence of a sequence $f_n(x)$, $n = 1, 2, \ldots$ [$f_n \in L^0(0, 1)$] to $f(x)$ in measure is equivalent to convergence in the metric (1), i.e., to the condition $\lim_{n \to \infty} \rho(f_n, f) = 0$.

III. Convergence almost everywhere (a.e.).

DEFINITION 1. If the series([1])

$$\sum_{n=1}^{\infty} f_n(x), \qquad x \in (0, 1), \tag{2}$$

---

([1]) Here and later, we always suppose that the terms of the series of functions under consideration, or of sequences, are measurable functions, finite almost everywhere.

converges a.e. to a function that is finite a.e., then the *majorant of the partial sums* of the series is the (a.e. finite, measurable) function

$$S^*(x) = \sup_{1 \le N < \infty} \left| \sum_{n=1}^{N} f_n(x) \right|. \tag{3}$$

DEFINITION 2. An O.N.S. $\Phi = \{\varphi_n(x)\}_{n=1}^{\infty}$, $x \in (0,1)$, is called a *convergence system* if every series of the form

$$\sum_{n=1}^{\infty} a_n \varphi_n(x), \qquad \sum_{n=1}^{\infty} a_n^2 < \infty, \tag{4}$$

converges a.e. on $(0,1)$.

Each convergence system generates, in a natural way, an operator, the majorant $S_{\Phi}^*$ of the partial sums, which acts from $l^2$ to $L^0(0,1)$.

In fact, for $\{a_n\} \in l^2$,

$$S_{\Phi}^*(\{a_n\}) = \sup_{1 \le N < \infty} \left| \sum_{n=1}^{N} a_n \varphi_n(x) \right|. \tag{5}$$

Most problems on the a.e. convergence of orthogonal series reduce to the study of properties of the operator $S_{\Phi}^*$ (for an arbitrary system $\Phi$ it is defined on a set of sequences $\{a_n\}$ (each with a finite number of nonzero terms), which is dense in $l^2$). Here we note the following proposition.

PROPOSITION 1. *In order for an O.N.S.* $\Phi = \{\varphi_n(x)\}_{n=1}^{\infty}$, $x \in (0,1)$, *to be a convergence system*

a) *it is necessary and sufficient that, for every sequence* $\{a_n\}_{n=1}^{\infty} \in l^2$, *the function*

$$\sup_{1 \le N < \infty} \left| \sum_{n=1}^{N} a_n \varphi_n(x) \right|$$

*is finite for almost all* $x \in (0,1)$; *and*

b) *it is sufficient that, for some* $p > 0$ *and some* $C_0 < \infty$, *and for every finite sequence* $\{a_n\}_{n=1}^{M}$, *the inequality*

$$\int_0^1 [S_{\Phi}^*(\{a_n\})]^p \, dx \le C_0 \left( \sum_{n=1}^{M} a_n^2 \right)^{p/2} \tag{6}$$

*is satisfied, where* $S_{\Phi}^*(\{a_n\})$ *is defined in* (5).

PROOF. a) We need to prove only the sufficiency. Suppose, on the contrary, that there is a series of the form (4) which diverges on a set of positive measure. This means that, for some $\varepsilon > 0$ and $\delta > 0$, there are sequences of integers $M_k$ and measurable integral-valued functions $N_k(x)$, with $M_k < N_k(x) < M_{k+1}$, such that for $k = 1, 2, \ldots$ the sets $G_k = \{x \in (0, 1) : |\sum_{n=M_k}^{N_k(x)} a_n \varphi_n(x)| > \varepsilon\}$ satisfy the inequality $m(G_k) > \delta$.

Define a sequence of numbers $\lambda_n$, $n = 1, 2, \ldots$, by setting $\lambda_n = \text{const} = C_k$ for $M_k \le n < M_{k+1}$, where the numbers $C_k \nearrow \infty$ increase so slowly that the sum $\sum_1^\infty a_n^2 \lambda_n^2$ is finite.

Then for the series $\sum_1^\infty a_n \lambda_n \varphi_n(x)$ the sum $\sum_1^\infty a_n^2 \lambda_n^2 < \infty$, and at the same time

$$\sup_{1 \le N < \infty} \left| \sum_{n=1}^N a_n \lambda_n \varphi_n(x) \right| \ge \frac{1}{2} \sup_{1 \le k < \infty} \left| \sum_{n=M_k}^{N_k(x)} a_n \lambda_n \varphi_n(x) \right| \ge \varepsilon \lim_{k \to \infty} C_k = \infty$$

for

$$x \in G \equiv \varlimsup_{k \to \infty} G_k \equiv \bigcap_{m=1}^\infty \bigcup_{k=m}^\infty G_k \qquad [m(G) \ge \delta],$$

which contradicts the assumption that every function of the form (5) is finite a.e.

b) Suppose, on the contrary, that (6) is satisfied but that for some sequence $\{a_n\}$, $\sum_1^\infty a_n^2 = 1$, we have

$$\sup_{1 \le N < \infty} \left| \sum_{n=1}^N a_n \varphi_n(x) \right| = \infty, \qquad x \in G, \ m(G) > 0.$$

It is then clear that for every constant $K$ we can find a function $N(x)$ such that $\sup_{x \in (0,1)} N(x) = M < \infty$ and

$$m \left\{ x \in (0, 1) : \left| \sum_{n=1}^{N(x)} a_n \varphi_n(x) \right| > K \right\} > \frac{m(G)}{2}.$$

But then, by the choice of $\{a_n\}_{n=1}^M$,

$$\int_0^1 [S_\Phi^*(\{a_n\})]^p \, dx \ge \frac{1}{2} K^p m(G) > C_0,$$

if the constant $K$ is sufficiently large. This inequality contradicts (6).

IV. Unconditional convergence in a linear metric space (in particular, in a Banach space).

DEFINITION 3. A series

$$\sum_{n=1}^\infty x_n \tag{7}$$

of elements of a linear metric space $X$ is said to be *unconditionally convergent* if, for every permutation $\sigma = \{\sigma(n)\}_{n=1}^\infty$ of the positive integers, the series

$$\sum_{n=1}^{\infty} x_{\sigma(n)}. \tag{7'}$$

converges in $X$.

It is easily seen that for any unconditionally convergent series the sum (7') is independent of $\sigma$.

THEOREM 1. *A necessary and sufficient condition for the unconditional convergence of a series* (7) *of elements of a complete linear metric space $X$ is that the series*

$$\sum_{n=1}^{\infty} \varepsilon_n x_n \tag{8}$$

*converges in $X$ for every set of numbers $\{\varepsilon_n\}$, $\varepsilon_n = \pm 1$, $n = 1, 2, \ldots$.*

PROOF. It is clear that the convergence of all the series (8) is equivalent to the convergence of all series

$$\sum_{n=1}^{\infty} \varepsilon_n' x_n, \qquad \varepsilon_n' = 0; 1, \ n = 1, 2, \ldots . \tag{8'}$$

Suppose that (7) diverges after a rearrangement $\sigma = \{\sigma(n)\}_{n=1}^{\infty}$ of its terms. This means that there are a number $\delta > 0$ and sequences $\{N_k\}_{k=1}^{\infty}$ and $\{M_k\}_{k=1}^{\infty}$, with $N_k \leq M_k < N_{k+1}$, such that

a) $\rho(\sum_{n=N_k}^{M_k} x_{\sigma(n)}, 0) \geq \delta$, $k = 1, 2, \ldots$ (here $\rho(x, 0)$ is the distance of $x$ from 0); and

b) if $A_k = \max\{\sigma(n): N_K \leq n \leq M_k\}$ and

$$B_k = \min\{\sigma(n): N_k \leq n \leq M_k\},$$

then $B_k < A_k < B_{k+1}$, $k = 1, 2, \ldots$.

Let $\varepsilon_n' = 1$ if $n \in \bigcup_{k=1}^{\infty} \{\sigma(m): N_k \leq m \leq M_k\}$, and $\varepsilon_n' = 0$ otherwise. Then it follows from a) and b) that, for $k = 1, 2, \ldots$,

$$\rho\left(\sum_{n=B_k}^{A_k} \varepsilon_n' x_n, 0\right) = \rho\left(\sum_{n=N_k}^{M_k} x_{\sigma(n)}, 0\right) \geq \delta,$$

i.e., there is a series of the form (8') that diverges in $X$.

Now suppose that (7) converges unconditionally. We shall prove that all series of the form (8'), and consequently those of the form (8), converge. In fact, if some series (8) diverged, we could find sequences $\{N_k\}_{k=1}^{\infty}$ and $\{M_k\}_{k=1}^{\infty}$ with $N_k \leq M_k < N_{k+1}$ such that

$$\rho\left(\sum_{n=N_k}^{M_k} \varepsilon_n' x_n, 0\right) \geq \delta > 0, \qquad k = 1, 2, \ldots .$$

Conversely, if $\{x_n\}$ is minimal, then since $x_k \notin X_k$ and $X_k$ is a closed subspace of $X$ ($k = 1, 2, \ldots$), we can find, by the Hahn-Banach theorem, a functional $y_k \in X^*$ that satisfies (11).

It is clear that if $X = L^p(0, 1)$ and $X^* = L^q(0, 1)$, $1 \le p < \infty$, $1/p + 1/q = 1$, then every O.N.S. $\{\varphi_n(x)\}_{n=1}^{\infty}$, where $\varphi_n \in L^p(0, 1) \cap L^q(0, 1)$ ($n = 1, 2, \ldots$), satisfies (11) with $x_n = \varphi_n(x)$ and $y_n = \varphi_n(x)$, $n = 1, 2 \ldots$.

Consequently the following proposition is a consequence of Theorem 2.

PROPOSITION 4. *Every O.N.S.* $\{\varphi_n(x)\}$, $x \in (0, 1)$, *with* $\varphi_n \in L^p(0, 1) \cap L^q(0, 1)$, $n = 1, 2, \ldots$, $1 \le p < \infty$, $1/p + 1/q = 1$, *is minimal in* $L^p(0, 1)$ *and is its own dual system.*

## §3. Fourier coefficients and partial sums

Let $\Phi = \{\varphi_n(x)\}_{n=1}^{\infty} \subset L^2(0, 1)$ be an O.N.S; let $f(x) \in L^1(0, 1)$ have the property that $f(x)\varphi_n(x) \in L^1(0, 1)$ for $n = 1, 2, \ldots$.

DEFINITION 10. The numbers

$$c_n(f) = c_n(f, \Phi) = \int_0^1 f(x)\varphi_n(x)\,dx, \qquad n = 1, 2 \ldots, \qquad (12)$$

are the *Fourier coefficients of* $f(x)$ *with respect to the system* $\Phi$, and the series

$$\sum_{n=1}^{\infty} c_n(f)\varphi_n(x) \qquad (13)$$

is the *Fourier series of* $f(x)$ *with respect to* $\Phi$.

It follows directly from Definition 10 that the $N$th partial sums of (13) satisfy the equation

$$S_N(f, t) \equiv \sum_{n=1}^{N} c_n(f)\varphi_n(t) = \int_0^1 f(x)K_N(x, t)\,dx, \qquad (14)$$

where $t \in (0, 1)$, and

$$K_N(x, t) = \sum_{n=1}^{N} \varphi_n(x)\varphi_n(t), \qquad x, t \in (0, 1), \ N = 1, 2, \ldots. \qquad (15)$$

DEFINITION 11. The function $K_N(x, t)$ defined in (15) is the *kernel of order* $N$ of the system $\Phi$, and the function

$$L_N(t) = \int_0^1 |K_N(x, t)|\,dx, \qquad t \in [0, 1], \ N = 1, 2, \ldots, \qquad (16)$$

is the $N$th *Lebesgue function of* $\Phi$.

The uniform boundedness of the Lebesgue functions $L_N(t)$, $N = 1, 2, \ldots$, is closely connected with the convergence of Fourier series in

the $L^1$ and $L^\infty$ metrics. More precisely, if we use the expressions for the norms of operators of the form (14) (see Proposition 1, §1), we find that
a) for every $t \in [0, 1]$

$$\sup_{f:\ \|f\|_C \leq 1} |S_N(f, t)| = L_N(t); \qquad (17)$$

b)

$$\sup_{f:\ \|f\|_1 \leq 1} \|S_N(f)\|_1 = \|L_N(t)\|_\infty. \qquad (18)$$

Let us also recall the fundamental facts about the Fourier series of functions in $L^2(0, 1)$. Let $f(x) \in L^2(0, 1)$, and let $S_N(x)$ be a polynomial in the system $\Phi$:

$$S_N(x) = \sum_{n=1}^{N} a_n \varphi_n(x). \qquad (19)$$

It follows easily from the orthonormality of $\Phi$ that

$$\|f - S_N\|_2^2 = \|f\|_2^2 - \sum_{n=1}^{N} c_n^2(f) + \sum_{n=1}^{N} (c_n(f) - a_n)^2, \qquad (20)$$

where, as before, $c_n(f)$ is the $n$th Fourier coefficient of $f(x)$.

The following proposition is an immediate consequence of (20).

PROPOSITION 5. a) *Among all sums* (19), *the one deviating least from* $f(x)$ *(in the* $L^2(0, 1)$ *norm) is the partial sum* (14) *of the Fourier series of* $f(x)$ *with respect to* $\Phi$; *here*

$$\|f - S_N(f)\|_2^2 = \|f\|_2^2 - \sum_{n=1}^{N} c_n^2(f). \qquad (21)$$

b) (*Bessel's inequality*). *For every function* $f \in L^2(0, 1)$ *we have*

$$\sum_{n=1}^{\infty} c_n^2(f) \leq \|f\|_2^2, \qquad (22)$$

We have the following theorems.

THEOREM 3 (RIESZ–FISCHER). *Let* $\Phi = \{\varphi_n(x)\}_{n=1}^{\infty}$ *be an O.N.S. and let* $\{c_n\}_{n=1}^{\infty} \in l^2$. *Then there is a function* $f \in L^2(0, 1)$ *such that* $c_n = c_n(f)$, $n = 1, 2, \ldots$, *and*

$$\sum_{n=1}^{\infty} c_n^2(f) = \|f\|_2^2. \qquad (23)$$

THEOREM 4. *Let* $\Phi = \{\varphi_n(x)\}_{n=1}^{\infty}$ *be an O.N.S. The following statements are equivalent*:

1) $\Phi$ *is complete in* $L^2(0, 1)$.

2) *The Fourier series, with respect to* $\Phi$, *of every function* $f \in L^2(0, 1)$ *converges to* $f$ *in the norm of* $L^2(0, 1)$.

3) *Equation* (23) (*Parseval's equation*) *holds for every* $f \in L^2(0, 1)$.

It follows immediately from Theorem 4 that if $\Phi$ is an O.N.S. which is complete in $L^2(0, 1)$, and $f$ and $g$ belong to $L^2(0, 1)$, then

$$\sum_{n=1}^{\infty} c_n(f)c_n(g) = \int_0^1 f(x)g(x)\,dx. \tag{23'}$$

It follows from Bessel's inequality (22) that the Fourier coefficients of a function in $L^2(0, 1)$ tend to zero as $n \to \infty$.

As for the Fourier coefficients of functions in $L^p(0, 1)$ with $1 \le p < 2$, it is easy to construct an example of a function $f \in L^p(0, 1)$, $1 \le p < 2$, and an O.N.S. $\Phi = \{\varphi_n(x)\}_{n=1}^{\infty}$, $x \in (0, 1)$, with $\varphi_n \in L^{\infty}(0, 1)$ for $n = 1, 2, \ldots$, such that $\lim_{n\to\infty} c_n(f) = +\infty$ (see, for example, §3.2). However, we have the following theorem.

THEOREM 5. *Let* $\Phi = \{\varphi_n(x)\}_{n=1}^{\infty}$ *be an O.N.S. of uniformly bounded functions*,

$$\|\varphi_n\|_{\infty} \le M < \infty, \qquad n = 1, 2, \ldots . \tag{24}$$

*Then* $\lim_{n\to\infty} c_n(f, \Phi) = 0$ *for every* $f \in L^1(0, 1)$.

PROOF Choose $\varepsilon > 0$ and find (using the completeness of $L^2(0, 1)$ in $L^1(0, 1)$), corresponding to a given $f(x)$, a function $g(x) \in L^2(0, 1)$ with $\|f - g\|_1 < M^{-1}\varepsilon$. Then

$$\varlimsup_{n\to\infty} |c_n(f, \Phi)| \le \varlimsup_{n\to\infty} |c_n(g, \Phi)| + \varlimsup_{n\to\infty} |c_n(f - g, \Phi)|. \tag{25}$$

Since $|c_n(f - g, \Phi)| \le \|f - g\|_1 \cdot \|\varphi_n\|_{\infty} \le M^{-1}\varepsilon M = \varepsilon$, it follows from (25) that

$$\varlimsup_{n\to\infty} |c_n(f, \Phi)| \le \varepsilon.$$

Since $\varepsilon$ is arbitrary, the conclusion of the theorem follows.

## §4. The basis property

DEFINITION 12. A system $\{x_n\}_{n=1}^{\infty}$ of elements of a Banach space $X$ is called a *basis* if for every element $x \in X$ there is a unique series

$$x \sim \sum_{n=1}^{\infty} a_n x_n, \qquad a_n = a_n(x) \in R^1, \ n = 1, 2, \ldots, \tag{26}$$

which converges to $x$ in the norm of $X$.

The following general result provides a criterion for a system to be a basis (for a proof see Appendix 1, §3).

THEOREM 6. *A necessary and sufficient condition for* $\{x_n\}_{n=1}^{\infty}$ *to be a basis of the Banach space* $X$ *is that the following three conditions are satisfied:*

a) $\{x_n\}$ *is complete in* $X$.

b) $\{x_n\}$ *is minimal.*

c) *There is a constant* $M > 0$ *such that, for all* $x \in X$,

$$\left\| \sum_{n=1}^{N} \langle x, y_n \rangle x_n \right\| \leq M \|x\|, \qquad N = 1, 2, \ldots, \tag{27}$$

*where* $\{y_n\}$ *is dual to* $\{x_n\}$ *(see Theorem 2).*

We have the following corollary of Theorem 6.

COROLLARY 1. *If* $\{x_n\}$ *is a basis for* $X$, *the functionals* $a_n(x)$ *[see (26)] are bounded linear functionals, defined by the equations* $a_n(x) = \langle x, y_n \rangle$, $x \in X$, $n = 1, 2, \ldots$, *where* $\{y_n\}$ *is conjugate to* $\{x_n\}$.

PROOF. Since $\sum_{m=1}^{\infty} a_m(x) x_m$ converges to $x$ in the $X$ norm, and the $y_n$, $n = 1, 2, \ldots$, are bounded linear functionals, we have, for $n = 1, 2, \ldots$,

$$\langle x, y_n \rangle = \lim_{N \to \infty} \left\langle \sum_{m=1}^{N} a_m(x) x_m, y_n \right\rangle = a_n(x),$$

as was to be proved.

THEOREM 7. *Let* $\{x_n\}_{n=1}^{\infty}$ *be a basis of the Banach space* $X$, *and* $\{y_n\}_{n=1}^{\infty}$ *the system dual to* $\{x_n\}$. *Then* $\{y_n\}$ *is a basis of* $Y$, *the closure of the linear span of* $\{y_n\}$ *(with the norm of the dual space* $X^*$ *of* $X$).

PROOF. The system $\{y_n\}$ is minimal (by the biorthogonality of $\{x_n\}$ and $\{y_n\}$), and complete in $Y$. Hence it is enough to prove (see Theorem 6) that, for each $y \in Y$,

$$\left\| \sum_{n=1}^{N} \langle y, x_n \rangle y_n \right\|_{X^*} \leq M \|y\|_{X^*}, \qquad N = 1, 2, \ldots \tag{28}$$

(here $x_n$ is considered as an element of $(X^*)^*$).

Let $y \in Y$ and a number $N$ be given. Then, by Theorem 6, for every $x \in X$ with $\|x\|_X = 1$,

$$\left| \left\langle x, \left( \sum_{n=1}^{N} \langle y, x_n \rangle y_n \right) \right\rangle \right| = \left| \sum_{n=1}^{N} \langle x, y_n \rangle \cdot \langle y, x_n \rangle \right|$$

$$= \left| \left\langle y, \left( \sum_{n=1}^{N} \langle x, y_n \rangle x_n \right) \right\rangle \right|$$

$$\leq \|y\|_{X^*} \cdot \left\| \sum_{n=1}^{N} \langle x, y_n \rangle x_n \right\| \leq M \cdot \|y\|_{X^*},$$

from which (28) immediately follows. This completes the proof of Theorem 7.

COROLLARY 2. *If the system $\{\varphi_n(x)\}$ is a basis in $L^p(0,1)$, $1 < p < \infty$, the dual system $\{\psi_n(x)\}$ is a basis in $L^q(0,1)$, where $1/p + 1/q = 1$.*

PROOF. By Theorem 7 it is sufficient to show that $\{\psi_n\}$ is complete in $L^q(0,1)$. But if this were not so, by Proposition 2 we could find a function $f \in L^p(0,1)$, $\|f\|_p > 0$, for which $\int_0^1 f(x)\psi_n(x)\,dx = 0$ for $n = 1, 2, \dots$. This contradicts the basis property of $\{\varphi_n\}$, since

$$f \overset{L^p}{=} \sum_{n=1}^{\infty} \left( \int_0^1 f(t)\varphi_n(t)\,dt \right) \varphi_n$$

by Corollary 1.

COROLLARY 3. *If the O.N.S. $\Phi = \{\varphi_n(x)\}$ is a basis in $L^p(0,1)$, and also*

$$\varphi_n \in L^p(0,1) \cap L^q(0,1), \qquad 1/p + 1/q = 1, \ 1 < p < \infty, \qquad (29)$$

*then $\Phi$ is a basis in $L^{p'}(0,1)$ for every $p' \in \Delta = [\min(p,q), \max(p,q)]$.*

PROOF. Since $\Phi$ is dual to itself (see Proposition 4), $\Phi$ is a basis in $L^q(0,1)$ by Corollary 2. Therefore $\Phi$ is complete and minimal in $L^p(0,1)$ and in $L^q(0,1)$, and therefore complete and minimal in $L^{p'}(0,1)$ for $p' \in \Delta$. Moreover, the partial-sum operator $S_N(f)$ ($N = 1, 2, \dots$) has bounded norm as an operator from $L^p(0,1)$ to $L^q(0,1)$ and from $L^q(0,1)$ to $L^p(0,1)$:

$$\|S_N(f)\|_p \leq M_1 \|f\|_p, \quad \|S_N(f)\|_q \leq M_2 \|f\|_q, \qquad N = 1, 2, \dots.$$

Hence by the interpolation theorem (see Theorem 1 in Appendix 1, §2) we obtain that $\|S_N(f)\|_{p'} \leq M \|f\|_{p'}$. All three of the conditions in Theorem 6 are verified; i.e., $\Phi$ is a basis in $L^{p'}(0,1)$ for $p' \in \Delta$.

Specializing Theorem 6 to $C(0,1)$ and using (17), we obtain the following result.

THEOREM 8. *A necessary and sufficient condition for an O.N.S.* $\Phi = \{\varphi_n(x)\}_{n=1}^{\infty} \subset C(0,1)$, *which is complete in* $C(0,1)$, *to be a basis in that space is that the Lebesgue functions of* $\Phi$ *are uniformly bounded*:

$$\|L_N(t)\|_C \leq M < \infty, \qquad N = 1, 2, \dots. \tag{30}$$

REMARK. It follows similarly from Theorem 6 and (18), taking account of Proposition 4, that an O.N.S. $\{\varphi_n(x)\}$ of bounded functions, which is complete in $L^1(0,1)$, is a basis in $L^1(0,1)$ if and only if

$$\|L_N(t)\|_\infty \leq M < \infty, \qquad N = 1, 2, \dots.$$

The next theorem follows easily from Theorem 8.

THEOREM 9. *If the O.N.S.* $\Phi = \{\varphi_n(x)\}_{n=1}^{\infty}$ *is a basis in* $C(0,1)$, *then* $\Phi$ *is also a basis in* $L^p(0,1)$ *for* $1 \leq p < \infty$.

PROOF. We verify that $\Phi$ satisfies the hypotheses of Theorem 6. The completeness of $\Phi$ in $L^p(0,1)$ is a direct consequence of its completeness in $C(0,1)$; and the minimality of $\Phi$ in $L^p(0,1)$, of Proposition 4.

To verify condition c), we use Hölder's inequality [see also (14) and (16)] to find that, when $f \in L^p(0,1)$,

$$|S_N(f,t)| \leq \int_0^1 |K_N(x,t)|^{1/q} |K_N(x,t)|^{1/p} |f(x)| \, dx$$

$$\leq \left\{ \int_0^1 |K_N(x,t)| \, dx \right\}^{1/q} \left\{ \int_0^1 |K_N(x,t)| \cdot |f(x)|^p \, dx \right\}^{1/p},$$

$$1/p + 1/q = 1. \tag{31}$$

From (31) and (30) we find that if $f \in L^p(0,1)$,

$$\int_0^1 |S_N(f,t)|^p \, dt \leq M^{p/q} \int_0^1 \int_0^1 |K_N(x,t)| |f(x)|^p \, dx \, dt$$

$$= M^{p/q} \int_0^1 |f(x)|^p \int_0^1 K_N(x,t) \, dt \, dx$$

$$\leq M^{p/q+1} \int_0^1 |f(t)|^p \, dt.$$

This completes the proof of Theorem 9.

We note that, as is shown by the Faber–Schauder system (see Chapter 6), the condition of orthogonality of the basis functions in Theorem 9 is essential.

DEFINITION 13. A basis $\{x_n\}_{n=1}^{\infty}$ of a Banach space $X$ is said to be *unconditional* if, for every permutation $\sigma = \{\sigma(n)\}_{n=1}^{\infty}$, the system $\{x_{\sigma(n)}\}_{n=1}^{\infty}$ is a basis for $X$.

Using Corollary 1 and Theorem 1 on unconditionally convergent series, we may assert that a necessary and sufficient condition for a basis $\{x_n\}$ to be unconditional is that for every sequence $\varepsilon = \{\varepsilon_n\}_{n=1}^{\infty}$, $\varepsilon_n = \pm 1$, for $n = 1, 2, \ldots$, and for every $x \in X$, the series

$$\sum_{n=1}^{\infty} \varepsilon_n \langle x, y_n \rangle x_n \equiv T_{\varepsilon}(x), \tag{32}$$

converges, where $\{y_n\} \subset X^*$ is the system dual to $\{x_n\}$.

THEOREM 10. *A necessary and sufficient condition that a system $\{x_n\}_{n=1}^{\infty}$ which is complete and minimal in $X$ should be an unconditional basis is that we can find a constant $M$ with the property that for every set of numbers $\{\varepsilon_n\}_{n=1}^{N}$ with $\varepsilon_n = \pm 1$ for $1 \le n \le N$, $N = 1, 2, \ldots$, the inequality*

$$\left\| \sum_{n=1}^{N} \varepsilon_n \langle x, y_n \rangle x_n \right\| \le M \|x\| \tag{33}$$

*is satisfied.*

PROOF. a) If (33) is satisfied, it follows from Theorem 6 that $\{x_n\}$ is a basis in $X$; it then remains to prove that (see Corollary 1) the series

$$\sum_{n=1}^{\infty} \varepsilon_n \langle x, y_n \rangle x_n$$

converges for every sequence $\{\varepsilon_n\}$, $\varepsilon_n = \pm 1$. Using (33), we have

$$\left\| \sum_{n=N}^{N'} \varepsilon_n \langle x, y_n \rangle x_n \right\| \le M \left\| \sum_{n=N}^{N'} \langle x, y_n \rangle x_n \right\| \le \delta_N,$$

for $N < N'$, where $\delta_N \to 0$ as $N \to \infty$.

b) Suppose on the contrary that $\{x_n\}$ is an unconditional basis in $X$, but that there is no constant $M$ for which all the conditions (33) are satisfied. Then we can find a sequence $1 < N_1 < N_2 < \cdots$ of integers, and sequences $\{a_n\}_{n=1}^{\infty}$ of coefficients and $\{\varepsilon_n\}_{n=1}^{\infty}$ of numbers, $\varepsilon_n = \pm 1$, such that, for $k = 1, 2 \ldots,$

$$1) \quad \left\| \sum_{n=N_k+1}^{N_{k+1}} a_n x_n \right\| = 1; \quad 2) \quad \left\| \sum_{n=N_k+1}^{N_{k+1}} \varepsilon_n a_n x_n \right\| \ge k^2. \tag{34}$$

Since the system $\{x_n\}$ in (34), 1) is a basis, we find that the series

$$\sum_{k=1}^{\infty} k^{-2} \sum_{n=N_k+1}^{N_{k+1}} a_n x_n$$

converges in $X$ to an element $x$, and moreover $\langle x, y_n \rangle = k^{-2} a_n$ for $N_k < n \leq N_{k+1}$. Consequently, using (34), 2), we see that the series $\sum_1^\infty \langle x, y_n \rangle x_n$ converges in $X$, whereas $\sum_1^\infty \varepsilon_n \langle x, y_n \rangle x_n$ diverges, which contradicts (see (32)) the fact that $\{x_n\}$ is an unconditional basis.

This completes the proof of Theorem 10.

We have the following corollary to Theorem 10.

COROLLARY 4. *A necessary and sufficient condition for a system* $\{x_n\}$ *which is complete and minimal in* $X$ *to be a basis is that the operators* $T_\varepsilon(x)$ *(see (32)) are defined for every sequence* $\varepsilon = \{\varepsilon_n\}^\infty$, $\varepsilon_n = \pm 1$, *and that the norms of these operators are bounded by a number independent of* $\varepsilon$. *Moreover,*

$$B\|x\| \leq \|T_\varepsilon(x)\| \leq A\|x\|, \qquad x \in X, \tag{35}$$

*where* $A > 0$ *and* $B > 0$ *are independent of* $x$ *and* $\varepsilon$.

The inequality $\|T_\varepsilon(x)\| \geq B\|x\|$ follows from the boundedness of $T_\varepsilon$ and the equation $T_\varepsilon^2(x) = x$.

Consider a biorthogonal system $\{\varphi_n(x), \psi_n(x)\}_{n=1}^\infty$, where $\varphi_n \in L^p(0, 1)$ and $\psi_n \in L^q(0, 1)$, $1 < p < \infty$, $1/p + 1/q = 1$. A fundamental role in the study of the unconditional basis property of $\{\varphi_n\}$ is played by the properties of the following operator: for $f \in L^p(0, 1)$ we set

$$P(f) = P(f, x) = \left\{ \sum_{n=1}^\infty [\langle f, \psi_n \rangle \varphi_n(x)]^2 \right\}^{1/2}. \tag{36}$$

The function (36) is sometimes called the *Paley function of* $\{\varphi_n\}$. In Chapter 2 we shall prove the following result by using (35).

THEOREM 11. *A necessary and sufficient condition for a system* $\Phi = \{\varphi_n(x)\}_{n=1}^\infty$, *which is complete and minimal in* $L^p(0, 1)$, *to be an unconditional basis in* $L^p(0, 1)$ *is that the function* (36) *is finite a.e. for every* $f \in L^p(0, 1)$ *and that the inequality*

$$B\|f\|_p \leq \|P(f)\|_p \leq A\|f\|_p \tag{37}$$

*is satisfied (where the constants* $A > 0$ *and* $B > 0$ *are independent of* $f$).

COROLLARY 5. *For* $p \neq 2$, *the space* $L^p(0, 1)$, $1 < p < \infty$, *has no unconditional and uniformly bounded orthonormal basis.*

PROOF. First let $p > 2$. Suppose, on the contrary, that there is an O.N.S. $\Phi = \{\varphi_n\}$ with $|\varphi_n(x)| \leq M$ for a.e. $x \in (0, 1)$ and $n = 1, 2, \ldots$, which is an unconditional basis in $L^p(0,1)$. But then by Bessel's inequality (see also Proposition 4) we find that

$$P(f,x) \leq M \left\{ \sum_{n=1}^{\infty} [\langle f, \varphi_n \rangle]^2 \right\}^{1/2} \leq M\|f\|_2,$$

a.e. on $(0,1)$, i.e., $\|P(f)\|_p \leq M\|f\|_2$. Then by (37) we find that

$$\|f\|_p \leq MB^{-1}\|f\|_2,$$

for every $f \in L^{(0,1)}$ $(2 < p < \infty)$; this is clearly impossible.

The case $1 < p < 2$ can be reduced to the case $p > 2$ by using Corollary 2.

In Chapters 3 and 5 we shall construct unconditional orthonormal bases in $L^P(0,1)$ for all $p$, $1 < p < \infty$. As for the spaces $L^1(0,1)$ and $C(0,1)$, we shall show (see Theorems 2.13 and 6.2) that no unconditional bases exist in these spaces.

# CHAPTER II

# Independent Functions and Their First Applications

## §1. Definition and construction of sequences of independent functions

DEFINITION 1. A set of real measurable functions $\{f_n(x)\}_{n=1}^N$ with domain $(0,1)$ is a *set of independent functions* if for every interval $I_n$, $n = 1,\ldots,N$, of the real line the following condition is satisfied:

$$m\{x \in (0,1): f_n(x) \in I_n, \ n = 1,\ldots,N\}$$
$$= \prod_{n=1}^{N} m\{x \in (0,1): f_n(x) \in I_n\}. \tag{1}$$

An infinite sequence of functions $\{f_n(x)\}_{n=1}^\infty$ is a *sequence* (or *system*) *of independent functions* (for short, an S.I.F.) if the set $\{f_n(x)\}_{n=1}^N$ is a set of independent functions for every $N = 1, 2, \ldots$.

There is an entirely analogous definition of a set of independent functions on any space $X$ with a measure $\mu$ for which $\mu(X) = 1$. In particular, just as in Definition 1, we can define independent functions on any space $X$ with a measure $\mu$ for which $\mu(X) = 1$. In particular, as in Definition 1 we can introduce independent functions on a set $G \subset R^q$ with $q$-dimensional Lebesgue measure $m(G) = 1$. If the measure of the set $G$ on which the functions $f_n(x)$ are defined is not 1 (but finite and positive), the definition of independence takes the following form: $\{f_n(x)\}_{n=1}^N$ is a set of independent functions if

$$m\{x \in G: f_n(x) \in I_n, \ n = 1,\ldots,N\}$$
$$= [m(G)]^{N+1} \prod_{n=1}^{N} m\{x \in G: f_n(x) \in I_n\}. \tag{2}$$

for every interval $I_n$, $n = 1,\ldots,N$.

We shall also need the following definition:

DEFINITION 2. *A sequence of measurable sets* $\{E_n\}_{n=1}^{\infty}$, $E_n \subset (0,1)$, *is a sequence of independent sets if the characteristic functions* $\{\chi_{E_n}(x)\}_{n=1}^{\infty}$ *of the sets form an* S.I.F.

Sequences of independent sets are much used in probability theory. In addition, the concept of independence plays a fundamental role in the theory of functions. Although orthogonal systems of independent functions form a very special and narrow subclass, they are indispensable in many problems that are discussed in this book. We especially often use properties of the Rademacher system, the simplest nontrivial S.I.F.

In this chapter we prove the propositions on S.I.F. that will be needed later, and give the initial applications of these propositions to the convergence of series of functions. We begin with explicit constructions of some classes of systems of independent functions.

THEOREM 1. *For* $n = 1, 2, \ldots$ *let there be given sequences* $\lambda_{s,n}$ *and* $\mu_{s,n}$, $s = 1, 2, \ldots$, *with* $\lambda_{s,n} \neq \lambda_{s',n}$ *for* $s \neq s'$, *and* $\mu_{s,n} \geq 0$ *with* $\sum_{s=1}^{\infty} \mu_{s,n} = 1$ *for all* $n$. *Then there is a sequence of independent piecewise constant(*[1]*) functions* $\{f_n(x)\}_{n=1}^{\infty}$, $x \in (0,1)$, *such that*

$$m\{x \in (0,1): f_n(x) = \lambda_{s,n}\} = \mu_{s,n}, \qquad s, n = 1, 2, \ldots. \tag{3}$$

The required sequence $\{f_n\}$ is constructed inductively, using the following notation: If $g(x)$ has been defined on the interval $I = (a, b)$, and $I' = (a', b')$ is another interval, we denote by $T_{I \to I'}(g, x)$ the function similar to $g(x)$, defined on $I'$:

$$T_{I \to I'}(g, x) = g\left(a + \frac{x - a'}{b' - a'}(b - a)\right), \qquad x \in I'. \tag{4}$$

We define $\{f_n\}$ only for almost all $x \in (0,1)$, since the property of independence is not affected by changing the functions on a set of measure zero. We first define the auxiliary functions

$$x \in \left(\sum_{p=1}^{s-1} \mu_{p,n}, \sum_{p=1}^{s} \mu_{p,n}\right), \qquad s = 1, 2, \ldots \tag{5}$$

(here $\sum_{p=1}^{0} \mu_{p,n} \equiv 0$). Then we set $f_1(x) = g_1(x)$, and if the set $\{f_n(x)\}_{n=1}^{N}$ of independent piecewise constant functions has already been defined, to construct $f_{N+1}(x)$ we consider the disjoint intervals $\{\delta_i\}$ of constancy of the function $f_N(x)$ $(m(\bigcup_i \delta_i) = 1)$ and set

$$F_{N+1}(x) = T_{(0,1) \to \delta_i}(g_{N+1}, x) \quad \text{for } x \in \delta_i, \ i = 1, 2 \ldots. \tag{6}$$

_____

[1]A function $f(x)$, $x \in (0,1)$, is *piecewise constant* if $f(x) = c_i$ for $x \in \delta_i$, $i = 1, 2, \ldots$, where $\{\delta_i\}$ is a finite or infinite set of pairwise disjoint intervals with $m(\bigcup_i \delta_i) = 1$.

It is clear that $f_{N+1}(x)$ is piecewise constant and that, for every interval $\delta_i$ of constancy of $f_N(x)$ and every interval $I_{N+1}$, we have

$$m\{x \in \delta_i: f_{N+1}(x) \in I_{N+1}\} = m(\delta_i)m\{x \in (0,1): f_{N+1}(x) \in I_{N+1}\}. \quad (7)$$

Notice also that it follows directly from the construction of $f_n(x)$ that

(∗) *The functions $f_n(x)$, $1 \le n \le N$, are constant on each interval of constancy of $f_{N+1}(x)$.*

From (7) and (∗) we find that for every set of intervals $I_n$, $n = 1, \ldots, N + 1$,

$$m \equiv m\{x \in (0,1): f_n(x) \in I_n, \ n = 1, \ldots, N + 1\}$$
$$= m\{x \in (0,1): f_n(x) \in I_n, \ n = 1, \ldots, N\}$$
$$\times m\{x \in (0,1): f_{N+1}(x) \in I_{N+1}\}. \quad (8)$$

If we now use the independence of the set $\{f_n\}_{n=1}^N$ (by the induction hypothesis), we find from (8) that

$$m = \prod_{n=1}^{N+1} m\{x \in (0,1): f_n(x) \in I_n\}.$$

Hence we have proved the independence of the set $\{f_n\}_{n=1}^{N+1}$, and consequently Theorem 1 is established.

The sequence of constant functions $f_n(x) = \lambda_n = \text{const}$ ($x \in (0,1); n = 1, 2, \ldots$) is a trivial example of the systems to which the construction in Theorem 1 leads if only one $\mu_{s,n}$ is different from zero for each $n$. The simplest nontrivial case of Theorem 1 is that in which, for all $n$, only two of the numbers $\mu_{s,n}$, $s = 1, 2, \ldots$, are different from zero: $\mu_{1,n} = \mu_{2,n} = \frac{1}{2}$. If we then take $\lambda_{1,n} = 1$, $\lambda_{2,n} = -1$, the construction in Theorem 1 leads to the S.I.F. that is most important for the theory of functions: the Rademacher system $\{r_n(x)\}_{n=1}^\infty$.

DEFINITION 3. For $n = 1, 2, \ldots$, the $n$th Rademacher function is defined by

$$r_n(x) = \begin{cases} 1 & \text{for } x \in ((i-1)/2^n, i/2^n) = \Delta_n^i, \ i \text{ odd} \\ -1 & \text{for } x \in ((i-1)/2^n, i/2^n) = \Delta_n^i, \ i \text{ even} \end{cases}$$
$$(i = 1, \ldots, 2^n). \quad (9)$$

In addition, it will be convenient to suppose in what follows that $r_0(x) = 1$ for $x \in (0,1)$ and that $r_n(i/2^n) = 0$ for $i = 0, 1, \ldots, 2^n$; $n = 0, 1, 2, \ldots$. Then we can give a more compact definition of the Rademacher functions by the formula

$$r_n(x) = \text{sgn} \sin 2^n \pi x, \quad x \in [0,1], \ n = 0, 1, \ldots. \quad (10)$$

COROLLARY 1. *The functions* $\{r_n(x)\}_{n=0}^{\infty}$, $x \in [0, 1]$, *form an S.I.F.*

In fact, as we have already noted, the Rademacher system is a special case of the systems constructed in Theorem 1.

The values of the functions $r_n(x)$ are directly connected with the binary expansion of $x$. In fact, let the binary-irrational $x \in (0, 1)$ be represented as an infinite binary expansion:

$$x = 0, \theta_1, \theta_2, \ldots, \quad \text{where } \theta_p = \theta_p(x) = 0 \text{ or } 1. \tag{11}$$

Then $x = \sum_1^{\infty} \theta_p 2^{-p}$, and therefore for every $n \geq 1$

$$\sum_{p=1}^{n} \theta_p 2^{-p} < x < \sum_{p=1}^{n} \theta_p 2^{-p} \sum_{p=n+1}^{\infty} 2^{-p} = \sum_{p=1}^{n} \theta_p 2^{-p} + 2^{-n}. \tag{12}$$

From the preceding inequality we have

$$\frac{\theta_n(x) + 2m}{2^n} < x < \frac{\theta_n(x) + 2m + 1}{2^n}, \tag{13}$$

where $m \geq 0$ is an integer. It follows from (9) and (13) that the equation $\theta_n(x) = 0$ holds for those and only those binary irrationals $x$ for which $r_n(x) = 1$.

In other words, we have

$$r_n(x) = 1 - 2\theta_n(x) = (-1)^{\theta_n(x)},$$

$$x \in (0, 1), \ x \neq \frac{i}{2^k}, \ 1 \leq i, \ k < \infty. \tag{14}$$

From the definition of independent functions, Corollary 1, and (14), the following corollary is immediate.

COROLLARY 2. *The functions* $\theta_n(x)$, $n = 1, 2, \ldots$, *defined by* (11) *form an S.I.F.*

Using Corollary 2, we can now construct a sequence $\{f_n(x)\}_{n=1}^{\infty}$ of independent uniformly distributed functions on $[0, 1]$, i.e., functions for which

$$m\{x \in (0, 1): f_n(x) \in I\} = m(I), \quad n = 1, 2, \ldots, \tag{15}$$

for every interval $I \subset (0, 1)$. To do this, we separate the positive integers in any way into an infinite number of disjoint sequences $\Lambda_n: \Lambda_n = \{k_{s,n}\}_{s=1}^{\infty}$, $n = 1, 2, \ldots$, and when $x \in (0, 1)$, $x \neq i/2^k$, we set

$$f_n(x) = 0, \theta_{k_{1,n}}(x)\theta_{k_{2,n}}(x), \ldots, \quad n = 1, 2, , \ldots, \tag{16}$$

i.e.,

$$f_n(x) = \sum_{s=1}^{\infty} 2^{-s} \theta_{k_{s,n}}(x), \quad n = 1, 2, \ldots, \tag{17}$$

where the $\theta_n(x)$ are defined in (11).

THEOREM 2. *The functions $f_n(x)$, $n = 1, 2, \ldots$, defined by (16), satisfy (15) and form an S.I.F.*

PROOF. The measurability of $f_n(x)$ follows immediately from (17). Let us prove that the functions $\{f_n\}$ are independent and simultaneously verify (15) for every interval $I \subset (0, 1)$. We choose a positive integer $N$ and consider a set of binary intervals

$$\omega_n = \left(\frac{i_n - 1}{2^{m_n}}, \frac{i_n}{2^{m_n}}\right) \subset (0, 1), \qquad 1 \leq i_n \leq 2^{m_n}, \; n = 1, \ldots, N.$$

Let the binary representation of $(i_n - 1)/2^{m_n}$ have the form

$$\frac{i_n - 1}{2^{m_n}} = \sum_{s=1}^{m_n} \varepsilon_{s,n} 2^{-s} \qquad (\varepsilon_{s,n} = 0; 1).$$

Then it follows from the definition of $f_n$ (see (16) and (17)) that $f_n(x) \in \omega_n$ if and only if $\theta_{k_{s,n}}(x) = \varepsilon_{s,n}$, $1 \leq s \leq m_n$. Therefore

$$m\{x \in (0, 1): f_n(x) \in \omega_n, n = 1, \ldots, N\}$$
$$= m\{x \in (0, 1): \theta_{k_{s,n}}(x) = \varepsilon_{s,n},$$
$$s = 1, \ldots, m_n; \; n = 1, \ldots, N\}. \tag{18}$$

But since the functions $\theta_n(x)$ are independent (see Corollary 2), if we recall that all the $k_{s,n}$ are different (by construction), we find that the right-hand side of (18) equals

$$\prod_{n=1}^{N} \prod_{s=1}^{m_n} \frac{1}{2} = \prod_{n=1}^{N} m(\omega_n).$$

Therefore

$$m\{x \in (0, 1): f_n(x) \in \omega_n, n = 1, \ldots, N\} = \prod_{n=1}^{N} m(\omega_n). \tag{19}$$

It follows at once from (19) that if, for each $n = 1, \ldots, N$, the set $E_n$ is a finite union of disjoint binary intervals:

$$E_n = \bigcup_{s=1}^{S_n} \omega_{s,n}, \tag{20}$$

then

$$m\{x \in (0, 1): f_n(x) \in E_n, \; n = 1, \ldots, N\} = \prod_{n=1}^{N} m(E_n). \tag{21}$$

In addition, we notice that for a given binary-rational number $i/2^k \in [0, 1]$ with binary representation

$$\frac{i}{2^k} = \sum_{s=1}^{\infty} \varepsilon_s 2^{-s} \qquad (\varepsilon_s = 0 \quad \text{for } s > k)$$

or

$$\frac{i}{2^k} = \sum_{s=1}^{\infty} \varepsilon_s' 2^{-s} \qquad (\varepsilon_s' = 1 \quad \text{for } s > k)$$

the equation $f_n(x) = i/2^k$ is satisfied only if $x$ belongs to the set

$$\{x \in [0, 1]: \theta_{k_{s,n}}(x) = \varepsilon_s, \ s = 1, 2, \dots\}$$
$$\cup \{x \in [0, 1]: \theta_{k_{s,n}}(x) = \varepsilon_s', \ s = 1, 2, \dots\},$$

i.e. (in view of the independence of the functions $\theta_n(x)$) on a set of measure zero. Therefore

$$m\{x \in (0, 1): f_n(x) \in \overline{E}_n, \ n = 1, \dots, N\}$$
$$= m\{x \in (0, 1): f_n(x) \in E_n, \ n = 1, \dots, N\}, \qquad (22)$$

where $\overline{E}$ is the closure of $E$.

Finally, let $I_n$, $n = 1, \dots, N$, be any set of intervals on the real axis. Since $0 \le f_n(x) \le 1$ for $x \in (0, 1)$ (see (16)), we have

$$m\{x \in (0, 1): f_n(x) \in I_n, \ n = 1, \dots, N\}$$
$$= m\{x \in (0, 1): f_n(x) \in I_n \cap [0, 1], \ n = 1, \dots, N\}. \qquad (23)$$

It is also clear that corresponding to every $\varepsilon > 0$ we can find sets $E_n$ and $E_n'$, $n = 1, \dots, N$, of the form (20), for which

$$E_n \subset I_n \cap [0, 1] \subset \overline{E}_n', \ m(E_n') - m(E_n) \le \varepsilon, \qquad n = 1, \dots, N. \qquad (24)$$

Using (21) and (22), we find from (24) that

$$\prod_{n=1}^{N} m\{I_n \cap [0, 1]\} - 2^N \varepsilon$$
$$\le m\{x \in (0, 1): f_n(x) \in I_n \cap [0, 1], \ n = 1, \dots, N\}$$
$$\le \prod_{n=1}^{N} m\{I_n \cap [0, 1]\} + 2^N \varepsilon,$$

which, since $\varepsilon$ in (23) is arbitrarily small, implies the equation

$$m\{x \in (0, 1): f_n(x) \in I_n, \ n = 1, \dots, N\} = \prod_{n=1}^{N} m(I_n \cap [0, 1]). \qquad (25)$$

Hence we have proved that (15) is satisfied (for $N = 1$) and that the functions $f_n$ are independent. This completes the proof of Theorem 2.

If we wish, in Theorem 1, to construct a sequence of independent functions the set of whose values is countable, we can start from (16) and construct an S.I.F. $\{\psi_n(x)\}$ for whose terms

$$m\{x \in (0, 1): \psi_n(x) > t\} = 1 - \lambda_n(t), \qquad t \in R^1, \; n = 1, 2, \ldots, \qquad (26)$$

where $\{\lambda_n(t)\}$ is any preassigned sequence of continuous distribution functions. More precisely, we have the following theorem

THEOREM 3. *Let there be given on the real axis a sequence of continuous strictly monotonic functions* $\{\lambda_n(t)\}_{n=1}^{\infty}$, *and, for* $n = 1, 2, \ldots$, *let* $\lim_{t \to -\infty} \lambda_n(t) = 0$, $\lim_{t \to +\infty} \lambda_n(t) = 1$, *and let* $\Lambda_n(t) = \lambda_n^{-1}(t)$ *be the inverse of* $\lambda_n(t)$. *Then the sequence* $\psi_n(x) = \Lambda_n(f_n(x))$, $x \in (0, 1)$, $n = 1, 2, \ldots$, *is an S.I.F. satisfying* (26).

LEMMA 1. *Let* $\{f_n(x)\}_{n=1}^{\infty}$, $x \in (0, 1)$, *be an S.I.F., and let* $(a_n, b_n)$ *be a finite or infinite interval that contains the range of the function* $f_n(x)$ *(i.e., $f_n(x) \in (a_n, b_n)$ for $x \in (0, 1)$), and let* $\Lambda_n(x)$ *be continuous and strictly monotonic on* $(a_n, b_n)$. *Then* $\{\Lambda_n(f_n(x))\}_{n=1}^{\infty}$ *is an S.I.F.*

PROOF. For $n = 1, 2, \ldots$, set $\delta_n = (\Lambda_n(a_n), \Lambda_n(b_n))$, and let $\{I_n\}_{n=1}^{\infty}$ be any set of real intervals; then

$$\begin{aligned}
E &= \{x \in (0, 1): \Lambda_n(f_n(x)) \in I_n, \; n = 1, \ldots, N\} \\
&= \{x \in (0, 1): \Lambda_n(f_n(x)) \in I_n \cap \delta_n, \; n = 1, \ldots, N\} \\
&= \{x \in (0, 1): f_n(x) \in \Lambda_n^{-1}(I_n \cap \delta_n), \; n = 1, \ldots, N\},
\end{aligned}$$

where $\Lambda_n^{-1}(c, d)$ is the interval $(c', d')$ such that $\Lambda_n(c') = c$, $\Lambda_n(d') = d$. Then, by the definition of independence,

$$\begin{aligned}
m(E) &= \prod_{n=1}^{N} m\{x \in (0, 1): f_n(x) \in \Lambda_n^{-1}(I_n \cap \delta_n)\} \\
&= \prod_{n=1}^{N} m\{x \in (0, 1): \Lambda_n(f_n(x)) \in I_n \cap \delta_n\} \\
&= \prod_{n=1}^{N} m\{x \in (0, 1): \Lambda_n(f_n(x)) \in I_n\}.
\end{aligned}$$

This completes the proof of Lemma 1. To establish Theorem 3 it remains only to verify (26). Using Theorem 2 (see (15)), we find that

$$\begin{aligned}
m\{x \in (0, 1): \Lambda_n(f_n(x)) > t\} \\
= m\{x \in (0, 1): \lambda_n(\Lambda_n(f_n(x))) > \lambda_n(t)\} \\
= m\{x \in (0, 1): f_n(x) > \lambda_n(t)\} = 1 - \lambda_n(t).
\end{aligned}$$

This completes the proof of Theorem 3.

An especially useful S.I.F. is $\{\xi_n\}_{n=1}^{\infty}$, constructed in Theorem 3 in the case when

$$\lambda_n(t) = \lambda(t) = \frac{1}{\sqrt{2\pi}} \int_{-\infty}^{t} e^{-y^2/2} dy, \qquad (27)$$

for $n = 1, 2, \ldots$, i.e., a sequence of independent and normally distributed functions.

It is clear that $\xi_n(x) \in L^p(0, 1)$ for every $p < \infty$ and $n = 1, 2, \ldots$. Moreover, we have the equations

$$\int_0^1 \xi_n(x)\,dx = 0, \qquad \int_0^1 \xi_n^2(x)\,dx = 1, \qquad n = 1, 2, \ldots, \qquad (28)$$

In fact, using the identity

$$\int_0^1 [\xi(x)]^p\,dx = -\int_{-\infty}^{\infty} t^p\,d\tilde{\lambda}_\xi(t) = -\int_{-\infty}^{\infty} t^p \tilde{\lambda}_\xi'(t)\,dt,$$

$$\xi \in L^p(0, 1), \ p = 1, 2; \ \tilde{\lambda}_\xi(t) = m\{x : \xi(x) > t\} \qquad (29)$$

(see Appendix 1, §1), we obtain, by integrating both sides by parts,

$$\int_0^1 \xi_n(x)\,dx = (2\pi)^{-1/2} \int_{-\infty}^{\infty} te^{-t^2/2}\,dt = 0,$$

$$\int_0^1 \xi_n^2(x)\,dx = (2\pi)^{-1/2} \int_{-\infty}^{\infty} t^2 e^{-t^2/2}\,dt$$

$$= -te^{-t^2/2}(2\pi)^{-1/2}|_{-\infty}^{\infty} + (2\pi)^{-1/2} \int_{-\infty}^{\infty} e^{-t^2/2}\,dt = 1.$$

## §2. Properties of systems of independent functions

Let us first notice a property of sets of independent functions which follows directly from Definition 1.

Let $\{f_n(x)\}_{n=1}^{N}$, $x \in G$, $G \subset R^1$, $0 < m(G) < \infty$, be a set of independent functions, and let $F_n$ ($n = 1, \ldots, N$) be a finite union of open, half-open, or closed intervals on the real axis; then

$$m\{x \in G : f_n(x) \in F_n, \ n = 1, \ldots, N\}$$

$$= [m(G)]^{-N+1} \prod_{n=1}^{N} m\{x \in G : f_n(x) \in F_n\}. \qquad (30)$$

Equation (30) can be extended to a wide class of sets $F_n$ by limiting processes; however, we do not dwell on this possibility, since a special case suffices for our purposes.

THEOREM 4. *For every set of independent functions* $\{f_n(x)\}_{n=1}^{\infty}$ *defined on the set* $G \subset R^1$, *with* $m(G) > 0$ *and* $f_n \in L^1(G)$, $n = 1, \ldots, N$, *the function* $\prod_{n=1}^{N} f_n(x)$ *also belongs to* $L^1(G)$ *and*

$$\int_G \prod_{n=1}^{N} f_n(x)\, dx = [m(G)]^{-N+1} \prod_{n=1}^{N} \int_G f_n(x)\, dx.$$

PROOF. First let $N = 2$. For each integer $k$ and each $\delta > 0$, we set

$$E_{k,\delta} = \{x \in G: f_1(x) \in [(k-1)\delta, k\delta)\}. \tag{31}$$

First let $f_1(x) \geq 0$ and $f_2(x) \geq 0$; then we have for each $\delta > 0$, by the definition of a Lebesgue integral,

$$\begin{aligned}
\int_G f_1(x) f_2(x)\, dx &= \sum_{k=1}^{\infty} \int_{E_{k,\delta}} f_1(x) f_2(x)\, dx \\
&= \sum_{k=1}^{\infty} k\delta \int_{E_{k,\delta}} f_2(x)\, dx + \rho,
\end{aligned}$$

where $|\rho| \leq \delta \|f_2\|_{L_2(G)}$. Consequently

$$\int_G f_1(x) f_2(x)\, dx = \lim_{\delta \to 0} \sum_{k=1}^{\infty} k\delta \int_{E_{k,\delta}} f_2(x)\, dx. \tag{32}$$

For $k = 1, 2, \ldots$, by equation (29),

$$\begin{aligned}
\int_{E_{k,\delta}} f_2(x)\, dx &= \int_0^{\infty} \lambda_{k,\delta}(t)\, dt, \\
\lambda_{k,\delta}(t) &= m\{x \in E_{k,\delta}: f_2(x) > t\};
\end{aligned} \tag{33}$$

here, since $f_1$ and $f_2$ are independent functions, we have, for $t > 0$ and $k = 1, 2, \ldots$,

$$\begin{aligned}
\lambda_{k,\delta}(t) &= \frac{m(E_{k,\delta})}{m(G)} \lambda_{f_2}(t), \\
\lambda_f(t) &= m\{x \in G: |f(x)| > t\}.
\end{aligned} \tag{34}$$

Since $\int_0^{\infty} \lambda_{f_2}(t)\, dt = \int_G f_2(x)\, dx$, we find from (33) and (34) that

$$\int_{E_{k,\delta}} f_2(x)\, dx = \frac{m(E_{k,\delta})}{m(G)} \int_G f_2(x)\, dx, \qquad k = 1, 2, \ldots. \tag{35}$$

Using the preceding equation, we find from (32) that for non-negative $f_1(x)$ and $f_2(x)$,

$$\begin{aligned}
\int_G f_1(x) f_2(x)\, dx &= \lim_{\delta \to 0} \sum_{k=1}^{\infty} k\delta\, m(E_{k,\delta}) \frac{1}{m(G)} \int_G f_2(x)\, dx \\
&= \frac{1}{m(G)} \int_G f_1(x)\, dx \int_G f_2(x)\, dx,
\end{aligned} \tag{36}$$

and then the finiteness of $\int_G f_1(x)f_2(x)\,dx$ follows from the finiteness of the integrals $\int_G f_1(x)\,dx$ and $\int_G f_2(x)\,dx$.

If now we are given independent summable functions $f_1(x)$ and $f_2(x)$ of arbitrary sign, the functions $|f_1(x)|$ and $|f_2(x)|$ are also independent (in fact, for every set of independent functions $f_n(x)$, $n = 1,\ldots,N$, the set $\{x\colon |f_n(x)| \in I_n, \ n = 1,\ldots,N\}$ has the form

$$\{x\colon f_n(x) \in I_n \cup (-I_n), \ n = 1,\ldots,N\},$$

where $-I_n = \{x\colon -x \in I_n\}$). It remains only to apply equation (30).

Consequently (see (36)) the integral $\int_G |f_1(x)f_2(x)|\,dx$ is finite.

Furthermore, we can argue as in the case when $f_1(x) \geq 0$ and $f_2(x) \geq 0$: from the equation

$$\int_G f_1(x)f_2(x) = \lim_{\delta \to 0} \sum_{k=-\infty}^{\infty} k\delta \int_{E_{k,\delta}} f_2(x)\,dx$$

and (35) (which is evidently valid also when $k < 1$) we find that

$$\int_G f_1(x)f_2(x)\,dx = [m(G)]^{-1} \int_G f_1(x)\,dx \int_G f_2(x)\,dx,$$

i.e., Theorem 4 is established for $N = 2$.

We now prove Theorem 4 for sets of $N > 2$ functions, assuming that it is known to be valid for $N - 1$ functions.

As before, let the set $E_{k,\delta}$ be defined by (31) for integral $k$ and $\delta > 0$; and also let $m(E_{k,\delta}) > 0$. Then $\{f_n(x)\}_{n=2}^{N}$ is clearly a set of independent functions on $E_{k,\delta}$, and we have (see (35))

$$\int_{E_{k,\delta}} \prod_{n=2}^{N} f_n(x)\,dx = [m(E_{k,\delta})]^{-N+2} \prod_{n=2}^{N} \int_{E_{k,\delta}} f_n(x)\,dx$$

$$= \frac{m(E_{k,\delta})}{[m(G)]^{N-1}} \prod_{n=2}^{N} \int_G f_n(x)\,dx, \qquad k = 0, \pm 1, \ldots. \tag{37}$$

Similarly,

$$\int_{\tilde{E}_{k,\delta}} \prod_{n=2}^{N} |f_n(x)|\,dx = \frac{m(\tilde{E}_{k,\delta})}{[m(G)]^{N-1}} \prod_{n=2}^{N} \int_G |f_n(x)|\,dx, \tag{37'}$$

where

$$\tilde{E}_{k,\delta} = \{x \in G\colon |f_1(x)| \in [(k-1)\delta, k\delta)\}, \qquad k = 1,2,\ldots, \ \delta > 0.$$

Taking $\delta = 1$ in (37'), we obtain

$$\int_G \prod_{n=1}^N |f_n(x)|\,dx = \sum_{k=1}^\infty \int_{\tilde E_{k,1}} \prod_{n=1}^N |f_n(x)|\,dx$$

$$\leq \sum_{k=1}^\infty k \int_{\tilde E_{k,1}} \prod_{n=2}^N |f_n(x)|\,dx$$

$$= [m(G)]^{1-N} \prod_{n=2}^N \int_G |f_n(x)|\,dx \sum_{k=1}^\infty km(\tilde E_{k,1}) < \infty,$$

i.e., $\prod_{n=1}^N |f_n(x)|$ is summable. Consequently the equation

$$\int_G \prod_{n=1}^N f_n(x)\,dx = \lim_{\delta \to 0} \sum_{k=-\infty}^\infty k\delta \int_{E_{k,\delta}} \prod_{n=2}^N f_n(x)\,dx$$

makes sense (see (32)); from this, it follows by (37) that

$$\int_G \prod_{n=1}^N f_n(x)\,dx = [m(G)]^{1-N} \prod_{n=2}^N \int_G f_n(x)\,dx \lim_{\delta \to 0} \sum_{k=-\infty}^\infty k\delta m(E_{k,\delta})$$

$$= [m(G)]^{1-N} \prod_{n=1}^N \int_G f_n(x)\,dx.$$

This completes the proof of Theorem 4.

COROLLARY 3. *An S.I.F.* $\{\psi_n(x)\}_{n=1}^\infty$, $x \in (0,1)$, *is an orthonormal system if it satisfies, for* $n = 1, 2, \dots$, *the conditions*

$$\int_0^1 \psi_n(x)\,dx = 0, \qquad \int_0^1 \psi_n^2(x)\,dx = 1.$$

We are going to investigate the properties of polynomials and series of O.N.S. that consist of independent functions $\{\psi_n(x)\}$. In this investigation we shall, in many cases, need to require that the functions $\psi_n(x)$ satisfy the additional requirement that they are uniformly bounded: $|\psi_n(x)| \leq M$, $x \in (0,1)$, $n = 1, 2, \dots$ (it is clear that $M \geq 1$, since

$$M \geq \|\psi_n\|_\infty \geq \|\psi_n\|_2 = 1).$$

THEOREM 5. *The following inequality holds for every set* $\{\psi_n(x)\}_{n=1}^N$ *of independent functions which satisfy*

$$\|\psi_n\|_2 = 1, \qquad \|\psi_n\|_\infty \leq M, \qquad \int_0^1 \psi_n(x)\,dx = 0$$

*for* $n = 1, \ldots, N$:

$$\lambda(t) = m \left\{ x \in (0,1) : \left| \sum_{n=1}^{N} a_n \psi_n(x) \right| > t \left( \sum_{n=1}^{N} a_n^2 \right)^{1/2} \right\} \le 2e^{-t^2/4M^2} \quad (38)$$

*for every* $t \ge 0$.

PROOF. Notice that

$$1 + \sum_{r=2}^{\infty} \frac{t^r}{r!} \le e^{t^2} \quad (39)$$

for every $t$. In fact, since $[(2j+1)!]^2 > \frac{1}{2}(2j)!(2j+2)!$, then, for $j \ge 1$, by using the inequality $|ab| \le \frac{1}{2}(a^2 + b^2)$ we obtain

$$\frac{|t^{2j+1}|}{(2j+1)!} \le \frac{\sqrt{2}|t^j t^{j+1}|}{[(2j)!(2j+2)!]^{1/2}} \le \frac{\sqrt{2}}{2} \left( \frac{t^{2j}}{(2j)!} + \frac{t^{2j+2}}{(2j+2)!} \right),$$

from which it follows that

$$1 + \sum_{r=2}^{\infty} \frac{t^r}{r!} = 1 + \sum_{j=1}^{\infty} \frac{t^{2j}}{(2j)!} + \sum_{j=1}^{\infty} \frac{t^{2j+1}}{(2j+1)!}$$

$$\le 1 + \sum_{j=1}^{\infty} \frac{t^{2j}}{(2j)!} + \frac{\sqrt{2}}{2} \cdot \frac{t^2}{2} + \sqrt{2} \sum_{j=2}^{\infty} \frac{t^{2j}}{(2j)!}$$

$$= 1 + \frac{1}{2} \left( 1 + \frac{\sqrt{2}}{2} \right) t^2 + (1 + \sqrt{2}) \sum_{j=2}^{\infty} \frac{t^{2j}}{(2j)!}$$

$$\le 1 + \sum_{j=1}^{\infty} \frac{t^{2j}}{j!} = e^{t^2}.$$

Let us now show that for every $z > 0$ and $n = 1, \ldots, N$,

$$\int_0^1 \exp\{z a_n \psi_n(x)\} \, dx \le \exp\{z^2 a_n^2 M^2\}. \quad (40)$$

In fact, if we expand $\exp\{z a_n \psi_n(x)\}$ in a Taylor series and use the equation $\int_0^1 \psi_n(x) \, dx = 0$, we obtain

$$\int_0^1 \exp\{z a_n \psi_n(x)\} \, dx = \int_0^1 \left[ 1 + \sum_{r=1}^{\infty} \frac{(z a_n \psi_n(x))^r}{r!} \right] dx$$

$$= 1 + \sum_{r=2}^{\infty} \int_0^1 \frac{(z a_n \psi_n(x))^r}{r!} \, dx$$

$$\le 1 + \sum_{r=2}^{\infty} \frac{|z a_n M|^r}{r!}$$

$$\le \exp\{z^2 a_n^2 M^2\}.$$

Let us set $P(x) = \sum_{n=1}^{\infty} a_n \psi_n(x)$ and

$$\lambda'(t) = m \left\{ x \in (0,1) : P(x) > t \left( \sum_{n=1}^{N} a_n^2 \right)^{1/2} \right\}, \qquad t \geq 0.$$

Since $e^x$ is monotonic, we can use Chebyshev's inequality to show that, for every $z > 0$,

$$\lambda'(t) = m \left\{ x \in (0,1) : \exp[zP(x)] > \exp \left[ zt \left( \sum_{n=1}^{N} a_n^2 \right)^{1/2} \right] \right\}$$

$$\leq \exp \left[ -tz \left( \sum_{n=1}^{N} a_n^2 \right)^{1/2} \right] \int_0^1 \exp[zP(x)] \, dx \qquad (41)$$

$$= \exp \left[ -tz \left( \sum_{n=1}^{N} a_n^2 \right)^{1/2} \right] \int_0^1 \prod_{n=1}^{N} \exp[za_n \psi_n(x)] \, dx.$$

By Theorem 4 (see also Lemma 1 and Theorem 3) the right-hand side of (41) is equal to

$$\exp \left[ -tz \left( \sum_{n=1}^{N} a_n^2 \right)^{1/2} \right] \prod_{n=1}^{N} \int_0^1 \exp[za_n \psi_n(x)] \, dx.$$

From this we find by using (40) that, for every $z > 0$,

$$\lambda'(t) \leq \exp \left\{ tz \left( \sum_{n=1}^{N} a_n^2 \right)^{1/2} + z^2 M^2 \sum_{n=1}^{N} a_n^2 \right\}. \qquad (42)$$

Taking $z = t[2M^2(\sum_{n=1}^{N} a_n^2)^{1/2}]^{-1}$ in (42), we obtain

$$\lambda'(t) \leq \exp \left[ -\frac{t^2}{2M^2} + \frac{t^2 M^2}{4M^4} \right] = \exp \left[ -\frac{t^2}{4M^2} \right]. \qquad (43)$$

If we now consider the set $\{-\psi_n(x)\}_{n=1}^{N}$ instead of $\{\psi_n(x)\}_{n=1}^{N}$, and apply (43), we find

$$\lambda''(t) \equiv m \left\{ x \in (0,1) : P(x) < -t \left( \sum_{n=1}^{N} a_n^2 \right)^{1/2} \right\}$$

$$\leq \exp \left[ -\frac{t^2}{4M^2} \right],$$

and consequently

$$\lambda(t) = \lambda'(t) + \lambda''(t) \leq 2 \exp \left[ -\frac{t^2}{4M^2} \right].$$

This completes the proof of Theorem 5.

We may notice that more precise estimates of the constants in Theorem 5 would let us obtain the inequality $\lambda(t) \leq 2\exp[-t^2/2M^2]$. However, for our purposes any inequality of the form $\lambda(t) \leq C\exp[-\gamma t^2]$, with $\gamma = \gamma(M) > 0$, would be sufficient, and for that reason we did not take the trouble to find precise constants in the estimates for $\lambda(t)$.

Theorem 5 leads to a number of corollaries concerning polynomials and a series involving S.I.F. $\{\psi_n(x)\}_{n=1}^{\infty}$, $x \in (0,1)$, whose components satisfy the conditions

$$\|\psi_n\|_2 = 1, \qquad \|\psi_n\|_{\infty} \leq M, \qquad \int_0^1 \psi_n(x)\,dx = 0,$$

$$n = 1, 2, \ldots. \quad (44)$$

THEOREM 6 (KHINCHIN'S INEQUALITY). *For all numbers $p > 2$ and $M \geq 1$ there exist constants $C_{p,M}$ such that, for every polynomial $P(x) = \sum_1^N a_n\psi_n(x)$ in an S.I.F. $\{\psi_n(x)\}_{n=1}^{\infty}$ that satisfies (44), the inequality*

$$\|P\|_p \leq C_{p,M}\|P\|_2 = C_{p,M}\left(\sum_{n=1}^N a_n^2\right)^{1/2}$$

*will be satisfied.*

PROOF. Without loss of generality we may suppose that $\|P\|_2 = (\sum_{n=1}^N a_n^2)^{1/2} = 1$. Let $\lambda(t) = m\{x \in (0,1): |P(x)| > t\}$. By Theorem 5, we have $\lambda(t) \leq 2\exp(-t^2/4M^2)$. Therefore (see Appendix 1, (2))

$$\|P\|_p = \left\{p\int_0^{\infty} t^{p-1}\lambda(t)\,dt\right\}^{1/p}$$

$$\leq \left\{2p\int_0^{\infty} t^{p-1}\exp\left(-\frac{t^2}{4M^2}\right)dt\right\}^{1/p}$$

$$= C_{p,M}.$$

THEOREM 7. *Every polynomial in an S.I.F. that satisfies (44) satisfies the inequalities*

a) $m\{x \in (0,1): |P(x)| \geq \frac{1}{2}\|P\|_2\} \geq c_M > 0$;

b) $\|P\|_p \geq \|P\|_1 \geq c_M\|P\|_2$ $\quad(c_M' > 0, \ 1 \leq p \leq \infty)$.

LEMMA 1. *Every function $f(x), x \in (0,1)$, with $f(x) \geq 0$, $\|f\|_1 = 1$, and $\|f\|_2 = K$, satisfies the inequality*

$$m\left\{x \in (0,1): f(x) \geq \frac{1}{4}\right\} \geq \frac{1}{K^2}.$$

PROOF. Let $Q = \{x \in (0,1): f(x) \geq 1/4\}$, $E = \{x \in (0,1): f(x) \geq 2K^2\}$. Then

$$K^2 \geq \int_E f^2(x)\,dx \geq 2K^2 \int_E f(x)\,dx,$$

i.e., $\int_E f(x)\,dx \leq \frac{1}{2} \int_{(0,1)\backslash E} f(x)\,dx \geq \frac{1}{2}$. Therefore

$$\frac{1}{2} \leq \int_{(0,1)\backslash E} f(x)\,dx = \int_{[(0,1)\backslash E]\cap Q} f(x)\,dx + \int_{[(0,1)\backslash E]\cap(0,1)\backslash Q]} f(x)\,dx$$
$$\leq 2K^2 m(Q) + \frac{1}{4},$$

from which we have $m(Q) \geq (8K^2)^{-1}$. This completes the proof of Lemma 1.

PROOF OF THEOREM 7. It is enough to prove part a), since part b) is an immediate corollary. By Theorem 6, we have $\|P\|_4 \leq C_{4,M}\|P\|_2$, and if we apply Lemma 1 to the function $P^2(x)\|P\|_2^{-2}$ (for $\|P\|_2 \neq 0$), we find that

$$m\{x \in (0,1): P^2(x)\|P\|_2^{-2} \geq \tfrac{1}{4}\} \geq (8C_{4,M}^4)^{-1},$$

from which a) follows with $c_M = (8C_{4,M}^4)^{-1}$. This completes the proof of the theorem.

By using Theorem 7 it is easy to obtain a necessary condition for the convergence in measure of the series

$$\sum_{n=1}^{\infty} a_n \psi_n(x) \qquad (45)$$

in an S.I.F. $\{\psi_n(x)\}$ that satisfies (44). This condition turns out to be equivalent to the convergence of the sum of squares of the coefficients of the series (45):

$$\sum_{n=1}^{\infty} a_n^2 < \infty. \qquad (46)$$

Inequality (46) is also a sufficient condition for the convergence both in measure and almost everywhere of series in any O.N.S. of independent functions. For convergence in measure this follows from the fact that under (46) a series (45), like any orthogonal series, converges in $L^2$, and *a fortiori* in measure.

The sufficiency of (46) for convergence a.e. of the series (45) will be proved below (see Theorem 9).

THEOREM 8. *If an S.I.F.* $\{\psi_n(x)\}_{n=1}^{\infty}$ *satisfies* (44), *and* (45) *converges in measure (a fortiori a.e.) on* (0, 1), *then* (46) *is satisfied. Moreover,*

$$m\left\{x \in (0,1): \left|\sum_{n=1}^{\infty} a_n \psi_n(x)\right| > \frac{1}{4}\left(\sum_{n=1}^{\infty} a_n^2\right)^{1/2}\right\} \geq c_M > 0. \qquad (47)$$

PROOF. Supposing that $\sum_{n=1}^{\infty} a_n^2 = \infty$, we can find an increasing sequence $\{N_k\}_{k=1}^{\infty}$ of integers such that

$$\sum_{n=N_k+1}^{N_{k+1}} a_n^2 > k^2, \qquad k = 1, 2, \ldots.$$

Then by Theorem 7 [part a)]

$$m\left\{x \in (0,1): \left|\sum_{n=N_k+1}^{N_{k+1}} a_n \psi_n(x)\right| > \frac{k}{2}\right\} \geq c_M > 0; \qquad k = 1, 2, \ldots,$$

which contradicts the convergence in measure of (45). Inequality (47) also follows from the inequality in part a) of Theorem 7 and from the fact that because the series (45) converges in measure, we have, for each $\alpha > 0$,

$$m\left\{x \in (0,1): \left|\sum_{n=1}^{\infty} a_n \psi_n(x)\right| > \alpha\right\}$$

$$\geq \varlimsup_{N \to \infty} m\left\{x \in (0,1): \left|\sum_{n=1}^{N} a_n \psi_n(x)\right| > 2\alpha\right\}.$$

THEOREM 9. *Every O.N.S. of independent functions,* $\Psi = \{\psi_n(x)\}_{n=1}^{\infty}$, $x \in (0,1)$, *is a convergence system (see Definition 1.2). Moreover, the majorant* $S_{\Psi}^{*}(\{a_n\})$ *of the partial sums (see* 1.5) *is a bounded operator from* $l^2$ *to* $L^2(0,1)$ :

$$\|S_{\Psi}^{*}(\{a_n\})\|_2 \leq 4\|\{a_n\}\|_{l^2}. \qquad (48)$$

REMARK. In contrast to Theorem 9, for the validity of Theorems 5–8 it was necessary to impose, besides the condition of orthonormality, supplementary conditions on the independent functions $\psi_n(x)$; for example, to suppose that $\|\psi_n\|_{\infty} \leq M$, $n = 1, 2, \ldots$. However, the requirement of uniform boundedness of the functions $\psi_n(x)$ can be weakened in each of these theorems. For example, in Theorem 8 it can be replaced by the condition $\|\psi_n\|_1 \geq c > 0$, $n = 1, 2, \ldots$, which, as is easily seen, is a necessary condition for the validity of Theorem 8.

LEMMA 1. *For every S.I.F.* $\Psi' = \{\psi'_n(x)\}_{n=1}^\infty$ *whose functions have only a finite number of values, and for all numbers* $\{a_n\}_{n=1}^N$,

$$\|S_{\Psi'}^*(\{a_n\})\|_2 \leq 4 \cdot \max_{1 \leq s \leq m \leq N} \left\|\sum_{n=s}^m a_n \psi'_n\right\|_2 \equiv 4M. \tag{49}$$

PROOF. Choose numbers $\{a_n\}_{n=1}^N$; we may also suppose, without loss of generality, that $M = 1$. Set

$$S^*(x) = S_{\Psi'}^*(\{a_n\}, x),$$

$$S_r(x) = \sum_{n=1}^r a_n \psi'_n(x), \qquad r = 1, \ldots, N.$$

We estimate the distribution function $\lambda(t) = m\{x \in (0,1): S^*(x) > t\}$. We first notice that by Chebyshev's inequality we have, for $1 \leq s \leq m \leq N$,

$$m\left\{x \in (0,1): \sum_{n=s}^m a_n \psi'_n(x) < -\frac{3}{2}\right\} \leq \frac{4}{9}M^2 < \frac{1}{2}. \tag{50}$$

For $t > 0$, set

$$G_t^+ = \left\{x \in (0,1): \max_{1 \leq r \leq N} S_r(x) > t\right\},$$

$$G_t^- = \left\{x \in (0,1): \min_{1 \leq r \leq N} S_r(x) < -t\right\}.$$

It is clear that

$$\lambda(t) \leq m(G_t^+) + m(G_t^-). \tag{51}$$

We represent the set $G_t^+$ in the form

$$G_t^+ = \bigcup_{q=1}^N G_t^+(q), \tag{52}$$

$$G_t^+(q) = \{x \in (0,1): S_r(x) \leq t, \ 1 \leq r \leq q, \ S_q(x) > t\}.$$

As we see from their definition (see (52)), the sets $G_t^+(q)$ for different values of $q$ are disjoint, and each $G_t^+(q)$ is a finite union of sets of the form $\{x \in (0,1): \psi'_n(x) = z_n, \ 1 \leq n \leq q\}$.

Using the independence of the functions $\psi'_n(x)$, we find that for every $z$ and $q = 1, \ldots, N-1$

$$m\left\{x \in G_t^+(q): \sum_{n=q+1}^N a_n \psi'_n(x) > z\right\}$$

$$= m[G_t^+(q)] \cdot m\left\{x \in (0,1): \sum_{n=q+1}^N a_n \psi'_n(x) > z\right\},$$

and consequently (see (50) and (52)),

$$m\left\{x \in G_t^+(q): S_N(x) > t - \frac{3}{2}\right\}$$

$$\geq m\left\{x \in G_t^+(q): \sum_{n=q+1}^N a_n \psi_n'(x) \geq -\frac{3}{2}\right\}$$

$$= m[G_t^+(q)]m\left\{x \in (0,1): \sum_{n=q+1}^N a_n \psi_n'(x) \geq -\frac{3}{2}\right\} \qquad (53)$$

$$> \frac{1}{2}m[G_t^+(q)].$$

If we add the inequalities (53) for $q = 1, \ldots, N$, we find that

$$m\left\{x \in G_t^+: S_N(x) > t - \tfrac{3}{2}\right\} > \tfrac{1}{2}m(G_t^+). \qquad (54)$$

If we consider the set $\{-\psi_n'\}$ instead of $\{\psi_n'\}$, we find from (54) that

$$m\left\{x \in G_t^-: S_N(x) < -t + \tfrac{3}{2}\right\} > \tfrac{1}{2}m(G_t^-). \qquad (55)$$

Combining (54) and (55) and using (51), we obtain

$$\lambda(t) \leq 2\lambda'(t - \tfrac{3}{2}), \qquad \lambda'(t) = m\left\{x \in (0,1): |S_N(x)| > t\right\}. \qquad (56)$$

By using equation (2) from Appendix 1, we deduce from (56) that

$$\|S^*\|_2^2 = 2\int_0^\infty t\lambda(t)\,dt$$

$$\leq 2\left[\int_0^{3/2} t\,dt + \int_{3/2}^\infty t\lambda(t)\,dt\right] \leq \frac{9}{4} + 4\int_{3/2}^\infty t\lambda'\left(t - \frac{3}{2}\right)dt$$

$$= \frac{9}{4} + 4\int_0^\infty t\lambda'(t)\,dt + 6\int_0^\infty \lambda'(t)\,dt$$

$$= \frac{9}{4} + 2\|S_N\|_2^2 + 6\|S_N\|_1 < 11M = 11.$$

This completes the proof of Lemma 1.

Retaining the notation used in the statement and proof of Lemma 1, we record the following inequality for later use:

$$\|S^*\|_p \leq C_p(M + \|S_N\|_p), \qquad 1 \leq p < \infty. \qquad (57)$$

In fact, if we suppose (without loss of generality) that $M = 1$, we find from (56) that

$$
\begin{aligned}
\|S^*\|_p^p &\leq 2p \int_0^\infty t^p \lambda' \left( t - \frac{3}{2} \right) dt \\
&\leq C_p \left[ \int_0^\infty \lambda'(t)\, dt + \int_0^\infty t^p \lambda'(t)\, dt \right] \leq C_p(\|S_N\|_1 + \|S_N\|_p^p) \\
&\leq C_p(1 + \|S_N\|_p^p) \leq C_p(M + \|S_N\|_p)^p.
\end{aligned}
$$

PROOF OF THEOREM 9. By Proposition 1.1, it is sufficient to establish (48) for every finite set $\{a_n\}_{n=1}^N$ such that $\sum_{n=1}^N a_n^2 = 1$. We approximate $\psi_n(x)$, $n = 1, \ldots, N$, by independent functions $\psi_n'(x)$ each of which has only finitely many values. For this purpose we introduce, for $-\infty < k < \infty$ and $0 < \delta < \infty$, sets $E_n(k, \delta) = \{x \in (0, 1) \colon \psi_n(x) \in [(k-1)\delta, k\delta)\}$, and for $n = 1, \ldots, N$ and $p = 1, 2, 3, \ldots$, we set

$$
\psi_n(x, p, \delta) = \begin{cases} k\delta, & \text{if } x \in E_n(k, \delta),\ -p < k \leq p, \\ -\delta p, & \text{if } \psi_n(x) < -\delta p, \\ \delta p, & \text{if } \psi_n(x) > \delta p. \end{cases}
$$

It is easily seen that, for all $p$ and $\delta$, the set $\Psi_{p,\delta} = \{\psi_n(x, p, \delta)\}_{n=1}^N$ is a set of independent functions and that $\|\psi_n(x, p, \delta) - \psi_n(x)\|_2 \to 0$ as $p \to \infty$ and $\delta \to 0$. Therefore

$$
\begin{aligned}
&\lim_{\substack{p \to \infty \\ \delta \to 0}} \max_{1 \leq s \leq m \leq N} \left\| \sum_{n=s}^m a_n \psi_n(x, p, \delta) \right\|_2 = 1, \\
&\lim_{\substack{p \to \infty \\ \delta \to 0}} \|S^*_{\Psi_{p,\delta}}(\{a_n\})\|_2 = \|S^*_\Psi(\{a_n\})\|_2.
\end{aligned} \tag{58}
$$

Then (48) for the set $\{a_n\}_{n=1}^N$ follows immediately from (58) and Lemma 1. This completes the proof of Theorem 9.

In the theory of functions, in particular, in order to construct orthonormal systems with some special property one often uses the following proposition on sequences of independent functions.

**BOREL–CANTELLI LEMMA.** *If* $\{E_n\}_{n=1}^{\infty}$, $E_n \subset (0,1)$, *is a sequence of independent sets such that* $\sum_{1}^{\infty} m(E_n) = \infty$ *and*

$$E_0 = \overline{\lim_{n \to \infty}} E_n \equiv \bigcap_{k=1}^{\infty} \bigcup_{n=k}^{\infty} E_n,$$

*then* $m(E_0) = 1$.

**PROOF.** It is sufficient to verify that $m(\bigcup_{n=k}^{\infty} E_n) = 1$ for $k = 1, \ldots$. However, for every $q > k$ we have

$$1 - m\left(\bigcup_{n=k}^{\infty} E_n\right) \leq 1 - m\left(\bigcup_{n=k}^{q} E_n\right) = m\left(\bigcap_{n=k}^{q} [(0,1)\backslash E_n]\right).$$

Since the sets $\{(0,1)\backslash E_n\}_{n=k}^{\infty}$ are also independent (see Definition 2), by Theorem 4 we have

$$m\left(\bigcap_{n=k}^{q} [(0,1)\backslash E_n]\right) = \int_0^1 \prod_{n=k}^{q} \chi_{(0,1)\backslash E_n}(x)\,dx = \prod_{n=k}^{q} (1 - m(E_n)).$$

It follows from the divergence of $\sum_{n=k}^{\infty} m(E_n)$ that

$$\lim_{q \to \infty} \prod_{n=k}^{q} (1 - m(E_n)) = 0,$$

and therefore

$$1 - m\left(\bigcup_{n=k}^{\infty} E_n\right) \leq \lim_{q \to \infty} m\left(\bigcap_{n=k}^{q} [(0,1)\backslash E_n]\right) = 0.$$

This completes the proof of the lemma.

The theorems previously proved in this section were of a general nature. However, the following result concerns a property which is fundamental in the theory of orthogonal series, a property possessed by sets of independent normally distributed functions (see §1, Theorem 3 and (27)).

**THEOREM 10.** *If* $\xi = \{\xi_n(x)\}_{n=1}^{N}$ *is a set of independent normally distributed functions for which* $\int_0^1 \xi_n(x)\,dx = 0$, $\int_0^1 \xi_n^2(x)\,dx = 1$, $n = 1, \ldots, N$, *and* $A = \{a_{m,n}\}_{m,n=1}^{N}$ *is an orthonormal matrix, then the function*

$$\xi_m'(x) = \sum_{n=1}^{N} a_{m,n}\xi_n(x), \qquad m = 1, \ldots, N,$$

*is also a set of normally distributed independent functions with*

$$\int_0^1 \xi_m'(x)\,dx = 0, \qquad \int_0^1 [\xi_m'(x)]^2\,dx = 1, \qquad m = 1, \ldots, N.$$

PROOF. For $m = 1, \ldots, N$ we have

$$\int_0^1 \xi_m'(x)\, dx = \sum_{n=1}^N a_{m,n} \int_0^1 \xi_n(x)\, dx = 0.$$

In addition, since $\{\xi_n(x)\}$ is an O.N.S., and $A$ is orthonormal, it is clear that the functions $\xi_m'(x)$, $m = 1, \ldots, N$, form an orthonormal set, and therefore $\int_0^1 [\xi_m'(x)]^2\, dx = 1$, $m = 1, \ldots, N$.

Let us now prove that for every set of intervals $I_m$ on the real axis, $m = 1, \ldots, N$, we have

$$m\{x \in (0,1): \xi_m'(x) \in I_m,\ m = 1, \ldots, N\}$$
$$= \prod_{n=1}^N (2\pi)^{-1/2} \int_{I_m} e^{-y^2/2} dy. \qquad (59)$$

Let $\overline{\xi}(x)$ and $\overline{\xi}'(x)$ denote the vectors

$$\overline{\xi}(x) = \{\xi_1(x), \ldots, \xi_N(x)\}, \qquad \overline{\xi}'(x) = \{\xi_1'(x), \ldots, \xi_N'(x)\},$$

and let $\pi_0$ be the parallelepiped

$$\pi_0 = \{y = \{y_m\}_{m=1}^N : y_m \in I_m,\ m = 1, \ldots, N\} \subset R^N$$

(we suppose that a standard basis $\{e_n\}_{n=1}^N$ has been given in $R^N$ and that the coordinates of vectors refer to this basis).

In addition, let $A^{-1}(\pi_0)$ be the preimage of $\pi_0$ under the rotation in $R^N$ defined by $A$ (i.e., under the rotation in which $\{y_n\}_{n=1}^N$ is transformed into the vector $\{y_m'\}_{m=1}^N$ with $y_m' = \sum_{n=1}^N a_{m,n} y_n$. Then

$$m\{x \in (0,1): \xi_m'(x) \in I_m,\ m = 1, \ldots, N\}$$
$$= m\{x \in (0,1): \overline{\xi}(x) \in A^{-1}(\pi_0)\}. \qquad (60)$$

For every parallelepiped $\pi$ with edges parallel to the coordinate axes, we have by hypothesis

$$m\{x \in (0,1): \overline{\xi}(x) \in \pi\} = (2\pi)^{N/2} \int_\pi \exp\left\{-\frac{1}{2} \sum_{n=1}^N y_n^2\right\} dy_1 \cdots dy_N,$$

consequently, we also have, for every set $P \subset R^N$ of the form

$$P = \bigcup_{s=1}^{s'} \overline{\pi}_s, \qquad \pi_s \cap \pi_q = \varnothing \text{ with } s \neq q \qquad (61)$$

(where $\pi_s$ is an open parallelepiped with edges parallel to the coordinate axes, and $\bar{\pi}_s$ is its closure), the equation

$$m\{x \in (0,1): \bar{\xi}(x) \in P\}$$

$$= (2\pi)^{-N/2} \int_P \exp\left\{-\frac{1}{2}\sum_{n=1}^{N} y_n^2\right\} dy_1 \ldots dy_N. \qquad (62)$$

But it is easily seen that for every $\varepsilon > 0$ we can find two sets $P'$ and $P''$ of the form (61) such that

a) $P' \subset A^{-1}(\pi_0) \subset P''$;     b) $m_N(P'') - m_N(P') \le \varepsilon$,

where $m_N$ is Lebesgue measure in $R^N$. Letting $\varepsilon$ tend to zero and using (60) and (62), we find that

$$m\{x \in (0,1): \xi'_m(x) \in I_m, \ m = 1,\ldots,N\}$$

$$= (2\pi)^{-N/2} \int_{A^{-1}(\pi_0)} \exp\left\{-\frac{1}{2}\sum_{n=1}^{N} y_n^2\right\} dy_1 \ldots dy_N. \qquad (63)$$

We make a change of variables in the integral (63) by introducing the variables $y'_m = \sum_{n=1}^{N}, \ldots, a_{m,n}y_n$, $m = 1, \ldots, N$, and using the invariance of $\exp\{-\frac{1}{2}\sum_{n=1}^{N} y_n^2\}$ under rotations in $R^N$. We find that the right-hand side of (63) equals

$$(2\pi)^{-N/2} \int_{\pi_0} \exp\left\{-\frac{1}{2}\sum_{m=1}^{N} (y'_m)^2\right\} dy'_1 \ldots dy'_N$$

$$= \prod_{m=1}^{N} (2\pi)^{-1/2} \int_{I_m} \exp\left\{-\frac{(y'_m)^2}{2}\right\} dy'_m.$$

This completes the proof of (59) and hence of Theorem 10.

## §3. Convergence for almost all choices of signs, and unconditional convergence

In this section we present applications of the results of §§1 and 2 to the convergence of series of functions. These results (see Theorems 11 and 12) are frequently used in the theory of orthogonal series.

DEFINITION 3. A series

$$\sum_{n=1}^{\infty} x_n, \qquad (64)$$

whose terms are elements of a linear metric space $X$ is said to *converge in X for almost all choices of signs* if the series

$$\sum_{n=1}^{\infty} r_n(t)x_n \qquad (65)$$

converges for almost all $t \in (0,1)$, in the metric of $X$ (here $\{r_n(t)\}$ is the Rademacher system (see §1)).

In exactly the same way we can define convergence of a series of functions almost everywhere for almost every choice of signs.

It is often very convenient to use the approach of applying random series (65) in order to study the properties of the series (64). For this reason, Definition 3 has become widely accepted and deserves careful study.

Definition 3 is directly connected with the definition of unconditional convergence of series in the space $X$ (see Theorem 1.1); but, as we shall see below, the study of convergence for almost all choices of signs is considerably simpler than for unconditional convergence.

THEOREM 11. *Let a series*

$$\sum_{n=1}^{\infty} f_n(x), \qquad x \in (0,1), \tag{66}$$

*be given. The following three conditions are equivalent:*

1) *the series* (66) *converges a.e. for almost every choice of signs;*
2) *the series* (66) *converges in measure for almost every choice of signs;*
3) *the sum* $\sum_{n=1}^{\infty} f_n^2(x)$ *is finite for almost all* $x \in (0,1)$.

PROOF. It is clear that 2) follows from 1) [1)⇒2)]. Let us prove that 2)⇒3) and then that 3)⇒1). Then Theorem 11 will be established.

Assume, on the contrary, that 2) is satisfied but

$$\sum_{n=1}^{\infty} f_n^2(x) = \infty \quad \text{for } x \in E, \; m(E) = \gamma > 0. \tag{67}$$

Then we can find a sequence $\{N_s\}_{s=1}^{\infty}$ of integers such that $1 < N_1 < N_2 < \cdots$ and

$$m(E_s) \equiv m \left\{ x \in (0,1): \sum_{n=N_s+1}^{N_{s+1}} f_n^2(x) > s \right\} > \frac{\gamma}{2}, \qquad s = 1,2,\ldots. \tag{68}$$

If we apply inequality a) of Theorem 7 to the polynomial

$$P_s(t;x) = \sum_{n=N_s+1}^{N_{s+1}} r_n(t) f_n(x)$$

in the Rademacher system (supposing that $x \in E_s$ is given), and use the fact that $\|P_s\|_2 > s^{1/2}$ for $x \in E_s$, by (68) we find that

$$m \left\{ t \in (0,1): |P_s(t,x)| > \tfrac{1}{2} s^{1/2} \right\} > c > 0, \qquad x \in E_s.$$

Therefore the planar measure

$$m_2 \left\{ (t,x) \in (0,1) \times (0,1) : |P_s(t,x)| > \frac{1}{2} s^{1/2} \right\}$$

exceeds

$$cm(E_s) > \frac{e\gamma}{2}. \tag{69}$$

It follows easily from (69) that if $G_s$ is the set

$$G_s = \left\{ t \in (0,1) : m \left\{ x \in (0,1) : |P_s(t,x)| > \frac{1}{2} s^{1/2} \right\} > \frac{c\gamma}{4} \right\},$$

then $m(G_s) \geq c\gamma/4$ for $s = 1, 2, \ldots$, and therefore $m(\overline{\lim}_{s \to \infty} G_s) \geq c\gamma/4$. However, by the very definition of the set $\overline{\lim}_{s \to \infty} G_s$, there are for each $t \in \overline{\lim}_{s \to \infty} G_s$ an infinite number of indices $s$ for which $t \in G_s$, i.e.,

$$m \left\{ x \in (0,1) : |P_s(t,x)| > \frac{1}{2} s^{1/2} \right\} > \frac{c\gamma}{4}.$$

Consequently for every $t \in \overline{\lim}_{s \to \infty} G_s$ the series

$$\sum_{n=1}^{\infty} r_n(t) f_n(x) \tag{70}$$

fails to converge in measure, in contradiction to condition 2). We have therefore proved that 2)$\Rightarrow$3).

Now let condition 3) be satisfied:

$$\sum_{n=1}^{\infty} f_n^2(x) < \infty \qquad \text{for } x \in E, \quad m(E) = 1. \tag{71}$$

Let us consider the set $F$ of points $(t,x)$ of the square $(0,1) \times (0,1)$ at which the series (70) does not converge. It follows from (71), by Theorem 9, that

$$m \{ t \in (0,1) : (t,x) \in F \} = 0$$

for every $x \in E$, and therefore by Fubini's theorem (see also (71))

$$\int_0^1 \int_0^1 \chi_F(t,x)\, dx\, dt = \int_0^1 \int_0^1 \chi_F(t,x)\, dt\, dx = \int_E \int_0^1 \chi_F(t,x)\, dt\, dx = 0,$$

i.e., $\int_0^1 \chi_F(t,x)\, dx = 0$ for a.e. $t \in (0,1)$. This implies the convergence of (70) a.e. (in $x$) for almost every choice of signs. This completes the proof of the implication 3)$\Rightarrow$1) and hence of Theorem 11.

The following corollary is an immediate consequence of Theorem 11 and Theorem 1 of Chapter 1.

COROLLARY 4. *If the series* (66) *converges unconditionally in measure* (*and* a fortiori *a.e.*), *then* $\sum_{n=1}^{\infty} f_n^2(x) < \infty$ *for almost all* $x \in (0,1)$.

The following result shows that the behavior of $(\sum_{n=1}^{\infty} f_n^2(x))^{1/2}$ serves also to describe convergence in $L^p(0,1)$, $1 \leq p < \infty$, for almost every choice of the signs in series of the form (66) with elements $f_n \in L^p(0,1)$.

THEOREM 12. *A necessary and sufficient condition for a series* (66) *with elements* $f_n \in L^p(0,1)$, $n = 1, 2, \ldots$, *to converge in* $L^p(0,1)$, $1 \leq p < \infty$, *for almost all choices of the signs, is that*

$$\left\{ \sum_{n=1}^{\infty} f_n^2(x) \right\}^{1/2} \in L^p(0,1). \tag{72}$$

LEMMA 1. *Let there be given a Rademacher series* $P(t) = \sum_{n=1}^{\infty} a_n r_n(t)$, $\sum_{n=1}^{\infty} a_n^2 < \infty$, *and let* $1 \leq p < \infty$. *The following inequalities hold:*
  a) $\|P^*(t)\|_p \leq C_p(\sum_{n=1}^{\infty} a_n^2)^{1/2}, P^*(t) = \sup_{1 \leq r < \infty} |\sum_{n=1}^{r} a_n r_n(t)|$;
  b) $\|P(t)\|_{L^p(E)} \geq \|P(t)\|_{L^1(E)} \geq c(\sum_{n=1}^{\infty} a_n^2)^{1/2}$
*for every set* $E \subset (0,1)$ *with* $m(E) > 1 - \delta$, *where* $c > 0$ *and* $\delta > 0$ *are positive absolute constants.*

Since the given series converges a.e. and in $L^p(0,1)$ (see Theorems 9 and 6), it is enough to prove Lemma 1 for the case when $P(t)$ is a polynomial $\sum_{n=1}^{N} a_n r_n(t)$. Then part a) is proved by first applying (57) to $P(t)$, and then applying Khinchin's theorem (Theorem 6). For the proof of part b), we recall that by Theorem 7

$$m(Q) \equiv m \left\{ t \in (0,1) : |P(t)| > \frac{1}{2} \left( \sum_{n=1}^{N} a_n^2 \right)^{1/2} \right\} > c' > 0.$$

Setting $\delta = \frac{1}{2} c'$, for any set $E \subset (0,1)$ with $m(E) > 1 - \delta$ we will have $m(E \cap Q) \geq \frac{1}{2} c'$ and

$$\|P\|_{L^1(E)} \geq \int_{E \cap Q} |P(t)| \, dt \geq \frac{c'}{2} \left( \sum_{n=1}^{N} a_n^2 \right)^{1/2} = c \left( \sum_{n=1}^{N} a_n^2 \right)^{1/2}.$$

This completes the proof of Lemma 1.

REMARK. Later (see Theorem 8.8) we shall prove that inequality a) of Lemma 1 holds not only for systems of independent functions but also for O.N.S. such that, for some $p > 2$, every polynomial $P(x)$ in the system satisfies the inequality $\|P\|_p \leq C\|P\|_2$.

PROOF OF THEOREM 12. To establish the sufficiency of condition (72) we assume the contrary, i.e., that (72) is satisfied, but for $t \in E$, $m(E) = \gamma > 0$,[2] the series (70) does not converge in $L^p(0,1)$. This means that there are sequences $\{N_k\}_{k=1}^\infty$ of integers and $\{M_k(t)\}_{k=1}^\infty$ of measurable integral-valued functions, with $N_k \leq M_k(t) < N_{k+1}$, $k = 1, \ldots, t \in (0,1)$, such that for $k = 1, 2, \ldots$,

$$\left\| \sum_{n=N_k}^{M_k(t)} r_n(t) f_n(x) \right\|_p \geq \rho > 0 \quad \text{for } t \in E_k, \ m(E_k) > \frac{\gamma}{2}.$$

Using Fubini's theorem, we deduce from the preceding inequalities that for $k = 1, 2, \ldots$

$$\int_0^1 \int_0^1 \left| \sum_{n=N_k}^{M_k(t)} r_n(t) f_n(x) \right|^p dt \, dx \geq \frac{\gamma}{2} \rho^p \equiv c > 0. \tag{73}$$

But by part a) of Lemma 1,

$$\int_0^1 \left| \sum_{n=N_k}^{M_k(t)} r_n(t) f_n(x) \right|^p dt \leq C_p \left( \sum_{n=N_k}^{N_{k+1}-1} f_n^2(x) \right)^{p/2} \tag{74}$$

for all $x \in (0,1)$. It follows from (73) and (74) that for $k = 1, 2, \ldots$

$$\int_0^1 \left( \sum_{n=N_k}^{N_{k+1}-1} f_n^2(x) \right)^{p/2} dx > c > 0.$$

However, this inequality contradicts (72), since it follows from (72) that

$$\lim_{N \to \infty} \left\| \left( \sum_{n=N}^\infty f_n^2 \right)^{1/2} \right\|_p = 0.$$

Hence the sufficiency of (72) is established; we now prove that it is necessary.

Let the series (70) converge in $L^p(0,1)$ for almost all $t \in (0,1)$. Then it evidently converges in measure for almost all $t \in (0,1)$, and consequently also converges almost everywhere, by Theorem 11. Therefore the series

---

[2]The measurability of $E$ follows from the facts that

$$E = \bigcup_{m=1}^\infty A_m, \quad \text{where } A_m = \bigcap_{s=1}^\infty \bigcup_{s<p \leq q} Q_{p,q,m}$$

and that $Q_{p,q,m} = \{t \in (0,1): \| \sum_{n=p}^q r_n(t) f_n(x) \|_p > m^{-1}\}$ consists of a finite number of binary intervals.

(70) converges for almost all points $(t, x)$ of the square $(0, 1) \times (0, 1)$, and its sum is measurable (as a function of two variables). Notice also that $\sum_{n=1}^{\infty} f_n^2(x) < \infty$ for almost all $x$ (see Theorem 11).

By hypothesis, the function

$$\psi(t) = \int_0^1 \left| \sum_{n=1}^{\infty} n = f_n(x) r_n(t) \right|^p dx$$

is finite for almost every $t \in (0, 1)$. Therefore we can find a set $E \subset (0, 1)$ with $m(E) > 1 - \delta$ ($\delta > 0$ is the constant in part b) of Lemma 1), and a constant $K$, such that $\psi(t) \leq K$ for $x \in E$.

Then by Fubini's theorem

$$\int_0^1 \int_E \left| \sum_{n=1}^{\infty} f_n(x) r_n(t) \right|^p dt\, dx = \int_E \psi(t)\, dt \leq K. \tag{74'}$$

If we estimate the inner integral on the left side of (74') by using part b) of Lemma 1 (remembering that the sum $\sum_{n=1}^{\infty} f_n^2(x)$ is finite for almost all $x$), we find that

$$\int_0^1 \left( \sum_{n=1}^{\infty} f_n^2(x) \right)^{p/2} dx \leq K.$$

This completes the proof of Theorem 12.

COROLLARY 5. *A sufficient condition for* (66) *to converge in* $L^p(0, 1)$, $1 \leq p < \infty$, *for almost all choices of the signs, is that*

$$\sum_{n=1}^{\infty} \|f_n\|_p^q < \infty; \qquad q = \min(2, p).$$

In fact, it is enough to show that

$$\left\| \left( \sum_{n=1}^{\infty} f_n^2 \right)^{1/2} \right\|_p \leq \left( \sum_{n=1}^{\infty} \|f_n\|_p^q \right)^{1/q}, \qquad q = \min(2, p). \tag{75}$$

For $1 \leq p \leq 2$, for every $x \in (0, 1)$,

$$\left( \sum_{n=1}^{\infty} f_n^2(x) \right)^{1/2} \leq \left( \sum_{n=1}^{\infty} |f_n(x)|^p \right)^{1/p}$$

(since $\|y\|_{l^r} \leq \|y\|_{l^p}$ if $1 \leq p \leq r$); consequently

$$\int_0^1 \left( \sum_{n=1}^{\infty} f_n^2(x) \right)^{p/2} dx \leq \int_0^1 \sum_{n=1}^{\infty} |f_n(x)|^p\, dx = \sum_{n=1}^{\infty} \|f_n\|_p^p.$$

For $2 < p < \infty$ we have

$$\left\|\left(\sum_{n=1}^{\infty} f_n^2\right)^{1/2}\right\|_p = \left\|\sum_{n=1}^{\infty} f_n^2\right\|_{p/2}^{1/2} \le \left(\sum_{n=1}^{\infty} \|f_n^2\|_{p/2}\right)^{1/2} = \left(\sum_{n=1}^{\infty} \|f_n\|_p^2\right)^{1/2}.$$

This establishes (75), and hence also Corollary 5.

Theorem 12 shows that the convergence of the series (66) in $L^p(0, 1)$, $1 \le p < \infty$, for almost all choices of the signs depends only on the behavior of $|f_n(x)|$ and not on the sign of $f_n(x)$, $n = 1, 2, \ldots$. For $p = \infty$, this is not the case. In fact, the series

$$\sum_{n=1}^{\infty} f_n^{(1)}(x), \qquad f_n^{(1)}(x) = a_n = \text{const}, \quad x \in (0, 1),$$

converges in $L^\infty(0, 1)$, for almost all choices of the signs, if and only if $\sum_{n=1}^{\infty} a_n^2 < \infty$ (see Theorems 8 and 9), although the series

$$\sum_{n=1}^{\infty} f_n^{(2)}(x), \qquad f_n^{(2)}(x) = a_n r_n(x), \quad x \in (0, 1),$$

converges in $L^\infty(0, 1)$ for almost all choices of signs only if $\sum_{n=1}^{\infty} |a_n| < \infty$, even though $|f_n^{(1)}(x)| = |f_n^{(2)}(x)|$ for almost all $x \in (0, 1)$.

The example $f_n^2(x) = a_n r_n(x)$, $n = 1, 2, \ldots$ (which can easily be modified to make $f_n^{(2)}(x)$ continuous), also shows that it is impossible to give a nontrivial condition (i.e., one better than the convergence of $\sum_{n=1}^{\infty} \|f_n\|_C$ that will ensure the uniform convergence of $\sum_{n=1}^{\infty} r_n(t)f_n(x)$ for almost all $t \in (0, 1)$. Instead of using information about the signs of the functions $f_n(x)$, we can in some cases obtain nontrivial propositions about the uniform convergence of this series for almost all $t \in (0, 1)$. For example, we shall show in §5 of Chapter 8 that if $f_n(x) = \int_0^x \varphi_n(y)\,dy$ for $n = 1, 2, \ldots$, where $\{\varphi_n\}_{n=1}^{\infty}$ is any O.N.S., then $\sum_{n=1}^{\infty} f_n(x)$ converges uniformly on $[0, 1]$ for almost every choice of signs (although the series $\sum_{n=1}^{\infty} \|f_n\|_C$ diverges for a wide class of O.N.S. $\{\varphi_n\}$, including the trigonometric system and the Haar system).

Later we shall sometimes use the following relation (which was actually applied in the proof of Theorem 12): for every set $\{f_n\}_{n=1}^{N} \subset L^p(0, 1)$ of functions, $1 \le p < \infty$,

$$c_p \left\|\left(\sum_{n=1}^{N} f_n^2\right)^{1/2}\right\|_p \le \left\{\int_0^1 \left\|\sum_{n=1}^{N} r_n(t)f_n\right\|_p^p dt\right\}^{1/p}$$

$$\le C_p \left\|\left(\sum_{n=1}^{N} f_n^2\right)^{1/2}\right\|_p, \tag{76}$$

where $C_p$ and $c_p > 0$ are numbers depending only on $p$. (The inequality (76) follows directly from Theorems 6 and 7, part b), and Fubini's theorem.)

We now prove, using (76), the theorem on unconditional bases that was stated in Chapter 1 (see Theorem 1.11).

1) Let $\{\varphi_n(x)\}_{n=1}^{\infty}$ be an unconditional basis in $L^p(0,1)$, $1 < p < \infty$, and $\{\psi_n(x)\}$ the conjugate system. Using Theorem 1.10 and Corollary 1.4, we find that for every function $f \in L^p(0,1)$ and every binary irrational $t \in (0,1)$ we have, for $N > N(f)$, the inequalities

$$b\|f\|_p \leq \left\| \sum_{n=1}^{N} r_n(t)\langle f, \psi_n \rangle \varphi_n \right\|_p \leq a\|f\|_p, \tag{77}$$

where the constants $b > 0$ and $a$ are independent of $f$. (The right-hand inequality in (77) holds for $N = 1, 2, \ldots$; the left-hand inequality holds for $N$ such that

$$\left\| f - \sum_{n=1}^{N} \langle f, \psi_n \rangle \varphi_n \right\| \leq \frac{1}{2}\|f\|_p.)$$

If we apply (76) when $f_n(x) = \langle f, \psi_n \rangle \varphi_n(x)$, $1 \leq n \leq N$, and then take limits as $N \to \infty$, we find from (77) that

$$m\|f\|_p \leq \left\| \left\{ \sum_{n=1}^{\infty} (\langle f, \psi_n \rangle \varphi_n)^2 \right\}^{1/2} \right\|_p \leq M\|f\|_p \qquad (m > 0). \tag{77'}$$

2) Let $\{\varphi_n(x)\}_{n=1}^{\infty}$ (with conjugate system $\{\psi_n(x)\}_{n=1}^{\infty}$) be a complete system, minimal in $L^p(0,1)$, such that (77') holds, and let there be given any set $\{\varepsilon_n\}_{n=1}^{N}$ with $\varepsilon_n = \pm 1$ for $n = 1, \ldots, N$. Then it follows from (77') that, for every $f \in L^p(0,1)$,

$$\left\| \sum_{n=1}^{N} \varepsilon_n \langle f, \psi_n \rangle \varphi_n \right\|_p \leq m^{-1} \left\| \left\{ \sum_{n=1}^{N} (\varepsilon_n \langle f, \psi_n \rangle \varphi_n)^2 \right\}^{1/2} \right\|_p$$

$$\leq m^{-1} \left\| \left\{ \sum_{n=1}^{\infty} (\langle f, \psi_n \rangle \varphi_n)^2 \right\}^{1/2} \right\|_p \leq M \cdot m^{-1} \|f\|_p,$$

and therefore (see Theorem 1.10) $\{\varphi_n(x)\}$ is an unconditional basis in $L^p(0,1)$.

This completes the proof of Theorem 1.11.

In Chapter 1 we have already mentioned that there are no unconditional bases in $L^1(0,1)$. We now prove this statement using (76).

THEOREM 13. *There is no unconditional basis in* $L^1(0, 1)$.

PROOF. We notice first that for every function $g(x) \in L^1(0, 1)$

$$
\begin{array}{ll}
\text{1) } \lim_{n \to \infty} \int_0^1 g(x) r_n(x)\, dx = 0; & (78) \\
\text{2) } \lim_{n \to \infty} \|g(x) + r_n(x) g(x)\|_1 = g(x)\|_1,
\end{array}
$$

where $\{r_n(x)\}$ is the Rademacher system.

Equation 1) in (78) follows from Theorem 1.5. In addition, from the definition of the Rademacher system (see (9)) we have

$$
\begin{aligned}
\|g + r_n g\|_1 - \|g\|_1 &= \sum_{i=1}^{2^n} \left[ \int_{(i-1)/2^n}^{i/2^n} |g(x)| \cdot |1 + r_n(x)|\, dx \right. \\
&\qquad\qquad\qquad \left. - \int_{(i-1)/2^n}^{i/2^n} |g(x)|\, dx \right] \\
&= \sum_{i=1}^{2^n} (-1)^{i-1} \cdot \int_{(i-1)/2^n}^{i/2^n} |g(x)|\, dx \\
&\le \int_0^{1-2^{-n}} ||g(x)| - |g(x + 2^{-n})||\, dx \le \omega_1(2^{-n}, |g|),
\end{aligned}
$$

where $\omega_1(\delta, |g|)$ is the integral modulus of continuity of $|g(x)|$, and $\lim_{\delta \to 0} \omega_1(\delta, g) = 0$ for every $g \in L^1(0, 1)$ (see Appendix 1, §1). Therefore part 2) of (78) follows from the preceding inequalities.

Now let us assume the contrary of the theorem: let $\{\varphi_n(x)\}$ be an unconditional basis in $L^1(0, 1)$, and let $\{\psi_n(x)\} \in L^\infty(0, 1)$ be the conjugate system, i.e., (see Corollary 1.1) for every $f \in L^1(0, 1)$ the series

$$
f(x) = \sum_{n=1}^{\infty} \langle f, \psi_n \rangle \varphi_n(x)
$$

converges unconditionally in $L^1(0, 1)$.

We now construct inductively a sequence of functions $g_s(x)$, $s = 1, 2, \ldots$:

$$
g_0(x) = 1, \qquad g_s(x) = \sum_{j=0}^{s-1} g_j(x) \cdot r_{k_s}(x), \qquad x \in (0, 1), \; s = 1, 2, \ldots,
$$

where the sequence of integers $k_s$, $s = 1, 2, \ldots$, increases so rapidly that

a) $\|g_s - u_s\|_1 \le 2^{-s-1}$, where $u_s$, $s = 0, 1, \ldots$, are "nonintersecting" polynomials in the basis $\{\varphi_n(x)\}$, i.e.,

$$
u_s(x) = \sum_{n=M_s}^{M'_s} a_n \varphi_n(x), \qquad M_s \le M'_s < M_{s+1}, \; s = 0, 1, \ldots;
$$

b)

$$\frac{1}{2} < \|g_s\|_1 = \left\|\sum_{j=0}^{s-1} g_j\right\|_1 = \left\|(1 + r_{k_{s-1}})\sum_{j=0}^{s-2} g_j\right\|_1 < 2, \qquad s = 2, 3, \ldots.$$

The possibility of constructing the sequence $\{k_s\}$ follows easily from (78) (in particular, we have used the fact that, by (78), 1), $\lim_{\delta\to\infty}\langle g \cdot r_s, \psi_n\rangle = 0$ for all $g \in L^1(0,1)$ and $n = 1, 2, \ldots$; and therefore we have

$$\lim_{s\to\infty}\left\|\sum_{n=1}^{N}\langle g \cdot r_s, \psi_n\rangle\varphi_n\right\|_1 = 0$$

for each given $N$).

It follows from inequalities a) and b) that

$$\left\|\sum_{j=0}^{s} u_j(x)\right\|_1 \leq 3, \qquad \|u_j(x)\|_1 \geq \frac{1}{4}, \ s, j = 0, 1, \ldots. \tag{79}$$

But since $\{\varphi_n(x)\}$ is an unconditional basis, there is a constant $K$ such that, for almost all $t \in (0,1)$ and $s = 0, 1, \ldots$,

$$\left\|\sum_{j=0}^{s} r_j(t)u_j(x)\right\|_1 = \left\|\sum_{j=0}^{s} r_j(t)\sum_{n=M_j}^{M_j'} a_n\varphi_n(x)\right\|_1$$
$$\leq K\left\|\sum_{j=0}^{s}\sum_{n=M_j}^{M_j'} a_n\varphi_n(x)\right\|_1 \leq 3K. \tag{80}$$

On the other hand, by (76) (see also (79))

$$\int_0^1 \left\|\sum_{j=0}^{s} r_j(t)u_j(x)\right\|_1 dt \geq c\left\|\left\{\sum_{j=0}^{s} u_j^2(x)\right\}^{1/2}\right\|_1$$
$$\geq c(s+1)^{-1/2}\sum_{j=0}^{s}\|u_j\|_1 \geq \frac{c}{4}(s+1)^{1/2}. \tag{81}$$

For sufficiently large $s$, inequalities (80) and (81) are mutually contradictory. This contradiction, resulting from the supposition that $\{\varphi_n\}$ is an unconditional basis in $L^1(0,1)$ establishes Theorem 13.

In Chapter 1 we mentioned the following theorem on series that converge a.e. for all choices of signs.

THEOREM 14. *If the series* (66) *has the property that the series*

$$\sum_{n=1}^{\infty} \varepsilon_n f_n(x) \tag{82}$$

*converges a.e. on* $(0,1)$ *for every sequence* $\{\varepsilon_n\}_{n=1}^{\infty}$, $\varepsilon_n = \pm 1$, *then every series of the form*

$$\sum_{n=1}^{\infty} \lambda_n f_n(x), \qquad \{\lambda_n\}_{n=1}^{\infty} \in l^{\infty} \tag{83}$$

*also converges a.e. on* $(0,1)$.

COROLLARY 6. *The unconditional convergence a.e. of the series* (66) *implies the unconditional convergence almost everywhere of* (83).

PROOF OF COROLLARY 6. It is clear that the series $\sum_{n=1}^{\infty} f_{\sigma(n)}(x)$ converges unconditionally almost everywhere for every permutation $\{\sigma(n)\}_{n=1}^{\infty}$. From this, it follows easily that every series of the form $\sum_{n=1}^{\infty} \varepsilon_n f_{\sigma(n)}(x)$, $\varepsilon_n = \pm 1$, $n = 1, 2, \ldots$, converges almost everywhere. It then remains only to apply Theorem 14 in order to obtain the convergence of $\sum_{n=1}^{\infty} \lambda_{\sigma(n)} f_{\sigma(n)}(x)$ almost everywhere.

The following lemma is needed for the proof of Theorem 14.

LEMMA 1. *Let there be given functions* $f_n \in L^2(0,1)$, $n = 1, \ldots, N$, *and numbers* $\lambda_n$ *with* $|\lambda_n| \le 1$ *for* $n = 1, \ldots, N$. *Then there is a set of numbers* $\varepsilon_n = \pm 1$, $n = 1, \ldots, N$, *such that*

$$\left\| \sum_{n=1}^{N} \lambda_n f_n - \sum_{n=1}^{N} \varepsilon_n f_n \right\|_2^2 \le \sum_{n=1}^{N} \|f_n\|_2^2. \tag{84}$$

PROOF. We may suppose without loss of generality that $|\lambda_n| < 1$, $n = 1, \ldots, N$ (in the contrary case we set $\varepsilon_n = \lambda_n$ for $|\lambda_n| = 1$ and prove the lemma for the remaining functions $f_n(x)$ and numbers $\lambda_n$). Let $\{g_n(t)\}_{n=1}^{N}$, $t \in (0,1)$, be a set of independent functions such that $g_n(t)$ takes, for $n = 1, \ldots, N$, only the two values $\lambda_n + 1$ and $\lambda_n - 1$, and that

$$\begin{aligned} m\{t \in (0,1) : g_n(t) = \lambda_n + 1\} &= (1 - \lambda_n)/2, \\ m\{t \in (0,1) : g_n(t) = \lambda_n - 1\} &= (1 + \lambda_n)/2. \end{aligned} \tag{85}$$

(The existence of such a set $\{g_n(t)\}$ was established in Theorem 1.) It follows from (85) that

$$\int_0^1 g_n(t)\, dt = 0, \qquad \int_0^1 g_n^2(t)\, dt = 1 - \lambda_n^2 \le 1, \quad n = 1, \ldots, N. \tag{86}$$

Since the functions $g_n(t)$, $n = 1, \ldots, N$, are pairwise orthogonal (see Corollary 3), we find from (86) that

$$
\int_0^1 \int_0^1 \left[ \sum_{n=1}^N g_n(t) f_n(x) \right]^2 dx \, dt = \int_0^1 \int_0^1 \left[ \sum_{n=1}^N g_n(t) f_n(x) \right]^2 dt \, dx
$$

$$
= \int_0^1 \sum_{n=1}^N f_n^2(x) \| g_n \|_2^2 \, dx
$$

$$
\leq \sum_{n=1}^N \| f_n \|_2^2 (1 - \lambda_n^2) \leq \sum_{n=1}^N \| f_n \|_2^2,
$$

and therefore there is a point $t_0 \in (0, 1)$ for which

$$
\left\| \sum_{n=1}^N g_n(t_0) f_n(x) \right\|_2^2 \leq \sum_{n=1}^N \| f_n \|_2^2. \tag{87}
$$

If we choose the set $\{\varepsilon_n\}_{n=1}^N$, $\varepsilon_n = \pm 1$, so that $g_n(t_0) = \lambda_n - \varepsilon_n$ for $n = 1, \ldots, N$ (see (85)), we can obtain the inequality (84) that we need from (87). This completes the proof of Lemma 1.

PROOF OF THEOREM 14. Let us suppose the contrary, i.e., that under the hypotheses of the theorem there is a series of the form (83), with $|\lambda_n| \leq 1$ for $n = 1, 2, \ldots$, which diverges at each point of a set $E \subset (0, 1)$ with $m(E) > 0$.

By Theorem 11, the convergence a.e. of all series of the form (82) implies that $\sum_{n=1}^\infty f_n^2(x) < \infty$ for almost all $x \in (0, 1)$. Therefore we can find a set $E_1 \subset (0, 1)$, with $m(E_1) > 1 - m(E)$, such that

$$
\sum_{n=1}^\infty n = \int_{E_1} f_n^2(x) \, dx < \infty.
$$

Set $E_2 = E \cap E_1$, and $g_n(x) = f_n(x) \chi_{E_1}(x)$, $n = 1, 2, \ldots$. Then it is clear that

a) the series $\sum_{n=1}^\infty \varepsilon_n g_n(x)$ converges a.e. on $(0, 1)$ for all choices of the signs $\varepsilon_n = \pm 1$, $n = 1, 2, \ldots$;

b) the series $\sum_{n=1}^\infty \lambda_n g_n(x)$ diverges for $x \in E_2$, $m(E_2) > 0$;

c) $\sum_1^\infty \| g_n \|_2^2 < \infty$.

We are going to show that a), b), and c) cannot all be satisfied. Hence we shall obtain a contradiction.

From b) there follows the existence of a number $\varepsilon_0 > 0$, sets $A_k$ of measure $m(A_k) \geq \gamma > 0$, $k = 1, 2, \ldots$, integers $N_1 < N_2 < \cdots < N_k < \cdots$, and functions $M_k(x)$, measurable on $(0, 1)$, with $N_k < M_k(x) \leq N_{k+1}$,

$x \in (0,1)$, $k = 1, 2, \ldots$, such that

$$\left| \sum_{n=N_k+1}^{M_k(x)} \lambda_n g_n(x) \right| > \varepsilon_0 \quad \text{for } x \in A_k; \ k = 1, 2, \ldots. \tag{88}$$

For $k = 1, 2, \ldots$, let $\delta_k = \sum_{n=N_k+1}^{N_{k+1}} \|g_n\|_2^2$. It follows from c) that

$$\lim_{k \to \infty} \delta_k = 0. \tag{89}$$

We fix $k$ and, for $N_k < n \leq N_{k+1}$ and $x \in (0,1)$, set

$$u_n(x) = \begin{cases} g_n(x) & \text{if } n \leq M_k(x), \\ 0 & \text{if } n > M_k(x). \end{cases} \tag{90}$$

The functions $u_n(x)$ are measurable and belong to $L^2(0,1)$, since $g_n(x)$ and $M_k(x)$ are measurable, $g_n \in L^2(0,1)$, and (see (88))

$$\left| \sum_{n=N_k+1}^{N_{k+1}} \lambda_n u_n(x) \right| > \varepsilon_0 \quad \text{for } x \in A_k. \tag{91}$$

By Lemma 1 there is a set $\{\varepsilon_n'\}_{n=N_k+1}^{N_{k+1}}$, $\varepsilon_n' = \pm 1$ such that

$$\left\| \sum_{n=N_k+1}^{N_{k+1}} \lambda_n U_n - \sum_{n=N_k+1}^{N_{k+1}} \varepsilon_n' u_n \right\|_2^2 \leq \sum_{n=N_k+1}^{N_{k+1}} \|u_n\|_2^2 \leq \sum_{n=N_k+1}^{N_{k+1}} \|g_n\|_2^2 = \delta_k.$$

If we apply Chebyshev's inequality to the preceding inequality and use (89), we find that for $k > k_0$

$$m(G_k) \equiv m\left\{ x \in (0,1): \left| \sum_{n=N_k+1}^{N_{k+1}} \lambda_n u_n(x) - \sum_{n=N_k+1}^{N_{k+1}} \varepsilon_n' u_n(x) \right| > \frac{\varepsilon_0}{2} \right\}$$

$$\leq \frac{4\delta_k}{\varepsilon_0^2} < \frac{1}{2}\gamma.$$

This implies that $m(A_k \backslash G_k) > \gamma/2$ for $k > k_0$. But for $x \in A_k \backslash G_k$ we have, by (91),

$$\left| \sum_{n=N_k+1}^{N_{k+1}} \varepsilon_n' u_n(x) \right| \geq \frac{\varepsilon_0}{2},$$

or equivalently (see (90)),

$$\left| \sum_{n=N_k+1}^{M_k(x)} \varepsilon_n' g_n(x) \right| \geq \frac{\varepsilon_0}{2}, \qquad x \in A_k \backslash G_k, \ k = k_0 + 1, \ldots.$$

Therefore the series $\sum_{n=1}^{\infty} \varepsilon'_n g_n(x)$, $\varepsilon'_n = \pm 1$, diverges at every point of the set $\overline{\lim}_{k\to\infty}(A_k\backslash G_k)$; since $m[\overline{\lim}_{k\to\infty}(A_k\backslash G_k)] \geq \gamma/2 > 0$, this contradicts condition a). This completes the proof of Theorem 14.

REMARK. There is an analog of Theorem 14 for convergence in measure. More precisely, it is easy to show, by using Lemma 1 and Corollary 4, that for a given sequence $\{f_n(x)\}_{n=1}^{\infty}$ the convergence of all series of the form (82) implies the convergence in measure of (83). (For convergence in any normed space, the analog of Theorem 14 can be proved very simply.

## §4. Random permutations

The applicability of the ideas used in §2 of this chapter is far from being restricted to systems of independent functions. In this section we discuss, in analogy with Theorem 9 of §2, a problem on the behavior of functions on the group $S(N)$ of permutations of the numbers $\{1,\ldots,N\}$.

Let us choose a set of numbers $\{a_n\}_{n=1}^{N}$. For $\sigma = \{\sigma(n)\}_{n=1}^{N} \in S(N)$ we set

$$S^*(\sigma) = \max_{1\leq p\leq N} |S_p(\sigma)|, \qquad S_p(\sigma) = S_p(\sigma, \{a_n\}) = \sum_{n=1}^{p} a_{\sigma(n)}. \tag{92}$$

Inequalities for $S^*(\sigma)$ and, in particular, inequalities for the measure of sets[3] $\{\sigma \in S(N): S^*(\sigma) > t\}$, $t > 0$, turn out to be useful in problems on rearrangements of orthogonal series. In particular, it will be shown at the end of this section (see Theorem 16), on the basis of these inequalities, that the terms of every orthogonal series with coefficients in $l^2$ can be rearranged in such a way that the resulting series is convergent a.e.

The following theorem shows that under the condition $\sum_{n=1}^{N} a_n = 0$ the behavior of the function (92) is closely connected with the behavior of a polynomial in the Rademacher system,

$$P(x) = \sum_{n=1}^{N} a_n r_n(x). \tag{93}$$

THEOREM 15. *There are absolute constants $C$ and $\gamma > 0$ such that, for $N = 1, 2, \ldots$, for every set of numbers $\{a_n\}_{n=1}^{N}$ with $\sum_{n=1}^{N} a_n = 0$,*

$$\mu\{\sigma \in S(N): S^*(\sigma) > t\} \leq Cm\{x \in (0,1): |P(x)| > \gamma t\}, \qquad t > 0.$$

We first introduce the following notation: if $\sigma = \{\sigma(n)\} \in S(N)$, we denote by $\overline{\sigma} = \{\overline{\sigma}(n)\}_{n=1}^{N}$ the permutation such that $\overline{\sigma}(n) = \sigma(N - n + 1)$,

---

[3] We suppose that we are given the natural measure $\mu = \mu_N$ on $S(N)$ for which the measure of each permutation $\sigma \in S(N)$ is equal to $1/N!$.

$n = 1, \ldots, N$. In addition, if $\{a_n\}_{n=1}^N$ is given, then for $\sigma \in S(N)$, $t <$ $S^*(\sigma)$, we set

$$N_t(\sigma) = \min\{p: |S_p(\sigma, \{a_n\})| > t\}.$$

LEMMA 1. *Let there be given a set of numbers* $\{b_n\}_{n=1}^m$ *with* $|\sum_{n=1}^m b_n| >$ $(\sum_{n=1}^m b_n^2)^{1/2}$, *and let* $1 \le p \le m/2$. *Then*

$$\mu\left\{\sigma \in S(m): |S_p(\sigma, \{b_n\})| \le \left|\sum_{n=1}^m b_n\right| \sqrt{\frac{7}{8}} \ge \frac{1}{7}\right\}.$$

PROOF OF LEMMA 1. It is easy to see that for all numbers $n$ and $k$, $1 \le n$, $k \le m$, $n \ne k$, we have the equations

$$\mu\{\sigma \in S(m): 1 \le \sigma(n) \le p\} = \frac{(m-1)!}{m!} p = \frac{p}{m},$$

$$\mu\{\sigma \in S(m): 1 \le \sigma(n), \sigma(k) \le p\} = \frac{(m-2)!(p-1)p}{m!} = \frac{(p-1)p}{(m-1)m},$$

for $p = 1, 2, \ldots$. Using these equations, we obtain

$$\frac{1}{m!} \sum_{\sigma \in S(m)} S_p^2(\sigma, \{b_n\}) = \frac{p}{m} \sum_{n=1}^m b_n^2 + \frac{2(p-1)p}{(m-1)m} \sum_{n \ne k} b_n b_k$$

$$= \left(\frac{p}{m} - \frac{(p-1)p}{(m-1)m}\right) \sum_{n=1}^m b_n^2$$

$$+ \frac{(p-1)p}{(m-1)m} \left(\sum_{n=1}^m b_n\right)^2$$

$$\le \frac{1}{2} \sum_{n=1}^m b_n^2 + \frac{1}{4} \left(\sum_{n=1}^m b_n\right)^2 \le \frac{3}{4} \left(\sum_{n=1}^m b_n\right)^2.$$

Therefore, by Chebyshev's inequality,

$$\mu\left\{\sigma \in S(m): |S_p(\sigma, \{b_n\})| > \left|\sum_{n=1}^m b_n\right| \sqrt{\frac{7}{8}}\right\}$$

$$\le \frac{3}{4} \left(\sum_{n=1}^m b_n\right)^2 \left[\sqrt{\frac{7}{8}} \sum_{n=1}^m b_n\right]^{-2} \le \frac{6}{7}.$$

PROOF OF THEOREM 15. We first prove that under the hypotheses of Theorem 15 we have the inequality

$$\mu\{\sigma \in S(N): S^*(\sigma) > t\} \le 14\mu\left\{\sigma \in S(N): |S_{[N/2]}(\sigma)| > \frac{t}{16}\right\},$$

$$t > \left(\sum_{n=1}^N a_n^2\right)^{1/2}, \quad (94)$$

and then we estimate the right-hand side of (94). Let $t > (\sum_{n=1}^{N} a_n^2)^{1/2}$ be given and

$$G_t' = \{\sigma \in S(N): S^*(\sigma) > t\}, \qquad G_t = \{\sigma \in G_t': N_t(\sigma) \leq [N/2]\}. \tag{95}$$

Notice that

$$\mu(G_t') \leq 2\mu(G_t). \tag{96}$$

In fact, it is clear that if $\sigma \in G_t'$ then $\overline{\sigma} \in G_t'$ also; but since (because $\sum_{n=1}^{N} a_n = 0$) when $\sigma \in G_t'$ we have $N_t(\sigma) + N_t(\overline{\sigma}) \leq N$, either $\sigma$ or $\overline{\sigma}$ is in $G_t$; this establishes (96).

We represent $G_t$ in the form

$$G_t = \bigcup_{q=1}^{[N/2]} G_t(q), \qquad G_t(q) = \{\sigma \in S(N): N_t(\sigma) = q\}. \tag{97}$$

It is clear that the sets $G_t(q)$ are disjoint for different values of $q$, and therefore

$$\mu(G_t) = \sum_{q=1}^{[N/2]} \mu(G_t(q)). \tag{98}$$

Moreover, for a permutation $\sigma = \{\sigma(n)\}_{n=1}^{N} \in G_t(q)$, let us denote by $\omega = \omega(\sigma)$ the set $\{\sigma(n)\}_{n=1}^{q}$, and set

$$G_t(q, \omega) = \{\sigma' = \{\sigma'(n)\}_{n=1}^{N} \in G_t(q): \sigma'(n) = \sigma(n), \ n = 1, \ldots, q\}.$$

It is easily seen that $G_t(q, \omega)$ contains all permutations $\{\sigma'(n)\} \in S(N)$ such that $\sigma'(n) = \sigma(n)$, $1 \leq n \leq q$, i.e.,

$$G_t(q, \omega) = \{\sigma' = \{\sigma'(n))\} \in S(N): \sigma'(n) = \sigma(n), \ n = 1, \ldots, q\}. \tag{99}$$

Since the sets $G_t(q, \omega)$ are disjoint, we have

$$\mu(G_t(q)) = \sum_{\omega} \mu(G_t(q, \omega)), \tag{100}$$

where the summation in (100) is over all distinct sets $\omega = \omega(\sigma) = \{\sigma(n)\}_{n=1}^{q}$ with $\sigma = \{\sigma(n)\}_{n=1}^{N} \in G_t(q)$.

Let us consider the set $G_t(q, \omega), 1 \leq q < [N/2]$, in more detail. Let $S = \sum_{n=1}^{q} a_{\sigma(n)}$; then $\sum_{n=q+1}^{N} a_{\sigma(n)} = -S$, and we have, by the construction of $G_t(q)$ and the choice of $t$ (see (94))

$$|S| > t > \left( \sum_{n=q+1}^{N} a_{\sigma(n)}^2 \right)^{1/2}. \tag{101}$$

In addition, if

$$\left| \sum_{n=q+1}^{[N/2]} a_{\sigma(n)} \right| \leq |S|\sqrt{\frac{7}{8}},$$

then

$$\left| \sum_{n=1}^{[N/2]} a_{\sigma(n)} \right| = \left| \sum_{n=1}^{q} a_{\sigma(n)} + \sum_{n=q+1}^{[N/2]} a_{\sigma(n)} \right|$$

$$\geq |S| \left( 1 - \sqrt{\frac{7}{8}} \right) \geq \frac{t}{16}. \tag{102}$$

But it follows from (99) and Lemma 1 (applied for $m = N - 1$, $p = [N/2] - q$, $b_n = a_{\sigma(n+q)}$; see also (101)) that when $1 \leq q \leq [N/2]$,

$$\mu[G_t(q, \omega)] \leq 7\mu \left\{ \sigma' \in G_t(q, \omega) : \left| \sum_{n=q+1}^{[N/2]} a_{\sigma'(n)} \right| \leq |S| \sqrt{\frac{7}{8}} \right\}.$$

Consequently, we find by using (102) that, for $1 \leq q \leq [N/2]$ and every set $\omega = \omega(\sigma)$, $\sigma \in G_t(q)$

$$\mu[G_t(q, \omega)] \leq 7\mu[G_t(q, \omega) \cap \{\sigma' \in S(N) : |S_{[N/2]}(\sigma)| > t/16\}]. \tag{103}$$

Combining the inequalities (102) for all different $\omega(\sigma)$, $\sigma \in G_t(q)$, $q = 1, 2, \ldots, [N/2]$, and using (96), (98), and (100), we find that

$$\mu(G_t') \leq 2\mu(G_t) \leq 14 \sum_{q=1}^{[N/2]} \mu \left\{ \sigma \in G_t(q) : |S_{[N/2]}(\sigma)| > \frac{t}{16} \right\}$$

$$\leq 14\mu \left\{ \sigma \in S(N) : |S_{[N/2]}(\sigma)| > \frac{t}{16} \right\}.$$

This completes the proof of (94). To complete the proof of Theorem 15, we now estimate

$$\mu \left\{ \sigma \in S(N) : |S_{[N/2]}(\sigma)| > \frac{t}{16} \right\}.$$

In the first place we observe that we may suppose that the theorem has been proved for $t \leq 16(\sum_{n=1}^{N} a_n^2)^{1/2}$, since in this case (by inequality a) of Theorem 7) we have

$$Cm \left\{ x \in (0, 1) : \left| \sum_{n=1}^{N} a_n r_n(x) \right| > \gamma t \right\} > 1$$

provided that $C$ is sufficiently large and $\gamma$ is sufficiently small. Moreover, we may suppose that $N > 16$, since for small $N$ the necessary estimate is easily obtained by appropriate choice of $C$ and $\gamma$.

For every $y > (\sum_{n=1}^{N} a_n^2)^{1/2}$ and $N > 16$, by using Lemma 1, for all $p$ such that $[N/2 - \sqrt{N}] \leq p \leq [N/2]$ we obtain

$$\mu\{\sigma \in S(N) : |S_{[N/2]}(\sigma)| > y\} \leq 7\mu \left\{ \sigma \in S(N) : |S_p(\sigma)| > y \left( 1 - \sqrt{\frac{7}{8}} \right) \right\}$$

$(\sqrt{N} < \frac{1}{2}(N/2)$ for $N > 16)$, and therefore for $N > 16$

$$\mu\{\sigma \in S(N): |S_{[N/2]}(\sigma)| > y\}$$

$$\leq 7(N-1)^{-1/2} \sum_{p=[N/2-\sqrt{N}]}^{[N/2]} \mu\left\{\sigma \in S(N): |S_p(\sigma)| > y\left(1 - \sqrt{\frac{7}{8}}\right)\right\}.$$

$$(104)$$

We shall estimate the right-hand side of (104) by using properties of the Rademacher system. It is easily seen that, for every $y > 0$,

$$\mu\{\sigma \in S(N): |S_p(\sigma)| > y\} = (C_N^p)^{-1} \sum_{\{\sigma(n)\}} 1, \tag{105}$$

where $C_N^p$ in (105) is the binomial coefficient, and the summation is over all sets $\{\sigma(n)\}_{n=1}^p$, $1 \leq \sigma(< \sigma(2) < \cdots < \sigma(p) \leq N$, such that $|\sum_{n=1}^p a_{\sigma(n)}| > y$.

But it follows from the most elementary properties of the Rademacher system that when $\sum_{n=1}^N a_n = 0$ the right-hand side of (105) is equal to

$$(C_N^p)^{-1} \cdot \text{card}\left\{\{\varepsilon_n\}_{n=1}^N, \varepsilon_n = 0; 1: \sum_{n=1}^N \varepsilon_n = p, \left|\sum_{n=1}^N \omega_n a_n\right| > y\right\}$$

$$= (C_N^p)^{-1} 2^N m\left\{x \in (0,1): \frac{1}{2}\sum_{n=1}^N (r_n(x)+1) = p, \left|\sum_{n=1}^N a_n r_n(x)\right| > 2y\right\}.$$

$$(106)$$

If we take into account that $C_N^p > C2^N/\sqrt{N}$ for $[N/2 - \sqrt{N}] \leq p \leq [N/2]$,[4] we find from (104)–(106) that

$$\mu\{\sigma \in S(N): |S_{[N/2]}(\sigma)| > y\}$$

$$\leq C \sum_{p=[N/2-\sqrt{N}]}^{[N/2]} m\left\{x \in (0,1): \sum_{n=1}^N (r_n(x)+1) = 2p,\right.$$

$$\left.\left|\sum_{n=1}^N a_n r_n(x)\right| > 2y\left(1 - \sqrt{\frac{7}{8}}\right)\right\}$$

$$\leq Cm\left\{x \in (0,1): \left|\sum_{n=1}^N a_n r_n(x)\right| > \frac{y}{10}\right\}, \qquad y > \left(\sum_{n=1}^N a_n^2\right)^{1/2}.$$

$$(107)$$

Inequality (107), applied for $y = t/16$ (see also (94)) completes the proof of Theorem 15.

If we apply Theorems 5 and 6 to estimate the polynomial (93), we immediately obtain the following corollary from Theorem 15.

---

[4] This follows easily from Stirling's formula.

COROLLARY 7. *The following inequalities are valid for all sets of numbers* $\{a_n\}_{n=1}^N$ *with* $\sum_{n=1}^N a_n = 0$ :

a) *For every* $t \geq 0$

$$\mu\left\{\sigma \in S(N): S^*(\sigma) > t\left(\sum_{n=1}^N a_n^2\right)^{1/2}\right\} \leq C \exp(-\gamma t^2), \qquad \gamma > 0;$$

b) *For every* $p$, $1 \leq p < \infty$,

$$\left\{\frac{1}{N!}\sum_{\sigma \in S(N)} [S^*(\sigma)]^p\right\}^{1/p} \leq C_p\left(\sum_{n=1}^N a_n^2\right)^{1/2}.$$

We also have the following corollary:

COROLLARY 8. *For* $1 \leq p < \infty$ *and all sets* $\{a_n\}_{n=1}^N$ *of numbers,*

$$\left\{\frac{1}{N!}\sum_{\sigma \in S(N)} [S^*(\sigma)]^p\right\}^{1/p} \leq C_p\left[\left|\sum_{n=1}^N a_n\right| + \left(\sum_{n=1}^N a_n^2\right)^{1/2}\right].$$

In fact, let us set $a_n' = a_n$ for $n = 1, 2, \ldots, N-1$ and $a_N' = a_N - \sum_{n=1}^N a_n$. Then $\sum_{n=1}^N a_n' = 0$ and, by part b) of Corollary 7,

$$
\begin{aligned}
\left\{\frac{1}{N!}\sum_{\sigma \in S(N)} [S^*(\sigma, \{a_n'\})]^p\right\}^{1/p} &\leq C_p\left(\sum_{n=1}^N (a_n')^2\right)^{1/2} \\
&\leq C_p'\left[\left|\sum_{n=1}^N a_n\right| + \left(\sum_{n=1}^N a_n^2\right)^{1/2}\right].
\end{aligned}
\tag{108}
$$

But, for every permutation $\sigma$,

$$S^*(\sigma, \{a_n\}) \leq S^*(\sigma, \{a_n'\}) + \left|\sum_{n=1}^N a_n\right|.$$

Hence the conclusion of Corollary 8 follows from (108).

Let us give an application of these results to a problem on the convergence of orthogonal series.

As will be shown later (see Corollary 3.6 and Theorem 9.1) there is an orthogonal series of the form

$$\sum_{n=1}^\infty a_n \varphi_n(x), \qquad \sum_{n=1}^\infty a_n^2 < \infty, \tag{109}$$

which diverges a.e. on $(0, 1)$. However, it follows at once from Theorem 11, proved in §3, that for almost every $t \in (0, 1)$ the series

$$\sum_{n=1}^{\infty} a_n r_n(t)\varphi_n(x), \qquad \sum_{n=1}^{\infty} a_n^2 < \infty, \qquad (110)$$

converges for almost every $x \in (0, 1)$. In other words, by changing the signs of the terms of the series (109), we can always make the transformed series converge a.e. The disadvantage of transforming a series by changing the signs of its terms is that the sum of the series (109) has nothing to do with the sum (in the sense of convergence in $L^2(0, 1)$) of the series (110). A different possible transformation of the series (109), which does not change its sum (since (109) converges unconditionally in $L^2(0, 1)$), is permutation of the terms of the series.

The following result shows that by rearranging the terms of the series it is also possible to obtain convergence a.e.

THEOREM 16. *For every orthogonal series of the form* (109) *there is a permutation* $\sigma = \{\sigma(n)\}_{n=1}^{\infty}$ *of the integers such that the series*

$$\sum_{n=1}^{\infty} a_{\sigma(n)}\varphi_{\sigma(n)}(x) \qquad (111)$$

*converges a.e. on* $(0, 1)$, *and moreover*

$$\int_0^1 \sup_{1 \leq N < \infty} \left[ \sum_{n=1}^{N} a_{\sigma(n)}\varphi_{\sigma(n)}(x) \right]^2 dx \leq C \left( \sum_{n=1}^{\infty} a_n^2 \right)^{1/2}. \qquad (112)$$

PROOF. We may suppose without loss of generality that $\sum_{n=1}^{\infty} a_n^2 = 1$. Since the series (109) converges in $L^2(0, 1)$, we can find a sequence $\{N_k\}_{k=1}^{\infty}$, $0 = N < N_2 < \cdots$, such that

$$\left\| \sum_{n=N_k+1}^{N_{k+1}} a_n \varphi_n \right\|_2 = \left( \sum_{n=N_k+1}^{N_{k+1}} a_n^2 \right)^{1/2} \leq 2^{-k+1}, \qquad k = 1, 2, \ldots. \qquad (113)$$

Then it is clear from Beppo Levi's theorem that the sequence $S_{N_k}(x)$ of partial sums of (109) converges a.e. on $(0, 1)$, and

$$\left\| \sup_{2 \leq k < \infty} |S_{N_k}(x)| \right\|_2 \leq \left\| \sum_{k=1}^{\infty} \left| \sum_{n=N_k+1}^{N_{k+1}} a_n \varphi_n \right| \right\|_2$$

$$\leq \sum_{k=1}^{\infty} \left\| \sum_{n=N_k+1}^{N_{k+1}} a_n \varphi_n \right\|_2 \leq 2. \qquad (114)$$

We can construct a permutation $\sigma$ such that the block of integers $N_k < n \leq N_{k+1}$ is a cycle, i.e., $\sigma((N_k, N_{k+1}]) = (N_k, N_{k+1}]$, $k = 1, 2, \ldots$.

Let us fix $k$ and estimate the number

$$I_k = \frac{1}{(N_{k+1} - N_k)!} \sum_{\sigma = \{\sigma(n)\}_{n=N_k+1}^{N_{k+1}}} \int_0^1 \max_{N_k < p \leq N_{k+1}} \left( \sum_{n=N_k+1}^{p} a_{\sigma(n)} \varphi_{\sigma(n)}(x) \right)^2 dx,$$

(115)

where the summation is over the set $S(N_{k+1} - N_k)$ of permutations of the set $\{N_{k+1}, \ldots, N_{k+1}\}$. If we interchange integration and summation in (115), and use Corollary 8 and (113), we obtain

$$I_k \leq C \int_0^1 \left\{ \left| \sum_{n=N_k+1}^{N_{k+1}} a_n \varphi_n(x) \right| + \left[ \sum_{n=N_k+1}^{N_{k+1}} a_n^2 \varphi_n^2(x) \right]^{1/2} \right\}^2 dx \leq C 2^{-2k},$$

$$k - 1, 2, \ldots. \quad (116)$$

It follows from (116) that for each $k = 1, 2, \ldots$ there is a permutation $\sigma_k = \{\sigma_k(n)\}_{n=N_k+1}^{N_{k+1}}$ of the block $(N_k, N_{k+1}]$ of integers, for which

$$\int_0^1 [\delta_k(x)]^2 dx \equiv \int_0^1 \left[ \max_{N_k < p \leq N_{k+1}} \sum_{n=N_k+1}^{p} a_{\sigma(n)} \varphi_{\sigma(n)}(x) \right]^2 dx \leq C 2^{-2k}.$$

(117)

We can now define the required permutation $\sigma = \{\sigma(n)\}_{n=1}^{\infty}$ by setting $\sigma(n) = \sigma_k(n)$ for $k = 1, 2, \ldots$, $n = N_k + 1, \ldots, N_{k+1}$.

From the convergence of the series

$$\sum_{k=1}^{\infty} \int_0^1 |\delta_k(x)| \, dx$$

(see (117)) and Beppo Levi's theorem, it follows that $\lim \delta_k(x) = 0$ for almost every $x \in (0, 1)$, and therefore (since the subsequence $S_{N_k}(x)$ converges a.e. (see (114)) the series (111) converges for almost all $x \in (0, 1)$.

Moreover, from (113) and (117) we obtain

$$\int_0^1 \sup_{1 \le N < \infty} \left[ \sum_{n=1}^{N} a_{\sigma(n)} \varphi_{\sigma(n)}(x) \right]^2 dx$$

$$\le \int_0^1 \left[ \sup_{2 \le k < \infty} |S_{N_k}(x)| + \sup_{1 \le k < \infty} \delta_k(x) \right]^2 dx$$

$$\le 2 \int_0^1 \sup_{2 \le k < \infty} S_{N_k}^2(x) \, dx + 2 \int_0^1 \sup_{1 \le k < \infty} \delta_k^2(x) \, dx$$

$$\le 8 + 2 \int_0^1 \sum_{k=1}^{\infty} \delta_k^2(x) \, dx \le C,$$

where $C$ is an absolute constant. This completes the proof of Theorem 16.

CHAPTER III

# The Haar System

In this chapter we study the Haar system, one of the classical orthonormal systems. It appears that the Haar system can be considered to be the simplest complete orthonormal system. Its simplicity and naturalness account for its great value in the theory of functions.

## §1. Definition; form of the partial sums

Before defining the Haar system, we introduce the standard notation for binary intervals, which will be used throughout the rest of the text. A *binary interval* is an interval of the form $((i-1)/2^k, i/2^k)$, where $i = 1, \ldots, 2^k$, $k = 0, 1, \ldots$.

For $n = 2^k + i$, $i = 1, \ldots, 2^k$, $k = 0, 1, \ldots$, we write

$$\Delta_n = \Delta_k^i = \left( \frac{i-1}{2^k}, \frac{i}{2^k} \right); \qquad \overline{\Delta}_n = \left[ \frac{i-1}{2^k}, \frac{i}{2^k} \right];$$

$$\Delta_1 = \Delta_0^0 = (0,1); \qquad \overline{\Delta}_1 = [0,1]. \tag{1}$$

If $\delta < (0,1)$ is any interval, we denote by $\delta^+$ and $\delta^-$ the left-hand and right-hand halves of $\delta$ (not including the midpoint). In particular $(n = 2^k + i)$,

$$\Delta_n^+ = (\Delta_k^i)^+ = \left( \frac{i-1}{2^k}, \frac{2i-1}{2^{k+1}} \right) = \Delta_{k+1}^{2i-1},$$

$$\Delta_n^- = (\Delta_k^i)^- = \left( \frac{2i-1}{2^{k+1}}, \frac{i}{2^k} \right) = \Delta_{k+1}^{2i}. \tag{2}$$

The intervals $\{\Delta_k^i\}_{i=1}^{2^k}$ will be called *intervals of the kth block*, $k = 0, 1, \ldots$ We note for later use the following simple properties of the family of binary intervals:

1) $\Delta_k^i \cap \Delta_k^j = \varnothing$ for $i \neq j$, $i, j = 1, \ldots, 2^k$, $k = 1, 2, \ldots$

2) If $\Delta_n$ and $\Delta_m$ are binary intervals and $\Delta_n \cap \Delta_m \neq \varnothing$, then either $\Delta_n \subset \Delta_m$ or $\Delta_m < \Delta_n$.

Property 1) is evident, and 2) follows because when $n = 2^k + i$ and $m = 2^l + j$, $k \geq 1$, the equation

$$\Delta_l^j = \left( \frac{j-1}{2^l}, \frac{j}{2^l} \right) = \left( \frac{2^{k-l}(j-1)}{2^k}, \frac{2^{k-l}j}{2^k} \right)$$

implies that either $\Delta_m \subset \Delta_n$ (if $2^{k-l}(j-1) < i \leq 2^{k-l}j$), or $\Delta_m \cap \Delta_n = \varnothing$ (if $i \leq 2^{k-l}(j-1)$ or $i > 2^{k-l}j$).

DEFINITION 1.   The *Haar system* is the system of functions $\chi = \{\chi_n(x)\}_{n=1}^{\infty}$, $x \in [0,1]$, where $\chi_1 \equiv 1$, and for $2^k < n \leq 2^{k+1}$, $k = 0, 1, \ldots, \chi_n(x)$ is defned as follows:

$$\chi_n(x) = \begin{cases} 0 & \text{for } x \notin \overline{\Delta}_n, \\ 2^{k/2} & \text{for } x \in \Delta_n^+, \\ -2^{k/2} & \text{for } x \in \Delta_n^-. \end{cases} \tag{3}$$

The values of $\chi_n(x)$ at points of discontinuity and at the endpoints of $[0,1]$ are specified so that $\chi_n(x) \in \mathscr{D}_{2^k}$, i.e., so that

$$\chi_n(x) = \lim_{\delta \to 0} \frac{1}{2}[\chi_n(x+\delta) + \chi_n(x-\delta)], \qquad x \in (0,1),$$

$$\chi_n(0) = \lim_{\delta \to +0} \chi_n(\delta), \qquad \chi_n(1) = \lim_{\delta \to +0} \chi_n(1-\delta). \tag{4}$$

The group of functions $\{\chi_n(x)\}_{n=2^k+1}^{2^{k+1}}$, $k = 0, 1, \ldots$, is called the *kth block*. It is often convenient to use, instead of the usual "single" indexing of the Haar functions, an indexing that shows directly in which block a given function lies; more precisely, to set, for $k = 0, 1, \ldots$; $i = 1, \ldots, 2^k$; $n = 2^k + i$,

$$\chi_k^{(i)}(x) = \chi_n(x), \qquad \chi_0^{(0)}(x) = 1, \; x \in [0,1]. \tag{5}$$

Then it is clear that the Haar system is the union of the blocks $\{\chi_k^{(j)}(x)\}_{i=1}^{2^k}$, $k = 0, 1, \ldots$ and the function $\chi_0^{(0)}(x)$.

It follows immediately from properties 1) and 2) that the Haar system is orthonormal. To show that it is complete, we use the following proposition.

PROPOSITION 1.   *For $N + 2^k$, $k = 0, 1, \ldots$, the linear span $G_N(\chi)$ of the functions $\{\chi_n(x)\}_{n=1}^N$ is $\mathscr{D}_N$, i.e.,*

$$G_N(\chi) \equiv \left\{ f(x) : f(x) = \sum_{n=1}^N a_n \chi_n(x) \right\} = \mathscr{D}_N,$$

$$N = 2^k, \; k = 0, 1, \ldots. \tag{6}$$

In fact, the Haar functions are linearly independent (since the Haar system is an O.N.S.), and therefore (6) follows from the fact that $G_N(\chi)$ and $\mathscr{D}_N$ are $N$-dimensional linear spaces, with (see (4)) $G_N(\chi) \subset \mathscr{D}_N$.

Since the set of functions $f \in \bigcup_{k=0}^{\infty} \mathscr{D}_{2^k}$ is dense in $L^p(0,1)$, $1 \le p < \infty$, it follows from Proposition 1 that the Haar system is complete in $L^p(0,1)$, $1 \le p < \infty$.

Let us find an expression for the partial sums $S_N(f, x)$ of the Fourier–Haar series of a function $f \in L^1(0,1)$ (see §1.3):

$$f(x) \sim \sum_{n=1}^{\infty} c_n(f)\chi_n(x),$$

where $c_n(f) = c_n(f, \chi)$ are the Fourier–Haar coefficients of $f(x)$, and by the definition of the Haar functions (see (3))

$$c_1(f) = \int_0^1 f(x)\,dx,$$

$$c_n(f) = 2^{k/2}\left[\int_{\Delta_n^+} f(x)\,dx - \int_{\Delta_n^-} f(x)\,dx\right], \qquad n = 2, 3, \ldots. \tag{7}$$

Let us show that, for $N = 2^k$, $k = 0, 1, \ldots,$

a)

$$S_N(f, x) \equiv \sum_{n=1}^{N} c_n(f)\chi_n(x) = 2^k \int_{\Delta_k^i} f(t)\,dt,$$

$$x \in \Delta_k^i, \ i = 1, \ldots, 2^k; \quad (8)$$

b)

$$S_N\left(f, \frac{i}{2^k}\right) = 2^{k-1}\left[\int_{\Delta_k^i} f(t)\,dt + \int_{\Delta_k^{i+1}} f(t)\,dt\right],$$

$$i = 1, \ldots, 2^k - 1, \quad (8')$$

$$S_N(f, 0) = 2^k \int_{\Delta_k^1} f(t)\,dt, \qquad S_N(f, 1) = 2^k \int_{\Delta_k^{2^k}} f(t)\,dt.$$

For $k = 0, 1, \ldots$ we define a function $m_k(f) \equiv m_k(f, x) \in \mathscr{D}_{2^k}$ by setting

$$m_k(f, x) = 2^k \int_{\Delta_k^i} f(t)\,dt; \qquad x \in \Delta_k^i, \ i = 1, \ldots, 2^k \tag{9}$$

(it is clear that the values of $m_k(f, x)$ at the points $i/2^k$, $i = 0, 1, \ldots, 2^k$, are determined by (9) and the condition $m_k(f) \in \mathscr{D}_{2^k}$).

Since $\chi_n(x)$, for $1 \le n \le N = 2^k$, and $m_k(f, x)$ are constant on each interval $\Delta_k^i$, $i = 1, \ldots, 2^k$, we have, for $n = 1, \ldots, N$,

$$
c_n(m_k(f)) = \sum_{i=1}^{2^k} \int_{\Delta_k^i} m_k(f, x) \chi_n(x)\, dx
$$

$$
= \sum_{i=1}^{2^k} \chi_n\left(\frac{2i-1}{2^{k+1}}\right) \int_{\Delta_k^i} f(t)\, dt = \sum_{i=1}^{2^k} \int_{\Delta_k^i} f(t) \chi_n(t)\, dt = c_n(f).
$$

Consequently

$$
S_n(f, x) = S_N(m_k(f), x), \qquad N = 2^k,\ k = 0, 1, \ldots
$$

But $m_k(f) \in \mathscr{D}_N$ and, by (6), $m_k(f) \in G_N(\chi)$. Hence

$$
S_N(f, x) = S_N(m_k(f), x) = m_k(f, x), \qquad x \in [0, 1],\ N = 2^k. \tag{10}
$$

From this, (8) and (8') follow.

For the partial sums $S_N(f, x)$ of index $N$ of the form $N = 2^k + i$, $i = 1, \ldots, 2^k - 1$, $k = 1, 2, \ldots$, we can obtain the following expression by using property 1) of binary intervals:

$$
S_N(f, x) = \begin{cases} S_{2^{k+1}}(f, x) & \text{for } x \in \left[0, \dfrac{i}{2^k}\right), \\[2ex] S_{2^k}(f, x) & \text{for } x \in \left(\dfrac{i}{2^k}, 1\right], \\[2ex] S_{2^k}(f, x) + c_N \chi_N(x) & \text{for } x = \dfrac{i}{2^k}. \end{cases} \tag{11}
$$

In concluding this section we note that we can deduce the following equation from (11) and (8): for $N = 2, 3, \ldots$,

$$
S_N(f, x) = \begin{cases} |\Delta_N^+|^{-1} \displaystyle\int_{\Delta_N^+} f(t)\, dt & \text{for } x \in \Delta_N^+, \\[2ex] |\Delta_N^-|^{-1} \displaystyle\int_{\Delta_N^-} f(t)\, dt & \text{for } x \in \Delta_N^-. \end{cases} \tag{11'}
$$

## §2. Inequalities for coefficients and theorems on the convergence of Fourier–Haar series

THEOREM 1. *The following inequalities are valid:*

$$
|c_n(f)| \le (2n)^{-1/2} \cdot \omega\left(\frac{1}{n}, f\right), \qquad n > 1, \tag{12}
$$

*if $f \in C(0, 1)$; and*

$$
|c_n(f)| \le n^{1/p - 1/2} \cdot \omega_p\left(\frac{1}{n}, f\right), \qquad n > 1, \tag{13}
$$

*if* $f \in L^p(0,1)$, $1 \le p < \infty$ (*here* $\omega(\delta, f)$ *and* $\omega_p(\delta, f)$ *are the moduli of continuity of* $f$ *in* $C$ *and* $L^p$ [*see Appendix* 1, §1]).

PROOF. Let $n = 2^k + i$, $i = 1, \ldots, 2^k$, $k = 0, 1, \ldots$ Then if $f \in C(0,1)$, according to (7) and (1) we have

$$|c_n(f)| = 2^{k/2} \left| \int_{(i-1)/2^k}^{(2i-1)/2^{k+1}} f(t)\, dt - \int_{(2i-1)/2^{k+1}}^{i/2^k} f(t)\, dt \right|$$

$$\le 2^{k/2} \int_{(i-1)/2^k}^{(2i-1)/2^{k+1}} |f(t) - f(t + 2^{-k-1})|\, dt$$

$$\le 2^{k/2} 2^{-k-1} \omega(2^{-k-1}, f) \le (2n)^{-1/2} \omega\left(\frac{1}{n}, f\right).$$

For $f \in L^p(0,1)$, $1 \le p < \infty$, we find by using Hölder's inequality that

$$|c_n(f)| \le 2^{k/2} \int_{(i-1)/2^k}^{(2i-1)/2^{k+1}} |f(t) - f(t + 2^{-k-1})|\, dt$$

$$\le 2^{k/2} \left\{ \int_{(i-1)/2^k}^{(2i-1)/2^{k+1}} |f(t) - f(t + 2^{-k-1})|^p\, dt \right\}^{1/p} \cdot 2^{-(k+1)(1-1/p)}$$

$$\le 2^{1/p-1} \cdot 2^{k(1/p-1/2)} \omega_p(2^{-k-1}, f) \le n^{1/p-1/2} \omega_p\left(\frac{1}{n}, f\right).$$

This completes the proof of Theorem 1.

REMARK. For $1 \le p < \infty$, inequality (13) does not ensure that the Fourier–Haar coefficients of a function $f \in L^p$ tend to zero; in fact, in this case $\overline{\lim}_{n \to \infty} c_n(f)$ may be $+\infty$. The reason for this is that $\lim_{n \to \infty} \|\chi_n\|_p = 0$ for $1 \le p < 2$. For example, it is easy to see that

$$f(x) = \sum_{k=1}^{\infty} a_k \chi_k^{(2)}(x) \in L^p(0,1) \qquad (1 \le p < 2)$$

if and only if

$$\sum_{k=1}^{\infty} |a_k|^p 2^{-(1/p-1/2)pk} < \infty$$

(since the supports of the functions $\chi_k^{(2)}(x)$, $k = 1, 2$, are disjoint, and consequently

$$\|f\|_p^p = \sum_{k=1}^{\infty} |a_k|^p \|\chi_k^{(2)}\|_p^p = \sum_{k=1}^{\infty} |a_k|^p 2^{-((1/p)-1/2)kp}).$$

THEOREM 2. *The Fourier–Haar series of every function $f(x) \in C(0,1)$ converges to $f(x)$, uniformly on $[0,1]$. Moreover,*

$$\rho_C(f,N) \equiv \|f - S_N(f)\|_C \leq 3\omega\left(\frac{1}{N},f\right), \qquad N \geq 1. \tag{14}$$

PROOF. It follows from $(10)$ and the evident inequality

$$\|m_k(f) - f\|_C \leq \omega(2^{-k},f), \qquad k = 0,1,\ldots,$$

that, when $N = 2^k$, $k = 0,1,\ldots,$

$$\rho_C(f,N) \leq \omega\left(\frac{1}{N},f\right). \tag{15}$$

If, however, $N = 2^k + i$, $i = 1,\ldots,2^k$, $k = 1,2,\ldots,$ then by $(11)$ and $(15)$

$$\rho_C(f,N) \leq \max\{\rho_C(f,2^{k+1}),\rho_C(f,2^k),\rho_C(f,2^k) + 2^{k/2}|c_N(f)|\}$$
$$\leq \omega(2^{-k},f) + 2^{k/2}|c_N(f)|. \tag{16}$$

Since

$$\omega(2^{-k},f) \leq 2\omega(2^{-k-1},f) \leq 2\omega(1/N,f),$$

the conclusion of Theorem 2 follows from $(16)$ and $(12)$.

THEOREM 3. *The Haar system is a basis in $L^p(0,1)$, $1 \leq p < \infty$. Moreover,*

$$\rho_p(f,N) \equiv \|f - S_N(f)\|_p \leq C_p\omega_p\left(\frac{1}{N},f\right), \qquad N = 1,2,\ldots$$
$$(C_p = 4^{1/p}(1 + 2^p)^{1/p}). \tag{17}$$

PROOF. The Haar system is complete in $L^p(0,1)$; it follows from Proposition 1.4 that it is minimal. Therefore (see Theorem 1.6 and also $(7)$ in Appendix 1), it is sufficient to establish $(17)$.

For $N = 2^k$, $k = 0, 1, \ldots$, we find from (10) by using Hölder's inequality that

$$
\begin{aligned}
\rho_p^p(f, N) &= \int_0^1 |f(x) - m_k(f, x)|^p \, dx \\
&= \sum_{i=1}^{2^k} 2^{kp} \int_{\Delta_k^i} \left| \int_{\Delta_k^i} [f(x) - f(t)] \, dt \right|^p dx \\
&\leq \sum_{i=1}^{2^k} 2^k \int_{\Delta_k^i} \int_{\Delta_k^i} |f(x) - f(t)|^p \, dt \, dx \\
&= 2^{k+1} \sum_{i=1}^{2^k} \iint_{\{x \in \Delta_k^i, t \in \Delta_k^i, x < t\}} |f(x) - f(t)|^p \, dt \, dx \\
&\leq 2^{k+1} \int_{1-2^{-k}}^1 \int_x^1 |f(x) - f(t)|^p \, dt \, dx \\
&\quad + 2^{k+1} \sum_{i=1}^{2^k-1} \int_{\Delta_k^i} \int_x^{x+2^{-k}} |f(x) - f(t)|^p \, dt \, dx \\
&= 2^{k+1} \int_{1-2^{-k}}^1 \int_0^{1-x} |f(x) - f(x+u)|^p \, du \, dx \\
&\quad + 2^{k+1} \sum_{i=1}^{2^k-1} \int_{\Delta_k^i} \int_0^{2^{-k}} |f(x) - f(x+u)|^p \, du \, dx \\
&= 2^{k+1} \int_0^{2^{-k}} \int_0^{1-u} |f(x) - f(x+u)|^p \, dx \, du \\
&\quad + 2^{k+1} \int_0^{2^{-k}} \int_0^{1-2^{-k}} |f(x) - f(x+u)|^p \, dx \, du \\
&\leq 2\omega_p^p(2^{-k}, f) + 2\omega_p^p(2^{-k}, f) = 4\omega_p^p(2^{-k}, f).
\end{aligned}
\tag{18}
$$

If now $N = 2^k + i$, $i = 1, \ldots, 2^k - 1$, $k = 1, 2, \ldots$, then by (11) and (18) (see also Appendix 1, (7))

$$
\begin{aligned}
\rho_p^p(f, N) &\leq \rho_p^p(f, 2^k) + \rho_p^p(f, 2^{k+1}) \\
&\leq 4[\omega_p^p(2^{-k}, f) + \omega_p^p(2^{-k-1}, f)] \leq 4(1 + 2^p)\omega_p^p\left(\frac{1}{N}, f\right).
\end{aligned}
$$

This completes the proof of Theorem 3.

THEOREM 4. *The Fourier–Haar series of every $f \in L^1(0,1)$ converges to* $f(x)$ *a.e. on* (0.1). *Furthermore, the majorants*

$$S^*(f,x) \equiv S^*_\chi(f,x) = \sup_{1 \le N < \infty} |S_N(f,x)| \tag{19}$$

*of the partial sums of the series satisfy the inequalities*
a) $m\{x \in (0.1): S^*(f,x) > y\} \le \frac{C}{y}\|f\|_1$;
b) $\|f\|_p \le \|S^*(f)\|_p \le C_p\|f\|_p$,   $f \in L^p(0,1)$, $1 < p \le \infty$.

PROOF. Inequality a) and the right-hand inequality in b) follow immediately from the theorem on the maximal function (see Appendix 1, Theorem 2), since it is evident that $|m_k(f,x)| \le M(f,x)$ for $k = 0,1,\dots$ and $x \in (0,1)$ (also see (9) and (10)).

In addition, it follows from the convergence of the Fourier–Haar series in the $L^1(0,1)$ metric, which was proved in Theorem 3, that for every $\varepsilon > 0$ and $N = 1,2,\dots$, the measure

$$m\left\{ x \in (0,1) : \sup_{M:N \le M} \left| \sum_{n=N}^M c_n(f)\chi_n(x) \right| > \varepsilon \right\}$$
$$\le \frac{C}{\varepsilon} \left\| \sum_{n=N}^\infty c_n(f)\chi_n(x) \right\|_1 \to 0$$

as $N \to \infty$; this also implies the convergence of the series $\sum_1^\infty c_n(f)\chi_n(x)$ to $f(x)$ a.e. Finally, it follows from the convergence of the series a.e. that $|f(x)| \le S^*(f,x)$, and therefore that the left-hand inequality in b) is valid.

Although it has many desirable properties, the Haar system has the drawback that it consists of discontinuous functions. The following two results provide a quantitative description of this phenomenon.

THEOREM 5. *The Fourier coefficients in the Haar system satisfy, for every* $f \in C(0,1)$, $f \ne \text{const}$, *the inequality*

$$\varlimsup_{n \to \infty} |c_n(f)|n^{3/2} > 0.$$

PROOF. Let us suppose the contrary, i.e.,

$$\varlimsup_{n \to \infty} |c_n(f_0)|n^{3/2} = 0, \qquad f_0 \in C(0,1), \ f_0 \ne \text{const}.$$

Let the sequence $n_k = 2^k + i_k$ $(1 \le i_k \le 2^k)$ have the property that

$$|c_{n_k}(f_0)| \equiv c_{n_k} = \max_{2^k < n \le 2^{k+1}} |c_n(f_0)|.$$

According to hypothesis, $2^{k/2}c_{n_k} = 2^{-k}\alpha_k$, where $\alpha_k \to 0$ as $k \to \infty$.

Since $\alpha_k$ tends to zero, we can find a number $k_0$ such that $\alpha_k < \alpha_{k_0}$ for $k > k_0$. Then

$$\sum_{k=k_0+1}^{\infty} 2^{k/2} c_{n_k} = \sum_{k=k_0+1}^{\infty} 2^{-k} \alpha_k < \alpha_{k_0} \sum_{k=k_0+1}^{\infty} 2^{-k} = c_{n_{k_0}} 2^{k_0/2}.$$

Hence, since $\Delta_k^i \cap \Delta_k^j = \varnothing$ for $i \neq j$ we find, for $x_0 = (2i_{k_0} - 1)/2^{k_0+1}$,

$$\sum_{n=n_{k_0}+1}^{\infty} |c_n(f_0)| \lim_{t \to 0} |\chi_n(x_0 + t) - \chi_n(x_0 - t)|$$

$$\leq \sum_{k=k_0+1}^{\infty} c_{n_k} \sum_{n=2^k+1}^{2^{k+1}} \lim_{t \to 0} |\chi_n(x_0 + t) - \chi_n(x_0 - t)|$$

$$= \sum_{k=k_0+1}^{\infty} c_{n_k} 2^{k/2+1} < 2^{k_0/2+1} c_{n_{k_0}}. \tag{20}$$

On the other hand,

$$\lim_{t \to 0} \sum_{n=1}^{n_{k_0}} |c_n(f_0)| \, |\chi_n(x_0 + t) - \chi_n(x_0 - t)|$$

$$= c_{n_{k_0}} \lim_{t \to 0} \chi_{n_{k_0}}(x_0 + t) - \chi_{n_{k_0}}(x_0 - t)| = 2^{k_0/2+1} c_{n_{k_0}}. \tag{21}$$

It follows from (20) and (21) that when $N > n_{k_0}$ the partial sums of the Fourier–Haar series of $f_0(x)$ are subject to the inequality

$$\lim_{t \to 0} |S_N(f_0, x_0 + t) - S_N(f_0, x_0 - t)| > \gamma > 0,$$

which contradicts the uniform convergence of $S_N(f_0, x)$ to $f_0(x)$.

This completes the proof of Theorem 5.

The example $f(x) \equiv x$, for which $|c_n(f)| \asymp n^{-3/2}$, shows the accuracy of our estimate. We also notice that it follows from Theorem 5 that it is impossible to have $\lim_{N \to \infty} N \|S_N(f) - f\|_\infty = 0$ for a continuous function $f(x) \neq$ const.

THEOREM 6. *If $f(x)$, $x \in [0, 1]$, is absolutely continuous, then*

$$\lim_{k \to \infty} 2^{k/2} \sum_{n=2^k+1}^{2^{k+1}} |c_n(f)| = \frac{1}{4} \int_0^1 |f'(x)| \, dx.$$

PROOF. Since $f(x)$ is absolutely continuous (also see (7)), we have, for $k = 1, 2, \ldots,$

$$I_k = 2^{k/2} \sum_{n=2^k+1}^{2^{k+1}} |c_n(f)|$$

$$= 2^k \sum_{i=1}^{2^k} \left| \int_{(i-1)/2^k}^{(2i-1)/2^{k+1}} [f(x) - f(x + 2^{-k-1})]\, dx \right|$$

$$= 2^k \sum_{i=1}^{2^k} \left| \int_{(i-1)/2^k}^{(2i-1)/2^{k+1}} \int_0^{2^{-k-1}} f'(x + u)\, du\, dx \right|. \tag{22}$$

Set

$$R_k = I_k - \sum_{i=1}^{2^k} \left| \frac{1}{2} \int_{(i-1)/2^k}^{(2i-1)/2^{k+1}} f'(x)\, dx \right|,$$

$$L_k = \sum_{i=1}^{2^k} \left| \frac{1}{2} \int_{(i-1)/2^k}^{(2i-1)/2^{k+1}} f'(x)\, dx \right| - \frac{1}{2} \sum_{i=1}^{2^k} \int_{(i-1)/2^k}^{(2i-1)/2^{k+1}} |f'(x)|\, dx.$$

Then (see (22))

$$|R_k| \le 2^k \sum_{i=1}^{2^k} \int_{(i-1)/2^k}^{(2i-1)/2^{k+1}} \int_0^{2^{-k-1}} |f'(x + u) - f'(x)|\, du\, dx$$

$$\le 2^k \int_0^{2^{-k-1}} \int_0^{1-2^{-k-1}} |f'(x + u) - f'(x)|\, dx\, du \le \omega_1(2^{-k-1}, f'). \tag{23}$$

Moreover, since we have for every $g \in L^1(a, b)$

$$\left| \int_a^b |g(x)|\, dx - \left| \int_a^b g(t)\, dt \right| \right| = \left| \int_a^b \left[ |g(x)| - (b-a)^{-1} \left| \int_a^b g(t)\, dt \right| \right] dx \right|$$

$$\le \int_a^b \left| g(x) - (b-a)^{-1} \int_a^b g(t)\, dt \right| dx$$

$$\le (b-a)^{-1} \int_a^b \int_a^b |g(x) - g(t)|\, dt\, dx,$$

we obtain the following estimate for $L_k$:

$$|L_k| \le 2^k \sum_{i=1}^{2^k} \int_{(i-1)/2^k}^{(2i-1)/2^{k+1}} \int_{(i-1)/2^k}^{(2i-1)/2^{k+1}} |f'(x) - f'(t)| \, dx \, dt$$

$$\le 2^{k+1} \int_0^{2^{-k-1}} \int_0^{1-2^{-k-1}} |f'(x+u) - f'(x)| \, dx \, du \le \omega_1(2^{-k-1}, f').$$
(24)

Since $\lim_{\delta \to 0} \omega_1(\delta, f') = 0$ (see Appendix 1, (7)) we find from (23) and (24) that

$$\lim_{k \to \infty} I_k = \lim_{k \to \infty} \frac{1}{2} \sum_{i=1}^{2^k} \int_{(i-1)/2^k}^{(2i-1)/2^{k+1}} |f'(x)| \, dx = \frac{1}{4} \int_0^1 |f'(x)| \, dx \quad (25)$$

(the right-hand equation in (25), which is valid for every $f' \in L^1(0, 1)$, has already been used in the proof of Theorem 2.13). This completes the proof of Theorem 6.

## §3. Unconditional convergence of Fourier–Haar series in $L^p(0, 1)$

In accordance with the general results of Chapter 1 (see Theorem 1.1 and Corollary 1.4) the study of unconditional convergence of Fourier-Haar series in $L^p(0, 1)$ reduces to the study of the operators $T_\varepsilon(f)$, where

$$T_\varepsilon(f, x) = \sum_{n=1}^{\infty} \varepsilon_n c_n(f) \chi_n(x), \qquad x \in (0, 1), \quad (26)$$

and $\varepsilon = \{\varepsilon_n\}_{n=1}^{\infty}$, $\varepsilon_n = \pm 1$, is a sequence of "signs."

It is clear that, since $\{\chi_n(x)\}$ is an O.N.S., the operators $T_\varepsilon(f)$ are isometric operators from $L^2(0, 1)$ to $L^2(0, 1)$, i.e.,

$$\|T_\varepsilon(f)\|_2 = \|f\|_2, \qquad f \in L^2(0, 1). \quad (27)$$

We are going to prove a number of propositions about the operators $T_\varepsilon$ and show, in particular, that (26) is a reasonable definition of $T_\varepsilon(f)$ for every $f \in L^1(0, 1)$.

THEOREM 7. *Let $f \in L^1(0, 1)$. Then for every sequence $\varepsilon = \{\varepsilon_n\}_{n=1}^{\infty}$, $\varepsilon_n = \pm 1$, the series (26) converges in measure to a function $T_\varepsilon(f)$ which is finite a.e. Moreover, for every $y > 0$,*

$$m\{x \in (0, 1): |T_\varepsilon(f, x)| > y\} \le 3/y \|f\|_1. \quad (28)$$

In the proof of Theorem 7 we shall need the following lemma on the operator $\widetilde{M}(f)$, which assigns to every $f \in L^1(0, 1)$ the function

$$\widetilde{M}(f, x) = \sup_{n:\Delta_n \ni x} |\Delta_n|^{-1} \int_{\Delta_n} |f(t)| \, dt. \quad (29)$$

It is easily seen (see (8) and (19)) that

$$\widetilde{M}(f,x) = S^*(|f|,x) \tag{30}$$

for binary-irrational $x \in (0,1)$.

LEMMA 1. *Let $f \in L^1(0,1)$, $y > \|f\|_1$, and let the set $G(y) = \{x \in (0,1) : \widetilde{M}(f,x) > y\}$ not be empty; then there are binary intervals $I_s$, $s = 1, 2, \ldots$, such that*

a) $G(y) = \bigcup I_s$, $I_s \cap I_{s'} = \varnothing$, $s \neq s'$, $s, s' = 1, 2, \ldots$;

b) $y < |I_s|^{-1} \int_{I_s}^s |f(t)| \, dt \leq 2y$, $s = 1, 2, \ldots$;

c) $\sum_s |I_s| = m[G(y)] \leq \frac{1}{y}\|f\|_1$.

PROOF OF LEMMA 1. It is clear that (see (9), (10)) $G(y) = \bigcup_0^\infty V_k$, where, for $k = 0, 1, \ldots$,

$$V_k = \left\{ x \in (0,1) : m_k(|f|,x) > y, x \neq \frac{i^k}{2}, \ i = 0, 1, \ldots, 2^k \right\}. \tag{31}$$

It is also clear tht if $V_k$ is not empty, it is the union of some collection of binary intervals of rank $k$. Therefore the set $W_k = V_k \backslash \bigcup_{s=0}^{k-1} V_s$ ($k = 1, 2, \ldots$), if not empty, also consists of binary intervals of rank $k$ (pairwise disjoint, in view of property 1). This means (since $W_k \cap W_{k'} = \varnothing$ $k \neq k'$), and $V_0 = \varnothing$ because $y > \|f\|_1$) that

$$G(y) = \bigcup_{k=1}^\infty V_k = \bigcup_{k=1}^\infty W_k = \bigcup_s I_s, \tag{32}$$

where $I_s$ are disjoint binary intervals (of rank $\geq 1$).

Let us show that inequality b) is satisfied for the intervals $I_s$. Let $\Delta = I_{s_0} \subset W_k$ be a binary interval of rank $k$ ($|\Delta| = 2^{-k}$) in the decomposition (32), and $\Delta'$ a binary interval of rank $k - 1$, containing $\Delta$, i.e.,

$$\Delta \subset \Delta', \qquad |\Delta'| = 2^{-k+1} = 2|\Delta|.$$

Since $\Delta \subset W_k \subset V_k \backslash V_{k-1}$, we find from (31) that

$$m_k(|f|,x) > y, \qquad m_{k-1}(|f|,x) \leq y, \quad x \in \Delta.$$

But by definition (see (9)) we have, for $x \in \Delta \subset \Delta'$,

$$m_k(|f|,x) = |\Delta|^{-1} \int_\Delta |f(t)| \, dt,$$

$$m_{k-1}(|f|,x) = |\Delta'|^{-1} \int_{\Delta'} |f(t)| \, dt.$$

Consequently

$$y < |\Delta|^{-1} \int_\Delta |f(t)|\, dt \le |\Delta|^{-1} \cdot |\Delta'| \cdot |\Delta'|^{-1} \int_{\Delta'} |f(t)|\, dt \le 2y.$$

Condition c) follows easily from a) and b):

$$m[G(y)] = \sum_s |I_s| \le \sum_s \frac{1}{y} \int_{I_s} |f(t)|\, dt \le \frac{1}{y}\|f\|_1.$$

This completes the proof of Lemma 1.

For the proof of Theorem 7, we have only to prove (28) for the case when $f(x)$ is an arbitrary polynomial in the Haar system. In fact, if (28) is satisfied for polynomials in the Haar system, we have, by Theorem 3, for every $f \in L^1(0,1)$ and every $\delta > 0$,

$$m\left\{x \in (0,1): \left|\sum_{n=M}^{M+N} \varepsilon_n c_n(f)\chi_n(x)\right| > \delta\right\} \le \frac{3}{\delta}\left\|\sum_{n=M}^{M+N} c_n(f)\chi_n\right\|_1 \to 0$$

when $M \to \infty$. This means that the partial sums $P_N(x)$ of (26) converge in measure to $T_\varepsilon(f,x)$, and therefore

$$m\{x \in (0,1): |T_\varepsilon(f,x)| > y\} \le \varliminf_{N\to\infty} m\{x \in (0,1): |P_N(x)| > y\}. \quad (33)$$

In addition, since $P_N(x) = T_\varepsilon(S_N(f), x)$, where $S_N(f)$, $N = 1,2,\ldots$, are the partial sums of the Fourier–Haar series of $f(x)$, then when we apply (28) to $S_N(f)$ we obtain

$$\varliminf_{N\to\infty} m\{x \in (0.1): |P_N(x)| > y\} \le \varliminf_{N\to\infty} \frac{3}{y}\|S_N(f)\|_1 = \frac{3}{y}\|f\|_1.$$

This implies (see (33)) that $f(x)$ satisfies (28).

Now let $g(x) = \sum_1^N c_n \chi_n(x)$ be an arbitrary polynomial in the Haar system and let $y > \|g\|_1$ (when $y \le \|g\|_1$ inequality (28) is evident). Applying Lemma 1 with $f(x) = g(x)$ and the number $y$, we consider the decomposition

$$g(x) = p(x) + h(x), \qquad x \in (0,1), \quad (34)$$

of $g(x)$, where

$$p(x) = \begin{cases} g(x) & \text{if } x \in (0,1)\backslash G(y), \\ |I_s|^{-1}\int_{I_s} g(t)\, dt & \text{if } x \in I_s,\ s = 1,2,\ldots, \end{cases} \quad (35)$$

$$h(x) = \begin{cases} 0 & \text{if } x \in (0,1)\backslash G(y), \\ g(x) - |I_s|^{-1}\int_{I_s} g(t)\, dt & \text{if } x \in I_s,\ s = 1,2,\ldots. \end{cases}$$

We note the following properties of $p(x)$ and $h(x)$:

I) $\|p\|_1 \le \|g\|_1$;

II) $p(x) \le 2y$ for a.e. $x \in (0,1)$;

III) $\int_{I_s} h(x)\, dx = 0$, $s = 1,2,\ldots$, $\operatorname{supp} h \subset G(y)$;

IV) $\operatorname{supp} T_\varepsilon(h) \subset G(y)$

[in IV) the function $T_\varepsilon(h)$ is determined, since $h \in L^\infty(0,1)$ and *a fortiori* $h \in L^2(0,1)$ (see II) and (27)); we recall that supp $f \equiv \{x \in (0,1) : f(x) \neq 0\}$].

Properties I) and III) follow immediately from the definitions of $p(x)$ and $h(x)$.

For $x \in G(y)$, inequality II) follows from inequality b) of Lemma 1; and for $x \notin G(y)$, from the following relation (see (30) and Theorem 4):

$$|p(x)| = |g(x)| \leq S^*(|g|,x) = \widetilde{M}(g,x) \quad \text{for almost all } x \notin G(y).$$

To establish property IV), it is enough to prove that for the Fourier–Haar coefficients $c_n(h)$ we have

$$c_n(h) = 0 \quad \text{if } \Delta_n \not\subset G(y).$$

By part a) of Lemma 1, and III),

$$
\begin{aligned}
c_n(h) &= \int_0^1 h(x)\chi_n(x)\,dx = \int_{G(y)\cap\Delta_n} h(x)\chi_n(x)\,dx \\
&= \sum_s \int_{I_s\cap\Delta_n} h(x)\chi_n(x)\,dx.
\end{aligned}
\tag{36}
$$

Moreover, since $I_n \not\subset G(y)$, by property 2) of binary intervals we have either $\Delta_n \cap I_s = \varnothing$, or $I_s \subset \Delta_n$, $I_s \neq \Delta_n$. If $I_s \subset \Delta_n$ and $I_s \neq \Delta_n$ it is clear that either $I_s \subset \Delta_n^+$ or $I_s \subset \Delta_n^-$; in either case, $\chi_n(x)$ is constant on $I_s$ and

$$\int_{I_s\cap\Delta_n} h(x)\chi_n(x)\,dx = c\int_{I_s} h(x)\,dx = 0$$

[see III)]. Consequently all terms of the sum (37) are zero. This establishes property IV).

From I) and II), using Chebyshev's inequality and (28), we find that, for $y > 0$,

$$
\begin{aligned}
m\{x \in (0,1) : |T_\varepsilon(p,x)| > y\} &\leq \frac{1}{y^2}\|T_\varepsilon(p)\|_2^2 = \frac{1}{y^2}\int_0^1 p^2(x)\,dx \\
&\leq \frac{2}{y}\int_0^1 |p(x)|\,dx.
\end{aligned}
$$

From this, by IV) (see also inequality c) of Lemma 1) there follows the inequality

$$
\begin{aligned}
m\{x \in (0,1) &: |T_\varepsilon(g,x)| > y\} \\
&\leq m[G(y)] + m\{x \in (0,1)\backslash G(y)|T_\varepsilon(p,x)| > y\} \leq (3/y)\|g\|_1.
\end{aligned}
$$

This completes the proof of Theorem 7.

Theorem 7 and Theorem 1.1 imply the following corollaries.

COROLLARY 1. *The Fourier–Haar series of every* $f(x) \in L^1(0,1)$ *converges unconditionally in measure.*

COROLLARY 2. *For each* $p$, $1 < p < \infty$, *there is a constant* $C_p > 0$ *such that*

$$\frac{1}{C_p}\|f\|_p \leq \|T_\varepsilon(f)\|_p \leq C_p\|f\|_p \tag{37}$$

*for every* $f \in L^p(0,1)$ *and every sequence* $\varepsilon = \{\varepsilon_n\}_{n=1}^\infty$ $\varepsilon_n = \pm 1$.

PROOF. According to Theorem 7, the linear operator $T_\varepsilon(f)$ is defined for all $f \in L^1(0,1)$; it has weak type $(1,1)$ (see (28)) a strong type $(2,2)$ (see (27)). Consequently, by Theorem 1 of Appendix 1, for $1 < p < 2$ there is a constant $C_p > 0$ such that

$$\|T_\varepsilon(f)\|_p \leq C_p\|f\|_p, \qquad f \in L^p(0,1). \tag{38}$$

If now $2 < p < \infty$, $1/p + 1/q = 1$, and $f \in L^p(0,1)$, then, first considering the partial sums $S_N(x)$ of the Fourier–Haar series of the function $f$, we find that for $N = 1, 2, \ldots$

$$\begin{aligned}
\|T_\varepsilon(S_N)\|_p &= \sup_{\|g\|_q \leq 1} \int_0^1 T_\varepsilon(S_N, x)g(x)\,dx \\
&= \sup_{\|g\|_q \leq 1} \sum_{n=1}^N \varepsilon_n c_n(f)c_n(g) \\
&= \sup_{\|g\|_q \leq 1} \int_0^1 S_N(x)T_\varepsilon(g, x)\,dx \\
&\leq \sup_{\|g\|_q \leq 1} \|S_N\|_p\|T_\varepsilon(g)\|_q \leq C_q\|S_N\|_p.
\end{aligned}$$

Hence, taking the limit as $N \to \infty$ and taking into account that $T_\varepsilon(S_N) \to T_\varepsilon(f)$ in measure and $S_N(x) \to f(x)$ in $L^p(0,1)$ (see Theorems 7 and 3), we obtain (38) also for $2 < p < \infty$. On the other hand, since $T_\varepsilon^2(f) = f$ we have $\|f\|_p = \|T_\varepsilon^2(f)\|_p \leq C_p\|T_\varepsilon(f)\|_p$. This completes the proof of Corollary 2.

If we use Corollaries 2 and 1.4, we see at once that the following theorem is valid.

THEOREM 8. *The Haar system is an unconditional basis in* $L^p(0,1)$ *for* $1 < p < \infty$.

In correspondence with the results of §4 of Chapter 1 (see Theorem 1.11), we use Theorems 7 and 8 to study properties of the operator that associates with $f \in L^1(0, 1)$ the function

$$P(f) = P(f, x) = \left\{ \sum_{n=1}^{\infty} c_n^2(f) \chi_n^2(x) \right\}^{1/2}, \tag{39}$$

which we call the *Paley function*.

THEOREM 9. *For every function $f \in L^1(0, 1)$, the sum* (39) *is finite a.e. on* $(0, 1)$; *moreover*,

1) $m\{x \in (0, 1): P(f, x) > y\} \leq (C/y)\|f\|_1$, $y > 0$;
2) $(1/C_p)\|f\|_p \leq \|P(f)\|_p \leq C_p\|f\|_p$,
*if $f \in L^p(0, 1)$, $1 < p < \infty$.*

PROOF. The finiteness a.e. of the sum (39) follows from inequality 1). It is enough to prove this inequality for polynomials in the Haar system (since $P(f, x) = \lim_{N \to \infty} P(S_N, x)$, $x \in (0, 1)$, and by Theorem 3 we have $\|S_N\|_1 \leq C\|f\|_1$; here $S_N$ is a partial sum of the Fourier–Haar series of $f(x)$).

Let $f(x) = \sum_{n=1}^{N} c_n(f)\chi_n(x)$. According to (28), for almost every $t \in (0, 1)$ we have

$$m\{x \in (0, 1) : F(x, t) > y\} \leq \frac{3}{y}\|f\|_1,$$

$$F(x, t) = \left| \sum_{n=1}^{N} r_n(t) c_n(f) \chi_n(x) \right|,$$

where $y > 0$ and $\{r_n(t)\}$ is the Rademacher system. Consequently

$$m_2\{(x, t) \in (0, 1) \times (0, 1) : F(x, t) > y\} \leq \frac{3}{y}\|f\|_1, \qquad y > 0. \tag{40}$$

On the other hand, for $x \in (0, 1)$ we have, by Theorem 2.7,

$$m\left\{ t \in (0, 1) : F(x, t) > \frac{1}{2}P(f, x) \right\} \geq c_1 > 0.$$

From this and (40) we obtain the inequality

$$m\left\{ x \in (0, 1) : \frac{1}{2}P(f, x) > y \right\} \cdot c_1 \leq \frac{3}{y}\|f\|_1, \qquad y > 0,$$

from which inequality 1) follows. Inequality 2) is a direct corollary of Theorems 8 and 1.11. This completes the proof of Theorem 9.

REMARK. The proof of inequality 1) of Theorem 9, is similar to the proof of Theorem 1.11 (see §2.3), and depends only on the property of the Haar system that an inequality (28) of weak type holds for every operator $T_\varepsilon$.

**THEOREM 10.** *There is a function $f_0(x) \in L^1(0,1)$ for which*
1) *the Paley function $P(f_0, x)$ is not summable;*
2) *the Fourier–Haar series*

$$\sum_{n=1}^{\infty} c_n(f_0)\chi_n(x) \tag{41}$$

*does not converge unconditionally in $L^1(0,1)$;*
3) *the majorant $S^*(f_0, x)$ of the partial sums of (41) is not summable.*

PROOF. For $n = 1, 2, \ldots$ we consider the polynomial

$$Q_n(x) = \chi_0^{(0)}(x) + \sum_{k=0}^{n} 2^{k/2}\chi_k^{(1)}(x)$$

in the Haar system. It is easily shown by induction on $n$ that

$$Q_n(x) = \begin{cases} 2^{n+1} & \text{for } x \in (0, 2^{-n-1}), \\ 0 & \text{for } x \in (2^{-n-1}, 1), \end{cases}$$

and consequently

$$\|Q_n\|_1 = 1, \qquad n = 1, 2, \ldots. \tag{42}$$

In addition,

a) $\|P(Q_n)\|_1 = \left\| \left\{ 1 + \sum_{k=0}^{n} 2^k [\chi_k^{(1)}(x)]^2 \right\}^{1/2} \right\|_1$

$$\geq \sum_{k=0}^{n} \int_{(\Delta_k^1)^-} |2^{k/2}\chi_k^{(1)}(x)| \, dx = \sum_{k=0}^{n} \frac{1}{2} = \frac{n+1}{2}, \qquad n = 1, 2, \ldots.$$

Now let

$$Q_n'(x) = \sum_{p=0}^{[n/2]} 2^p \chi_{2p}^{(1)}(x);$$

then evidently

$$\|Q_n'\|_1 \geq \sum_{p=0}^{[n/2]} \|Q_n'\|_{L^1((\Delta_{2p}^1)^-)},$$

but for $x \in (\Delta_{2p}^1)^-$ we have $\chi_{2s}^{(1)}(x) = 0$ if $s > p$, and therefore

$$|Q_n'(x)| \geq \left| 2^{2p} - \sum_{k=0}^{p-1} 2^k \chi_{2k}^{(1)}(x) \right| \geq 2^{2p} - \sum_{k=0}^{p-1} 2^{2k} \geq \frac{1}{2} 2^{2p},$$

$$x \in (\Delta_{2p}^1)^-, \ 1 \leq p \leq [n/2]. \tag{43}$$

It follows from (43) that

b)    $$\|Q_n'\|_1 \geq \sum_{p=1}^{[n/2]} \frac{1}{2} 2^{2p} \cdot 2^{-2p-1} = \frac{1}{4}\left[\frac{n}{2}\right], \qquad n = 2, 3, \ldots;$$

c)    $$\|S^*(Q_n)\|_1 \geq \int_0^1 \frac{1}{2} \max_{1 \leq k \leq n} |2^{k/2} \chi_k^{(1)}(x)|\, dx$$

$$\geq \frac{1}{2} \sum_{k=0}^n \int_{(\Delta_k^1)^-} |2^{k/2} \chi_k^{(1)}(x)|\, dx \geq \frac{1}{4}(n+1),$$

$$n = 1, 2, \ldots.$$

Set

$$f_0(x) = \sum_{s=1}^\infty 3^{-s} Q_{4^s}(x) = \frac{1}{2} + \sum_{k=0}^\infty \left(\sum_{s:4^s \geq k} 3^{-s}\right) \chi_k^{(1)}(x).$$

The function $f_0(x)$ satisfies the requirements of Theorem 10. In fact, $\|f_0\|_1 \leq \sum_{s=1}^\infty 3^{-s} = \frac{1}{2}$, by (42). In addition, if we use the non-negativity of the coefficients of $Q_n$ and inequality a), we obtain

$$\|P(f_0)\|_1 \geq \sup_{1 \leq s < \infty} 3^{-s}\|P(Q_{4^s})\|_1 \geq \frac{1}{2} \sup_{1 \leq s < \infty} \left(\frac{4}{3}\right)^s = +\infty.$$

Similarly, by using inequalities b) and c), we can verify that $S^*(f_0) \notin L^1(0,1)$ and that the subseries

$$\sum_{p=1}^\infty c_{2p,1}(f_0)\chi_{2p}^{(1)}(x)$$

of (41) diverges in $L^1(0,1)$; therefore (see Theorem 1.1) series (41) diverges in $L^1(0,1)$ after a permutation of its terms.

This completes the proof of Theorem 10.

We showed previously (see Theorems 4 and 9) that when $1 < p < \infty$

$$\text{a)} \quad \|P(f)\| \asymp \|f\|_p; \qquad \text{b)} \quad \|S^*(f)\|_p \asymp \|f\|_p. \tag{44}$$

When $p = 1$, both relations in (44) fail (see Theorem 10). However, we have the following theorem.

THEOREM 11. *There are positive constants $C$ and $c$ such that*

$$c\|P(f)\|_1 \leq \|S^*(f)\|_1 \leq C\|P(f)\|_1$$

*for every $f \in L^1(0,1)$.*

In the proof of Theorem 11 and of some other results in this chapter, a fundamental role is played by the following property of the Haar system.

LEMMA 1. *For $j = 1, 2, \ldots$ let $A_j$ be a set of intervals of the form*

$$A_j = \{(0,1), \varnothing, \Delta_n^+, \Delta_n^- : n = 1, \ldots, j\}, \qquad A_0 = \{(0,1), \varnothing\}$$

*and let $\mathscr{F}_j$ be the family of sets representable in the form of a union of some number of intervals in $A_j$ ($j = 0, 1, \ldots$). Also let $\tau(x)$, $x \in [0,1]$, be a function taking the values $0, 1, \ldots, \infty$ and such that*

$$e_j \equiv \{x \in (0,1) : \tau(x) = j\} \in \mathscr{F}_j, \qquad j = 0, 1, \ldots. \tag{45}$$

*Then there is a sequence $\{\varepsilon_n\}_{n=1}^\infty$, $\varepsilon_n = 0$ or $1$, such that*

$$\varepsilon_n \chi_n(x) = \begin{cases} \chi_n(x) & \text{if } n \le \tau(x), \\ 0 & \text{if } n > \tau(x), \end{cases} \qquad x \in (0,1) \backslash R_2, \; n = 1, 2, \ldots, \tag{46}$$

*where $R_2 = \{\{i/2^k\}_{i=0}^{2^k}\}_{k=0}^\infty$ and consequently, for arbitrary numbers $c_n$, $n = 1, 2, \ldots$,*

$$\sum_{n=1}^{\tau(x)} c_n \chi_n(x) = \sum_{n=1}^\infty \varepsilon_n c_n \chi_n(x), \qquad x \in (0,1) \backslash R_2 \tag{47}$$

*(here $\sum_{n=1}^0 \equiv 0$).*

PROOF. We first notice the following property of the family $\mathscr{F}_j$, $j = 0, 1, \ldots$.

If the interval

$$\Delta \in \mathscr{F}_j, \; n > j, \text{ and } \Delta \cap \Delta_n \ne \varnothing, \text{ then } \Delta_n \subset \Delta. \tag{48}$$

In fact, let $\Delta = \Delta_s^+$ (or $\Delta_s^-$), $1 \le s \le j$, and $\Delta \cap \Delta_n \ne \varnothing$; then if $\Delta_n \not\subset \Delta$, by property 2) of binary intervals we have $\Delta_s^+ = \Delta \subset \Delta_n$; moreover, $\Delta_s^+ \ne \Delta_n$, and therefore we also have $\Delta_s \subset \Delta_n$, i.e., $s \ge n$. This contradicts the hypothesis $s \le j$, $j < n$.

Now let $\tau(x)$ satisfy (45). Then the right-hand part of (46) is equal to

$$\chi_{E_n}(x)\chi_n(x), \qquad E_n = \{x \in (0,1) : \tau(x) \ge n\}, \; n = 1, 2, \ldots, \tag{49}$$

and, as usual, $\chi_{E_n}(x)$ is the characteristic function of $E_n$.

Let us show that

$$\chi_{E_n}(x) = a_n = \text{const} \quad \text{for } x \in \Delta_n, \; n = 1, 2, \ldots \; (a_n = 0 \text{ or } 1).$$

Then Lemma 1 will be proved, since by (49) equation (46) will be satisfied with $\varepsilon_n = a_n$, $n = 1, 2, \ldots$.

Suppose, on the contrary, that there are two points $x_1$ and $x_2$ of the interval $\Delta_n$ for which

$$\tau(x_1) \ge n, \qquad \tau(x_2) = j < n.$$

This means that $x_2 \in e_j$, $x_1 \notin e_j$, i.e.,

$$e_j \cap \Delta_n \neq \varnothing, \qquad e_j \not\supset \Delta_n, \quad j < n,$$

which contradicts property (48) of sets in $\mathscr{F}_j$ (see also (45)). This completes the proof of Lemma 1.

PROOF OF THEOREM 11. For $n = 1, 2, \ldots$, set $\chi_n^* = |\Delta_n|^{-1}\chi_{\Delta_n}(x)$. It is clear that, for almost all $x \in (0, 1)$ we have $\chi_n^*(x) = \chi_n^2(x)$, and hence for almost all $x$ we have

$$P(f, x) = \left\{ \sum_{n=1}^{\infty} c_n^2(f)\chi_n^*(x) \right\}^{1/2}.$$

In addition, for $N = 1, 2, \ldots$, set

$$P_N(f, x) = \left\{ \sum_{n=1}^{N} c_n^2 \chi_n^*(x) \right\}^{1/2} \qquad (c_n = c_n(f)),$$

$$S_N^*(f, x) = \sup_{1 \leq m \leq N} \left| \sum_{n=1}^{m} c_n \chi_n(x) \right|.$$

Consider, for a given $t > 0$, the function

$$\tau(x) = \begin{cases} +\infty & \text{if } P(f, x) \leq t, \\ \inf\{N : P_{n+1}(f, x) > t\} & \text{if } P(f, x) > t. \end{cases} \tag{50}$$

For $j = 0, 1, 2, \ldots$ the set $e_j = \{x \in (0, 1) : \tau(x) = j\}$ is either empty or coincides with $\Delta_{j+1} \in \mathscr{F}_j$. Hence by using Lemma 1 we can define a sequence $\{\varepsilon_n\}_{n=1}^{\infty}$, $\varepsilon_n = 0$ or 1, for which (46) is satisfied.

Let

$$g(x) = \sum_{n=1}^{\tau(x)} c_n \chi_n(x) = \sum_{n=1}^{\infty} \varepsilon_n c_n \chi_n(x), \qquad x \in (0, 1)\backslash R_2.$$

Then it follows from (47) and (50) that

a) $P(g, x) \leq t$ for $x \in (0, 1)\backslash R_2$;

b) $S^*(f, x) = S^*(g, x)$ if $P(f, x) \leq t$, $x \in (0, 1)\backslash R_2$;

c) $P(f, x) = P(g, x)$ if $P(f, x) \leq t$, $x \in (0, 1)\backslash R_2$.

From a)–c) we find, by using Chebyshev's inequality and the equivalence $\|S^*(g)\|_2 \asymp \|P(g)\|_2$, $g \in L^2$ (see (44)), that

$$
\begin{aligned}
m\{x \in (0,1) &: S^*(f,x) > t, P(f,x) \le t\} \\
&\le m\{x \in (0,1) : S^*(g,x) > t\} \\
&\le t^{-2}\|S^*(g)\|_2^2 \\
&\le Ct^{-2}\|P(g)\|_2^2 = Ct^{-2} \int_{\{x:P(f,x)>t\}} P^2(g,x)\,dx \\
&\quad + Ct^{-2} \int_{\{x:P(f,x)\le t\}} P^2(g,x)\,dx \\
&\le Cm\{x \in (0,1) : P(f,x) > t\} + Ct^{-2} \int_{\{x:P(f,x)\le t\}} P^2(f,x)\,dx.
\end{aligned}
$$

Consequently,

$$
\begin{aligned}
\lambda_{S^*(f)}(t) &\equiv m\{x \in (0,1) : S^*(f,x) > t\} \\
&\le m\{x \in (0,1) : P(f,x) > t\} \\
&\quad + m\{x \in (0,1) : S^*(f,x) > t, P(f,x) \le t\} \\
&\le (C+1)\lambda_{P(f)}(t) + Ct^{-2} \int_{\{x:P(f,x)\le t\}} P^2(f,x)\,dx.
\end{aligned}
$$

From the preceding inequality we obtain, by using the equation

$$
\int_0^1 |f(x)|^p\,dx = p \int_0^\infty t^{p-1}\lambda_f(t)\,dt,
$$
$$
\lambda_f(t) = m\{x \in (0,1) : |f(x)| > t\}, \quad (51)
$$

(see Appendix 1, (2)),

$$
\begin{aligned}
\|S^*(f)\|_1 &= \int_0^\infty \lambda_{S^*(f)}(t)\,dt \le (C+1)\|P(f)\|_1 \\
&\quad + C \int_0^\infty t^{-2} \int_{\{x:P(f,x)\le t\}} P^2(f,x)\,dx\,dt \\
&= (C+1)\|P(f)\|_1 + C \int_0^1 P^2(f,x) \int_{P(f,x)}^\infty t^{-2}\,dt\,dx \\
&= C'\|P(f)\|_1.
\end{aligned}
$$

To prove the converse inequality we consider, with a fixed $t > 0$, the function

$$
\tau'(x) = \begin{cases} +\infty & \text{if } \sup_{N:x\in\Delta_N} \|S_N(f)\|_{C(\Delta_N)} \le t, \\ \inf\{N : \|S_{N+1}(f)\|_{C(\Delta_{N+1})} > t, \ x \in \Delta_{N+1}\}, \\ & \hspace{2cm} \text{in the opposite case.} \end{cases} \quad (52)
$$

It is easily seen that for $j = 0, 1, 2, \ldots$ the set $e_j' = \{x \in (0,1) : \tau'(x) = j\}$ is either empty or identical with $\Delta_{j+1}$, i.e., in all cases $e_j' \in \mathscr{F}_j$. We apply Lemma 1 to $\tau'(x)$ and set

$$g'(x) = \sum_{n=1}^{\tau'(x)} c_n \chi_n(x) = \sum_{n=1}^{\infty} \varepsilon_n' c_n \chi_n(x) \qquad (\varepsilon_n' = 0; 1). \tag{53}$$

Then, by (46) and (52), we shall have

    a')   $S^*(g', x) \leq t$ for $x \in (0,1) \backslash R_2$;

    b')   $S^*(f, x) = S^*(g', x)$

          for $x \in E \equiv \{x \in (0,1) : \tau'(x) = +\infty\} \backslash R_2$;   (53')

    c')   $P(f, x) = P(g', x)$ for $x \in E$;

    d')   $E \subset \{x \in (0,1) : S^*(f, x) \leq t\}$.

We also show that

$$m[(0,1) \backslash E] \leq 2m\{x \in (0,1) : S^*(f, x) > t\}. \tag{54}$$

    In fact,

$$m[(0,1) \backslash E] = m\{x \in (0,1) : \tau'(x) < +\infty\} = \sum_{j=0}^{\infty} m(e_j').$$

Moreover, as we showed above, if $e_j' \neq \varnothing$ then $e_j' = \Delta_{j+1}$ and $\|S_{j+1}(f)\|_{C(\Delta_{j+1})} > t$ (see (52)). But $S_{j+1}(f, x)$ is constant on the halves $\Delta_{j+1}^+$ and $\Delta_{j+1}^-$ of $\Delta_{j+1}$; hence either $\Delta_{j+1}^+$ or $\Delta_{j+1}^-$ is a subset of $\{x \in (0,1) : S^*(f, x) > t\}$. Hence, since the sets $e_j'$ are disjoint, we obtain (54).

From conditions a')–d') we obtain, by using Chebyshev's inequality and (44),

$$m\{x \in E : P(f, x) > t\} = m\{x \in E : P(g', x) > t\}$$
$$\leq t^{-2}\|P(g')\|_2^2 \leq Ct^{-2}\|S^*(g')\|_2^2$$
$$= Ct^{-2}\int_E [S^*(g', x)]^2\, dx$$
$$+ Ct^{-2}\int_{(0,1) \backslash E} [S^*(g', x)]^2\, dx$$
$$\leq Ct^{-2}\int_{\{x : S^*(f,x) \leq t\}} [S^*(f, x)]^2\, dx$$
$$+ Cm[(0,1) \backslash E].$$

It follows from the preceding inequality and (54) that

$$\lambda_{P(f)}(t) = \{x \in (0,1) : P(f,x) > t\}$$
$$\leq m[(0,1)\backslash E] + m\{x \in E : P(f,x) > t\}$$
$$\leq C'\lambda_{S^*(f)}(t) + Ct^{-2}\int_{\{x:S^*(f,x)\leq t\}} [S^*(f,x)]^2\,dx$$

and, by (51),

$$\|P(f)\|_1 = \int_0^\infty \lambda_{P(f)}(t)\,dt$$
$$\leq C'\|S^*(f)\|_1 + C\int_0^\infty t^{-2}\int_{\{x:S^*(f,x)\leq t\}} [S^*(f,x)]^2\,dx\,dt$$
$$= C'\|S^*(f)\|_1 + C\int_0^1 [S^*(f)]^2\int_{S^*(f,x)}^\infty t^{-2}\,dt\,dx \leq C''\|S^*(f)\|_1.$$

This completes the proof of Theorem 11.

COROLLARY 3. *The Haar system is an unconditional basis in the space $L^*(0,1)(^1)$ of functions $f \in L^1(0,1)$ for which the norm*

$$\|f\|_* = \|S^*(f)\|_1 \tag{55}$$

*is finite.*

In fact, let $f \in L^*(0,1)$, and let $\varepsilon = \{\varepsilon_n\}_{n=1}^\infty$ be any sequence of numbers $\varepsilon_n = \pm 1$. Then by Theorem 11

$$\left\|\sum_{n=1}^\infty \varepsilon_n c_n(f)\chi_n(x)\right\|_* \asymp \int_0^1 \left\{\sum_{n=1}^\infty c_n^2(f)\chi_n^2(x)\right\}^{1/2} dx \asymp \|f\|_*$$

(here the constants in the equivalence relations are independent of $\varepsilon$ and $f$). Hence it follows by Corollary 1.4 that the Haar system is an unconditional basis in $L^*(0,1)$.

From Theorem 11 we can derive the following corollary.

COROLLARY 4. *A necessary and sufficient condition for the Fourier–Haar series of a function $f \in L^1(0,1)$ to converge unconditionally in $L^1(0,1)$ is that $f \in L^*(0,1)$.*

PROOF. If $f \in L^*(0,1)$, then its Fourier–Haar series converges unconditionally in $L^*(0,1)$ (see Corollary 3), and therefore also in $L^1(0,1)$ (since, for every $g \in L^1(0,1)$, $S^*(g,x) \geq |g(x)|$ for almost all $x \in (0,1)$ (see Theorem 4), and therefore $\|g\|_1 \leq \|g\|_*$).

---

$(^1)$It is easy to verify that $L^*(0,1)$, with the norm (55), is a Banach space.

Now let $f \in L^1(0, 1)$ and let its Fourier–Haar series converge unconditionally in $L^1(0, 1)$. In this case, by Theorem 1.1 the series

$$\sum_{n=1}^{\infty} r_n(t) c_n(f) \chi_n(x),$$

converges in $L^1(0, 1)$ for almost all $t \in (0, 1)$, and therefore (see Theorem 2.12) $f \in L^*(0, 1)$. This completes the proof of Corollary 4.

THEOREM 12. *Let $T$ be a linear operator from $L^1(0, 1)$ to $L^0(0, 1)$, such that*

a) *for every $f \in L^*(0, 1)$ (see (55))*

$$m\{x \in (0, 1) : |T(f, x)| > t\} \le \frac{M_1}{t} \|f\|_*, \qquad t > 0; \qquad (56)$$

b) *for every $f \in L^q(0, 1)$ ($q$ is a number satisfying the inequality $1 < q < \infty$)*

$$m\{x \in (0, 1) : |T(f, x)| > t\} \le M_2 \frac{\|f\|_q{}^q}{t}, \qquad t > 0. \qquad (57)$$

*Then for each $p$, $1 < p < q$, $T$ is a bounded operator from $L^p(0, 1)$ to $L^p(0, 1)$, i.e.,*

$$\|T(f)\|_p \le M \|f\|_p, \qquad f \in L^p(0, 1).$$

REMARK. Theorem 12 sharpens a theorem of Marcinkiewicz (see Appendix 1, Theorem 1), and shows that the condition of being of weak type $(1, 1)$ can be replaced by the somewhat less restrictive condition (56). We shall use Theorem 12 only in Chapter 6, but its proof is directly connected with Theorem 11, and for that reason it is included in this section.

PROOF. Let there be given a number $p$, $1 < p < q$, and a function $f \in L^p(0, 1)$. Take a number $t > \|f\|_1$ and consider the functions $\tau'(x)$ and $g'(x)$, and the sets $e'_j$, $j = 0, 1, \ldots$, and $E$, defined in the proof of Theorem 11 for each $t > 0$ (see (52) and (53)). We noticed that $e'_j$ either is empty, or coincides with $\Delta_{j+1}$. Therefore (see (53) and (11)), if $j > 0$ and $x \in e'_j \ne \emptyset$, we have

$$g'(x) = \sum_{n=1}^{j} c_n(f) \chi_n(x) = [m(e'_j)]^{-1} \int_{e'_j} f(y) \, dy. \qquad (58)$$

In addition, it follows at once from the definition of $\tau'(x)$ (see (52)) that $e'_0$ is empty if $t > \|f\|_1$. Consequently (also see (53))

$$g'(x) = \begin{cases} f(x) & \text{if } x \in E, \\ [m(e'_j)]^{-1} \int_{e'_j} f(y) \, dy & \text{if } x \in e'_j. \end{cases} \qquad (59)$$

From a'), b'), and c') and (54) it follows (see the proof of Theorem 11) that (also see (53))

a)   $\|g'\|_\infty \leq t$;

b)   $G(t) = \{x \in (0,1)\backslash R_2 : S^*(f,x) > t\} \subset \bigcup_{j=1}^{\infty} e'_j$,

$$E \subset (0,1)\backslash G(t); \quad (60)$$

c)   $\sum_{j=1}^{\infty} m(e'_j) \leq 2m[G(t)]$.

Let $g''(x) = f(x) - g'(x)$; then by (59)

$$\text{supp } g'' \subset \bigcup_{j=1}^{\infty} e'_j, \qquad \int_{e'_j} g''(x)\,dx = 0, \qquad j = 1,2,\ldots. \quad (61)$$

Since $T$ is a linear operator,

$$\begin{aligned}
\lambda(t) &= m\{x \in (0,1) : |T(f,x)| > t\} \\
&\leq m\left\{x \in (0,1) : T(g',x)| > \frac{t}{2}\right\} \\
&\quad + m\left\{x \in (0,1) : |T(g'',x)| > \frac{t}{2}\right\} \\
&= \lambda_1(t) + \lambda_2(t).
\end{aligned} \quad (62)$$

Using (57) and (60), a) and c), we obtain

$$\begin{aligned}
\lambda_1(t) &= m\left\{x \in (0,1) : |T(g',x)| > \frac{t}{2}\right\} \\
&\leq \frac{2^q M_2}{t^q} \int_0^1 |g'(x)|^q\,dx \\
&\leq 2^q M_2 m[(0,1)\backslash E] + \frac{2^q M_2}{t^q} \int_E |f(x)|^q\,dx \\
&\leq 2^q M_2 \left[2m[G(t)] + t^{-q} \int_{(0,1)\backslash G(t)} |S^*(f,x)|^q\,dx\right],
\end{aligned}$$

From this, using Theorem 4 and (51), we obtain

$$\int_{\|f\|_1}^{\infty} t^{p-1}\lambda_1(t)\,dt \le M'\left\{\int_0^{\infty} t^{p-1}m[G(t)]\,dt\right.$$

$$+ \int_0^{\infty} t^{-q+p-1}\int_{(0,1)\backslash G(t)} |S^*(f,x)|^q\,dx\,dt\bigg\}$$

$$\le M'\left\{\|S^*(f)\|_p^p + \int_0^1 |S^*(f,x)|^q\int_{S^*(f,x)}^{\infty} t^{-q+p-1}\,dt\,dx\right\}$$

$$\le 2M'\|S^*(f)\|_p^p \le M\|f\|_p^p. \tag{63}$$

Moreover, by (60), a) and (61) we have $S(g'',x) \le S^*(f,x)+t$ for almost all $x \in (0,1)$ and $\operatorname{supp} S^*(g'') \subset \bigcup_1^{\infty} e_j'(^2)$ and therefore (see (56) and (60),b) and c))

$$\lambda_2(t) = m\left\{x \in (0,1) : |T(g''x,)| > \frac{t}{2}\right\}$$

$$\le 2M_1 t^{-1}\int_{\bigcup_{j=1}^{\infty} e_j'} S^*(g'',x)\,dx$$

$$\le |2M_1 t^{-1}\left\{\int_{G(t)} S^*(f,x)\,dx + \int_{\bigcup_{j=1}^{\infty} e_j'\backslash G(t)} S^*(f,x)\,dx + \int_{\bigcup_{j=1}^{\infty} e_j'} t\,dx\right\}$$

$$\le 2M_1 t^{-1}\int_{G(t)} S^*(f,x)\,dx + 6M_1 m[G(t)].$$

Applying the inequality just obtained for $\lambda_2(t)$, and again using (51) and Theorem 4, we find that

$$\int_{\|f\|_1}^{\infty} t^{p-1}\lambda_2(t)\,dt \le M'\left\{\int_0^{\infty} t^{p-1}m[G(t)]\,dt\right.$$

$$+ \int_0^{\infty} t^{p-2}\int_{G(t)} S^*(f,x)\,dx\,dt\bigg\}$$

$$\le M''\left\{\|S^*(f)\|_p^p + \int_0^1 S^*(f,x)\int_0^{S^*(f,x)} t^{p-2}\,dt\,dx\right\}$$

$$\le M\|f\|_p^p. \tag{64}$$

---

($^2$)Here we used the fact that $c_n(g'') = 0$ by (61), if $\Delta_n \not\subset \bigcup_j e_j'$; note also that we used essentially the same fact in the proof of Theorem 7 (see (36)).

It follows at once from (62)–(64) that

$$\|T(f)\|_p^p = p \int_0^\infty t^{p-1} \lambda(t)\, dt$$

$$\leq p \int_0^{\|f\|_1} t^{p-1}\, dt + p \int_{\|f\|_1}^\infty t^{p-1}[\lambda_1(t) + \lambda_2(t)]\, dt$$

$$\leq M'(\|f\|_1^p + \|f\|_p^p)$$

$$\leq M\|f\|_p^p.$$

This completes the proof of Theorem 12.

## §4. Convergence almost everywhere and in measure of Fourier series in the Haar system

THEOREM 13. *A necessary and sufficient condition for a series*

$$\sum_{n=1}^\infty a_n \chi_n(x) \tag{65}$$

*in the Haar system to converge almost everywhere on a set* $E \subset (0,1)$, $m(E) > 0$, *is that the sum of the series*

$$\sum_{n=1}^\infty a_n^2 \chi_n^2(x) \tag{66}$$

*is finite for almost all* $x \in E$.

REMARK. The proofs of Theorems 11 and 13 are quite complicated; consequently, to save space, we shall refer back to the proof of Theorem 11.

PROOF. SUFFICIENCY. Let the sum be finite for almost every $x \in E$. Then

$$F(x) = \sum_{n=1}^\infty a_n^2 \chi_n^*(x) < \infty \tag{67}$$

for almost all $x \in E$, and

$$\chi_n^*(x) = |\Delta_n|^{-1} \chi_{\Delta_n}(x), \qquad n = 1, 2, \ldots.$$

Take $\varepsilon \in (0, 1)$, and then take $k$ so that

$$m(G_k) < \varepsilon, \qquad G_k = \{x \in E : F(x) > k\}. \tag{68}$$

As in (50), we define the function

$$\tau(x) = \begin{cases} +\infty & \text{if } F(x) \leq k, \\ \inf\{N : F_{N+1}(x) > k\} & \text{if } F(x) > k, \end{cases} \tag{69}$$

where $F_N(x) = \sum_{n=1}^{N} a_n^2 \chi_n^*(x)$ is a partial sum of (67). Since, for $j = 1, 2, \ldots$, the set $\{x \in (0, 1) : \tau(x) = j\}$ is either empty or identical with $\Delta_{j+1} \in \mathscr{F}_j$, by Lemma 1 for Theorem 11 we can find a sequence $\varepsilon_n = 0$ or $1$, $n = 1, 2, \ldots$, for which

$$\varepsilon_n \chi_n(x) = \begin{cases} \chi_n(x) & \text{if } n \leq \tau(x), \\ 0 & \text{if } n > \tau(x) \end{cases} \tag{70}$$

for $n = 1, 2, \ldots$ and $x \in (0, 1) \backslash R_2$. Then (see (69) and (47))

$$\sum_{n=1}^{\infty} \varepsilon_n^2 a_n^2 \chi_n^2(x) = \sum_{n=1}^{\tau(x)} a_n^2 \chi_n^*(x) \leq k; \qquad x \in (0, 1) \backslash R_2.$$

Integrating the inequality, we find that $\sum_{n=1}^{\infty} \varepsilon_n^2 a_n^2 \leq k < \infty$, i.e., the series

$$\sum_{n=1}^{\infty} \varepsilon_n a_n \chi_n(x) \tag{71}$$

is the Fourier–Haar series of a function in $L^2(0, 1)$ and consequently (see Theorem 4) converges a.e. in $(0, 1)$. But when $x \in E \backslash (G_k \cup R_2)$ the series (71) coincides with (65) (see (47) and (69)), and therefore (65) converges for almost all $x$ in the set

$$E \backslash G_k, \qquad m(E \backslash G_k) \geq m(E) - \varepsilon$$

(see (68)). Since $\varepsilon$ can be arbirarily small, it follows that (65) converges for almost all $x \in E$.

*Necessity.* We choose $\varepsilon \in (0, 1)$ and find a closed set $G$ such that

$$G \subset E, \qquad m(E \backslash G) \leq \frac{\varepsilon}{3}. \tag{72}$$

Let us consider the series

$$\sum_{n=1}^{\infty} c_n \chi_n(x), \tag{73}$$

which is obtained from (65) if we omit (i.e., set $c_n = 0$) all terms $a_n \chi_n(x)$ whose supports lie in one of the complementary intervals of $G$. It is easy to verify that (73) converges a.e. in $(0, 1)$, and consequently, for sufficiently large $M$,

$$m \left\{ x \in (0, 1) : \sup_{1 \leq N < \infty} |S_N(x)| > M \right\} < \frac{\varepsilon}{3}, \tag{74}$$

where $S_N(x)$, $N = 1, 2, \ldots$, are the partial sums of (73).

As in (52) we define the function

$$\tau'(x) = \begin{cases} +\infty & \text{if } \sup_{N : x \in \Delta_N} \|S_N\|_{C(\Delta_N)} \leq M, \\ \inf\{N : \|S_{N+1}\|_{C(\Delta_{N+1})} > M, \ x \in \Delta_{N+1}\}, \\ \qquad\qquad\qquad\qquad\qquad \text{in the contrary case.} \end{cases} \tag{75}$$

In the proof of Theorem 11 it was shown that $\{x \in (0,1) : \tau'(x) = j\} \in \mathcal{F}_j$ for $j = 0, 1, \ldots$, and that (see (54))

$$m\left\{x \in (0,1) : \tau'(x) < \infty\right\} \leq 2m\left\{x \in (0,1) : \sup_{1 \leq N < \infty} |S_N(x)| > M\right\}. \quad (76)$$

Using Lemma 1 from Theorem 11, we find a sequence $\varepsilon_n = 0$ or $1$; $n = 1, 2, \ldots$, for which, when $x \in (0,1) \backslash R_2$,

$$\varepsilon_n \chi_n(x) = \begin{cases} \chi_n(x) & \text{if } n \leq \tau'(x), \\ 0 & \text{if } n > \tau'(x). \end{cases} \quad (77)$$

Then the series

$$g(x) = \sum_{n=1}^{\tau'(x)} c_n \chi_n(x) = \sum_{n=1}^{\infty} \varepsilon_n c_n \chi_n(x),$$

just as for (73), converges a.e. on $(0,1)$; moreover (see (75))

$$\left\| \sum_{n=1}^{N} \varepsilon_n c_n \chi_n(x) \right\|_{\infty} \leq M, \qquad N = 1, 2, \ldots. \quad (78)$$

It follows from (78) that $(\sum_{n=1}^{\infty} \varepsilon_n^2 c_n^2)^{1/2} \leq M$, and consequently also

$$\sum_{n=1}^{\infty} \int_0^1 \varepsilon_n^2 c_n^2 \chi_n^2(x)\,dx \leq M^2.$$

From the preceding inequality we obtain, using Beppo Levi's theorem,

$$\sum_{n=1}^{\infty} \varepsilon_n^2 c_n^2 \chi_n^2(x) < \infty \qquad \text{for almost all } x \in (0,1),$$

and therefore (see (77)),

$$\sum_{n=1}^{\infty} c_n^2 \chi_n^2(x) < \infty \quad \text{for almost all } x \in B \equiv \{x : \tau'(x) = +\infty\}.$$

But the series (65) and (73) coincide at the points of $G$, and consequently

$$\sum_{n=1}^{\infty} a_n^2 \chi_n^2(x) < \infty \quad \text{for almost all } x \in G \cap B. \quad (79)$$

Because $m[(0,1) \backslash B] \leq 2\varepsilon/3$ and $m(G \cap B) \geq m(G) - 2\varepsilon/3 \geq m(E) - \varepsilon$ (where $\varepsilon > 0$ may be taken to be arbitrarily small) (see (74), (76), (72)), (79) implies the finiteness of the sum (66) for almost all $x \in E$. This completes the proof of Theorem 13.

COROLLARY 5. *The following propositions are equivalent for every series* (65) *of Haar functions and every set* $E \subset (0, 1)$ *with* $m(E) > 0$ :
a) *the series* (65) *converges for almost all* $x \in E$;
b) *the series*

$$\sum_{n=1}^{\infty} \varepsilon_n a_n \chi_n(x)$$

*converges for arbitrary "signs"* $\varepsilon_n = \pm 1$, $n = 1, 2, \ldots$, *and for almost every* $x \in E$;
c) *the sum* (66) *is finite for almost every* $x \in E$;
d) *the series* (65) *converges in measure on* $E$ *for almost every choice of signs*;
e) *the series* (65) *converges unconditionally in measure on* $E$.

PROOF. The validity of the implications a)⇔c) and b)⇔c) follows from Theorem 13, and the equivalence of c) and d) follows from Theorem 2.11; finally, e)⇒d) and $b$) ⇒e) follow from Theorem 2.11 via Theorem 1.1.

The Haar system has the property that for every function $f(x)$ that is measurable and a.e. finite, there is a series (65) that converges to $f(x)$ almost everywhere. (We shall obtain this proposition in Chapter 10 as a corollary of a general result.)

On the other hand, the following theorem shows that there is no Haar series whose partial sums $S_N(x)$ satisfy

$$\lim_{N \to \infty} S_N(x) = +\infty \quad \text{for } x \in E, \ m(E) > 0.$$

THEOREM 14. *If the partial sums of* (65) *satisfy the condition*

$$\overline{\lim_{N \to \infty}} \sum_{n=1}^{N} a_n \chi_n(x) < +\infty, \qquad x \in E, \ m(E) > 0,$$

*then* (65) *converges a.e. on* $E$ (*to a function which is finite a.e.*).

REMARK. Theorem 14 will be needed in Chapter 10 (see Theroem 10.4).

PROOF. We may suppose without loss of generality that $a_1 = 0$. Take $\varepsilon > 0$, and let $G \subset E$ be a closed set such that $m(G) > m(E) - \varepsilon$. Consider the series

$$\sum_{n=1}^{\infty} a_n^{(1)} \chi_n(x),  \tag{80}$$

which is obtained from (65) if we omit (i.e., set $a_n^{(1)} = 0$) the terms $a_n \chi_n(x)$ whose supports are in the complement of $G$. It is clear that

$$a_n^{(1)} \chi_n(x) = a_n \chi_n(x); \qquad n = 1, 2, \ldots, \ x \in G.  \tag{81}$$

Moreover, at each point $x \in [0, 1] \backslash G$ only a finite number of terms of (80) are different from zero. Consequently

$$\sup_{1 \leq N < \infty} S_N^{(1)}(x) \equiv \sup_{1 \leq N < \infty} \sum_{n=1}^{N} a_n^{(1)} \chi_n(x) = M(x) < \infty, \qquad x \in [0, 1].$$

It follows from this that for every $\delta > 0$ there is a number $M = M_\delta > 0$ such that

$$m\{x \in (0, 1) : M(x) > M\} < \delta. \tag{82}$$

For $x \in (0, 1)$, set

$$\tau(x) = \begin{cases} +\infty & \text{if } \sup_{N:x\in\Delta_N} \|S_N^{(1)}\|_{C(\Delta_N)} \leq M, \\ \inf\{N : \|S_{N+1}^{(1)}\|_{C(\Delta_{N+1})} > M, \quad x \in \Delta_{N+1}\}, \\ & \text{in the contrary case.} \end{cases} \tag{83}$$

It is easily seen that for $j = 0, 1, \ldots$ the set $e_j = \{x \in (0, 1) : \tau(x) = j\}$ is either empty or identical with $\Delta_{j+1} \in \mathscr{F}_j$. Therefore by using Lemma 1 from Theorem 11, we can find numers $\varepsilon_n = 0$ or 1 for $n = 1, 2, \ldots$, for which

$$\varepsilon_n \chi_n(x) = \begin{cases} \chi_n(x) & \text{if } n \leq \tau(x), \\ 0 & \text{if } n > \tau(x), \end{cases}$$

for $x \in (0, 1) \backslash R_2$ and $n = 1, 2, \ldots$. Then if we set $a_n^{(2)} = \varepsilon_n a_n$ for $n = 1, 2, \ldots$, we have for $n = 1, 2, \ldots$

$$a_n^{(2)} \chi_n(x) = a_n^{(1)} \chi_n(x)$$

for almost every $x \in G_1 = \{x : \tau(x) = +\infty\}$. (84)

Let us estimate the measure of $G_1$. If $e_j$ is, for some $j$, not empty, then $e_j = \Delta_{j+1}$ and, by (83), we have $\|S_{j+1}^{(1)}\|_{C(\Delta_{j+1})} > M$, so that $S_{j+1}^{(1)}(t)$ is constant on the intervals $\Delta_{j+1}^+$ and $\Delta_{j+1}^-$. This implies that one of these intervals is a subset of $\{x \in (0, 1) : M(x) > M\}$ (see (81)). Therefore (see (82)) $m(\bigcup_{j=0}^{\infty} e_j) \leq 2\delta$ and

$$m(G_1) = 1 - m\left(\bigcup_{j=0}^{\infty} e_j\right) \geq 1 - 2\delta. \tag{85}$$

By the construction of $\tau(x)$ (see (83)) it is clear that for almost all $x \in (0, 1)$

$$\sup_{1 \leq N < \infty} S_N^{(2)}(x) \equiv \sup_{1 \leq N < \infty} \sum_{n=1}^{N} a_n^{(2)} \chi_n(x)$$

$$= \sup_{1 \leq N < \infty} \sum_{n=1}^{\min\{N, \tau(x)\}} a_n^{(1)} \chi_n(x) \leq M. \tag{86}$$

Furthermore, let $B = B_\delta = 2M\delta^{-1}$. We obtain

$$\tau'(x) = \begin{cases} +\infty & \text{if } \inf_{N:x\in\Delta_N}[\min_{t\in\Delta_N} S_N^{(2)}(t)] \geq -B, \\ \inf\{N : \min_{t\in\Delta_{N+1}} S_{N+1}^{(2)}(t) < -B, \ x \in \Delta_{N+1}\} \\ & \hspace{2cm} \text{in the contrary case.} \quad (87) \end{cases}$$

Arguing as in the construction of the series $\sum_{n=1}^\infty a_n^{(2)}\chi_n(x)$, we find numbers $\varepsilon_n' = 0$ or $1$, $n = 1, 2, \ldots$, such that if we set $a_n^{(3)} = \varepsilon_n' a_n^{(2)}$, $n = 1, 2, \ldots$, we have

$$\sum_{n=1}^\infty a_n^{(3)}\chi_n(x) = \sum_{n=1}^{\tau'(x)} a_n^{(2)}\chi_n(x), \qquad x \in (0,1)\backslash R_2,$$

and consequently

$$a_n^{(3)}\chi_n(x) = a_n^{(2)}\chi_n(x)$$
$$\text{for almost every } x \in G_2 \equiv \{x : \tau'(x) = +\infty\}. \quad (88)$$

Moreover (see (87) and (86)), for almost every $x \in (0,1)$ we have

a) $\qquad \inf_{1\leq N<\infty} \sum_{n=1}^N a_n^{(3)}\chi_n(x) > -B;$

$$(89)$$

b) $\qquad \sup_{1\leq N<\infty} \sum_{n=1}^N a_n^{(3)}\chi_n(x) \leq M.$

It follows from (89) that $\sum_{n=1}^\infty (a_n^{(3)})^2 \leq \max\{M^2, B^2\} < \infty$. Therefore, by Theorem 4, the series $\sum_{n=1}^\infty a_n^{(3)}\chi_n(x)$ converges almost everywhere on $(0,1)$ to a function $g(x)$ in $L^2(0,1)$. Therefore (see (81), (84), and (88)) the series (65) converges to $g(x)$ for almost all $x \in G \cap G_1 \cap G_2$.

We now estimate the measure of $G_2$ and show that

$$m(G_2) \geq 1 - \delta. \quad (90)$$

For this purpose we again use a frequently applied property of functions of the form (87): for each $j$, $j = 0, 1, \ldots$, the set $e_j' = \{x \in \tau'(x) = j\}$ is either empty or identical with $\Delta_{j+1}$, and then the inequality

$$S(x) = \sum_{n=1}^{\tau'(x)+1} a_n^{(2)}\chi_n(x) < -B$$

holds in one of $\Delta_{j+1}^+$ or $\Delta_{j+1}^-$. From this it follows that

$$m\{x \in (0,1) : S(x) < -B\} \geq \frac{1}{2} m\{x \in (0,1) : \tau'(x) < \infty\}. \quad (91)$$

In addition (taking account of the equations $a_1^{(3)} = a_1^{(2)} = a_1^{(1)} = 0$) we have

$$
\int_0^1 S(x)\,dx = \int_0^1 \sum_{n=1}^{\tau'(x)} a_n^{(2)} \chi_n(x)\,dx + \sum_{n=1}^{\infty} \int_{e'_{n-1}} a_n^{(2)} \chi_n(x)\,dx
$$

$$
= \int_0^1 \sum_{n=1}^{\infty} a_n^{(3)} \chi_n(x)\,dx = 0
$$

(92)

(we have used the fact that $\int_{\Delta_n} \chi_n(x)\,dx = 0$; notice also that, by (86) and (87), we have $|S(x)| \le M + B$ for almost all $x \in (0,1)$).

Since $S(x) < M$ for almost all $x \in (0,1)$ (see (86)), we find from (92) that

$$
0 = \int_0^1 S(x)\,dx \le M - Bm\{x \in (0,1) : S(x) < -B\},
$$

i.e., $m\{x \in (0,1) : S(x) < -B\} \le M/B = \delta/2$, from which (90) follows by virtue of (91).

Applying inequalities (85) and (90), we find

$$
m(G \cap G_1 \cap G_2) \ge m(G) - 3\delta \ge m(E) - \varepsilon - 3\delta \qquad (G \subset E); \qquad (93)
$$

moreover, it was shown above that (65) converges for almost all $x \in G \cap G_1 \cap G_2$. Since the numbers $\varepsilon$ and $\delta$ in (93) can be taken arbitrarily small, it follows that (65) converges for almost all $x \in E$. This completes the proof of Theorem 14.

## §5. Absolute convergence almost everywhere and unconditional convergence almost everywhere for Haar series

The following result shows that absolute convergence a.e. and unconditional convergence a.e. are equivalent for Haar series.

THEOREM 15. *A necessary and sufficient condition for the series*

$$
\sum_{n=1}^{\infty} a_n \chi_n(x) \qquad (94)
$$

*to converge unconditionally a.e. on a set* $E \subset (0,1)$, $m(E) > 0$, *is that the sum*

$$
\sum_{n=1}^{\infty} |a_n \chi_n(x)|
$$

*is finite for almost all* $x \in E$.

LEMMA 1. *For every polynomial of the form*

$$
\sum_{n=M}^{N} a_n \chi_n(x), \qquad 1 < M < N,
$$

*there is a permutation* $\{\sigma(n)\}_{n=M}^{N}$ *of the numbers* $M, M+1, \ldots, N,$ *such that for every* $x \in [0,1]$

$$\max_{M \leq p \leq q \leq N} \left| \sum_{n=p}^{q} a_{\sigma(n)} \chi_{\sigma(n)}(x) \right| \geq \frac{1}{4} \sum_{n=M}^{N} |a_n \chi_n(x)|.$$

PROOF OF LEMMA 1. Let $t_n \in (0,1)$, $n = 2, 3, \ldots$, be the centers of the intervals $\Delta_n$. Then

$$\begin{aligned} \chi_n(x) &\geq 0 \quad \text{for } 0 \leq x \leq t_n \chi_n(x) \\ &\leq 0 \quad \text{for } t_n \leq x \leq 1, \; n = 2, 3, \ldots \end{aligned} \tag{95}$$

First we consider the case when all the coefficients $a_n$, $n = M, M+1, \ldots, N$, have the same sign, say $a_n \geq 0$, $n = M, \ldots, N$.

We choose the permutation $\{\sigma(n)\}_{n=M}^{N}$ so that the numbers $t_{\sigma(n)}$ are arranged in increasing order:

$$0 < t_{\sigma(M)} < t_{\sigma(M+1)} < \cdots < t_{\sigma(N)} < 1$$

(note that in view of property 2) of binary intervals, all the $t_n$ are different). Then (see (95)):

1) if $x \in [0, t_{\sigma(M)})$ then $\chi_{\sigma(n)}(x) \geq 0$ for $n = M, M+1, \ldots, N$;

2) if $x \in [t_{\sigma(M+i)}, t_{\sigma(M+i+1)})$, $0 \leq i < N - M$, then $\chi_{\sigma(n)}(x) \leq 0$ for $M \leq n \leq M + i$ and $\chi_{\sigma(n)}(x) \geq 0$ for $M + i < n \leq N$;

3) if $x \in [t_{\sigma(N)}, 1]$ then $x_{\sigma(n)}(x) \leq 0$ for $n = M, M+1, \ldots, N$.

Remembering that the $a_n$ are nonnegative, we find from 1), 2), and 3) that

$$\sup_{M \leq p \leq q \leq N} \left| \sum_{n=p}^{q} a_{\sigma(n)} \chi_{\sigma(n)}(x) \right| \geq \frac{1}{2} \sum_{n=M}^{N} |a_n \chi_n(x)|; \qquad x \in [0,1]. \tag{96}$$

If now the signs of the coefficients $a_n$ are arbitrary and $S$ is the number of nonnegative coefficients ($1 \leq S \leq N - M$), the permutation $\{\sigma(n)\}_{n=M}^{N}$ is constructed so that

a) $a_{\sigma(n)} \geq 0$ for $M \leq n < M + S$; $a_{\sigma(n)} < 0$ for $M + S \leq n \leq N$;

b) in each of the blocks $\{\sigma(n)\}_{n=M}^{M+S-1}$ and $\{\sigma(n)\}_{n=M+S}^{N}$ the numbers $\sigma(n)$ are ordered in the direction of increase of $t_n$, i.e., $t_{\sigma(n)} < t_{\sigma(m)}$ for $M \leq n < m < M + S$ and for $M + S \leq n < m \leq M.$

Then for $x \in [0, 1]$, by (96),

$$\max_{M \le p \le q \le N} \left| \sum_{n=p}^{q} a_{\sigma(n)} \chi_{\sigma(n)}(x) \right|$$

$$\ge \max \left\{ \frac{1}{2} \sum_{n=M}^{M+S-1} |a_{\sigma(n)} \chi_{\sigma(n)}(x)|, \frac{1}{2} \sum_{n=M+S}^{N} |a_{\sigma(n)} \chi_{\sigma(n)}(x)| \right\}$$

$$\ge \frac{1}{4} \sum_{n=M}^{N} |a_n \chi_n(x)|.$$

PROOF OF THEOREM 15. We need only prove the necessity. Suppose the contrary: the series (94) converges unconditionally almost everywhere on $E$, but

$$\sum_{n=1}^{\infty} |a_n \chi_n(x)| = +\infty \quad \text{for } x \in F \subset E, \ m(f) > 0.$$

Then there are sets $A_k \subset F$, $m(A_k) > m(F) - 1/k$, $k = 1, 2, \ldots$, and an increasing sequence of integers $N_k$, $k = 1, 2, \ldots$ ($N_1 = 0$), such that

$$\sum_{n=N_k+1}^{N_{k+1}} |a_n \chi_n(x)| > 1 \quad \text{for } x \in A_k, \ k = 1, 2, \ldots$$

If we now apply Lemma 1 to the polynomial $\sum_{n=N_k+1}^{N_{k+1}} a_n \chi_n(x)$, we find a permutation $\{\sigma(n)\}_{n=N_k+1}^{N_{k+1}}$ of the numbers $N_k + 1, N_k + 2, \ldots, N_{k+1}$ such that

$$\max_{N_k < p \le q \le N_{k+1}} \left| \sum_{n=p}^{q} a_{\sigma(n)} \chi_{\sigma(n)}(x) \right| \ge \frac{1}{4}, \quad x \in A_k, \ k = 1, 2, \ldots.$$

But then the series $\sum_{n=1}^{\infty} a_{\sigma(n)} \chi_{\sigma(n)}(x)$ diverges on the set $\overline{\lim}_{k \to \infty} A_k$, i.e., for almost all $x \in F$, which contradicts the unconditional convergence of (94) a.e. on $E$. This completes the proof of Theorem 15.

Theorem 15 substantially simplifies the study of unconditional convergence a.e. for Haar series.

Since the series

$$\sum_{k=1}^{\infty} \sum_{i=1}^{2^k} \frac{1}{k 2^{k/2}} \chi_k^{(i)}(x) \tag{97}$$

with coefficients in $l^2$ does not converge absolutely at any binary irrational $x$, we have the following corollary to Theorem 15.

COROLLARY 6. *The Haar system is not an unconditional convergence system (see Definition* 1.5).

The example of the series (97) lets us, when we take account of Theorem 15 and Corollary 5, obtain the following general proposition, which is fundamental for further work.

COROLLARY 7. *There is a series* $\sum_1^\infty f_n(x)$, $x \in (0, 1)$, *of functions, which is not unconditionally convergent a.e., although all series of the form*

$$\sum_{n=1}^\infty \varepsilon_n f_n(x), \qquad \varepsilon_n = \pm 1, \; n = 1, 2, \ldots,$$

*converge a.e. on* $(0, 1)$.

Corollary 6 can be improved (see Theorems 16 and 17).

THEOREM 16. *There is a function* $f \in L^\infty(0, 1)$ *whose Fourier–Haar series diverges a.e. on* $(0.1)$ *after a rearrangement of its terms.*

REMARK. In Chapter 9 we shall show, using Theorem 16, that for every C.O.N.S. there is a continuous function whose Fourier series with respect to this system diverges a.e. after a rearrangement of its terms.

PROOF OF THEOREM 16. For $m = 2, 3, \ldots,$ set

$$P_m(x) = m^{-4} \sum_{k=m^8+1}^{m^8+m^7} \sum_{i=1}^{2^k} 2^{-k/2} \chi_k^{(i)}(x) \equiv \sum_{n=1}^\infty c_n(P_m)\chi_n(x). \tag{98}$$

For a given $m$, we can argue in the proof of Theorem 11 to find numbers $\varepsilon_n = 0$ or 1, $n = 1, 2, \ldots,$ such that (see (53) and (52))

$$Q_m(x) \equiv \sum_{n=1}^{\tau_n(x)} c_n(P_m)\chi_n(x) = \sum_{n=1}^\infty \varepsilon_n c_n(P_m)\chi_n(x), \tag{99}$$

where

$$\tau_m(x) = \begin{cases} +\infty \quad \text{if } \sup_{N:x\in\Delta_N} \|S_N(P_m)\|_{C(\Delta_N)} \leq 1, \\ \inf\{N : \|S_{N+1}(P_m)\|_{C(\Delta_{N+1})} > 1, \; x \in \Delta_{N+1}\} \\ \hspace{6cm} \text{in the contrary case.} \end{cases}$$

Moreover (see (54))

$$m\{x \in (0, 1) : \tau_m(x) < \infty\} \leq 2m\{x \in (0, 1) : S^*(P_m, x) > 1\}. \tag{100}$$

It is clear that $Q_m(x)$, $m = 2, 3, \ldots,$ are Haar polynomials with no similar terms (see (98) and (99)). We notice some of their properties.

By the construction of $\tau_m(x)$,

$$\text{a) } \|Q_m\|_\infty \leq 1; \qquad \text{b) } \|Q_m\|_2 \leq \|P_m\|_2 \leq m^{-1/2}. \tag{101}$$

If $x \in E_m \equiv \{x \in (0,1) : \tau(x) = \infty\}$, then

$$\sum_{n=1}^\infty |c_n(Q_m)\chi_n(x)| = \sum_{n=1}^\infty |c_n(P_m)\chi_n(x)| = m^3, \qquad m = 2, 3, \ldots. \tag{102}$$

Finally, it follows from (100), (44), and the Chebyshev inequality that

$$m[(0,1)\backslash E_m] \leq 2m\{x \in (0,1) : S^*(P_m, x) > 1\}$$

$$\leq 2\|S^*(P_m)\|_2^2 \leq C\|P_m\|_2^2 \leq Cm^{-1}, \qquad m = 2, 3, \ldots. \tag{103}$$

Let us show that the function

$$f(x) = \sum_{m=2}^\infty m^{-2} Q_m(x)$$

satisfies the requirements of Theorem 16. In fact, by (101), a), we have $\|f\|_\infty \leq \sum_{m=2}^\infty m^{-2}$; and by (102) (since the $Q_m$ have no similar terms)

$$\sum_{n=1}^\infty |c_n(f)\chi_n(x)| \geq m^{-2} \sum_{n=1}^\infty |c_n(Q_m)\chi_m(x)| = m,$$

$$x \in E_m, \ m = 2, 3, \ldots,$$

i.e.,

$$\sum_{n=1}^\infty |c_n(f)\chi_n(x)| = \infty \tag{104}$$

from $x \in \overline{\lim}_{m\to\infty} E_m$. But it follows from (103) that $m(\overline{\lim}_{m\to\infty} E_m) = 1$, and therefore (see the proof of Theorem 15 and (104)) the Fourier–Haar series of $f(x)$ diverges a.e. after a rearrangement of its terms. This completes the proof of Theorem 16.

DEFINITION 2. A sequence $\{\omega(n)\}$ $(1 \leq \omega(1) \leq \omega(2) \leq \cdots)$ is a *Weyl multiplier for unconditional convergence a.e. of series in the O.N.S.* $\{\varphi_n(x)\}_{n=1}^\infty$ if the condition

$$\sum_{n=1}^\infty a_n^2 \omega(n) < \infty \tag{105}$$

guarantees the unconditional convergence a.e. of the series $\sum_1^\infty a_n \omega_n(x)$.

THEOREM 17. *A necessary and sufficient condition for the sequence* $\{\omega(n)\}$, $1 \leq \omega(1) \leq \omega(2) \leq \cdots$, *to be a Weyl multiplier for unconditional convergence a.e. of series of Haar functions is that*

$$\sum_{n=1}^\infty \frac{1}{n\omega(n)} < \infty. \tag{106}$$

PROOF. If the sum (106) is finite and the series (105) converges for a sequence $\{\omega(n)\}$, $1 \leq \omega(1) \leq \omega(2) \leq \cdots$, then, since $\|\chi_n\|_1 \leq 2n^{-1/2}$, $n = 1, 2, \ldots$, we can use Cauchy's inequality to obtain

$$\sum_{n=1}^{\infty} \int_0^1 |a_n \chi_n(x)| \, dx \leq 2 \sum_{n=1}^{\infty} \frac{|a_n|}{\sqrt{n}}$$

$$= 2 \sum_{n=1}^{\infty} |a_n| \sqrt{\omega(n)} \cdot \frac{1}{\sqrt{n\omega(n)}}$$

$$\leq 2 \left\{ \sum_{n=1}^{\infty} a_n^2 \omega(n) \right\}^{1/2} \left\{ \sum_{n=1}^{\infty} \frac{1}{n\omega(n)} \right\}^{1/2} < \infty.$$

Consequently the series (94) converges absolutely a.e. on $(0,1)$, by Beppo Levi's theorem.

Suppose, on the other hand, that (106) is not satisfied. Then

$$\sum_{k=1}^{\infty} \omega^{-1}(2^k) \geq \sum_{k=1}^{\infty} 2^{-k} \sum_{n=2^k+1}^{2^{k+1}} \omega^{-1}(n) \geq \sum_{n=1}^{\infty} \frac{1}{n\omega(n)} = \infty$$

and we can find a sequence $\{q_k\}_{k=1}^{\infty}$, $\lim_{k \to \infty} q_k = \infty$, such that

$$\sum_{k=1}^{\infty} \frac{1}{\omega(2^{k+1})q_k} = \infty, \qquad \sum_{k=1}^{\infty} \frac{1}{\omega(2^{k+1})q_k^2} < \infty. \tag{107}$$

Consider the series

$$\sum_{n=1}^{\infty} a_n \chi_n(x) \equiv \sum_{k=1}^{\infty} \sum_{i=1}^{2^k} \frac{1}{2^{k/2} \omega(2^{k+1}) q_k} \chi_k^{(i)}(x). \tag{108}$$

Using (107), we find that

a) for almost all $x \in (0,1)$

$$\sum_{n=1}^{\infty} |a_n \chi_n(x)| \geq \sum_{k=1}^{\infty} \frac{1}{\omega(2^{k+1})q_k} = \infty;$$

b)

$$\sum_{n=1}^{\infty} a_n^2 \omega(n) \leq \sum_{k=1}^{\infty} \frac{1}{\omega(2^{k+1})q_k^2} < \infty,$$

and therefore (see Theorem 15 and Definition 2) $\{\omega(n)\}$ is not a Weyl multiplier for unconditional convergence a.e. of Haar series.

This completes the proof of Theorem 17.

In concluding this section we present a sufficient condition for the absolute convergence a.e. of the Fourier–Haar series of $f(x)$, formulated in terms of the integral modulus of continuity of $f(x)$.

THEOREM 18. *If $f \in L^1(0, 1)$ and*

$$\omega_1(\delta, f) = O(|\log\delta|^{-1/2-\varepsilon}) \qquad (\delta \to 0), \qquad (109)$$

*for some $\varepsilon > 0$, then the Fourier–Haar series of $f$ converges absolutely a.e.*

PROOF. For $k = 0, 1, \ldots$ and $x \in (0, 1)$, set

$$F_k(x) = \sum_{i=1}^{2^k} |c_{k,i}(f)\chi_k^{(i)}(x)| = \sum_{n=2^k+1}^{2^{k+1}} |c_n(f)\chi_n(x)|.$$

It is clear that, for $x \in (0, 1)\backslash R_2$,

$$F_k^2(x) = \sum_{i=1}^{2^k} c_{k,i}^2(f)[\chi_k^{(i)}(x)]^2.$$

Consequently, for $x \in (0, 1)\backslash R_2$,

$$\begin{aligned}
\sum_{k=1}^{\infty} F_k(x) &= \sum_{q=0}^{\infty} \sum_{k=2^q}^{2^{q+1}-1} F_k(x) \\
&\leq \sum_{q=0}^{\infty} 2^{q/2} \left\{ \sum_{k=2^q}^{2^{q+1}-1} F_k^2(x) \right\}^{1/2} \\
&\leq \sum_{q=0}^{\infty} 2^{q/2} \left\{ \sum_{k=2^q}^{\infty} F_k^2(x) \right\}^{1/2} \\
&= \sum_{q=0}^{\infty} 2^{q/2} P(g_q, x),
\end{aligned} \qquad (110)$$

where $g_q(x) = f(x) - \sum_{n=1}^{2^{2^q}} c_n(f)\chi_n(x)$, and $P(g_q, x)$ is the Paley function of $g_q(x)$ (see (39)).

Consider the sets

$$E_q = \{x \in (0, 1) : P(g_q, x) > 2^{-(q/2)(1+\varepsilon)}\}, \qquad q = 0, 1, \ldots. \qquad (111)$$

It follows from Theorem 9 and (17) that

$$m(E_q) \leq 2^{(q/2)(1+\varepsilon)}\|g_q\|_1 \leq 2^{(q/2)(1+\varepsilon)}\omega_1(2^{-2^q}, f) \leq C2^{-(q/2)\varepsilon}.$$

Therefore $\sum_{q=0}^{\infty} m(E_q) < \infty$ and consequently for almost every $x \in (0, 1)$ there is a number $q(x)$ such that $x \notin E_q$ for $q > q(x)$. But then, for almost all $x \in (0, 1)$, we obtain (see (110) and (111))

$$\sum_{k=1}^{\infty} F_k(x) \leq \sum_{q=0}^{q(x)} 2^{q/2} P(g_q, x) + \sum_{q=q(x)+1}^{\infty} 2^{q/2} P(g_q, x)$$

$$\leq \sum_{q=0}^{q(x)} 2^{q/2} P(g_q, x) + \sum_{q=q(x)+1}^{\infty} 2^{-(q/2)\varepsilon} < \infty.$$

This completes the proof of Theorem 18.

COROLLARY 8. *The Fourier-Haar series converges absolutely a.e. for every continuous function $f(x)$ with modulus of continuity $\omega(\delta, f) = O(|\log \delta|^{-1/2-\varepsilon})$ $(\delta \to 0, \varepsilon > 0)$.*

REMARK. We can prove, as an analog of Theorem 16, that Corollary 8 (and therefore Theorem 18) cannot be essentially improved, i.e., there is a function $f \in C(0, 1)$ with $\omega(\delta, f) \leq (|\log \delta|^{-1/2})$ $(\delta \to 0)$ whose Fourier-Haar series diverges absolutely for almost every $x \in (0, 1)$.

## §6. Transformations of the Haar system

As we shall show in Chapters 8, 9, and 10, the Haar system is very useful in solving a number of problems in the theory of general orthogonal series. More precisely, it turns out to be possible to extend many propositions that were first proved for the Haar system to a wide class of complete orthonormal systems. In this brief section we present a preliminary survey of extensions of this kind.

Let there be given a family of measurable sets

$$\mathscr{E} = \{\{E_k^i\}_{i=1}^{2^k}\}_{k=0}^{\infty}, \tag{112}$$

where $E_0^1 = [0, 1]$, $E_k^i \subset [0, 1]$ for $k > 0$, and the following conditions are satisfied:

   I) $m(E_k^i) = 2^{-k}$;
   II) $E_k^i = E_{k+1}^{2i-1} \cup E_{k+1}^{2i}$;
   III) $E_k^i \cap E_k^j = \varnothing$ for $i \neq j$.

An example of a family of sets that satisfies conditions I)–III) is the family of binary intervals (with inclusion of the endpoints, as appropriate).

PROPOSITION 2. *For every family $\mathscr{E}$ with properties I)–III) there is a measurable function $u(x)$, $x \in (0, 1)$, such that*

   1) *for every measurable set $G \subset [0, 1]$, the set $u^{-1}(G) \equiv \{x \in (0, 1) : u(x) \in G\}$ is measurable and $m[u^{-1}(G)] = m(G)$;*

   2)

$$\chi_{E_k^i}(x) = \chi_{\Delta_k^i}(u(x)),$$

*for* $k = 0, 1, \ldots$; $i = 1, \ldots, 2^k$; *and almost all* $x \in (0, 1)$, *where, as usual,* $\chi_E(x)$ *is the characteristic function of the set* $E$.

PROOF. For $k = 0, 1, \ldots$, we set

$$u_k(x) = \frac{i - 1}{2^k} \quad \text{for } x \in E_k^i, \; i = 1, \ldots, 2^k.$$

It follows from property II) of the family $\mathscr{E}$ that $0 \le u_k(x) \le u_{k+1}(x) \le 1$ for every $x \in [0, 1]$ and $k = 0, 1, \ldots$. Therefore the function

$$u(x) = \lim_{k \to \infty} u_k(x)$$

is defined for all $x \in [0, 1]$, and is measurable (as a pointwise limit of measurable functions). It follows directly from the definition of $u(x)$ that, for every interval $\Delta \subset (-\infty, \infty)$,

$$m\{x \in (0, 1) : u(x) \in \Delta\} = \lim_{k \to \infty} m\{x \in (0, 1) : u_k(x) \in \Delta\}$$
$$= m[\Delta \cap (0, 1)],$$

and therefore, for $\Delta \subset (0, 1)$,

$$m\{x \in (0, 1) : u(x) \in \Delta\} = m\{x \in (0, 1) : u(x) \in \overline{\Delta}\}$$
$$= m(\Delta), \tag{113}$$

where $\overline{\Delta}$ is the closure of $\Delta$.

It follows at once from (113) that

$$m[u^{-1}(V)] = m(V) \tag{114}$$

for every open or closed set $V \subset [0, 1]$.

Moreover, it is easily seen that if $u(x) \in \Delta_k^i$ then $x \in E_k^i$, i.e.,

$$u^{-1}(\Delta_k^i) \subset E_k^i, \qquad k = 0, 1, \ldots, \; i = 1, \ldots, 2^k. \tag{115}$$

From (114) and (115), taking account of the equation $m(E_k^i) = 2^{-k}$ (see 1)), we obtain $m[E_k^i \backslash u^{-1}(\Delta_k^i)] = 0$, which establishes property 2) of $u(x)$.

Now let $G \subset (0, 1)$ be any measurable set. We fix $\varepsilon > 0$ and find an open set $V$ satisfying

$$G \subset V \subset (0, 1), \qquad m(G) + \varepsilon \ge m(V).$$

Then $u^{-1}(G) \subset u^{-1}(V)$ and (see (114))

$$m^*[u^{-1}(G)] \le m[u^{-1}(V)] \le m(G) + \varepsilon$$

($m^*(E)$ denotes the outer measure of $E$). From this it follows, since $\varepsilon > 0$ is arbitrary, that

$$m^*[u^{-1}(G)] \le m(G).$$

On the other hand, if $m_*$ denotes the inner measure, we have

$$m_*[u^{-1}(G)] = 1 - m^*[[0,1]\backslash u^{-1}(G)]$$
$$= 1 - m^*[u^{-1}([0,1]\backslash G)]$$
$$= 1 - m^*[u^{-1}((0,1)\backslash G)] \geq 1 - m[(0,1)\backslash G]$$
$$= m(G).(^3)$$

Consequently $u^{-1}(G)$ is a measurable set and equation 1) holds. This completes the proof of Proposition 2.

For every family $\mathscr{E}$ with properties I)–III), we define a system of functions $\{\chi_n(x,\mathscr{E})\}_{n=1}^{\infty}$ on $(0,1)$ by setting

$$\chi_1(x,\mathscr{E}) \equiv \chi_0^{(0)}(x,\mathscr{E}) \equiv 1,$$

$$\chi_n(x,\mathscr{E}) \equiv \chi_k^i(x,\mathscr{E}) = \begin{cases} 2^{k/2} & \text{if } x \in E_{k+1}^{2i-1}, \\ -2^{k/2} & \text{if } x \in E_{k+1}^{2i}, \\ 0 & \text{if } x \in (0,1)\backslash E_k^i, \end{cases} \qquad (116)$$

where $n = 2^k + i$, $k = 0,1,\ldots$; $i = 1,\ldots,2^k$.

We say that the system $\{\chi_n(x,\mathscr{E})\}_{n=1}^{\infty}$ is a *system of Haar type*. It follows at once from conditions I)–III) that every system of Haar type is an orthonormal system. Moreover, it follows from Proposition 2 that, for every polynomial $P(x) = \sum_1^N a_n \chi_n(x,\mathscr{E})$, the equation

$$P(x) = P'(u(x)) \quad \text{for a.e. } x \in (0,1)$$

is satisfied, where

$$P'(x) = \sum_{n=1}^{N} a_n \chi_n(x) \qquad (117)$$

and that the distribution functions of $P(x)$ and $P'(x)$ are the same.

It is clear from what we have said that many results about the Haar system can be carried over to systems $\{\chi_n(x,\mathscr{E})\}$. In particular, we notice the following corollary.

COROLLARY 9. *If $\{\chi_n(x,\mathscr{E})\}$ is any system of Haar type, there is a function $g \in L^{\infty}(0,1)$ for which*

$$g(x) \overset{L^2}{=} \sum_{n=1}^{\infty} \int_0^1 \chi_n(t,\mathscr{E})g(t)\,dt\chi_n(x,\mathscr{E}) \qquad (118)$$

---

$(^3)$We used the fact that, by (114), $m\{x \in (0,1) : u(x) = t\} = 0$ for every $t$. Notice also that in applications of Proposition 2 (see Chapters 9 and 10) the sets $E_k^i$ have such simple structure that their measurability is completely obvious.

*and the series* (118) *diverges unboundedly a.e., after some rearrangement of its terms.*

PROOF. Let

$$\sum_{n=1}^{\infty} c_n \chi_n(x) \equiv \sum_{n=1}^{\infty} c_n(f) \chi_n(x) \qquad \left( \sum_{n=1}^{\infty} c_n^2 < \infty \right) \qquad (119)$$

be the Fourier series of the function $f \in L^{\infty}(0,1)$ constructed in Theorem 16. Then, for some permutation $\{\sigma(n)\}_{n=1}^{\infty}$ of the integers, the series

$$\sum_{n=1}^{\infty} c_{\sigma(n)} \chi_{\sigma(n)}(x)$$

diverges unboundedly for $x \in E$, $m(E) = 1$. This means that for $x \in u^{-1}(E)$, i.e., almost everywhere (see Proposition 2), the series

$$\sum_{n=1}^{\infty} c_{\sigma(n)} \chi_{\sigma(n)}(u(x)) = \sum_{n=1}^{\infty} c_{\sigma(n)} \chi_{\sigma(n)}(x, \mathscr{E}).$$

diverges. Therefore if we set

$$g(x) \overset{L^2}{=} \sum_{n=1}^{\infty} c_n \chi_n(x, \mathscr{E}),$$

and use the fact that $g \in L^{\infty}(0,1)$ (since, by (117) and (8), the partial sums of the series (119) do not exceed $\|f\|_{\infty}$ for almost all $x \in (0,1)$), we find that $g(x)$ satisfies the requirements of Corollary 9.

REMARK. Since the series (119) converges in $L^p(0,1)$ for every $p \in [1, \infty)$ (see Theorem 3), it follows, by (117), that the series (118) has the same property.

## CHAPTER IV

# Some Results on the Trigonometric and Walsh Systems

In this chapter we prove a number of results about the properties of the most fundamental orthogonal system: the trigonometric system. For the most part, these theorems are classical and can be found in monographs on the theory of trigonometric series. We also present some comparatively recent results. We naturally make no attempt to provide a complete survey of this material. The necessity of considering the properties of the trigonometric system was forced on the authors by our principle of making the exposition self-contained, without references to the specialized literature. Many results of the theory of general orthogonal series are generalizations or analogs of propositions of the trigonometric system. In addition, in a number of problems the trigonometric system behaves in the best possible way, and by its example verifies the definitive nature of some of the general theorems proved in this book.

In §1 we state, without proofs, some facts that are presented in university courses.

At the end of the chapter we give the definition, and prove some properties, of another classical O.N.S., the Walsh system.

### §1. Properties of the partial sums of Fourier series, Fourier coefficients, and Fejér means

The trigonometric system is the system

$$\left\{ \frac{1}{2}, \ \cos nx, \ \sin nx, \ n = 1, 2, \ldots, \ x \in [-\pi, \pi] \right\}. \tag{1}$$

The trigonometric system is complete in $L^2(-\pi, \pi)$ and orthogonal. The functions in the system (1) are not normalized in $L^2(-\pi, \pi)$. The system

$$\left\{ \frac{1}{\sqrt{2\pi}}, \ \frac{1}{\sqrt{\pi}} \cos nx, \ \frac{1}{\sqrt{\pi}} \sin nx, \ n = 1, 2, \ldots, \ x \in [-\pi, \pi] \right\}$$

is orthonormal.

To each $f \in L^1(-\pi, \pi)$ there corresponds its Fourier series:

$$\frac{a_0(f)}{2} + \sum_{n=1}^{\infty} a_n(f) \cos nx + b_n(f) \sin nx, \tag{2}$$

where

$$
\begin{aligned}
a_n(f) &= \frac{1}{\pi} \int_{-\pi}^{\pi} f(x) \cos nx \, dx, && n = 0, 1, \ldots, \\
b_n(f) &= \frac{1}{\pi} \int_{-\pi}^{\pi} f(x) \sin nx \, dx, && n = 1, 2, \ldots.
\end{aligned} \tag{3}
$$

The partial sums of the series (2) satisfy, for $N = 1, 2, \ldots$, the equation

$$
\begin{aligned}
S_N(x) = S_N(f, x) &= \frac{a_0(f)}{2} + \sum_{n=1}^{N} a_n(f) \cos nx + b_n(f) \sin nx \\
&= \frac{1}{\pi} \int_{-\pi}^{\pi} f(x + t) D_N(t) \, dt,
\end{aligned} \tag{4}
$$

where

$$D_N(t) = \frac{1}{2} + \sum_{n=1}^{N} \cos nt = \frac{\sin(N + \frac{1}{2})t}{2 \sin \frac{t}{2}}, \qquad t \in [-\pi, \pi], \ N = 1, 2, \ldots,$$

is called the *Dirichlet kernel*. If we also set $D_0(t) \equiv \frac{1}{2}$, we have

$$S_0(f, x) = \frac{a_0(f)}{2} = \frac{1}{\pi} \int_{-\pi}^{\pi} f(x + t) D_0(t) \, dt.$$

By (4), the Lebesgue functions $L_N(x)$ of the trigonometric system (see Definition 1.11) are independent of $x$, and equal to

$$L_N(x) = L_N = \frac{1}{\pi} \int_{-\pi}^{\pi} |D_N(t)| \, dt, \qquad x \in [-\pi, \pi]. \tag{5}$$

The system of complex-valued functions $\{e^{inx}\}_{n=-\infty}^{\infty}$, $x \in [-\pi, \pi]$, is also called the trigonometric system.

To a function $f \in L^1(-\pi, \pi)$ there corresponds the series

$$\sum_{n=-\infty}^{\infty} c_n(f) e^{inx}, \qquad c_n(f) = \frac{1}{2\pi} \int_{-\pi}^{\pi} f(x) e^{-inx} \, dx, \tag{6}$$

which is sometimes more convenient than the seies (2), although, of course, the partial sums of (2) and (6) are the same:

$$\sum_{n=-N}^{N} c_n(f) e^{inx} = S_N(f, x),$$

where $S_N(f, x)$ is defined in (4).

Later we shall simplify the expression for the partial sums of (2), but first we notice some properties of Fourier coefficients.

THEOREM 1. *The Fourier coefficients of a function $f \in L^1(-\pi, \pi)$ satisfy*

$$\max\{|a_n|(f)|, |b_n(f)|\} \leq \frac{1}{4\pi}\omega_1^{(2)}\left(\frac{\pi}{n}, f\right) \leq \frac{1}{2\pi}\omega_1\left(\frac{\pi}{n}, f\right),$$

$$n = 1, 2, \ldots, \quad (7)$$

*where $\omega_1^{(2)}(\delta, f)$ and $\omega_1(\delta, f)$ are the integral moduli of continuity of $f(x)$, of second and first order, respectively (see Appendix 1, (6')).*

PROOF. From (3) we find, since $\cos n(x \pm \frac{\pi}{2}) = -\cos nx$,

$$a_n(f) = -\frac{1}{\pi}\int_{-\pi}^{\pi} f(x)\cos n\left(x \pm \frac{\pi}{n}\right) dx$$

$$= -\frac{1}{\pi}\int_{-\pi}^{\pi} f\left(x \pm \frac{\pi}{n}\right) \cos nx\, dx,$$

and therefore

$$a_n(f) = \frac{1}{4\pi}\int_{-\pi}^{\pi}\left[2f(x) - f\left(x + \frac{\pi}{n}\right) - f\left(x - \frac{\pi}{n}\right)\right]\cos nx\, dx,$$

$$|a_n(f)| \leq \frac{1}{4\pi}\omega_1^{(2)}\left(\frac{\pi}{n}, f\right).$$

The coefficients $b_n$ can be estimated similarly. This completes the proof of Theorem 1.

COROLLARY 1. *Let $f \in L^1(-\pi, \pi)$, $g \in L^\infty(-\pi, \pi)$. Then the Fourier coefficients of $F_x(t) = f(x + t)g(t)$ tend to zero, uniformly for $x \in [-\pi, \pi]$.*

PROOF. by (7), it is enough to prove that $\omega_1(\delta, F_x) \to 0$ as $\delta \to 0$, uniformly in $x$. Let $\|g\|_\infty = M$; then, for each $\varepsilon > 0$, if we represent $f(x)$ in the form

$$f(x) = f_1(x) + f_2(x), \qquad \|f_1\|_\infty = B < \infty, \quad \|f_2\|_1 \leq \frac{\varepsilon}{4M},$$

we have

$$\int_{-\pi}^{\pi} |F_x(t+h) - F_x(t)| t \leq \int_{-\pi}^{\pi} |f(x+t)| \cdot |g(t+h) - g(t)| dt$$

$$+ \int_{-\pi}^{\pi} |f(x+t+h) - f(x+t)| \cdot |g(t+h)| dt$$

$$\leq \int_{-\pi}^{\pi} |f_1(x+t)| \cdot |g(t+h) - g(t)| dt$$

$$+ \int_{-\pi}^{\pi} |f_2(x+t)| \cdot |g(t+h) - g(t)| dt$$

$$+ M\omega_1(h, f) \leq B\omega_1(h, g) + \frac{\varepsilon}{2} + M\omega_1(h, f)$$

$$\leq \varepsilon,$$

if $h$ is sufficiently small. This completes the proof of Corollary 1.

We now show that the partial sums of every function $f \in L^1(-\pi, \pi)$ satisfy, as $N \to \infty$,

$$S_N(f, x) = \frac{1}{\pi} \int_{-\pi}^{\pi} \frac{f(x+t) \sin Nt}{t} dt + o(1), \tag{8}$$

where $o(1)$ tends to zero uniformly for $x \in [-\pi, \pi]$.

In fact, for $t \in [-\pi, \pi]$ and $N = 1, 2, \ldots$, let us set

$$S_N^*(f, x) = \frac{1}{\pi} \int_{-\pi}^{\pi} f(x+t) D_N^*(t) \, dt, \qquad D_N^*(t) = \frac{\sin Nt}{t},$$

then we have

$$D_N(t) - D_N^*(t) = g(t) \sin Nt + \tfrac{1}{2} \cos Nt, \tag{9}$$

where

$$g(t) = \frac{1}{2 \tan \frac{t}{2}} - \frac{1}{t}, \qquad \|g\|_{L^\infty(-\pi, \pi)} = A < \infty.$$

Therefore, by Corollary 1, as $N \to \infty$ we have

$$S_N(f, x) - S_N^*(f, x) = \frac{1}{\pi} \int_{-\pi}^{\pi} f(x+t) \left[ g(t) \sin Nt + \frac{1}{2} \cos Nt \right] dt \to 0$$

uniformly for $x \in [-\pi, \pi]$; this establishes (8).

If we set $f(x) \equiv 1$ in (8), we obtain

$$1 = \frac{1}{\pi} \int_{-\pi}^{\pi} \frac{\sin Nt}{t} dt + o(1),$$

and consequently (again see (8)) if $f \in L^1(-\pi, \pi)$, $x \in [-\pi, \pi]$,

$$S_N(f, x) - f(x) = \frac{1}{\pi} \int_{-\pi}^{\pi} [f(x+t) - f(x)] \frac{\sin Nt}{t} dt + f(x)[o(1)] \tag{10}$$

(in (10), as in (8), $o(1)$ tends to zero uniformly on $[-\pi, \pi]$).

Using (9), we easily obtain the following expression for the Lebesgue constants (see (5)):

$$L_N = \frac{1}{\pi} \int_{-\pi}^{\pi} |D_N(t)| \, dt = \frac{4}{\pi^2} \ln N + O(1) \qquad \text{as } N \to \infty, \qquad (11)$$

from which there follows the next theorem (see Theorem 1.8 and the remark following it).

THEOREM 2. *The trigonometric system is not a basis in the spaces* $C(-\pi, \pi)$ *and* $L^1(-\pi, \pi)$.

In contrast to the partial sums of the Fourier series of a continuous $f(x)$, their arithmetic means always converge uniformly to $f(x)$. More precisely, if we consider the sequence

$$\sigma_N(x) = \sigma_N(f, x) = \frac{1}{N+1} \sum_{\nu=0}^{n} S_\nu(f, x), \qquad N = 0, 1, \ldots, \qquad (12)$$

then $\sigma_N(f, x) \to f(x)$ uniformly for $x \in [-\pi, \pi]$ provided that $f \in C(-\pi, \pi)$ (also see §4 for a stronger result).

The means $\sigma_N(f, x)$ are the Fejér means. Here (see (4)) for every $f \in L^1(-\pi, \pi)$

$$\sigma_N(f, x) = \frac{1}{\pi} \int_{-\pi}^{\pi} f(x+t) \frac{1}{N+1} \sum_{\nu=0}^{N} D_\nu(t) \, dt$$

$$= \frac{1}{\pi} \int_{-\pi}^{\pi} f(x+t) K_N(t) \, dt, \qquad (12')$$

where $K_N(t)$ is the Fejér kernel,

$$K_N(t) = \frac{1}{N+1} \sum_{\nu=0}^{N} D_\nu(t) = \frac{2}{N+1} \left[ \frac{\sin \frac{1}{2}(N+1)t}{2 \sin \frac{t}{2}} \right]^2, \qquad (13)$$

$$N = 0, 1, \ldots.$$

Notice that (see (13) and (12')) for $N = 1, 2, \ldots$

a) $K_N(t) \geq 0, \quad t \in [-\pi, \pi];$ \qquad b) $\dfrac{1}{\pi} \displaystyle\int_{-\pi}^{\pi} K_N(t) \, dt = 1.$ \qquad (14)

It follows from (14) that for $f \in C(-\pi, \pi)$ and $x \in [-\pi, \pi]$

$$|\sigma_N(f, x)| \leq \frac{1}{\pi} \int_{-\pi}^{\pi} |f(x+t)| K_N(t) \, dt \leq \|f\|_{C(-\pi,\pi)}, \qquad N = 0, 1, \ldots. \quad (15)$$

If, on the other hand, $f \in L^p(-\pi, \pi)$, $1/p + 1/q = 1$, $1 \le p < \infty$, then by Hölder's inequality

$$|\sigma_N(f, x)| = \frac{1}{\pi} \left| \int_{-\pi}^{\pi} f(x+t) K_N^{1/p}(t) K_N^{1/q}(t) \, dt \right|$$

$$\le \frac{1}{\pi} \left[ \int_{-\pi}^{\pi} |f(x+t)|^p K_N(t) \, dt \right]^{1/p} \cdot \pi^{1/q}$$

and consequently

$$\|\sigma_N(f)\|_p^p \le \frac{1}{\pi} \int_{-\pi}^{\pi} \int_{-\pi}^{\pi} |f(x+t)|^p K_N(t) \, dt \, dx$$

$$= \frac{1}{\pi} \int_{-\pi}^{\pi} K_N(t) \int_{-\pi}^{\pi} |f(x+t)|^p \, dx \, dt$$

$$\le \|f\|_p^p \frac{1}{\pi} \int_{-\pi}^{\pi} K_N(t) \, dt = \|f\|_p^p. \tag{16}$$

From (16) it is easy to obtain the convergence in the $L^p(-\pi, \pi)$ norm, $1 \le p < \infty$, of $\sigma_N(f, x)$ to $f(x) \in L^p(-\pi, \pi)$.

It should be noticed that in proving (15) and (16) we used only the property $\|K_N\|_1/\pi \le 1$ of the kernel $K_N(t)$. More precisely, we actually proved that for every $Q(t)$ with $\|Q\|_1 \le 1$

$$\|F\|_p \le \|f\|_p, \qquad F(x) = \int_{-\pi}^{\pi} f(x+t) Q(t) \, dt,$$

$$f \in L^p(-\pi, \pi), \ 1 \le p \le \infty. \tag{16'}$$

## §2. Best approximation. Vallée–Poussin means

DEFINITION 1. Let $f \in C(-\pi, \pi)$. The *best approximation to $f(x)$ by trigonometric polynomials of order $N$ in $C(-\pi, \pi)$* is defined to be

$$E_N = E_N(f) = \inf_T \|T - f\|_{C(-\pi, \pi)}, \qquad N = 0, 1, \ldots,$$

where the infimum is taken over all

$$T(x) = \frac{a_0}{2} + \sum_{n=1}^{N} a_n \cos nx + b_n \sin nx.$$

Similarly, if $f \in L^p(-\pi, \pi)$, $1 \le p < \infty$, then

$$E_N^{(p)} = E_N^{(p)}(f) = \inf_T \|T - f\|_{L^p(-\pi, \pi)}, \qquad N = 0, 1, \ldots,$$

is the *best approximation to $f(x)$ in $L^p(-\pi, \pi)$ by trigonometric polynomials of order $N$.*

The following theorem connects the smoothness of $f(x)$ with the behavior of its best approximations.

THEOREM 3. *There is an absolute constant $C(^1)$ such that*
a) *for every $f \in C(-\pi, \pi)$*

$$E_N(f) \leq C\omega\left(\frac{1}{N}, f\right), \qquad N = 1, 2, \ldots; \tag{17}$$

b) *if $f \in L^p(-\pi, \pi)$, $1 < p < \infty$,*

$$E_N^{(p)}(f) \leq C\omega_p\left(\frac{1}{N}, f\right), \qquad N = 1, 2, \ldots. \tag{18}$$

In the proof of Theorem 3 we use the Vallée–Poussin means, which are very useful in many problems about the approximation of functions by trigonometric polynomials.

DEFINITION 2. The *Vallée–Poussin means* of a function $f \in L^1(-\pi, \pi)$ are the polynomials

$$V_N(x) = V_N(f, x) = \frac{1}{N}\sum_{\nu=N}^{2N-1} S_\nu(f, x), \qquad N = 1, 2, \ldots. \tag{19}$$

It is easily seen that the Vallée–Poussin and Fejér means are connected by the following equation (see (12) and (19)):

$$V_N(f, x) = 2\sigma_{2N-1}(f, x) - \sigma_{N-1}(f, x), \qquad N = 1, 2, \ldots. \tag{20}$$

From (20) it follows immediately from (15) and (16) that

a) $\qquad \|V_N(f)\|_C \leq 3\|f\|_C, \qquad N = 1, 2, \ldots, \ f \in C(-\pi, \pi);$
b) $\quad \|V_N(f)\|_p \leq 3\|f\|_p, \qquad N = 1, 2, \ldots, \ f \in L^p(-\pi, \pi), \ 1 \leq p < \infty.$
$$\tag{21}$$

Another useful property of the Vallée–Poussin means is that they satisfy

$$V_N(T_m, x) = T_m(x), \qquad x \in [-\pi, \pi] \tag{22}$$

for every trigonometric polynomial of order $m$, with $N \geq m$.

Properties (21) and (22) are sufficient for us to obtain the following proposition.

PROPOSITION 1. *For every $f \in C(-\pi, \pi)$ we have*

$$E_{2N-1}(f) \leq \|V_N(f) - f\|_C \leq 4E_N(f). \tag{23}$$

*If $f \in L^p(-\pi, \pi)$, $1 \leq p < \infty$, then*

$$E_{2N-1}^{(p)}(f) \leq \|V_N(f) - f\|_p \leq 4E_N^{(p)}(f). \tag{24}$$

PROOF. Since $V_N(f)$ is a trigonometric polynomial of order $2N - 1$, we need only establish the right-hand inequalities in (23) and (24). Let

---

$(^1)$We do not discuss the size of $C$.

$f \in C(-\pi, \pi)$, $N = 1, 2, \ldots$, and let $T_N(x)$ be the polynomial of best approximation (of order $\leq N$) of $f$;

$$\|f - T_N\|_C = E_N(f)$$

(the existence of such a polynomial is easily established by a compactness argument).

Then, using (21), part a) and (23), we obtain

$$\|V_N(f) - f\|_C \leq \|V_N(f) - T_N\|_C + \|T_N - f\|_C$$
$$= \|V_N(f - T_N)\|_C + E_N(f) \leq 4E_N(f),$$

which establishes (23). Inequality (24) is proved in the same way.

Using the representation (12') for the Fejér means, and (20), we obtain the following expression for the Vallée–Poussin means:

$$V_N(f, x) = \frac{1}{\pi} \int_{-\pi}^{\pi} f(x + t) \frac{\cos Nt - \cos 2Nt}{N(2\sin\frac{t}{2})^2} \, dt, \qquad (25)$$

since (see (13))

$$2K_{2N-1}(t) - K_{N-1}(t) = \frac{4\sin^2 Nt}{2N(2\sin\frac{t}{2})^2} - \frac{2\sin^2\frac{N}{2}t}{N(2\sin\frac{t}{2})^2}$$

$$= \frac{2\cos^2\frac{N}{2}t - 2\cos^2 Nt}{N(2\sin\frac{t}{2})^2} = \frac{\cos Nt - \cos 2Nt}{N(2\sin\frac{t}{2})^2}.$$

The function

$$P_N(t) = \frac{\cos Nt - \cos 2Nt}{N(2\sin\frac{t}{2})^2}$$

is the *Vallée–Poussin kernel of order* $N$ ($N = 1, 2, \ldots$).

The kernel $P_N(t)$ is an even trigonometric polynomial of order $2N - 1$ (see (13) and (20)); moreover (see (14)),

$$\frac{1}{\pi} \int_{-\pi}^{\pi} P_N(t) \, dt = 1, \qquad \frac{1}{\pi} \int_{-\pi}^{\pi} |P_N(t)| \, dt \leq 3, \qquad N = 1, 2, \ldots \quad (26)$$

we shall use the following property of $P_N(t)$: for $y \in [-\pi, \pi]$ it satisfies

$$|Q_N(y)| \equiv \left| \int_{|y|}^{\pi} P_N(t) \, dt \right| \leq 10 \min\left(1, \frac{1}{N^2 y^2}\right), \qquad N = 1, 2, \ldots . \quad (27)$$

In fact, since $f(t) = (2\sin(t/2))^{-2}$ decreases monotonically on $(0, \pi)$ and $\int_a^b g(t) \, dt \leq 3/N$, where $g(t) = \cos Nt - \cos 2Nt$ and $a$ and $b$ are arbitrary on $[0, \pi]$, we find, by using the second mean value theorem, that

$$N \cdot Q_N(y) = \int_{|y|}^{\pi} f(t) g(t) \, dt = f(|y|) \int_{|y|}^{\xi} g(x) \, dx, \qquad \xi \in (|y|, \pi),$$

$$N|Q_N(y)| \leq \left(2\sin\frac{y}{2}\right)^{-2} \frac{3}{N} \leq 3\left(\frac{\pi}{2}\right)^2 \frac{1}{y^2 N} < \frac{10}{y^2 N},$$

which, together with the inequality $|Q_N(y)| \le 3\pi$ (see (26)) establishes (27). It follows from (27) that

$$\|Q_N\|_1 = 2\left[\int_0^{1/N} 10\,dy + \int_{1/N}^{\pi} \frac{10}{N^2 y^2}\,dy\right] \le \frac{60}{N}, \qquad N = 1, 2, \ldots. \quad (28)$$

PROOF OF THEOREM 3. We can prove (17) and (18) simultaneously (with the same constant $C \le 44$), writing $\|f\|_\infty \equiv \|f\|_{C(-\pi,\pi)}$ and denoting by $\omega_p(\delta, f)$, $p = \infty$, the ordinary modulus of continuity of $f \in C(-\pi, \pi)$.

We shall show that the required approximation to $f(x)$ is given (for every $p$) by its Vallée–Poussin means $V_m(f)$, where $m = [(N+1)/2]$ ($V_m(f)$ is a polynomial of order $\le N$).

We first approximate $f(x)$ by mean-values of the form

$$I_\delta(f, x) = \frac{1}{2\delta}\int_{-\delta}^{\delta} f(x+t)\,dt, \qquad \delta > 0.$$

We have the inequality

$$\|f - I_\delta(f)\|_p = \left\{\int_{-\pi}^{\pi}\left|\frac{1}{2\delta}\int_{-\delta}^{\delta}[f(x+t)-f(x)]\,dt\right|^p dx\right\}^{1/p}$$

$$\le \frac{1}{(2\delta)^{1/p}}\left\{\int_{-\pi}^{\pi}\int_{-\delta}^{\delta}|f(x+t)-f(x)|^p\,dt\,dx\right\}^{1/p}$$

$$\le \frac{1}{(2\delta)^{1/p}}\left\{2\int_0^{\delta}\omega_p^p(t,f)\,dt\right\}^{1/p} \le \omega_p(\delta, f) \quad (29)$$

(here we used the inequality $\|f\|_{L^1(E)} \le [m(E)]^{1-p^{-1}}\|f\|_{L^p(E)}$, which is an immediate consequence of Hölder's inequality). Then the derivative of $I_\delta(f)$ can be estimated as follows:

$$\|I_\delta'(f)\|_p = \left\|\frac{f(x+\delta)-f(x-\delta)}{2\delta}\right\|_p$$

$$\le \frac{1}{2\delta}\omega_p(2\delta, f) \le \frac{1}{\delta}\omega_p(\delta, f). \quad (30)$$

(Note that when $p = \infty$ inequalities (29) and (30) follow directly from the definitions of the moduli of continuity.)

Let $N$ be given. Set $\delta = 1/N$ and $g(x) = I_{1/N}(f, x)$. Then (with $m = [(N+1)/2]$)

$$\|f - V_m(f)\|_p \le \|f - g\|_p + \|g - V_m(g)\|_p + \|V_m(f-g)\|_p. \quad (31)$$

It follows from (29) and (21) that

$$\|f - g\|_p + \|V_m(f-g)\|_p \le 4\omega_p\left(\frac{1}{N}, f\right), \quad (32)$$

and consequently it remains only to estimate the middle term on the right of (31). Integrating by parts and using the evenness of $P_m(t)$, we obtain

$$
\begin{aligned}
V_m(g,x) - g(x) &= \frac{1}{\pi} \int_{-\pi}^{\pi} [g(x+t) - g(x)] P_m(t)\, dt \\
&= \frac{1}{\pi} \int_0^\pi [g(x+t) + g(x-t) - 2g(x)] P_m(t)\, dt \\
&= -\frac{1}{\pi} [g(x+t) + g(x-t) - 2g(x)] Q_m(t) \big|_0^\pi \\
&\quad + \frac{1}{\pi} \int_0^\pi Q_m(t)[g'(x+t) - g'(x-t)]\, dt \\
&= \frac{1}{\pi} \int_{-\pi}^{\pi} Q_m(t) g'(x+t)\, \mathrm{sgn}\, t\, dt,
\end{aligned}
$$

and therefore (see (16′), (28), and (30)),

$$
\|g - V_m(g)\|_p \le \frac{1}{\pi} \|Q_m\|_1 \|g'\|_p \le 40\omega_p\left(\frac{1}{N}, f\right). \tag{33}
$$

Combining (32) and (33), and using (31), we obtain

$$
\|f - V_m(f)\|_p \le 44\omega_p\left(\frac{1}{N}, f\right), \qquad N = 1,2,\ldots \left(m = \left[\frac{N+1}{2}\right]\right). \tag{34}
$$

This completes the proof of Theorem 3.

REMARK. By a proof similar to that of (34) we can show that if the $2\pi$-periodic function $f(x)$ has a $k$th derivative $f^{(k)}(x) \in C(-\pi, \pi)$ ($k = 1, 2, \ldots$), then

$$
\|f - V_{[(N+1)/2]}(f)\|_\infty \le \frac{C_k}{N^k} \omega\left(\frac{1}{N}, f^{(k)}\right), \qquad N = 1, 2, \ldots.
$$

## §3. Convergence of trigonometric series in $L^p$ and almost everywhere

It will be shown in Chapter 5 (see Theorem 5.3) that if $g \in L^p(-\pi, \pi)$ with $1 \le p \le \infty$ then the function

$$
\tilde{g}(x) = \frac{1}{\pi} \int_{-\pi}^{\pi} \frac{g(t)}{2\tan\frac{x-t}{2}}\, dt \equiv \lim_{\varepsilon \to 0} \frac{1}{\pi} \int_{\varepsilon \le |t-x| \le \pi} \frac{g(t)}{2\tan\frac{x-t}{2}}\, dt \tag{35}
$$

exists, is finite for almost all $x \in [-\pi, \pi]$, and satisfies

$$
\|g\|_p \le C_p \|g\|_p \qquad \text{if } 1 < p < \infty. \tag{36}
$$

We use (36) in this section to prove that the trigonometric system is a basis in $L^p(-\pi, \pi)$, $1 < p < \infty$.

THEOREM 4. *The trigonometric system is a basis in $L^p(-\pi, \pi)$ $1 < p <$ $\infty$. Moreover, if $f \in L^p(-\pi, \pi)$ then*

$$\|S_N(f) - f\|_p \le C_p E_N^{(p)}(f). \tag{37}$$

PROOF. It is clear that the trigonometric system is complete (see, for example, Theorem 3) and minimal in $L^p(-\pi, \pi)$; therefore, to show that it is a basis it is enough to show (by Theorem 1.6) that when $1 < p < \infty$

$$\|S_N(f)\|_p \le C_p\|f\|_p, \qquad f \in L^p(-\pi, \pi), \ N = 1, 2, \dots. \tag{38}$$

If we use the equation

$$\sin\left(N + \frac{1}{2}\right) t = \sin Nt \cos\frac{t}{2} + \cos Nt \sin\frac{t}{2},$$

we can write the partial sums $S_N(f, x)$ (see (4)) in the form

$$
\begin{aligned}
S_N(f, x) &= \frac{1}{\pi} \int_{-\pi}^{\pi} f(t) D_N(t - x)\, dt \\
&= \frac{1}{\pi} \int_{-\pi}^{\pi} \frac{f(t)\sin N(t - x)}{2\tan\frac{t-x}{2}}\, dt + \frac{1}{2\pi}\int_{-\pi}^{\pi} f(t)\cos N(t - x)\, dt \\
&= \frac{\cos Nx}{\pi} \int_{-\pi}^{\pi} \frac{f(t)\sin Nt}{2\tan\frac{t-x}{2}}\, dt - \frac{\sin Nx}{\pi}\int_{-\pi}^{\pi}\frac{f(t)\cos Nt}{2\tan\frac{t-x}{2}}\, dt \\
&\quad + \frac{1}{2\pi}\int_{-\pi}^{\pi} f(x + t)\cos Nt\, dt, \tag{39}
\end{aligned}
$$

where the two integrals on the right of (39) are to be interpreted as in (35), i.e., as principal values.

From (39), it follows by (36) and (16) that

$$\|S_N(f)\|_p \le C_p(\|f(t)\cos Nt\|_p + \|f(t)\sin Nt\|_p) + \|f\|_p \le C_p\|f\|_p.$$

This establishes (38). To obtain (37), we choose a polynomial $T_N$ of order $N$ such that $\|f - T_n\|_p = E_N^{(p)}(f)$. Since $S_N(T_N, x) = T_N(x)$, we find from (38) that

$$
\begin{aligned}
\|S_N(f) - f\|_p &\le \|S_N(f - T_N)\|_p + \|T_N - f\|_p \\
&\le (C_p + 1)E_N^{(p)}(f).
\end{aligned}
$$

This completes the proof of Theorem 4. From (37) and Theorem 3 we obtain the following corollary.

COROLLARY 2. *For every $f \in L^p(-\pi, \pi)$, $1 < p < \infty$, we have the inequalities*

$$\|S_N(f) - f\|_p \le C_p \omega_p(\tfrac{1}{N}, f), \qquad N = 1, 2, \dots.$$

As we noticed earlier (see Theorem 2), Theorem 4 fails for $p = 1$. This follows from the inequality (see (11))

$$\|D_N\|_1 \geq c \ln N, \qquad N = 1, 2, \ldots. \tag{40}$$

The following result yields a wide generalization of (40) and shows that no change in the orders of the terms in the trigonometric system can essentially decrease the rate of growth of the Lebesgue functions.

THEOREM 5. *There is an absolute constant $c_0$ such that, for every sequence of integers $1 \leq n_1 < n_2 < \cdots$ and arbitrary complex numbers $d_k$, $k = 1, 2, \ldots$, we have the inequality*

$$\int_0^1 \left| \sum_{k=1}^n d_k e^{2\pi i n_k t} \right| dt \geq c_0 \sum_{k=1}^N \frac{|d_k|}{k}, \qquad N = 1, 2, \ldots. \tag{41}$$

COROLLARY 3. *For every set of positive integers $1 \leq n_1 < \cdots < n_N$, we have the inequality*

$$\|P(t)\|_{L^1(-\pi,\pi)} \geq c \ln N, \qquad P(t) = \sum_{k=1}^N \cos n_k t, \tag{42}$$

*where $c > 0$ is an absolute constant.*

In fact, by applying (41), we obtain

$$\|P\|_{L^1(-\pi,\pi)} = 2\pi \int_0^1 \left| \sum_{k=1}^N \frac{1}{2}(e^{2\pi i n_k t} + e^{-2\pi i n_k t}) \right| dt$$

$$= \pi \int_0^1 \left| e^{2\pi i (n_N+1)t} \sum_{k=1}^N (e^{2\pi i n_k t} + e^{-2\pi i n_k t}) \right| dt$$

$$\geq \pi c_0 \sum_{k=1}^{2N} \frac{1}{k} \geq c \ln N.$$

PROOF OF THEOREM 5. We first notice three simple facts:
I) *Let $f \in L^2(0,1)$ be a real function with Fourier series*

$$f(t) \sim \sum_{n=-\infty}^{\infty} c_n(f) e^{2\pi i n t}, \qquad c_n(f) = \int_0^1 f(t) e^{-2\pi i n t} \, dt$$

*and*

$$h(t) = c_0(f) + 2 \sum_{n=-\infty}^{\infty} c_n(f) e^{2\pi i n t};$$

*then $\operatorname{Re} h(t) = f(t)$.*

In fact, since $\mathrm{Re}[(a + ib)(c + id)] = ac - bd$, we have

$$\mathrm{Re}\, h(t) = c_0(t) + 2 \sum_{n=-\infty}^{\infty} (\mathrm{Re}\, c_n \cdot \cos 2\pi nt - \mathrm{Im}\, c_n \cdot \sin 2\pi nt),$$

but $c_n(f) = \overline{c_{-n}(f)}$, $n = \pm 1, \pm 2, \ldots$, since $f(t)$ is real, and therefore

$$\mathrm{Re}\, h(t) = c_0(f) + \sum_{n=-\infty}^{-1} (\mathrm{Re}\, c_n \cdot \cos 2\pi nt - \mathrm{Im}\, c_n \cdot \sin 2\pi nt)$$

$$+ \sum_{n=1}^{\infty} (\mathrm{Re}\, c_n \cdot \cos 2\pi nt - \mathrm{Im}\, c_n \cdot \sin 2\pi nt) = \mathrm{Re}\, f(t) = f(t).$$

II) *For all complex-valued functions $f, g \in L^2(0, 1)$ we have the equation*

$$\int_0^1 f(t)\overline{g(t)}\, dt = \sum_{n=-\infty}^{\infty} c_n(f)\overline{c_n(g)}. \tag{43}$$

In fact, by Parseval's equation

$$\int_0^1 f(t)\overline{g(t)}\, dt = \sum_{n=-\infty}^{\infty} c_n(f)c_{-n}(\overline{g}),$$

but $\overline{c_n(g)} = c_{-n}(\overline{g})$, and we obtain (43).

III) *If $\mathrm{Re}\, u \geq 0$ then $|e^{-u} - 1| \leq 2|u|$.*

In fact, $|e^{-u}| = e^{-\mathrm{Re}\, u} \leq 1$, and therefore when $|u| \leq 1$ we have $|e^{-u} - 1| \leq |e^{-u}| + 1 \leq 2|u|$; if, however, $|u| < 1$, then

$$|e^{-u} - 1| \leq \sum_{n=1}^{\infty} \frac{|u|^n}{n!} \leq |u| \sum_{n=1}^{\infty} \frac{1}{n!} \leq 2|u|.$$

We now turn to the proof of the theorem; without loss of generality, we may suppose that $d_k \neq 0$ for $k = 1, 2, \ldots$. We define $\alpha$ and $\beta$ by $\alpha = \frac{1}{2}$, $\beta = \ln 2$; then $e^{-\beta} + \alpha = 1$. We divide the sequence $\{n_k\}_{k=1}^{\infty}$ into blocks $\{\Lambda_s\}_{s=1}^{\infty}$ so that $\Lambda_s$ contains $100^s$ terms, and so that if $n_k \in \Lambda_s$ and $n_{k'} \in \Lambda_{s'}$, and $n_k < n_{k'}$ then $s \leq s'$.

For $s = 1, 2, \ldots$, set

$$f_x(t) = 100^{-s} \sum_{n_k \in \Lambda_s} (\mathrm{sgn}\, d_k)e^{2\pi i n_k t}, \qquad t \in [0, 1], \tag{44}$$

where $\mathrm{sgn}\, d \equiv d/|d|$ if $d \neq 0$ and $\mathrm{sgn}\, 0 = 0$.

It is clear that for $s = 1, 2, \ldots$,

$$\|f_s\|_C \leq 1, \qquad \|f_s\|_2 = 100^{-s/2}. \tag{45}$$

For each $N$, $N = 1, 2, \ldots$, we construct a function $F_N(t)$ for which $\|F_N\|_C \leq 1$ and

$$I_N = \left| \int_0^1 \sigma_N(t)\overline{F_N(t)}\, dt \right| \geq \frac{1}{400} \sum_{k=1}^{N} \frac{|d_k|}{k}, \qquad (46)$$

where

$$\sigma_N(t) = \sum_{k=1}^{N} d_k e^{2\pi i n_k t}.$$

From this we at once obtain (41) (with $c_0 = 400^{-1}$), since

$$I_N \leq \|F_N\|_C \int_0^1 |\sigma_N(t)|\, dt.$$

For $s = 1, 2, \ldots$, set

$$h_s(t) = c_0(|f_s|) + 2 \sum_{n=-\infty}^{-1} c_n(|f_s|)e^{2\pi i n t}, \qquad t \in [0, 1].$$

We notice the following properties of $h_s(t)$, $s = 1, 2, \ldots$:
a) $\operatorname{Re} h_s(t) = |f_s(t)| \geq 0$, $\quad t \in [0, 1]$, $s = 1, 2, \ldots$ (see I)).

b)
$$\|h_s\|_2 = \left\{ c_0^2(|f_s|) + 4 \sum_{n=-\infty}^{-1} c_n^2(|f_s|) \right\}^{1/2}$$

$$\leq 2\|f_x\|_2 \leq \frac{2}{100^{s/2}} \text{ (see (45))}. \qquad (47)$$

c) $c_n(h_s) = 0$ for $n > 0$, $s = 1, 2, \ldots$.
d) The Fourier series of $h_s(t)$, $s = 1, 2, \ldots$, converges absolutely (since $|f_s(t)| \in \operatorname{Lip} 1$, property d) follows from Corollary 5, to be proved below).
Finally we define the sequence $\{g_s(t)\}_{s=1}^{\infty}$ by setting

$$g_1(t) = \alpha f_1(t), \qquad g_{s+1}(t) = e^{-\beta h_{s+1}(t)} g_s(t) + \alpha f_{s+1}(t),$$
$$t \in [0, 1], \ s = 1, 2, \ldots.$$

From the functions $g_s(t)$ we are going to select the function $F_N(t)$ that we need (see (46)). First we prove that

$$\|g_s\|_C \leq 1, \qquad s = 1, 2, \ldots. \qquad (48)$$

By (45) we have $\|g_1\|_C \leq 1$. Supposing that (48) has already been verified for $g_{s-1}(t)$, we shall have (see (47,a))

$$|g_s(t)| \leq \exp\{-\beta h_s(t)\}| + \alpha|f_s(t)|$$
$$= \exp\{-\beta \operatorname{Re} h_s(t)\} + \alpha|f_s(t)|$$
$$= \exp\{-\beta|f_s(t)|\} + \alpha|f_s(t)| \leq 1, \qquad t \in [0, 1].$$

The last inequality follows because $\psi(x) = \exp(-\beta x) + \alpha x$ is convex on $[0, 1]$ (because $\psi''(x) = \beta^2 \exp(-\beta x) > 0$), $\psi(0) = \psi(1) = 1$, and $|f_s(t)| \leq 1$ for $t \in [0, 1]$ (see (45)).

Now we show by induction that, for $s = 1, 2, \ldots,$

$$g_s(t) = \alpha \sum_{l=1}^{s} f_l(t) \exp\left\{ - \sum_{m=l+1}^{s} \beta h_m(t) \right\} \tag{49}$$

(we take $\sum_{m=s+1}^{s} \equiv 0$). Equation (49) is evident for $s = 1$.

Supposing that (49) has already been verified for $g_{s-1}$, we obtain

$$g_s = \exp\{-\beta h_s\} g_{s-1} + \alpha f_s$$

$$= \exp\{-\beta h_s\} \alpha \sum_{l=1}^{s-1} f_l \exp\left\{ - \sum_{m=l+1}^{s-1} \beta h_m \right\}$$

$$+ \alpha f_s = \alpha \sum_{l=1}^{s-1} f_l \exp\left\{ - \sum_{m=l+1}^{s} \beta h_m \right\} + \alpha f_s,$$

which establishes (49) for all $s$.

The function $F_N(t)$ that satisfies (46) is constructed as follows: for a given $N = 1, 2, \ldots,$ we determine a number $s = s(N)$ so that $n_N \in \Lambda_s$, and set

$$F_N(t) = g_s(t), \qquad t \in [0, 1]. \tag{50}$$

We shall estimate the Fourier coefficients of $F_N(t)$ and show that

$$|c_n(F_N) - \alpha c_n(f_j)| \leq (\tfrac{\alpha}{2})|c_n(f_j)|, \qquad n \in \Lambda_j, \ 1 \leq j \leq s. \tag{51}$$

The required inequality (46) will follow at once from (51).

In fact, it follows from (51) that (see II) and (44))

$$I_N = \left| \sum_{n=-\infty}^{\infty} c_n(\sigma_N)\overline{c_n(F_N)} \right| = \left| \sum_{k=1}^{N} c_{n_k}(\sigma_N)\overline{c_{n_k}(F_N)} \right|$$

$$= \left| \sum_{j=1}^{s} \sum_{n_k \in \Lambda_j, \ k \leq N} d_k \overline{c_{n_k}(F_N)} \right|$$

$$= \left| \alpha \sum_{j=1}^{s} \sum_{n_k \in \Lambda_j, k \leq N} \{d_k \overline{c_{n_k}(f_j)} + c_k[\overline{c_{n_k}(F_N)} - \alpha c_{n_k}(f_j)]\} \right|$$

$$\geq \alpha \sum_{j=1}^{s} \sum_{n_k \in \Lambda_j, k \leq N} 100^{-j}|d_k| - \frac{\alpha}{2} \sum_{j=1}^{s} \sum_{n_k \in \Lambda_j, k \leq N} |d_k| \, |c_{n_k}(f_j)|$$

$$= \frac{\alpha}{2} \sum_{j=1}^{s} 100^{-j} \sum_{n_k \in \Lambda_j, \ k \leq N} |d_k|,$$

and, remembering that $k \geq 100^{j-1}$ for $n_k \in \Lambda_j$, we obtain

$$|I_N| \geq \frac{\alpha}{2} \sum_{j=1}^{s} \sum_{n_k \in \Lambda_j, \, k \leq N} \frac{|d_k|}{100k} = \frac{1}{400} \sum_{k=1}^{N} \frac{|d_k|}{k}.$$

Consequently it remains only to prove (51). Let the numbers $j$, $1 \leq j \leq s$, and $n \in \Lambda_j$, be given. Set

$$E_1(t) = \exp\left\{ -\sum_{m=l+1}^{s} \beta h_m(t) \right\}, \qquad l = 1, \ldots, s;$$

we are going to estimate the $n$th Fourier coefficient of $f_l(t)E_l(t)$ and then use (49).

Let us show that

$$c_n(f_l E_l) = 0 \quad \text{for } l < j. \tag{52}$$

In fact, by (47),c), we have $c_k(\psi_l) = 0$ for $k > 0$ and $l = 1, \ldots, s$, where $\psi_1(t) = -\sum_{m=l+1}^{s} \beta h_m(t)$. Consequently, for every $p$, $p = 0, 1, \ldots$, and $l = 1, 2, \ldots$, we have $c_k((\psi_l)^p) = 0$ if $k > 0$, and

$$c_k(E_l) = c_k(\exp \psi_l) = \sum_{p=0}^{\infty} \frac{1}{p!} \int_0^1 \psi_l^p(t) e^{-2\pi i k t} \, dt = 0, \qquad k > 0. \tag{53}$$

(Notice that since $\psi_l(t) \in C(-\pi, \pi)$ (see (47),d)), the series $\sum_{p=0}^{\infty} \psi^{p(t)}/p!$ converges uniformly in $t$ and can be integrated term by term.) Since $c_k(f_l) = 0$ for $k \notin \Lambda_l$ (see (44)), from (53) we at once obtain (52).

Moreover, it is clear that

$$c_n(f_l E_l) = c_n(f_l) + c_n(f_l[E_l - 1]), \qquad l = j, j+1, \ldots, s, \tag{54}$$

and since $c_n(f_l) = 0$ if $l > j$ (because $n \in \Lambda_j$), we have

$$c_n(f_l E_l) = c_n(f_l[E_l - 1]), \qquad l = j+1, \ldots, s. \tag{55}$$

Using (52), (54), and (55), we deduce from (49) and (50) that

$$c_n(F_N) = c_n(g_s) = \alpha \sum_{l=1}^{s} c_n(f_l E_l) = \alpha c_n(f_j) + \alpha \sum_{l=j}^{s} c_n(f_l[E_l - 1]) \tag{56}$$

($n \in \Lambda_j$, $s = s(N)$). Since $|c_n(f_j)| = 100^{-j}$ (see (44)), it is clear from (56) that (51) will be established if we verify that

$$Q \equiv \sum_{l=j}^{s} |c_n(f_l[E_l - 1])| \leq \frac{1}{2} 100^{-j}. \tag{57}$$

Using Cauchy's inequality (see also (45)), we obtain

$$|c_n(f_l[E_l - 1])| \leq \|f_l(E_l - 1)\|_1$$
$$\leq \|f_l\|_2\|E_l - 1\|_2 \leq 100^{-l/2}\|E_l - 1\|_2. \tag{58}$$

In addition, since $\operatorname{Re} h_m(t) \geq 0$ for $m = 1, 2, \ldots$ (see (47),a)), we have

$$\|E_l - 1\|_2 \equiv \left\|\exp\left\{-\sum_{m=l+1}^{s} \beta h_m(t)\right\} - 1\right\|_2$$
$$\leq 2\beta \left\|\sum_{m=l+1}^{s} h_m(t)\right\|_2 \leq 4\beta \sum_{m=l+1}^{s} 100^{-m/2} \tag{59}$$

(here we used inequality III) and (47),b)).
It follows from (58) and (59) that

$$Q \leq 4\beta \sum_{l=j}^{\infty} 10^{-l} \sum_{m=l+1}^{\infty} 10^{-m} \leq \frac{1}{2}100^{-j}.$$

Hence (57) and (51), and consequently Theorem 5, are established.

REMARK. We shall show in Chapter 9 that a weaker version of Corollary 3 remains valid for all O.N.S. $\{\varphi_n(x)\}_{n=1}^{\infty}$, $x \in (0, 1)$, with $|\varphi_n(x)| \leq M$, $n = 1, 2, \ldots$, $x \in (0, 1)$. More precisely, we shall prove that for such systems

$$\frac{1}{N}\sum_{p=1}^{N} \left\|\sum_{n=1}^{p} \varphi_n(x)\right\|_1 \geq c_M \ln N > 0, \qquad N = 2, 3, \ldots.$$

Moreover, as is shown by the Walsh system $\{w_n(x)\}$ (see §5), for which $|w_n(x)| = 1$ for almost all $x \in (0, 1)$, $n = 1, 2, \ldots$, and

$$\left\|\sum_{n=1}^{2^k} w_n(x)\right\|_1 = 1 \quad \text{for } k = 1, 2, \ldots,$$

the norms $\|\sum_1^{N_k} \varphi_n(x)\|_1$ can be bounded for a subsequence of the indices $\{N_k\}_{k=1}^{\infty}$.

We now turn to the question of the convergence of trigonometric series almost everywhere. Here the most natural question remained open up to 1966, when Lennart Carleson proved the following deep theorem.

THEOREM C. *The trigonometric system is a convergence system (see Definition 1.2).*

Before Carleson's work it was not even known whether the Fourier series of every continuous function necessarily converges at even one point.

Unfortunately, the proof of Theorem C is too complicated to be presented here. We shall, however, prove a result that was the best one available for forty years before Carleson's work.

THEOREM 6. *A sufficient condition for the convergence almost everywhere of the series*

$$\sum_{n=1}^{\infty} a_n \cos nx + b_n \sin nx \tag{60}$$

*is that the sum*

$$\sum_{n=1}^{\infty} (a_n^2 + b_n^2) \cdot \ln n < \infty$$

*is finite.*

We shall obtain Theorem 6 as a corollary of a theorem to be proved in Chapter 8, which is a general result on the connection between convergence almost everywhere of series in an orthogonal system and estimates for the Lebesgue constants of the system (see Theorem 8.4).

Theorem C can be extended to the spaces $L^p(-\pi, \pi)$ with $p > 1$: the Fourier series of every function in $L^p(-\pi, \pi)$, $p > 1$, converges a.e. On the other hand, we have the following theorem.

THEOREM K. *There is a function $f(x) \in L^1(-\pi, \pi)$ whose Fourier series [see (2)] diverges for almost all $x \in (-\pi, \pi)$.*

Theorem K will be obtained in Chapter 9 as a consequence of a more general result (see Theorem 9.10).

## §4. Uniform and absolute convergence of Fourier series

As we have already seen (see Theorem 2, and also equations 1.(17) and (11)), there are continuous $2\pi$-periodic functions whose Fourier series diverge at a point. At the same time, there are many conditions on a function $f(x)$ that guarantee the uniform convergence of the series (2). One such condition is easily obtained from Theorem 3.

THEOREM 7. *If the modulus of continuity of $f(x) \in C(-\pi, \pi)$ satisfies the condition $\omega(\delta, f) = o(\ln^{-1}(1/\delta))$ as $\delta \to 0$, then the series (2) converges to $f(x)$, uniformly on $[-\pi, \pi]$.*

LEMMA 1. *For every $f(x) \in C(-\pi, \pi)$, the inequality*

$$\|f - S_N(f)\|_C \leq C \ln N \cdot E_N(f), \qquad N = 2, 3, \ldots, \tag{61}$$

*is satisfied; here $C$ is an absolute constant.*

In fact, let $T_N$ be the polynomial of best approximation (of order $N$) to $f(x)$: $\|f - T_N\| = E_N(f)$. Then, by using 1.(17) and the evident equation $S_N(T_N, x) \equiv T_N(x)$ (also see (11)), we find that

$$\|f - S_N(f)\|_C \leq \|f - T_N\|_C + \|S_N(T_N - f)\|_C$$
$$\leq (1 + L_N)E_N(f) \leq C \ln N \cdot E_N(f), \qquad N = 2, 3, \ldots$$

The conclusion of Theorem 7 follows directly from Lemma 1 and Theorem 3 (see (17)).

REMARK. Theorem 7 is best possible in the sense that there is a function $f \in C(-\pi, \pi)$ with $\omega(\delta, f) = O(\ln^{-1}(1/\delta))$ as $\delta \to 0$, for which the series (2) diverges at $x = 0$.

THEOREM 8. *For every* $f \in C(-\pi, \pi)$

$$\frac{1}{N+1} \sum_{\nu=0}^{N} |f(x) - S_\nu(f, x)| \leq \frac{C}{N+1} \sum_{\nu=0}^{N} E_\nu(f),$$
$$x \in [-\pi, \pi], \ N = 0, 1, \ldots \quad (62)$$

(*C is an absolute constant*).

The following corollary is an immediate consequence of Theorem 8.

COROLLARY 4. *For every* $f \in C(-\pi, \pi)$,

$$\lim_{N \to \infty} \frac{1}{N+1} \sum_{\nu=0}^{N} |f(x) - S_\nu(f, x)| = 0$$

*uniformly for* $x \in [-\pi, \pi]$.

The property of the partial sums of the Fourier series of a continuous function, described in Corollary 4, is known as *uniform strong summability* of the Fourier series. The phrase "strong summability" is used because Theorem 8 strengthens Fejér's theorem on the uniform convergence of the means $\sigma_N(f, x)$ to $f(x)$.

PROOF OF THEOREM 8. Let us first show that, for every function $g \in C(-\pi, \pi)$,

$$\frac{1}{N+1} \sum_{\nu=0}^{N} |S_\nu(g, x)| \leq C\|g\|_C, \qquad x \in [-\pi, \pi], \ N = 0, 1, \ldots. \quad (63)$$

It follows from (8) that it is sufficient to prove inequality (63) for $S^*_\nu(g,x)$ (see (8), (9)) instead of $S_\nu(x)$. For a given $x \in [-\pi, \pi]$, we have

$$\frac{1}{N+1} \sum_{\nu=0}^{N} |S^*_\nu(g,x)| = \frac{1}{N+1} \sum_{\nu=0}^{N} \varepsilon_\nu \frac{1}{\pi} \int_{-\pi}^{\pi} g(x+t) \frac{\sin \nu t}{t} \, dt$$

$$= \frac{1}{(N+1)\pi} \int_{-\pi}^{\pi} g(x+t) \sum_{\nu=0}^{N} \varepsilon_\nu \frac{\sin \nu t}{t} \, dt$$

$$\leq \frac{1}{(N+1)\pi} \|g\|_C \int_{-\pi}^{\pi} \left| \sum_{\nu=0}^{N} \varepsilon_\nu \frac{\sin \nu t}{t} \right| \, dt, \qquad (64)$$

where $\varepsilon_\nu = \operatorname{sgn} S^*_\nu(g,x)$, $\nu = 0, \ldots, N$.

Moreover, if we use Cauchy's inequality and Parseval's equation, we find that

$$\int_{-\pi}^{\pi} \left| \sum_{\nu=0}^{N} \varepsilon_\nu \frac{\sin \nu t}{t} \right| \, dt \leq \int_{|t| \leq 1/(N+1)} \sum_{\nu=0}^{N} \nu \, dt$$

$$+ \left\{ \int_{1/(N+1) \leq |t| \leq \pi} \left[ \sum_{\nu=0}^{N} \varepsilon_\nu \sin \nu t \right]^2 \, dt \right\}^{1/2}$$

$$\times \left\{ \int_{1/(N+1) \leq |t| \leq \pi} \frac{dt}{t^2} \right\}^{1/2}$$

$$\leq C(N+1),$$

from which (63) follows by (64).

If we apply (63) with $N = 2k$ to $f(x) - T_k(f,x)$, where $T_k(f)$ is the polynomial of order $k$ of best approximation to $f(x)$, and use the fact that $S_\nu(T_k) = T_k$ for $\nu \geq k$, we obtain

$$\sum_{\nu=k}^{2k} |f(x) - S_\nu(f,x)| = \sum_{\nu=k}^{2k} |f(x) - T_k(x) - S_\nu(f - T_k, x)|$$

$$\leq C(2k+1)\|f - T_k\|_C$$

$$= C(2k+1)E_k(f), \qquad k = 0, 1, \ldots. \qquad (65)$$

It follows from (65) that, for $r = 1, 2, \ldots$,

$$\sum_{\nu=0}^{2^r} |f(x) - S_\nu(f,x)| \leq C E_0(f) + C' \sum_{s=0}^{r-1} 2^s E_{2^s}(f)$$

$$\leq C \sum_{\nu=0}^{2^{r-1}} E_\nu(f) \leq C \sum_{\nu=0}^{2^r} E_\nu(f). \qquad (66)$$

Inequality (62) with an arbitrary $N$ follows easily from (66), since, because the numbers $E_\nu(f)$ decrease monotonically, we have

$$\sum_{\nu=0}^{\infty} E_\nu(f) \le 2 \sum_{\nu=0}^{2^r} E_\nu(f)$$

for $2^r \le N < 2^{r+1}$. This completes the proof of Theorem 8.

The following result shows that if we transform a given continuous function by a suitable change of variable we can always obtain a function whose Fourier series converges uniformly.

THEOREM 9. *For every function* $f \in C(-\pi, \pi)$ *there is a function* $\tau(x)$, *continuous on* $(-\pi, \pi)$, *such that*

$$-\pi = \tau(-\pi) < \tau(x) < \tau(y) < \tau(\pi) = \pi$$
$$\text{for } -\pi < x < y < \pi \tag{67}$$

*and the Fourier series of the composite function* $F(x) = f(\tau(x))$ *converges uniformly on* $[-\pi, \pi]$.

PROOF. We shall prove in Chapter 6 (see Corollary 6.2) that for every $f \in C(-\pi, \pi)$ we can always find a function $\tau(x)$ satisfying (67) such that the second-order integral modulus of continuity of $F(x) = f(\tau(x))$ satisfies $\omega_1^{(2)}(\delta, F) = o(\delta)$ as $\delta \to 0$. But then by (7)

$$|a_n(F)| + |b_n(F)| = o\left(\frac{1}{n}\right) \quad \text{as } n \to \infty. \tag{68}$$

Hence Theorem 9 will be established if we verify that the Fourier series of every function $F \in C(-\pi, \pi)$ with Fourier coefficients of the form (68) converges uniformly on $[-\pi, \pi]$.

We use the evident equation (see (12))

$$S_N(F, x) - \sigma_N(F, x) = \sum_{n=1}^{N} \frac{n}{N+1}[a_n(F) \cos nx + b_n(F) \sin nx]. \tag{69}$$

It follows from (69) that

$$\|S_N(F) - \sigma_N(F)\|_C \le \frac{1}{N+1} \sum_{n=1}^{N} n[|a_n(F)| + |b_n(F)|] = o(1) \quad \text{as } N \to \infty,$$

and since the means $\sigma_N(F, x)$ converge uniformly to $F(x)$, the same is true for the partial sums $S_N(F, x)$. This completes the proof of Theorem 9.

We now consider the question of the uniform convergence of trigonometric series for almost every choice of signs (see Definition 2.3). The following theorem is often useful.

THEOREM 10. *If the coefficients of the series* (60) *satisfy*

$$\sum_{n=1}^{\infty}(a_n^2 + b_n^2)\ln^{1+\varepsilon} n < \infty, \tag{70}$$

*with some* $\varepsilon > 0$, *then the series* (60) *converges for almost all choices of the signs of its terms.*

PROOF. To begin with, let us introduce the notation

$$\|f\|_U = \sup_{1\le N<\infty} \|S_N(f)\|_{C(-\pi,\pi)},$$

where, as usual, $S_N(f)$ is the $N$th partial sum of the Fourier series of $f$.

The proof is based on an inequality for the distribution function of an arbitrary polynomial in the Rademacher system, obtained in Chapter 2 (see Theorem 2.5):

$$m\left\{t \in (0,1): \left|\sum_{n=1}^{N} a_n r_n(t)\right| > y \left(\sum_{n=1}^{N} a_n^2\right)^{1/2}\right\} \le 2e^{-y^2/4},$$
$$y \ge 0, \qquad N = 1,2,\ldots. \tag{71}$$

From (71) we can deduce that, for all numbers $a_n$ and $b_n$, and for $n = 1,\ldots,N$,

$$\mu \equiv m\left\{t \in (0,1): \left\|\sum_{n=1}^{N}(a_n \cos nx + b_n \sin nx)r_n(t)\right\|_U\right.$$
$$\left. > 8 \left(\sum_{n=1}^{N} a_n^2 + b_n^2\right)^{1/2} \ln^{1/2} N\right\} \le \frac{200}{N}. \tag{72}$$

For this purpose we notice that every trigonometric polynomial of degree $\le N$ satisfies

$$\|P\|_{C(-\pi,\pi)} \le 2 \max_{-50N^2\le k\le 50N^2} \left|P\left(\frac{2\pi k}{100N^2}\right)\right|. \tag{73}$$

In fact, if $P'(x)$ is the derivative of $P(x)$, then

$$\|P'\|_c \le N \sum_{n=1}^{N}(|a_n| + |b_n|) \le N(2N)^{1/2}\left[\sum_{n=1}^{N}(a_n^2 + b_n^2)\right]^{1/2}$$
$$= N^{3/2}(2\pi)^{1/2}\|P\|_2 \le 2\pi N^{3/2}\|P\|_C,$$

and if $x_0 \in (-\pi,\pi)$ is a point such that

$$|P(x_0)| = \|P\|_C \quad \text{and} \quad x_0 \in \left[\frac{2\pi(k-1)}{100N^2}, \frac{2\pi k}{100N^2}\right]$$

then by Lagrange's theorem

$$\left| P(x_0) - P\left(\frac{2\pi k}{100N^2}\right) \right| \le \frac{2\pi}{100N^2} \|P'\|_C$$

$$\le \frac{(2\pi)^2}{100N^{1/2}} |P(x_0)| \le \frac{1}{2}|P(x_0)|.$$

Therefore $|P(2\pi k/100N^2)| \ge \frac{1}{2}\|P\|_C$, which establishes (73).

Using (73) and (71) and the inequality $|a\cos x + b\sin x| \le (a^2+b^2)^{1/2}$), we obtain

$$\mu \le m\left\{ t \in (0,1) : \max_{\substack{k,M:1\le M\le N,\\ -50N^2 \le k \le 50N^2}} \left| \sum_{n=1}^{M} \left( a_n \cos\frac{2\pi k}{100N^2} + b_n\sin\frac{2\pi k}{100N^2} \right) r_n(t) \right| \right.$$

$$\left. \ge 4\left(\sum_{n=1}^{N} a_n^2 + b_n^2\right)\ln^{1/2} N \right\}^{1/2}$$

$$\le 100N^3 \max_{M,k} m\left\{ t \in (0,1) : \left| \sum_{n=1}^{M} \left( a_n\cos\frac{2\pi k}{100N^2} + b_n\sin\frac{2\pi k}{100N^2} \right) r_n(t) \right| \right.$$

$$\left. \ge 4\left(\sum_{n=1}^{N} a_n^2 + b_n^2\right)^{1/2} \ln^{1/2} N \right\}$$

$$\le 100N^3 \cdot 2N^{-4} = 200 \cdot N^{-1};$$

this establishes (72).

Let $F(x,t)$ denote the $L^2$-sum of the series

$$\sum_{n=1}^{\infty} (a_n\cos nx + b_n\sin nx)r_n(t), \tag{74}$$

and, for $k = 0,1,\ldots$, set

$$Q_k(x,t) = \sum_{n=2^{2^k}}^{2^{2^{k+1}}-1} (a_n\cos nx + b_n\sin nx)r_n(t), \tag{75}$$

$$d_k = \left( \sum_{n=2^{2^k}}^{2^{2^{k+1}}-1} a_n^2 + b_n^2 \right)^{1/2}.$$

By (72) we have, for each $k = 0,1,\ldots$,

$$\mu_k \equiv m\{ t \in (0,1) : \|Q_k(x,t)\|_U > 8d_k 2^{(k+1)/2} \} \le 200\, 2^{-2^k},$$

and consequently

$$\sum_{k=0}^{\infty} \mu_k < \infty. \tag{76}$$

It follows from (76) that, for almost all $t \in (0, 1)$, for some index $k(t)$ onward,

$$\|Q_k(x, t)\|_U \le 8 d_k 2^{(k+1)/2}, \qquad k \ge k(t).$$

Therefore for almost every $t \in (0, 1)$ (see (70) and (75))

$$\sum_{k=0}^{\infty} \|Q_k(x, t)\|_U = \sum_{k=0}^{k(t)-1} \|Q_k(x, t)\|_U + \sum_{k=k(t)}^{\infty} \|Q_k(x, t)\|_U$$

$$\le A(t) + 16 \sum_{k=k(t)}^{\infty} d_k 2^{k/2}$$

$$= A(t) + \sum_{k=k(t)}^{\infty} d_k 2^{k(1/2+\varepsilon/2)} \cdot 2^{-k\varepsilon/2}$$

$$\le A(t) + \left\{ \sum_{k=k(t)}^{\infty} d_k^2 2^{k(1+\varepsilon)} \right\}^{1/2} \left\{ \sum_{k=k(t)}^{\infty} 2^{-\varepsilon k} \right\}^{1/2}$$

$$< \infty. \tag{77}$$

(In (77) $A(t)$ is a number depending on $t$, and $\varepsilon$ is the same as in (70).) Since

$$\left\| \sum_{n=2^{2p}}^{\infty} (a_n(\cos nx + b_n \sin nx) r_n(t) \right\|_U \le \sum_{k=p}^{\infty} \|Q_k\|_U, \qquad p = 0, 1, \dots,$$

the conclusion of Theorem 10 follows from (77).

From the inequality (72), which we used to prove Theorem 10, it is easy to deduce that

$$\int_0^1 \left\| r_1(t) + \sum_{n=1}^{N} r_{2n}(t) \cos nx + r_{2n+1}(t) \sin nx \right\|_C dt \le C(N \ln N)^{1/2},$$

$$N = 2, 3, \dots. \tag{78}$$

Moreover, in (72), and therefore in (78), the Rademacher functions can be replaced by an orthonormal set of independent, normally distributed, functions $\{\xi_n(t)\}_{n=1}^{2N+1}$, $t \in (0, 1)$ (see §2.1). This is because in the proof of (72) we used only a single property of the Rademacher functions, namely the exponential estimate (71), which is also valid for the functions $\{\xi_n(t)\}$

(see Theorem 2.10). Therefore

$$\int_0^1 \|G_N(x,t)\|_{C(-\pi,\pi)}\, dt \le C(N \ln N)^{1/2}, \qquad N = 2, 3, \dots, \tag{79}$$

where

$$G_N(x,t) = \frac{1}{\sqrt{2}} \xi_1(t) + \sum_{n=1}^{N} \xi_{2n}(t) \cos nx + \xi_{2n+1}(t) \sin x \tag{80}$$

[the choice of the coefficient of $\xi_1(t)$ in (80) is different from that used in (78); this is evidently unimportant]. It is useful to observe that the inequalities in (78) and (79) are of the best possible order. This can be verified very simply for $\{\xi_n(t)\}$.

PROPOSITION 2. *There is an absolute constant $c > 0$ such that*

$$\int_0^1 \|G_N(x,t)\|_{C(-\pi,\pi)}\, dt \ge c(N \ln N)^{1/2}, \qquad N = 2, 3, \dots, \tag{81}$$

*where $G_N(t)$ is defined in* (80).

PROOF. Choose an integer $N$ and consider the set of points $x_k = -\pi + 2\pi k/(2N+1)$, $k = 1, \dots, 2N+1$, and the matrix $A_{2N+1} = \{a_{k,m}\}_{k,m=1}^{2N+1}$ with

$$a_{k,2n} = \left(\frac{2}{2N+1}\right)^{1/2} \cos nx_k, \qquad a_{k,2n+1} = \left(\frac{2}{2N+1}\right)^{1/2} \sin nx_k,$$

$$n = 1, \dots, N, \ a_{k,1} = (2N+1)^{-1/2}, \ k = 1, \dots, 2N+1.$$

Then $A_{2N+1}$ is orthonormal. In fact, for $1 \le k, k' \le 2N+1$,

$$\sum_{m=1}^{2N+1} a_{k,m} a_{k',m} = \frac{1}{2N+1} + \frac{2}{2N+1} \sum_{n=1}^{N} \cos n(x_k - x_{k'})$$

$$= \frac{2}{2N+1} D_N(x_k - x_{k'}) = \delta_{k,k'}$$

(here $D_N(x)$ is the Dirichlet kernel (see (4)), and $\delta_{k,k'} = 1$ when $k = k'$, $\delta_{k,k'} = 0$ for $k \ne k'$).

Using the orthonormal property of $A_{2N+1}$ and Theorem 2.10, we find that the functions

$$f_k(t) = \left(\frac{2}{2N+1}\right)^{1/2} G_N(x_k, t), \qquad t \in (0, 1), \ k = 1, \dots, 2N+1,$$

also form a set of normally distributed independent functions. Therefore (see equation 2.(27)),

$$m(E_N) \equiv m \left\{ t \in (0,1) : \|G_N(x,t)\|_C < \left( \frac{2N+1}{2} \ln N \right)^{1/2} \right\}$$

$$\leq m \left\{ t \in (0,1) : \max_{1 \leq k \leq 2N+1} |f_k(t)| < (\ln N)^{1/2} \right\}$$

$$= \prod_{k=1}^{2N+1} \left( 1 - \frac{2}{\sqrt{2\pi}} \int_{\ln^{1/2} N}^{\infty} e^{-z^2/2} \, dz \right) \tag{82}$$

$$\leq \left( 1 - \frac{\gamma}{N} \right)^{2N+1} \leq \rho < 1,$$

where $\gamma \in (0,1)$ and $\rho < 1$ is an absolute constant. Proposition 2 follows from (82), since

$$\int_0^1 \|G_N(x,t)\|_C \, dt \geq \left( \frac{2N+1}{2} \right)^{1/2} \ln^{1/2} N \cdot m((0,1)\backslash E_N)$$

$$\geq C(N \ln N)^{1/2}.$$

Although, as we noted above, inequality (78) is of the best possible order, there are choices of signs $\varepsilon_n = \pm 1$, $n = 1, \ldots, 2N+1$, for which the norm

$$\left\| \varepsilon_1 + \sum_{n=1}^{N} \varepsilon_{2n} \cos nx + \varepsilon_{2n+1} \sin nx \right\|_C$$

does not exceed $CN^{1/2}$.[2]

In fact, we have the following theorem.

THEOREM 11. *There is a sequence of numbers* $\{\varepsilon_n\}_{n=0}^{\infty}$ *with* $\varepsilon_n = \pm 1$ *for which*

$$\left\| \sum_{n=0}^{N} \varepsilon_n e^{inx} \right\|_{C(-\pi,\pi)} \leq 5N^{1/2}, \qquad N = 0, 1, \ldots.$$

LEMMA 1. *Let* $\{z_n\}_{j=1}^{\infty}$ *be a sequence of complex numbers of modulus unity. Set* $P_0 = Q_0 = 1$ *and*

$$P_{j+1} = P_j + z_{j+1} Q_j, \qquad Q_{j+1} = P_j - z_{j+1} Q_j, \qquad j = 0, 1, \ldots. \tag{83}$$

*Then, for* $j = 1, 2, \ldots$, *we can represent* $P_j$ *and* $Q_j$ *as sums:*

$$P_j = \sum_{\{\alpha_s\}} \varepsilon(\{\alpha_s\}) \prod_{s=1}^{j} z_s^{\alpha_s}, \qquad Q_j = \sum_{\{\alpha_s\}} \varepsilon'(\{\alpha_s\}) \prod_{s=1}^{j} z_s^{\alpha_s}, \tag{84}$$

---

[2]Since $\| \sum_{n=1}^{N} \varepsilon_{2n} \cos nx + \varepsilon_{2n+1} \sin nx \|_2 = (2N/\pi)^{1/2}$, this inequality is of the best possible order.

*where both sums are over all sets $\{\alpha_s\}_{s=1}^j$ of zeros and ones, and $\varepsilon(\{\alpha_s\})$ and $\varepsilon'(\{\alpha_s\})$ depend on $\{\alpha_s\}$ but have only the values $+1$ and $-1$. Moreover,*

$$|P_j|^2 + |Q^j|^2 = 2^{j+1}. \tag{85}$$

The first part of the lemma is easily verified by induction. To prove (85), we observe that it is evidently satisfied for $j = 0$; and since

$$\begin{aligned}
|P_{j+1}|^2 + |Q_{j+1}|^2 &= (P_j + z_{j+1}Q_j)(\overline{P}_j + \bar{z}_{j+1}\overline{Q}_j) \\
&\quad + (P_j - z_{j+1}Q_j)(\overline{P}_j - \bar{z}_{j+1}\overline{Q}_j) \\
&= 2(|P_j|^2 + |Q_j|^2), \qquad j = 0, 1, \ldots,
\end{aligned}$$

it follows that $|P_j|^2 + |Q_j|^2 = 2^{j+1}$ for $j = 1, 2, \ldots$. This completes the proof of Lemma 1.

PROOF OF THEOREM 11. Set $z_j = z_j(x) = \exp(i2^{j-1}x)$, $j = 1, 2, \ldots$, $x \in (-\pi, \pi)$; and let $P_j = P_j(x)$ and $Q_j = Q_j(x)$, $j = 0, 1, \ldots$, be defined as in Lemma 1 (see (83)). If the number $n = 0, 1, 2, \ldots$, is represented in the form $n = \alpha_1 + 2\alpha_2 + \cdots + 2^{j-1}\alpha_j$, where $\alpha_s = 0$ or $1$ for $s = 1, \ldots, j-1$ and $\alpha_j = 1$ (this representation is unique), we can set (see (84))

$$\varepsilon_n = \varepsilon(\{\alpha_s\}_{s=1}^j), \qquad \varepsilon_n' = \varepsilon'(\{\alpha_s\}_{s=1}^j).$$

By (84), $P_j(x)$ and $Q_j(x)$ are trigonometric polynomials of the form

$$P_j(x) = \sum_{n=0}^{2^j-1} \varepsilon_n e^{inx}, \qquad Q_j(x) = \sum_{n=0}^{2^j-1} \varepsilon_n' e^{inx},$$

and moreover (see (85)) we have $\|P_j\|_C \le 2^{(j+1)/2}$, $\|Q_j\|_C \le 2^{(j+1)/2}$.

It is easy to see from the construction of $P_j(x)$ and $Q_j(x)$ (see, in particular, (83)) that if $p = r \cdot 2^j$, $q = (r+1)2^j - 1$, $r = 0, 1, \ldots$; $j = 1, 2, \ldots$, then

$$\text{either } \left| \sum_{n=p}^q \varepsilon_n e^{inx} \right| = |P_j(x)|, \quad \text{or } \left| \sum_{n=p}^q \varepsilon_n e^{inx} \right| = |Q_j(x)|.$$

Consequently we always have

$$\left\| \sum_{n=r2^j}^{(r+1)2^j-1} \varepsilon_n e^{inx} \right\|_C \le 2^{(j+1)/2}. \tag{86}$$

Now let a number $N$ be given with $2^{p-1} < N < 2^p$, $p > 1$. We express $N$ in base 2: $N = \sum_{s=1}^p \alpha_s 2^{s-1}$, where $\alpha_s = 0$ or $1$ ($\alpha_p = 1$). Also let

$$N_j = \sum_{s=p-j+1}^p \alpha_s 2^{s-1}, \qquad j = 1, \ldots, p, \ N_0 = 0,$$

then we have the equation

$$\sum_{n=0}^{N} \varepsilon_n e^{inx} = \sum_{j:\alpha_{p-j+1}\neq 0} \sum_{n=N_{j-1}}^{N_j-1} \varepsilon_n e^{inx} + \varepsilon_N e^{iNx}. \tag{87}$$

Since $N_{j-1}$ is divisible by $2^{p-j}$ and $N_j - N_{j-1} = \alpha_{p-j+1}2^{p-j} = 2^{p-j}$ (if $\alpha_{p-j+1} \neq 0$), we may use (86) to estimate the inner sums in (87). It follows that

$$\left| \sum_{n=0}^{N} \varepsilon_n e^{inx} \right| \leq 1 + \sum_{j=1}^{p-1} 2^{(j+1)/2} \leq \frac{2^{(p+1)/2}}{\sqrt{2}-1} < 5N^{1/2}.$$

This completes the proof of Theorem 11.

REMARK. By a similar argument using (86), we can show that the sequence $\{\varepsilon_n\}$ constructed in Theorem 11 satisfies, for all $M$ and $N$ ($0 \leq M < N$),

$$\left\| \sum_{n=M}^{N} \varepsilon_n e^{inx} \right\|_C \leq C(N-M)^{1/2}.$$

We conclude this section by discussing the absolute convergence of Fourier series. The following theorem gives a sufficient condition for absolute convergence i.e., the convergence of

$$\sum_{n=1}^{\infty} |a_n(f)| + |b_n(f)|, \tag{88}$$

THEOREM 12. *Let $f \in L^2(-\pi,\pi)$ and let its modulus of continuity in $L^2(-\pi,\pi)$ satisfy the condition*

$$\sum_{n=1}^{\infty} \frac{\omega_2(\frac{1}{n}, f)}{\sqrt{n}} < \infty;$$

*then the sum of the series* (88) *is finite.*[3]

PROOF. By Theorem 3

$$\sum_{n=1}^{\infty} E_n^{(2)}(f) n^{-1/2} < \infty, \tag{89}$$

moreover, by Parseval's theorem

$$E_n^{(2)}(f) = \left\{ \frac{1}{\pi} \sum_{k=n+1}^{\infty} \rho_k^2 \right\}^{1/2},$$

$$\rho_k = \rho_k(f) = [a_k^2(f) + b_k^2(f)]^{1/2}, \ k = 1, 2, \ldots. \tag{90}$$

---

[3]Consequently, $f(x)$ agrees a.e. on $[-\pi, \pi]$ with a continuous function.

Therefore

$$\sum_{n=1}^{\infty} \left( \sum_{k=n}^{\infty} \rho_k^2 \right)^{1/2} n^{-1/2} < \infty,$$

and we obtain

$$\frac{1}{2} \sum_{n=1}^{\infty} |a_n(f)| + |b_n(f)| \le \sum_{n=1}^{\infty} \rho_n = \sum_{n=1}^{\infty} \sum_{k=1}^{n} \frac{\rho_n}{n}$$

$$= \sum_{k=1}^{\infty} \sum_{n=k}^{\infty} \frac{\rho_n}{n} \le \sum_{k=1}^{\infty} \left( \sum_{n=k}^{\infty} \rho_n^2 \right)^{1/2} \left( \sum_{n=k}^{\infty} \frac{1}{n^2} \right)^{1/2}$$

$$\le C \sum_{k=1}^{\infty} \left( \sum_{n=k}^{\infty} \rho_n^2 \right)^{1/2} k^{-1/2} < \infty.$$

This completes the proof of Theorem 12.
We notice the following two corollaries of Theorem 12.

COROLLARY 5. *Let $f \in C(-\pi, \pi)$ and let $\omega(\delta, f) = O(\delta^\alpha)$ $(\delta \to 0)$ for some $\alpha > \frac{1}{2}$; then the Fourier series of $f$ converges absolutely.*

The conclusion of Corollary 5 follows immediately from Theorem 12 and the inequality $\omega_2(\delta, f) \le (2\pi)^{1/2} \omega(\delta, f)$.

COROLLARY 6. *Let $f \in C(-\pi, \pi)$ be of bounded variation and in addition let $\omega(\delta, f) = O(\log(1/\delta))^{-\gamma}$ $(\delta \to 0)$ for some $\gamma > 2$. Then the Fourier series of $f$ converges absolutely.*

PROOF. The $L^1$ modulus of continuity of a function $f(x)$ of bounded variation satisfies $\omega_1(\delta, f) = O(\delta)$ $(\delta \to 0$ (see Appendix 1, Proposition 2). Hence, for $\delta < \delta_0 < 1$,

$$[\omega_2(\delta, f)]^2 = \sup_{0 < h \le \delta} \|f(x + h) - f(x)\|_2^2$$

$$\le \sup_{0 < h \le \delta} \|f(x + h) - f(x)\|_C \cdot \sup_{0 < h \le \delta} \|f(x + h) - f(x)\|_1$$

$$\le C \frac{\delta}{\log^\gamma(\frac{1}{\delta})}.$$

Therefore $\sum_{n=1}^{\infty} n^{-1/2} \omega_2(1/n, f) < \infty$, and Corollary 6 follows from Theorem 12.

REMARK. The conclusions of Corollaries 5 and 6 are precise, in the sense that it is not possible to replace $\alpha$ by $\frac{1}{2}$ or $\gamma$ by 2. This will follow immediately from a theorem on general orthonormal systems which will be proved in Chapter 9 (see Theorem 9.11 and Corollary 9.2).

## §5. The Walsh system.
### Definition and some properties

In this section we use the expression of a number $x$ in base 2: for $x \geq 0$,

$$x = \sum_{k=-\infty}^{\infty} \theta_k(x)2^{-k}, \quad \text{where } \theta_k = 0 \text{ or } 1. \tag{91}$$

The coefficients $\theta_k(x)$ are uniquely determined if we make the additional assumption that when $x$ is a binary rational we consider the expansion (1) with a finite number of nonzero coefficients.

It is clear that when $x \in (0,1)$ the coefficients $\theta_k(x) = 0$ when $k \leq 0$. In Chapter 2 we saw the connection between $\theta_k(x)$, $k = 1, 2, \ldots$, and the Rademacher system, and showed that $\{\theta_k(x)\}_{k=1}^{\infty}$ forms a sequence of independent functions on $(0,1)$ (see 2.(14) and Corollary 2.2).

If $n$ is a positive integer,

$$n = \sum_{k=0}^{\infty} \theta_{-k(n)}2^k = \sum_{k=0}^{k(n)} \theta_{-k(n)}2^k, \quad k(n) = [\log_2 n], \tag{92}$$

$$\theta_{-k(n)}(n) = 1.$$

DEFINITION 3. The *Walsh system* is the system $W = \{w_n(x)\}_{n=0}^{\infty}$, $x \in [0,1]$, where $w_0(x) \equiv 1$ and, for $n \geq 1$,

$$w_n(x) = \prod_{k=0}^{\infty}[r_{k+1}(x)]^{\theta_{-k(n)}} = r_{k(n)+1}(x) \prod_{k=0}^{k(n)-1} [r_{k+1}(x)]^{\theta_{-k(n)}}, \tag{93}$$

where $r_k(x)$, $k = 1, 2, \ldots$, are the Rademacher functions.

The Walsh system is orthonormal, and complete in $L^p(0,1)$, $1 \leq p < \infty$. The pairwise orthogonality of $w_n(x)$ follows directly from the fact that that the Rademacher system is an S.I.F. (see Theorem 2.4). To establish the completeness of $W$ in $L^p(0,1)$, $0 \leq p < \infty$, we observe that it follows from the definition (93) that the functions $w_n(x)$, $0 \leq n < 2^k$, are constant on each interval $((i-1)/2^k, i/2^k)$, $1 \leq i \leq 2^k$. Therefore after alteration at a finite number of points the functions $w_n(x)$, $0 \leq n < 2^k$, are elements of the space $\mathscr{D}_{2^k}$, and since they are pairwise orthogonal they form a basis in $\mathscr{D}_{2^k}$. Therefore

$$\left\{ P(x) : P(x) = \sum_{n=0}^{2^k-1} a_n w_n(x) \right\} \cong \mathscr{D}_{2^k}, \quad k = 0, 1, \ldots \tag{94}$$

(here $\cong$ means that $P(x) \in \mathscr{D}_{2^k}$ after alteration at a finite number of points). Since the piecewise constant functions are dense in $L^p(0,1)$, we immediately obtain from (94) the completeness of the Walsh system in $L^p(0, 1)$, $1 \leq p < \infty$.

The following property of the Walsh functions follows easily from the definition (see (93)): for $k \geq 0$, $i = 0, 1, \ldots, 2^k - 1$,

$$w_{2^k+i}(x) = r_{k+1}(x) \cdot w_i(x) = w_{2^k}(x)w_i(x), \qquad x \in [0, 1]. \qquad (95)$$

For the study of the Walsh system, we introduce the operator $\dotplus$ for adding numbers in base 2: if $x, y \in [0, \infty)$, $x = \sum_{k=-\infty}^{\infty} \theta_k(x)2^{-k}$, $y = \sum_{k=-\infty}^{\infty} \theta_k(y)2^{-k}$, we set

$$x \dotplus y = \sum_{k=-\infty}^{\infty} |\theta_k(x) - \theta_k(y)|2^{-k}. \qquad (96)$$

It is easy to see that 1) if $x \geq 0$ and $y \geq 0$ are integers, then $x \dotplus y$ is also a nonnegative integer; and 2) if $x, y \in [0, 1)$ then $x \dotplus y \in [0, 1)$.

The following properties of $\dotplus$ follow directly from the definition (96): if $x, y, z \in [0, 1)$, then

a) $(x \dotplus y) \dotplus z = x \dotplus (y \dotplus z)$, $\qquad x \dotplus y = y \dotplus x$, $x \dotplus x = 0$;

b) $|(x \dotplus y) - x| \leq y$.

In addition, because the functions $\theta_n(x)$, $n = 1, 2, \ldots$, $x \in (0, 1)$, are independent, it is easy to deduce that

c) for every $f \in L^1(0, 1)$ the function

$$f_y(x) = f(x \dotplus y), \qquad x \in (0, 1), \qquad (97)$$

with $y \in [0, 1)$ fixed, is equimeasurable with $f(x)$ on $[0, 1)$, and consequently

$$\int_0^1 f_y(x)\, dx = \int_0^1 f(x)\, dx. \qquad (98)$$

Finally, we show by induction that

d) for every integer $N \geq 1$ with $N = 2^{k_1} + \cdots + 2^{k_s}$, $k_1 > \cdots > k_s \geq 0$, we have the equation

$$\{N \dotplus j\}_{j=0}^{N-1} = \bigcup_{l=1}^{s} Q_{k_l}, \qquad (99)$$

where $Q_k = \{n : 2^k \leq n < 2^{k+1}\}$, $k = 0, 1, \ldots$.

First notice that for $k = 0, 1, \ldots$; $0 \leq j \leq 2^k$,

$$2^k \dotplus j = 2^k + j, \qquad (100)$$

from which (99) follows for $n$ of the form $2^k$, $k = 0, 1, \ldots$ Suppose now that (99) has been verified for $N \leq 2^k$ ($k = 1, 2, \ldots$); we prove it for $2^k < N \leq 2^{k+1} - 1$. Let

$$N = 2^{k_1} + \cdots + 2^{k_s} = 2^k + i, \qquad i = 2^{k_2} + \cdots + 2^{k_s} \quad (k_1 = k).$$

Then

$$\{N \dotplus j\}_{j=0}^{N-1} = \{(2^k + i) \dotplus j\}_{j=0}^{2^k-1} \cup \{(2^k + i) \dotplus (2^k + p)\}_{p=0}^{i-1}$$

$$\equiv I_1 \cup I_2. \tag{101}$$

It is easily seen that if $j_1 \neq j_2$, $0 \leq j_1, j_2 \leq 2^k - 1$, then $i \dotplus j_1 \neq i \dotplus j_2$, and moreover $0 \leq i \dotplus j_1 \leq 2^k - 1$, $0 \leq i \dotplus j_2 \leq 2^k - 1$. Hence it follows that $\{i \dotplus j\}_{j=0}^{2^k-1} = \{j\}_{j=0}^{2^k-1}$. Using (100) and a), we obtain

$$I_1 = \{(2^k + i) \dotplus j\}_{j=0}^{2^k-1} = \{2^k \dotplus i(i \dotplus j)\}_{j=0}^{2^k-1}$$

$$= \{2^k + j\}_{j=0}^{2^k-1} = Q_k = Q_{k_1}. \tag{102}$$

In addition, since $(2^k + i) \dotplus (2^k + p) = i \dotplus p$ $(0 \leq i, p \leq 2^k - 1)$, then by the inductive hypothesis we obtain

$$I_2 = \{i \dotplus p\}_{p=0}^{i-1} = \bigcup_{l=2}^{s} Q_{k_l}. \tag{103}$$

It follows from (101)–(103) that (99) holds for $2^k < N \leq 2^{k+1} - 1$, and therefore for all $N$.

The following equation clarifies the connection between the operation $\dotplus$ and properties of the Walsh functions:

$$w_n(x) \cdot w_m(x) = w_{n \dotplus m}(x), \qquad n, m \geq 0, \ x \in (0,1)\backslash R_2, \tag{104}$$

where $R_2$ is the set of binary-rational points of $[0, 1]$.

Equation (104) follows directly from the definition of $w_n(x)$ (see (93)). In fact, since $|r_k(x)| = 1$ for $x \in (0,1)\backslash R_2$, $k = 0, 1, \ldots$, we obtain

$$w_n(x) \cdot w_m(x) = \prod_{k=0}^{\infty} [r_{k+1}(x)]^{\theta_{-k(n)} + \theta_{-k(m)}}$$

$$= \prod_{k=0}^{\infty} [r_{k+1}(x)]^{|\theta_{-k(n)} - \theta_{-k(m)}|} = w_{n \dotplus m}(x).$$

We now show that when $n \geq 0$

$$w_n(x)w_n(y) = w_n(x \dotplus y) \quad \text{if } \{x, y, x \dotplus y\} \subset (0,1)\backslash R_2. \tag{105}$$

Since $r_k(x) = (-1)^{\theta_k(x)}$ for $k = 1, 2, \ldots$, and $x \in (0,1)\backslash R_2$ (see 2.(14)) we have, by (93),

$$w_n(x)w_n(y) = \prod_{k=0}^{\infty} (-1)^{\theta_{k+1}(x)\theta_{-k(n)}} \cdot (-1)^{\theta_{k+1}(y)\theta_{-k(n)}}$$

$$= \prod_{k=0}^{\infty} (-1)^{\theta_{-k(n)}[\theta_{k+1}(x) + \theta_{k+1}(y)]}, \qquad x, y \in (0,1)\backslash R_2, \tag{106}$$

and since

$$(-1)^{[\theta_{k+1}(x)+\theta_{k+1}(y)]} = (-1)^{[\theta_{k+1}(x)-\theta_{k+1}(y)]} = (-1)^{\theta_{k+1}(x\dotplus y)}$$

(see (96)), we obtain (105) from (106).

Each function $f \in L^1(0,1)$ has a Fourier series in the Walsh system:

$$f \sim \sum_{n=0}^{\infty} c_n(f)w_n(x), \qquad c_n(f) = \int_0^1 f(x)w_n(x)\,dx. \qquad (107)$$

Let us consider some properties of the coefficients and partial sums of the series (107). We shall need the following proposition.

PROPOSITION 3. *If* $f \in L^1(0,1)$ *and* $y \in (0,1)\backslash R_2$, *then (see (97))*

$$c_n(f_y) = w_n(y) \cdot c_n(f), \qquad n = 0, 1, \ldots .$$

PROOF. Using properties a) and c) of $\dotplus$ we see that, for $n = 0, 1, \ldots,$

$$c_n(f_y) = \int_0^1 f(x \dotplus y)w_n(x)\,dx$$

$$= \int_0^1 f((x \dotplus y) \dotplus y)w_n(x \dotplus y)\,dx = \int_0^1 f(x)w_n(x \dotplus y)\,dx,$$

and since (105) is satisfied for almost all $x \in (0,1)$ when $y$ is fixed in $(0,1)\backslash R_2$, we obtain

$$c_n(f_y) = w_n(y)\int_0^1 f(x)w_n(x)\,dx = w_n(y)c_n(f).$$

COROLLARY 7. *The Fourier–Walsh coefficients of a function* $f \in C(0,1)$ *satisfy*

$$|c_n(f)| \le \tfrac{1}{2}\omega(\tfrac{1}{n}, f), \qquad n = 1, 2, \ldots$$

PROOF. Let $n = 2^k + i$ $(k = 0, 1, \ldots; \ i = 0, 1, \ldots, 2^k - 1)$ and $y \in 2^{-k-1}, 2^{-k})\backslash R_2$. We see from (93) then that $w_n(y) = -1$ and, by Proposition 3,

$$c_n(f_y) = -c_n(f), \qquad 2^k \le n < 2^{k+1}, \qquad y \in (2^{-k-1}, 2^{-k})\backslash R_2. \qquad (108)$$

Therefore (also see b)) for each $y \in (2^{-k-1}, 2^{-k})\backslash R_2$

$$|c_n(f)| = \frac{1}{2}|c_n(f) - c_n(f_y)|$$

$$\le \frac{1}{2}\int_0^1 |f(x) - f(x \dotplus y)|\,dx \le \frac{1}{2}\omega(y, f),$$

from which it follows at once that $|c_n(f)| \le \tfrac{1}{2}\omega(2^{-k-1}, f) \le \tfrac{1}{2}\omega(\tfrac{1}{n}, f)$.

In a similar way we can prove the following result on the absolute convergence of the series of Fourier–Walsh coefficients of $f(x)$.

THEOREM 13. *Let $f \in C(0,1)$ and let $\omega(\delta, f) = O(\delta^{\alpha})$ ($\delta \to 0$) with some $\alpha > \frac{1}{2}$; then*

$$\sum_{n=0}^{\infty} |c_n(f)| < \infty.$$

PROOF. Let $k = 0, 1, \ldots$, $y \in (2^{-k-1}, 2^{-k})$, and $g_y(x) = f(x) - f(x \dotplus y) = f(x) - f_y(x)$ [$x \in (0,1)$]. Then by (108) we have $c_n(g_y) = 2c_n(f)$ for $2^k \le n < 2^{k+1}$, and consequently

$$\delta_k \equiv \sum_{n=2^k}^{2^{k+1}-1} c_n^2(f) = \frac{1}{4} \sum_{n=2^k}^{2^{k+1}-1} c_n^2(g_y)$$

$$\le \frac{1}{4} \int_0^1 g_y^2(x) \, dx = \frac{1}{4} \int_0^1 [f(x) - f(x \dotplus y)]^2 \, dx$$

(here we used Bessel's inequality). From the preceding relation, taking account of inequality b), we deduce that

$$\delta_k \le \tfrac{1}{4} \omega^2(y, f) \le \omega^2(2^{-k}, f), \qquad k = 0, 1, \ldots,$$

and therefore, by Cauchy's inequality,

$$\sum_{n=2^k}^{2^{k+1}-1} |c_n(f)| \le 2^{k/2} \delta_k^{1/2} \le 2^{k/2} \omega(2^{-k}, f), \qquad k = 0, 1, \ldots. \tag{109}$$

If we add the inequalities (109) and use the hypothesis of the theorem, according to which $\omega(2^{-k}, f) \le C 2^{-k\alpha}$, $\alpha > \frac{1}{2}$, we obtain

$$\sum_{n=1}^{\infty} |c_n(f)| \le C \sum_{k=0}^{\infty} 2^{k/2} \cdot 2^{-k\alpha} < \infty.$$

This completes the proof of Theorem 13.

We now consider the Lebesgue functions of the Walsh system:

$$L_N(x) = \int_0^1 |K_N(x, y)| \, dy \qquad K_N(x, y) = \sum_{n=0}^{N-1} w_n(x) w_n(y),$$

$$N = 1, 2, \ldots.$$

It follows from (105) that, for $x \in (0,1) \backslash R_2$, and for almost all $y \in (0,1)$,

$$K_N(x, y) = \sum_{n=0}^{N-1} w_n(x \dotplus y), \qquad N = 1, 2, \ldots.$$

If we set $D_N^W(t) = \sum_{n=0}^{N-1} w_n(t)$, $t \in (0,1)$, then (see (97) and (98))

$$L_N(x) = \int_0^1 |D_N^W(x \dotplus y)| \, dy = \int_0^1 |D_N^W(t)| \, dt = L_N, \qquad x \in (0,1) \backslash R_2.$$

Therefore the Lebesgue functions of the Walsh system are independent of $x$ for $x \in (0,1)\backslash R_2$. (Notice also that when $x \in R_2$ we have $w_n(x) = 0$ for $n > N(x)$, and therefore at a point $x \in R_2$ the Lebesgue functions satisfy $L_N(x) = L_{N(x)}(x)$ for $n > N(x)$.)

By (95) we have, for $t \in (0,1)$ and $k \geq 1$,

$$r_k(t)D_{2^{k-1}}^W(t) = \sum_{n=0}^{2^{k-1}-1} r_k(t)w_n(t)$$

$$= \sum_{n=0}^{2^{k-1}-1} w_{2^{k-1}+n}(t) = \sum_{n\in Q_{k-1}} w_n(t). \tag{110}$$

Therefore

$$D_{2^k}^W(t) = D_{2^{k-1}}^W(t) + \sum_{n\in Q_{k-1}} w_n(t) = D_{2^{k-1}}^W(t)[1 + r_k(t)].$$

From this, since $D_1^W(t) = w_0(t) \equiv 1$, we obtain

$$D_{2^k}^W(t) = \prod_{s=1}^{k}[1 + r_s(t)], \qquad t \in (0,1). \tag{111}$$

It follows immediately from the definition of the Rademacher functions (see Definition 2.3) and equation (111) that, for $k = 0, 1, \ldots$,

$$D_{2^k}^W(t) = \begin{cases} 2^k & \text{if } t \in (0, 2^{-k}), \\ 0 & \text{if } t \in (2^{-k}, 1), \end{cases} \quad t \neq i/2^k, \ i = 0, 1, \ldots, 2^k, \tag{112}$$

and for $x \in (0,1)\backslash R_2$

$$L_{2^k}(x) = L_{2^k} = \int_0^1 |D_{2^k}^W(t)|\, dt = 1, \qquad k = 0, 1, \ldots. \tag{113}$$

We now estimate $L_N$ for general values of $N$. Let $N = 2^{k_1} + \cdots + 2^{k_s}$, $k_1 > \cdots > k_s \geq 0$ ($N > 1$). Then, by (104) and (99) (also see (110)) we have, for $t \in (0,1)\backslash R_2$,

$$w_N(t)D_N^W(t) = \sum_{n=0}^{N-1} w_N(t)w_n(t) = \sum_{n=0}^{N-1} w_{N+n}(t)$$

$$= \sum_{l=1}^{s} \sum_{n\in Q_{k_l}} w_n(t) = \sum_{l=1}^{s} r_{k_l+1}(t)D_{2^{k_l}}^W(t). \tag{114}$$

It follows from (113) and (114) that

$$L_N = \int_0^1 |D_N^W(t)|\,dt = \int_0^1 |w_N(t)D_N^W(t)|\,dt$$

$$\leq \sum_{l=1}^{s} \int_0^1 |D_{2^{k_l}}^W(t)|\,dt = s,$$

and since $s \leq [\log_2 N] + 1$, we have established the inequality

$$L_N \leq 1 + [\log_2 N], \qquad N = 1, 2, \ldots. \tag{115}$$

On the other hand,

$$\varlimsup_{N \to \infty} \frac{L_N}{\log_2 N} > 0. \tag{116}$$

Let us prove (116). For $p = 1, 2, \ldots$, set $N_p = \sum_{k=0}^{p} 2^{2k}$. Then, using the equation (see (112)) $r_{k+1}(t)D_{2^k}^W(t) = 2^{k/2}\chi_k^{(1)}(t)$, where $\{\chi_k^{(i)}\}$ is the Haar system and $t \in (0,1) \backslash R_2$, we find from (114) that

$$w_{N_p}(t)D_{N_p}^W(t) = \sum_{k=0}^{p} r_{2k+1}D_{2^{2k}}^W(t) = \sum_{k=0}^{p} 2^k \chi_{2k}^{(1)}(t), \qquad t \in (0,1) \backslash R_2. \tag{117}$$

We showed in Chapter 3 (see 3.(44)) that $\|\sum_{k=0}^{p} 2^k \chi_{2k}^{(1)}\|_1 \geq p/4$, and therefore

$$L_{N_p} = \int_0^1 |w_{N_p}(t)D_{N_p}^W(t)|\,dt \geq \frac{p}{4} > \frac{\log_2 N_p}{10} \qquad \text{for } p > p_0.$$

This establishes (116) (for a generalization of (116) see §9.3).

We should note that using the connection between the Walsh and Haar systems greatly simplifies the study of convergence questions for Walsh series. The following equation is often very useful: for $f \in L^1(0,1)$ and $k = 0, 1, \ldots,$

$$S_{2^k}(f, W, x) \equiv \sum_{n=0}^{2^k - 1} c_n(f)w_n(x) = S_{2^k}(f, \chi, x), \qquad x \neq \frac{i}{2^k}, \ 0 \leq i \leq 2^k. \tag{118}$$

Here the left-hand side contains the partial sums of the Fourier–Walsh series, and the right-hand side contains the partial sums of the Fourier–Haar series.

Since the sets of polynomials of order $2^k$ are the same for both the Haar and Walsh systems (see (94) and Proposition 3.1), equation (118) for $f \in L^2(0,1)$ follows directly from Proposition 1.5 (see Chapter 1). If, for any $f \in L^1(0,1)$, we then choose a sequence of functions $\{f_j\} \subset L^2(0,1)$, $\|f_j - f\|_1 \to 0$ as $f \to \infty$, we obtain

$$S_{2^k}(f, W, x) = \lim_{j \to \infty} S_{2^k}(f_j, W, x)$$
$$= \lim_{j \to \infty} S_{2^k}(f_j, \chi, x) = S_{2^k}(f, \chi, x)$$

$(x \neq i/2^k, 0 \le i \le 2^k)$, which establishes (118) in the general case.

By means of (115) and (118) we can establish the following theorem.

THEOREM 14. *For $f \in C(0,1)$ and $x \in (0,1)\backslash R_2$ we have the inequality*

$$|S_N(f,x) - f(x)| \le C \cdot \omega\left(\frac{1}{N}, f\right) \ln N, \qquad N = 2, 3, \ldots,$$

*where $S_N(f) = \sum_{n=0}^{N-1} c_n(f) w_n(x)$ are the partial sums of* (107), *and $C$ is an absolute constant.*

PROOF. Let $N = 2^k + i$, $k = 1, 2, \ldots$, $0 \le i < 2^k$. For $x \in (0,1)\backslash R_2$, it follows from (118) that

$$|S_N(f,x) - f(x)| = S_N(f,x) - S_{2^k}(f,x) + S_{2^k}(f,x) - f(x)$$
$$= S_N(f - S_{2^k}(f), x) + S_{2^k}(f, \chi, x) - f(x). \quad (119)$$

By Theorem 3.2

$$\|S_{2^k}(f, \chi, x) - f(x)\|_C \le 3\omega\left(\frac{1}{2^k}, f\right). \quad (120)$$

If we then use (115) and 1.(17), we find that when $x \in (0,1)\backslash R_2$

$$|S_N(f - S^{2^k}(f), x)| \le L_N \|f - S^{2^k}(f)\|_\infty \le 2\log_2 N \|f - S^{2^k}(f, \chi, x)\|_\infty$$
$$\le 6\log_2 N\omega(2^{-k}, f). \quad (121)$$

Since $\omega(2^{-k}, f) \le 2\omega(2^{-k-1}, f) \le 2\omega(1/N, f)$, we obtain the conclusion of Theorem 14 by combining (120) and (121) (also see (119)).

COROLLARY 8. *If $f \in C(0,1)$ and $\omega(\delta, f) = o(\ln^{-1}(1/\delta))$ as $\delta \to 0$, then the Fourier-Walsh series of $f(x)$ converges to $f(x)$, uniformly on $(0,1)\backslash R_2$.*

We next consider the convergence of Fourier–Walsh series in $L^p(0,1)$. We first observe that, by (116) (also see the remark after theorem 1.8), the Walsh system is not a basis in $L^1(0,1)$, i.e., there is a function $f \in L^1(0,1)$ for which (107) does not converge in $L^1(0,1)$. The situation is different in the spaces $L^p(0,1)$ with $1 < p < \infty$. We have the following theorem.

THEOREM 15. *The Walsh system is a basis in $L^p(0, 1)$, $1 < p < \infty$.*

PROOF. Let $g \in L^p(0, 1)$. We consider the grouped Walsh series:

$$c_0(g)w_0(x) + \sum_{k=0}^{\infty} E_k(g, x), \qquad E_k(g, x) = \sum_{n \in Q_k} c_n(g)w_n(x). \qquad (122)$$

By (118), $E_k(g, x)$ is a polynomial in the Haar system:

$$E_k(g, x) = S_{2^{k+1}}(g, \chi, x) - S_{2^k}(g, \chi, x) = \sum_{n=2^k+1}^{2^{k+1}} d_n(g)\chi_n(x),$$

where the $d_n(g)$ are the Fourier–Haar coefficients of $g(x)$. Hence, recalling that

$$E_k^2(g, x) = \sum_{n=2^k+1}^{2^{k+1}} d_n^2(g)\chi_n^2(x),$$

and using Theorem 3.9, we find that, for every set of integers $0 \le k_1 < \cdots < k_s$,

$$\left\| \sum_{l=1}^{s} E_{k_l}(g, x) \right\|_p \le C_p \left\| \left\{ \sum_{l=1}^{s} E_{k_l}^2(g, x) \right\}^{1/2} \right\|_p$$

$$\le C_p \left\| \left\{ \sum_{n=1}^{\infty} d_n^2(g)\chi_n^2(x) \right\}^{1/2} \right\|_p \le C_p' \|g\|_p. \qquad (123)$$

We use (123) to prove the following inequality: for every $f \in L^p(0, 1)$, $1 < p < \infty$,

$$\|S_N(f, W, x)\|_p \le C_p \|f\|_p, \qquad N = 1, 2, \ldots. \qquad (124)$$

Since the Walsh system is minimal (see Proposition 1.4) and complete in $L^p(0, 1)$, the conclusion of Theorem 15 will follow directly from (124) (by Theorem 1.6).

Let $N \ge 1$ be given, and $N = 2^{k_1} + \cdots + 2^{k_s}$, $k_1 > \cdots > k_s \ge 0$. Let $g(x) = w_N(x) \cdot f(x)$. It follows from (104) that, for $n = 0, 1, \ldots$,

$$c_n(f) = \int_0^1 f(x)w_N(x)w_N(x)w_n(x)\, dx$$

$$= \int_0^1 g(x)w_{N+n}(x)\, dx = c_{N+n}(g).$$

According to (99) and (104) (also see (122)) we obtain

$$w_N(x)S_N(f,W,x) = \sum_{n=0}^{N-1} c_n(f)w_N(x)w_n(x)$$

$$= \sum_{n=0}^{N-1} c_{N \dotplus n}(g)w_{N \dotplus n}(x)$$

$$= \sum_{l=1}^{s} \sum_{n \in Q_{k_l}} c_n(g)w_n(x) = \sum_{l=1}^{s} E_{k_l}(g,x).$$

Using the last equation and (123), we find that

$$\|S_N(f,W,x)\|_p = \|w_N(x)S_N(f,W,x)\|_p \le C_p'\|g\|_p = C_p'\|f\|_p.$$

This establishes (124), and therefore Theorem 15.

REMARK. As we have seen, many propositions about the convergence of trigonometric series are also valid for Fourier–Walsh series. The existence of analogous properties for the trigonometric and Walsh systems is not accidental, and is due to a large extent to the fact that both systems are systems of group characters. For lack of space, we shall not discuss the details here.

CHAPTER V

# The Hilbert Transform and Some Function Spaces

## §1. The Hilbert transform

DEFINITION 1. The *Hilbert transform* of a function $f(x) \in L^1_{1/(1+|x|)}(R^1)$ is

$$T(f,x) = \int_{-\infty}^{\infty} \frac{f(t)}{x-t} dt \equiv \int_{\{t:\ |t-x|>\delta\}} \frac{f(t)}{x-t} dt. \tag{1}$$

We denote by $L^1_{1/(1+|x|)}(R^1)$ the set of functions $f(x)$ such that

$$\int_{-\infty}^{\infty} \frac{|f(x)|}{1+|x|} dx < \infty.$$

It is easily seen that $L^p(R^1) \subset L^1_{1/(1+|x|)}(R^1)$ for every $p$, $1 \leq p < \infty$.

There is a similar definition for the Hilbert transform of a measure $\mu$ that has bounded variation on $R^1$:

$$T(\mu,x) = \lim_{\delta \to +0} \int_{\{t:\ |t-x|>\delta\}} \frac{1}{x-t} d\mu(t).$$

The properties of the Hilbert transform, to which we devote this section, play a fundamental role in problems about the convergence of trigonometric series, and consequently in the study of many function spaces. In Chapters 8 and 10 we shall show that properties of the Hilbert transform are also very useful in the theory of general orthogonal series.

We are going to show that when $f \in L^p(R^1)$, $1 \leq p < \infty$, the limit in (1) exists and is finite for almost all $x \in R^1$. However, we notice first that Definition 1 has the following consequences:

a) if $g(x) = f(x+a)$, $a \in R^1$, then

$$T(g,x) = \lim_{\delta \to +0} \int_{\{t:\ |x-t|>\delta\}} \frac{f(t+a)}{x-t} dt = T(f,x+a); \tag{2}$$

b) if $g'(x) = f(\lambda x)$, $\lambda \in R^1$, then

$$T(g',x) = (\operatorname{sgn}\lambda)[T(f,\lambda,x)]. \tag{3}$$

145

In fact, if $\lambda = 0$ then $g'(x) = \text{const}$ and, by (1), $T(g', x) \equiv 0$. For $\lambda \neq 0$

$$T(g', x) = \lim_{\delta \to +0} (\text{sgn}\,\lambda) \int_{\{z:\,|\lambda x - z| > |\lambda\delta|\}} \frac{f(z)}{\lambda x - z}\,dz = (\text{sgn}\,\lambda) T(f, \lambda x).$$

The example of the characteristic function of the interval $(0, 1)$, namely $\chi_{(0,1)}(x)$, for which

$$T(\chi_{(0,1)}, x) = \int_0^1 \frac{dt}{x - t} = \ln\left|\frac{x}{x - 1}\right|, \qquad x \in R^1, \quad x \neq 0, 1, \qquad (4)$$

shows that the Hilbert transform of a function in $L^1(R^1)$ need not belong to the same space. However, we have the following theorem.

THEOREM 1. *For every $f \in L^1(R^1)$ and almost all $x \in R^1$, the limit* (1) *exists and is finite. Moreover, the operator $T: f \to T(f)$ is of weak type* $(1, 1)$, *i.e.,*

$$m\{x \in R^1: |T(f, x)| > y\} \leq \frac{64}{y}\|f\|_{L^1(R^1)}, \qquad y > 0. \qquad (5)$$

LEMMA 1 (fundamental lemma). *Let $-\infty < a_1 < a_2 < \cdots < a_n < \infty$; let $\delta_a$ be a measure concentrated at the point $a \in R^1$ with $\delta_a(a) = 1$, and let*

$$\mu = \sum_{j=1}^n \mu_j \delta_{a_j} \qquad (\mu_j > 0, j = 1, \ldots, n).$$

*Then for each $y > 0$*

$$m\{x \in R^1: T(\mu, x) > y\} = \frac{1}{y} \sum_{j=1}^n \mu_j$$

$$= m\{x \in R^1: T(\mu, x) < -y\}. \qquad (6)$$

PROOF. It is easily seen that the function

$$T(\mu, x) = \sum_{j=1}^n \frac{\mu_j}{x - a_j}, \qquad x \neq a_j, \; j = 1, \ldots, n, \qquad (7)$$

is monotone decreasing on each interval $(a_j, a_{j+1})$, $j = 0, 1, \ldots, n$ (where $a_0 = -\infty$, $a_{n+1} = +\infty$), and moreover that
a) $\lim_{x - a_j \to +0} T(\mu, x) = +\infty$, $\lim_{x - a_j \to -0} T(\mu, x) = -\infty, j = 1, \ldots, n$;
b) $T(\mu, x) < 0$ for $x \in (-\infty, a_1)$, $\lim_{x \to -\infty} T(\mu, x) = 0$;
c) $T(\mu, x) > 0$ for $x \in (a_n, +\infty)$, $\lim_{x \to +\infty} T(\mu, x) = 0$.
Consequently, in each interval $(a_j, a_{j+1})$, $j = 1, \ldots, n$, there is a unique point $x_j$ at which $T(\mu, x_j) = y$ and

$$m(E) \equiv m\{x \in R^1: T(\mu, x) > y\} = \sum_{j=1}^n (x_j - a_j) = \sum_{j=1}^n x_j - \sum_{j=1}^n a_j. \qquad (8)$$

Since the numbers $x_j$ are the roots of the equation $T(\mu, x) = y$, and therefore (see (7)) also of the equation

$$\prod_{j=1}^{n}(x - a_j) - \frac{1}{y}\sum_{j=1}^{n}\mu_j \prod_{\substack{1 \leq i \leq n \\ i \neq j}} (x - a_i) = 0, \tag{9}$$

by Vieta's theorem the sum $\sum_{j=1}^{n}(-x_j)$ is equal to the coefficient of $x^{n-1}$ in the polynomial (9), i.e.,

$$-\sum_{j=1}^{n} x_j = -\sum_{j=1}^{n} a_j - \frac{1}{y}\sum_{j=1}^{n}\mu_j,$$

whence it follows by (8) that $m(E) = (1/y)\sum_{j=1}^{n}\mu_j$. The second equation in (6) is verified in just the same way. This completes the proof of Lemma 1.

LEMMA 2. *Let $E \subset R^1$ be a compact set and $\{I_\alpha\}$ a covering of $E$ by intervals $I_\alpha$. Then we can select from $\{I_\alpha\}$ a finite set of pairwise disjoint intervals $\{I_j\}_{j=1}^{n}$ such that $\sum_{j=1}^{n} |I_j| \geq \frac{1}{2}m(E)$.*

PROOF (also see Lemma 1 in Appendix 1). Since $E$ is compact we can reduce $\{I_\alpha\}$ to a finite covering $\{I_s\}_{s=1}^{N}$, $I_s = (a_s, b_s)$, $a_1 \leq \cdots \leq a_N$. As we can easily see, we may suppose (by deleting, if necessary, some intervals from $\{I_s\}_{s=1}^{N}$) that each point of the real line belongs to at most two intervals $I_s$. Then the sets $\{I_{2s-1}\}$, $1 \leq 2s - 1 \leq N$, and $\{I_{2s}\}$, $1 \leq 2s \leq N$, consist of pairwise disjoint intervals. Since $\sum_{s=1}^{N} |I_s| \geq m(E)$, one of these sets satisfies the requirements of the lemma.

For $f \in L^1(R^1)$ and $\delta > 0$, let

$$T_\delta(f, x) = \int_{\{t:\ |x-t|>\delta\}} \frac{f(t)}{x - t}\, dt \quad \text{and} \quad T^*(f, x) = \sup_{\delta > 0} |T_\delta(f, x)|. \tag{10}$$

LEMMA 3. *For every function $f \in L^1(R^1)$ the following inequality is satisfied for $y > 0$:*

$$m\{x \in R^1\colon T^*(f, x) \geq y\} \leq \frac{64}{y}\|f\|_{L^1(R^1)}.$$

PROOF. First let $f(x) \geq 0$, $x \in R^1$, and let $y$ be given. For $\delta > 0$ set

$$E_\delta^+ = \left\{x \in R^1\colon \sup_{\varepsilon \geq \delta} T_\varepsilon(f, x) \geq y\right\},$$

$$E_\delta^- = \left\{x \in R^1\colon \inf_{\varepsilon \geq \delta} T_\varepsilon(f, x) \leq -y\right\}.$$

It is easily seen that the sets $E_\delta^+$ and $E_\delta^-$ are compact. In fact, their bound-edness follows because, for every $x$ and $A > 0$,

$$|T_\varepsilon(f,x)| \leq \frac{1}{\varepsilon} \int_{x-A}^{x+A} |f(t)|\, dt + \frac{1}{A}\|f\|_{L^1(R^1)}$$

(see (10)), and therefore $|T_\varepsilon(f,x)| < y$ for sufficiently large $x$ and $\varepsilon \geq \delta > 0$.

For each point $x \in E_\delta^+$ there is an interval $I = (x-\varepsilon, x+\varepsilon)$, $\varepsilon = \varepsilon(x) \geq \delta$, for which

$$\int_{R^1\setminus I} \frac{f(t)}{x-t}\, dt \geq y.$$

Since $E_\delta^+$ is covered by the intervals $I$, we can, using the compactness of $E_\delta^+$ and Lemma 2, choose a finite number of pairwise disjoint intervals $\{I_j\}_{j=1}^n$ such that

a)    $$m(E_\delta^+) \leq 2\sum_{j=1}^n |I_j|;$$

(11)

b)    $$\int_{R^1\setminus I_j} \frac{f(t)}{c(I_j)-t}\, dt \geq y, \qquad j = 1,\ldots,n$$

(here $c(I)$ is the center of the interval $I$).

By using the uniform continuity of $(x-t)^{-1}$ on the set $\{t: |x-t| \geq \delta\}$, it is easy to see that for each $\rho > 0$ there is a number $\alpha = \alpha(\rho) > 0$ such that if we divide the real line into intervals $\omega_s = ((s-1)\alpha, s\alpha)$, $s = 0, \pm 1, \pm 2, \ldots$, we will have, for $j = 1,\ldots,n$,

$$\left| \int_{R^1\setminus I_j} \frac{f(t)}{c(I_j)-t}\, dt - \sum_{s:\, c(\omega_s)\notin I_j} \frac{1}{c(I_j)-c(\omega_s)} \int_{\omega_s} f(t)\, dt \right| < \rho y. \qquad (12)$$

Let

$$g(x) = \sum_{s=-\infty}^\infty \frac{\int_{\omega_s} f(t)\, dt}{x - c(\omega_s)}, \qquad g_j(x) = \sum_{s:\, c(\omega_s)\in I_j} \frac{\int_{\omega_s} f(t)\, dt}{x - c(\omega_s)}.$$

It is clear that the function

$$g(x) - g_j(x) = \sum_{s:\, c(\omega_s)\notin I_j} \frac{1}{x - c(\omega_s)} \int_{\omega_s} f(t)\, dt$$

decreases monotonically on $I_j$, $j = 1,\ldots,n$. Moreover, by inequalities (11),b) and (12), $g(x) - g_j(x) > y(1 - \rho)$ for $x = c(I_j)$, and therefore, on the entire left-hand half of $I_j$,

$$g(x) - g_j(x) > y(1 - \rho) \qquad (x \in I_j, x \leq c(I_j)). \qquad (13)$$

It follows from (13) that, for $j = 1, \ldots, n$,

$$|I_j|/2 \leq m\left\{x \in I_j: g(x) > \frac{1}{2}y(1 - \rho)\right\}$$
$$+ m\left\{x \in I_j: g_j(x) < -\frac{1}{2}y(1 - \rho)\right\}. \tag{14}$$

Combining inequalities (14) for $j = 1, \ldots, n$, and using the fact that for all $\alpha > 0$ we have

$$m\{x \in I_j: g(x) > \alpha\} = \lim_{N \to \infty} m\left\{x \in I_j: \sum_{s=-N}^{N} \frac{\int_{\omega_s} f(t)\, dt}{x - c(\omega_s)} > \alpha\right\},$$

we obtain

$$\frac{1}{2}\sum_{j=1}^{n}|I_j| \leq \varlimsup_{N \to \infty} m\left\{x \in R^1: \sum_{s=-N}^{N} \frac{\int_{\omega_s} f(t)\, dt}{x - c(\omega_s)} > \frac{y}{2}(1 - \rho)\right\}$$
$$+ \sum_{j=1}^{n} m\left\{x \in R^1: g_j(x) < -\frac{y}{2}(1 - \rho)\right\}. \tag{15}$$

It follows directly from (15) and Lemma 1 that

$$\frac{1}{2}\sum_{j=1}^{n}|I_j| \leq \frac{2}{y(1 - \rho)}\int_{R^1} f(t)\, dt + \frac{2}{y(1 - \rho)}\int_{R^1} f(t)\, dt.$$

Letting $\rho$ approach zero in this inequality (see also (11),a)), we obtain

$$m(E_\delta^+) \leq \frac{16}{y}\int_{-\infty}^{\infty} f(t)\, dt. \tag{16}$$

Similarly, we can verify that

$$m(E_\delta^-) \leq \frac{16}{y}\int_{-\infty}^{\infty} f(t)\, dt. \tag{16'}$$

It follows from (16) and (16') (which hold for all $\delta > 0$) that, for every nonnegative function $f \in L^1(R^1)$,

$$m\{x \in R^1: T^*(f, x) \geq y\} \leq \frac{32}{y}\|f\|_{L^1(R^1)}, \quad y > 0. \tag{17}$$

If now the function $f \in L^1(R^1)$ is of arbitrary sign, we represent it in the form $f(x) = f^+(x) - f^-(x)$, where

$$f^+(x) = \begin{cases} f(x) & \text{if } f(x) \geq 0, \\ 0 & \text{if } f(x) < 0, \end{cases} \quad f^-(x) = f^+(x) - f(x),$$

$$\|f^+\|_{L'(R')} + \|f^-\|_{L'(R')} = \|f\|_{L'(R')},$$

and use

$$\{x \in R^1 : T^*(f, x) \geq y\} \subset \left\{x \in R^1 : T^*(f^+, x) \geq \frac{y}{2}\right\}$$
$$\cup \left\{x \in R^1 : T^*(f^-, x) \geq \frac{y}{2}\right\}.$$

Then the conclusion of Lemma 3 follows from (17).

PROOF OF THEOREM 1. Let us show that the finite limit (see (1) and (10))

$$T(f, x) = \lim_{\delta \to +0} T_\delta(f, x)$$

exists for almost all $x \in R^1$. We represent $f(x)$ in the form $f(x) = f_1(x) + f_2(x)$, where $f_1$ is a piecewise constant function, and $\|f_2\|_{L^1(R^1)} \leq \varepsilon$ ($\varepsilon > 0$). Then

$$\varlimsup_{\delta, \delta' \to +0} |T_\delta(f, x) - T_{\delta'}(f, x)| \leq \varlimsup_{\delta, \delta' \to +0} |T_\delta(f_1, x) - T_{\delta'}(f_1, x)|$$
$$+ \varlimsup_{\delta, \delta' \to 0} |T_\delta(f_2, x) - T_{\delta'}(f_2, x)|. \quad (18)$$

Since $f_1$ is piecewise constant, the first term on the right-hand side of (18) is zero for almost all $x \in R^1$. Moreover, by Lemma 3, for every $y > 0$,

$$m \left\{ x \in R^1 : \varlimsup_{\delta, \delta' \to +0} |T_\delta(f_2, x) - T_{\delta'}(f_2, x)| > y \right\}$$
$$\leq m \left\{ x \in R^1 : T^*(f_2, x) > \frac{y}{2} \right\} \leq \frac{128\varepsilon}{y}.$$

Since the number $\varepsilon > 0$ in the preceding inequality can be taken arbitrarily small, we have, for every $y > 0$,

$$m \left\{ x \in R^1 : \varlimsup_{\delta, \delta' \to +0} |T_\delta(f, x) - T_{\delta'}(f, x)| > y \right\} = 0,$$

i.e., the limit (1) exists and is finite for almost all $x \in R^1$. But then (5) follows directly from Lemma 3. This completes the proof of Theorem 1.

REMARK. Theorem 1 can also be proved by a method similar to that used in proving the inequalities of weak type in Theorem 3.7.

The following corollaries are immediate consequences of Theorem 1 (see (5)).

COROLLARY 1. *If the sequence of functions* $f_n \in L^1(R^1)$, $n = 1, 2, \ldots$, *converges to* $f(x)$ *in the norm of* $L^1(R^1)$, *then the sequence* $T(f_n, x)$ *converges in measure to* $T(f, x)$.

COROLLARY 2. *For every* $f \in L^p(R^1)$, $1 \leq p < \infty$, *the Hilbert transform (see (1)) exists and is finite for almost every* $x \in R^1$.

In fact, for $N = 0, \pm 1, \pm 2, \ldots$ let

$$f_N(t) = \begin{cases} f(t) & \text{for } t \in (N, N+1), \\ 0 & \text{for } t \notin (N, N+1); \end{cases}$$

then, for $x \in (N, N+1)$,

$$T(f, x) = T(f_N, x) + \int_{N+1}^{\infty} \frac{f(t)}{x-t} \, dt + \int_{-\infty}^{N} \frac{f(t)}{x-t} \, dt. \qquad (19)$$

Applying Hölder's inequality, we obtain

$$\left| \int_{N+1}^{\infty} \frac{f(t)}{x-t} \, dt \right| + \left| \int_{-\infty}^{N} \frac{f(t)}{x-t} \, dt \right|$$

$$\leq 2\|f\|_{L^p(R^1)} \cdot \left\{ \int_{R^1 \setminus (N, N+1)} |x-t|^{-q} \, dt \right\}^{1/q} < \infty \qquad (20)$$

for all $x \in (N, N+1)$ (here $1/p + 1/q = 1$). Since $f_N \in L^1(R^1)$, $N = 0, \pm 1, \ldots$, and therefore, by Theorem 1, the Hilbert transform $T(f_N, x)$ exists and is finite for almost all $x \in R^1$, Corollary 2 follows from (19) and (20).

The following result shows that the Hilbert transform is a bounded operator on $L^p(R^1)$ for $1 < p < \infty$.

THEOREM 2. *The inequality*

$$\|T(f)\|_{L^p(R^1)} \leq C_p \|f\|_{L^p(R^1)} \qquad (21)$$

*is satisfied for every $f \in L^p(R^1)$ ($1 < p < \infty$), with $C_p$ depending only on $p$.*

We divide the proof of Theorem 2 into several steps. First we observe that $\|T(\chi_{(0,1)})\|_{L^2(R^1)} = c_0 < \infty$ (see (4)), and consequently (see (2) and (3)) the characteristic function $\chi_\omega(x)$ of every interval $\omega \subset R^1$ satisfies the equation

$$\|T(\chi_\omega)\|_{L^2(R^1)} = c_0 |\omega|^{1/2} = c_0 \|\chi_\omega\|_{L^2(R^1)}. \qquad (22)$$

We now prove (21) for $p = 2$ for "simple" functions $f(x)$, i.e., for functions of the form

$$f(x) = a_i \quad \text{for } x \in \omega_i = (ih, (i+1)h), \ h > 0, \ i = 0, \pm 1, \pm 2, \ldots, \qquad (23)$$

where only finitely many numbers $a_i$ are different from zero. We need the following lemma.

LEMMA 1. *There is an absolute constant $C > 0$ such that for every fini-tary([1]) sequence $\{a_i\}_{i=-\infty}^{\infty}$ we have the inequalities*

$$\left\{ \sup_{\{b_j\}:\ \sum_{j=-\infty}^{\infty} b_j^2 = 1} \sum_{\substack{i,j=-\infty \\ i \neq j}}^{\infty} \frac{a_i b_j}{(j-i)^s} \right\}^2 = \sum_{j=-\infty}^{\infty} \left[ \sum_{\substack{i=-\infty \\ i \neq j}}^{\infty} \frac{a_i}{(j-i)^s} \right]^2$$

$$\leq C \sum_{i=-\infty}^{\infty} a_i^2, \qquad s = 1, 2. \quad (24)$$

PROOF. For $s = 2$ we have

$$\left| \sum_{\substack{i,j=-\infty \\ i \neq j}}^{\infty} \frac{a_i b_j}{(j-i)^2} \right| = \left| \sum_{\substack{k=-\infty \\ k \neq 0}}^{\infty} \frac{1}{k^2} \sum_{i=-\infty}^{\infty} a_i b_{i+k} \right|,$$

whence, since

$$\left| \sum_{i=-\infty}^{\infty} a_i b_{i+k} \right| \leq \left( \sum_{i=-\infty}^{\infty} a_i^2 \right)^{1/2} \left( \sum_{i=-\infty}^{\infty} b_{i+k}^2 \right)^{1/2} = \left( \sum_{i=-\infty}^{\infty} a_i^2 \right)^{1/2}$$

for every $k$, we obtain (24) for $s = 2$.

For $s = 1$ we use the equation

$$I \equiv \sum_{j=-\infty}^{\infty} \left( \sum_{\substack{i=-\infty \\ i \neq j}}^{\infty} \frac{a_i}{j-i} \right)^2 = \sum_{j=-\infty}^{\infty} \sum_{\substack{\nu,\mu=-\infty \\ \nu \neq j, \mu \neq j}}^{\infty} \frac{a_\nu a_\mu}{(j-\nu)(j-\mu)}$$

$$= \sum_{\nu,\mu=-\infty}^{\infty} a_\nu a_\mu \sum_{\substack{j=-\infty \\ j \neq \nu,\mu}}^{\infty} \frac{1}{(j-\nu)(j-\mu)}. \quad (25)$$

Since

$$\frac{1}{(j-\nu)(j-\mu)} = \frac{1}{\nu-\mu} \left[ \frac{1}{j-\nu} - \frac{1}{j-\mu} \right]$$

for $j \neq \mu$, $j \neq \nu$, $\mu \neq \nu$, we find that the coefficient of $a_\nu a_\mu$ ($\nu \neq \mu$) on the right-hand side of (25) equals

$$\frac{1}{\nu-\mu} \sum_{\substack{j=-\infty \\ j \neq \nu,\mu}}^{\infty} \left[ \frac{1}{j-\nu} - \frac{1}{j-\mu} \right] = \frac{2}{(\nu-\mu)^2}.$$

---

([1])That is, $a_i = 0$ if $|i|$ is sufficiently large.

Consequently (by (24), already proved for $s = 2$),

$$|I| \leq \sum_{\mu=-\infty}^{\infty} a_\mu^2 \sum_{\substack{j=-\infty \\ j \neq \mu}}^{\infty} \frac{1}{(j-\mu)^2} + 2 \left| \sum_{\substack{\mu,\nu=-\infty \\ \mu \neq \nu}}^{\infty} \frac{a_\nu a_\mu}{(\nu-\mu)^2} \right|$$

$$\leq C \sum_{\nu=-\infty}^{\infty} a_\nu^2.$$

This completes the proof of Lemma 1.

REMARK. If we consider the matrix

$$H_N = \{h_{i,j}\}_{i,j=1}^N \quad \text{with } h_{i,j} = \begin{cases} \dfrac{1}{i-j} & \text{for } i \neq j, \\ 0 & \text{for } i = j, \end{cases}$$

$$N = 2, 3, \ldots$$

(the Hilbert matrix), it then follows from Lemma 1 that

$$\|H_N\| \leq C, \qquad N = 2, 3, \ldots \tag{26}$$

(the definition of the norm of a matrix was given at the beginning of the book). At the same time, as is easily verified,

$$\|\{|h_{i,j}|\}_{i,j=1}^N\| \asymp \ln N. \tag{27}$$

Inequalities (26) and (27) are frequently used in the theory of orthogonal series.

Digressing from the proof of Theorem 2, we also present an approach to inequalities for the norms of matrices $H$ of the form

$$H = \{h_{s,j}\}_{s,j=-\infty}^{\infty}, \qquad h_{s,j} = c_{s-j}, \tag{28}$$

by using properties of the trigonometric system.

LEMMA 2. *Let* $\{c_j\}_{j=-\infty}^{\infty}$ *be a sequence of complex numbers such that* $\sum_{j=-\infty}^{\infty} |c_j|^2 < \infty$ *and in addition*

$$f(x) \overset{L^2}{=} \sum_{j=-\infty}^{\infty} c_j e^{ijx} \in L^{\infty}(-\pi, \pi). \tag{29}$$

*Then the norm of the matrix* (28) *satisfies*

$$\|H\| = \sup_{\{b_s\}:\, \sum_{s=-\infty}^{\infty} |b_s|^2 \leq 1} \left\{ \sum_{j=-\infty}^{\infty} \left| \sum_{s=-\infty}^{\infty} b_s c_{s-j} \right|^2 \right\}^{1/2} \leq \|f\|_{L^{\infty}(-\pi,\pi)}.$$

PROOF. Let $\sum_{s=-\infty}^{\infty} |b_s|^2 \leq 1$ and

$$g(x) = \sum_{s=-\infty}^{\infty} b_s e^{-isx},$$

$$\int_{-\pi}^{\pi} |g(x)|^2 \, dx = 2\pi \sum_{s=-\infty}^{\infty} |b_s|^2 \leq 2\pi. \tag{30}$$

Since the series (29) and (30) converge in $L^2(-\pi, \pi)$, we have, for $j = 0, \pm 1, \pm 2, \ldots,$

$$\int_{-\pi}^{\pi} g(x)f(x)e^{ijx} \, dx = \lim_{N \to \infty} \sum_{s=-N}^{N} \sum_{\nu=-\infty}^{\infty} \int_{-\pi}^{\pi} b_s e^{-isx} c_\nu e^{i(\nu+j)x} \, dx$$

$$= 2\pi \sum_{s=-\infty}^{\infty} b_s c_{s-j}.$$

Consequently (see (30)),

$$\sum_{j=-\infty}^{\infty} \left| \sum_{s=-\infty}^{\infty} b_s c_{s-j} \right|^2 = \sum_{j=-\infty}^{\infty} \left| \frac{1}{2\pi} \int_{-\pi}^{\pi} g(x)f(x)e^{ijx} \, dx \right|^2$$

$$= \frac{1}{2\pi} \int_{-\pi}^{\pi} |g(x)f(x)|^2 \, dx$$

$$\leq \frac{1}{2\pi} \|f\|_{L^\infty(-\pi,\pi)}^2 \cdot \int_{-\pi}^{\pi} |g(x)|^2 \, dx \leq \|f\|_{L^\infty(-\pi,\pi)}^2,$$

and Lemma 2 is established.

We notice a corollary of Lemma 2, which will be needed in Chapters 8 and 10:

Let the matrix $H_N(A)$ ($N = 2, 3, \ldots; A = 1, \ldots, N - 1$) have the form

$$H_N(A) = \{h_{s,j}\}_{s,j=1}^{N},$$

$$h_{s,j} = \begin{cases} \dfrac{1}{s-j} & \text{if } 0 < |s - j| \leq A, \\ 0 & \text{if } s = j \text{ or } |s - j| > A; \end{cases}$$

then there is an absolute constant $C$ such that

$$\|H_N(A)\| \leq C, \qquad N = 2, 3, \ldots, \quad A = 1, \ldots, N - 1. \tag{31}$$

In fact, it is clear that

$$\|H_N(A)\| \leq \|H'(A)\|,$$
$$H'(A) = \{h_{s,j}\}_{s,j=-\infty}^{\infty} \equiv \{c_{s-j}\}_{s,j=-\infty}^{\infty},$$

where $c_j = 1/j$ if $0 < |j| \le A$, and $c_j = 0$ otherwise. Since

$$\frac{1}{2}\left\|\sum_{\substack{j=-A \\ j\neq 0}}^{A} \frac{1}{j} e^{ijx}\right\|_{L^\infty(-\pi,\pi)} = \left\|\sum_{j=1}^{A} \frac{1}{j} \sin jx\right\|_{L^\infty(-\pi,\pi)}$$

$$\le \sup_{0\le x\le\pi}\left|\int_0^x \sum_{j=1}^{A} \cos jt\,dt\right|$$

$$\le \frac{\pi}{2} + \sup_{0\le x\le\pi}\left|\int_0^x D_A(t)\,dt\right|$$

$$\le C' + \sup_{0\le x\le\pi}\left|\int_0^x \frac{\sin At}{t}\,dt\right| \le C$$

($D_A(t)$ is the Dirichlet kernel of order $A$; for details see §4.1, in particular 4.(4) and 4.(9)); then (31) follows from Lemma 2.

We return to the proof of Theorem 2.

Let $f(x)$ be a simple function of the form (23) and let $x \in \omega_j$, $j = 0, \pm 1, \pm 2, \ldots$ (see (23)). Then $f(x) = \sum_{i=-\infty}^{\infty} a_i \chi_{\omega_i}(x)$ and

$$T(f,x) = \sum_{i:\,|i-j|\le 1} a_i T(\chi_{\omega_i}, x) + \sum_{i:\,|i-j|>1} a_i T(\chi_{\omega_i}, x)$$

$$\equiv S_1(x) + S_2(x).$$

It follows directly from (22) that

$$\int_{\omega_j} S_1^2(x)\,dx \le C(a_{j-1}^2 + a_j^2 + a_{j+1}^2)h. \tag{32}$$

Moreover, since

$$S_2(x) = \sum_{i:\,|i-j|>1} a_i \int_{\omega_i} \frac{dt}{x-t}$$

$$= \sum_{i:\,|i-j|>1} \frac{a_i}{j-i} + \sum_{i:\,|i-j|>1} a_i \left(\int_{\omega_i} \frac{dt}{x-t} - \frac{1}{j-i}\right),$$

if we use the inequalities

$$\left|\int_{\omega_i} \frac{dt}{x-t} - \frac{1}{j-i}\right| = \left|\int_{\omega_i} \left[\frac{1}{x-t} - \frac{1}{(j-i)h}\right]dt\right|$$

$$\le \int_{\omega_i} \left|\frac{(jh-x)-(ih-t)}{(x-t)(j-i)h}\right|dt \le \frac{C}{(j-i)^2},$$

with $|i - j| > 1$ and $x \in \omega_j$, we obtain

$$
\int_{\omega_j} S_2^2(x)\,dx \le Ch \left( \sum_{i:\,|i-j|>1} \frac{a_i}{j-i} \right)^2 + Ch \left( \sum_{i:\,|i-j|>1} \frac{|a_i|}{(j-i)^2} \right)^2
$$

$$
\le C'h \left[ \left( \sum_{\substack{i=-\infty \\ i \ne j}}^{\infty} \frac{a_i}{j-i} \right)^2 + \left( \sum_{\substack{i=-\infty \\ i \ne j}}^{\infty} \frac{|a_i|}{(j-i)^2} \right)^2 + a_{j-1}^2 + a_{j+1}^2 \right].
$$

$$(33)$$

Adding (32) and (33), we obtain

$$
\int_{R^1} T^2(f,x)\,dx
$$

$$
= \sum_{j=-\infty}^{\infty} \int_{\omega_j} T^2(f,x)\,dx
$$

$$
\le Ch \left[ \sum_{j=-\infty}^{\infty} a_j^2 + \sum_{j=-\infty}^{\infty} \left( \sum_{\substack{i=-\infty \\ i \ne j}}^{\infty} \frac{a_i}{j-i} \right)^2 + \sum_{j=-\infty}^{\infty} \left( \sum_{\substack{i=-\infty \\ i \ne j}}^{\infty} \frac{|a_i|}{(j-i)^2} \right)^2 \right],
$$

from which it follows by Lemma 1 (see also (23)) that

$$
\|T(f)\|_{L^2(R^1)} \le Ch \sum_{j=-\infty}^{\infty} a_j^2 = C\|f\|_{L^2(R^1)}^2. \tag{34}
$$

Therefore we have verified (21) for simple functions when $p = 2$. The transition to arbitrary functions $f \in L^2(R^1)$ can be made by a standard argument. In fact, first let $f(x)$ be such that $f(x) = 0$ for $|x| > N$. Then we choose a sequence of simple functions $\{f_n(x)\}_{n=1}^{\infty}$ for which

$$
\lim_{n \to \infty} \|f - f_n\|_{L^2(R^1)} = 0, \qquad \lim_{n \to \infty} \|f - f_n\|_{L^1(R^1)} = 0.
$$

Then, by (34) and Corollary 1, we will have
  a) $\|T(f_n)\|_{L^2(R^1)} \le C\|f_n\|_{L^2(R^1)}$,     $n = 1, 2, \ldots$;
  b) $T(f_n, x) \to T(f, x)$ in measure on $R^1$.[2]
  Hence it follows by Fatou's theorem that

$$
\|T(f)\|_{L^2(R^1)} \le \varliminf_{n \to \infty} \|T(f_n)\|_{L^2(R^1)} \le C\|f\|_{L^2(R^1)}.
$$

---

[2]It follows from b) that $T(f_{n_i}, x) \to T(f, x)$ for almost all $x \in R^1$, for some subsequence $\{n_i\}$.

Finally, let $f(x)$ be any function in $L^2(R^1)$. Since for each $x \in R^1$ we have

$$T(f, x) = \lim_{N \to \infty} T(f \cdot \chi_{(-N,N)}, x)$$

(see the proof of Corollary 2), and since $\|T(f \cdot \chi_{(-N,N)})\|_{L^2(R^1)} \leq C\|f\|_{L^2(R^1)}$, if we use Fatou's theorem again we obtain

$$\|T(f)\|_{L^2(R^1)} \leq C\|f\|_{L^2(R^1)}, \qquad f \in L^2(R^1). \tag{35}$$

This completes the proof of (21) for $p = 2$.[3]

Since the Hilbert transform operator has weak type $(1,1)$ and strong type $(2,2)$ (see (5) and (35)), we can apply Marcinkiewicz's interpolation theorem (see Appendix 1, Theorem 1) to obtain (21) for $1 < p < 2$. It remains to consider the case $2 < p < \infty$, where, as we saw in the case $p = 2$, we may restrict the proof of (2) to simple functions $f(x)$.

We first verify that

$$\int_{R^1} T(f, x)g(x)\, dx = -\int_{R^1} f(x)T(g, x)\, dx \tag{36}$$

for all simple functions $f(x)$ and $g(x)$.

Since $T$ is a continuous operator on $L^2(R^1)$, we may suppose, without loss of generality, that the intervals of constancy of $f(x)$ (see (23)) are the same as those of $g(x)$. If we then use the linearity of $T$, we can verify (36) for the case when $f = \chi_\omega, g = \chi_{\omega'}$, where $\omega$ and $\omega'$ are intervals on the line, either disjoint or identical. If $\omega \cap \omega' = \varnothing$, then

$$\int_{R^1} T(f, x)g(x)\, dx = \int_{\omega'} \int_\omega \frac{1}{x - t}\, dt\, dx,$$

and since the function, of two variables, $(x - t)^{-1}\chi_{\omega'}(x)\chi_\omega(t) \in L^1(R^2)$, we have, by Fubini's theorem,

$$\int_{R^1} T(f, x)g(x)\, dx = -\int_\omega \int_{\omega'} \frac{1}{t - x}\, dx\, dt$$
$$= -\int_{R^1} f(x)T(g, x)\, dx.$$

If $\omega = \omega' = (a, b)$, then

$$\int_{R^1} T(f, x)g(x)\, dx = \int_\omega T(\chi_\omega, x)\, dx = 0$$

(since it is easy to see from the definition (see (1)) that

$$T\left(\chi_\omega, \frac{a+b}{2} + z\right) = -T\left(\chi_\omega, \frac{a+b}{2} - z\right), \qquad z \in R^1).$$

Therefore (36) is established.

---

[3] It can be shown that $\|T(f)\|_{L^2(R^1)} = \pi\|f\|_{L^2(R^1)}$ for every $f \in L^2(R^1)$.

Let $2 < p < \infty$, and let $g(x)$ be a simple function with norm $\|g\|_{L^q(R^1)} \le 1$ $(1/p + 1/q = 1)$.

Then $1 < q < 2$, and by using (36) and (21) (already proved for $1 < p < 2$) we find that

$$\left| \int_{R^1} T(f,x)g(x)\,dx \right| = \left| \int_{R^1} f(x)T(g,x)\,dx \right|$$

$$\le \|f\|_{L^p(R^1)} \|T(g)\|_{L^q(R^1)} \le C_q \|f\|_{L^p(R^1)}.$$

From the last inequality we obtain $\|T(f)\|_{L^p(R^1)} \le C_q \|f\|_{L^p(R^1)}$ by using the fact that $\|F\|_{L^p(R^1)} = \sup \int_{R^1} F(x)g(x)\,dx$, for every $F(x)$ $(x \in R^1)$, where the supremum is taken over all simple functions $g(x)$ with $\|g\|_{L^q(R^1)} \le 1$. This completes the proof of Theorem 2.

For $A > 0$, let (see (1) and (10))

$$T_A'(f,x) = T(f,x) - T_A(f,x) = \int_{\{t:\, |x-t|\le A\}} \frac{f(t)}{x-t}\,dt. \tag{37}$$

We notice the following properties of $T_A'(f)$, which follow from Theorems 1 and 2.

a) If $f \in L^1(R^1)$ then $(y > 0)$

$$m\{x \in R^1 : |T_A'(f,x)| > y\} \le \frac{2^8}{y} \|f\|_{L^1(R^1)}. \tag{38}$$

b) If $f \in L^p(R^1)$, $1 < p < \infty$, then

$$\|T_A'(f)\|_{L^p(R^1)} \le C_{p,A} \|f\|_{L^p(R^1)}, \tag{39}$$

where the constant $C_{p,A}$ depends only on $p$ and $A$.

In fact, since

$$\{x \in R^1 : |T_A'(t,x)| > y\} \subset \left\{ x \in R^1 : |T(f,x)| > \frac{y}{2} \right\}$$

$$\cup \left\{ x \in R^1 : |T_A(f,x)| > \frac{y}{2} \right\},$$

the inequality (38) follows immediately from (5) and Lemma 3 (see Theorem 1). For $f \in L^p(R^1)$, $1 < p < \infty$, we find, by applying Hölder's inequality, that for every $x \in R^1$

$$|T_A(f,x)| \le \left\{ \int_{\{t:\, |x-t|>A\}} |f(t)|^p\,dt \right\}^{1/p} \left\{ \int_{\{t:\, |x-t|>A\}} |x-t|^{-q}\,dt \right\}^{1/q}$$

$$\le C_{p,A} \|f\|_{L^p(R^1)}$$

$(1/p + 1/q = 1)$. Then it follows from Theorem 2 that

$$\|T_A'(f)\|_{L^p(R^1)} \le \|T(f)\|_{L^p(R^1)} + \|T_A(f)\|_{L^p(R^1)} \le C_{p,A} \|f\|_{L^p(R^1)}.$$

We note that the constant $C_{p,A}$ in (39) can be taken to be independent of $A$, although this is not essential for our purposes. We shall use (38) and (39) to investigate the periodic analog of the Hilbert transform: the operator of forming the conjugate function.

DEFINITION 2. If $f \in L^1(-\pi, \pi)$, its *conjugate function* is

$$\tilde{f}(x) = \frac{1}{\pi} \int_{-\pi}^{\pi} \frac{f(t)}{2 \tan \frac{x-t}{2}} dt = \frac{1}{\pi} \int_{\{t: |x-t| \leq \pi\}} \frac{f(t)}{2 \tan \frac{x-t}{2}} dt,$$
$$x \in [-\pi, \pi], \quad (40)$$

where the integral in (40) is a *principal value*, i.e., the limit as $\varepsilon \to 0$ of the integrals

$$\int_{\{t: \varepsilon \leq |x-t| \leq \pi\}} \frac{f(t)}{2 \tan \frac{x-t}{2}} dt.$$

THEOREM 3. *For every function $f \in L^1(-\pi, \pi)$ the conjugate function $\tilde{f}$ exists, is finite almost everywhere on $(-\pi, \pi)$, and has the following properties:*

a) $m\{x \in (-\pi, \pi): |\tilde{f}(x)| > y\} \leq (C_1/y)\|f\|_{L^1(-\pi,\pi)}$, $\quad y > 0$.
b) *If $f \in L^p(-\pi, \pi)$, $1 < p < \infty$, then $\tilde{f} \in L^p(-\pi, \pi)$, and*

$$\|\tilde{f}\|_{L^p(-\pi,\pi)} \leq C_p \|f\|_{L^p(-\pi,\pi)}.$$

PROOF. It is clear that, for $x \in (-\pi, \pi)$,

$$\tilde{f}(x) = \int_{\{t: |x-t| \leq \pi\}} \frac{f_1(t)}{2 \tan \frac{x-t}{2}} dt,$$
$$f_1(t) = \frac{1}{\pi} f(t) \chi_{[-2\pi,2\pi]}(t), \quad t \in R^1. \quad (41)$$

Let

$$K(x, t) = \frac{1}{2 \tan \frac{x-t}{2}} - \frac{1}{x-t}.$$

Then (see (37))

$$\tilde{f}(x) = T'_\pi(f_1, x) + \int_{\{t: |x-t| \leq \pi\}} K(x,t) f_1(t) dt$$
$$\equiv T'_\pi(f_1, x) + G(x). \quad (42)$$

It is easily seen that $|K(x,t)| \leq C_0 < \infty$ for $|x - t| < \pi$. Therefore

$$|G(x)| \leq C_0 \|f_1\|_{L^1(R^1)} = 2C_0 \|f\|_{L^1(-\pi,\pi)},$$

and it follows from (38) that, when $1 < p < \infty$,

$$\|\tilde{f}\|_{L^p(-\pi,\pi)} \leq C'_p \cdot \|f\|_{L^p(-\pi,\pi)} + 2C_0 \|f\|_{L^1(-\pi,\pi)}$$
$$\leq C_p \|f\|_{L^p(-\pi,\pi)}.$$

Moreover, inequality a) is evidently valid for $y \leq 4C_0\|f\|_{L^1(-\pi,\pi)}$ if $C_1(4C_0)^{-1}$ $\geq 2\pi$. For $y > 4C_0\|f\|_{L^1(-\pi,\pi)}$ if we use

$$\{x \in (-\pi, \pi): |\tilde{f}(x)| > y\} \subset \{x \in R^1: |T'_\pi(f_1, x)| > y - 2C_0\|f\|_{L^1(-\pi,\pi)}\},$$

and then apply (38), we obtain

$$m\{x \in (-\pi, \pi): |\tilde{f}(x)| > y\} \leq \frac{2^8\|f_1\|_{L^1(R^1)}}{y - 2C_0\|f\|_{L^1(-\pi,\pi)}}$$

$$\leq \frac{2^{10}}{y}\|f\|_{L^1(-\pi,\pi)},$$

i.e., inequality a) is always satisfied, with $C_1 = \max(2^{10}, 8\pi C_0)$. This completes the proof of Theorem 3.

## §2. The spaces $\operatorname{Re}\mathscr{H}^1$ and BMO

In this section we study the space $\operatorname{Re}\mathscr{H}^1$, an important function space that consists of the functions $f \in L^1(-\pi, \pi)$ for which the conjugate function $\tilde{f}$ is also summable. The study of the space $\operatorname{Re}\mathscr{H}^1$ will be continued in the next chapter (see §6.5) by using the properties of Franklin's orthogonal system. Here we use essentially only the material of Appendix 2, where there is a collection of facts from the theory of functions of a complex variable.

Let $\operatorname{Re}\mathscr{H}^p$ ($1 \leq p < \infty$) be the space of functions $f(x)$ which are the boundary functions of functions in the space $H^p$ (see §2 of Appendix 2):

$$f(x) = \lim_{r \to 1} \operatorname{Re} F(re^{ix}) \quad \text{for almost all } x \in (-\pi, \pi), \ F \in H^p. \tag{43}$$

It is shown in Appendix 2, §2, that

$$\operatorname{Re}\mathscr{H}^p = \{f \in L^p(-\pi, \pi): \tilde{f} \in L^p(-\pi, \pi)\}, \qquad 1 \leq p \leq \infty, \tag{44}$$

and that $\operatorname{Re}\mathscr{H}^p$ is a Banach space with norm

$$\|f\|_{\operatorname{Re}\mathscr{H}^p} = \|f\|_{L^p(-\pi,\pi)} + \|\tilde{f}\|_{L^p(-\pi,\pi)}; \tag{45}$$

moreover, if $\operatorname{Im} F(0) = 0$ in (43) then

$$\|F(z)\|_{H^p} \leq \|f(x)\|_{\operatorname{Re}\mathscr{H}^p} \leq \sqrt{2}\|F(z)\|_{H^p} \quad (1 \leq p \leq \infty). \tag{46}$$

Theorem 3 shows that when $1 < p < \infty$ the space $\operatorname{Re}\mathscr{H}^p$ is identical with $L^p(-\pi, \pi)$ and

$$\|f\|_{L^p(-\pi,\pi)} \leq \|f\|_{\operatorname{Re}\mathscr{H}^p} \leq (1 + C_p)\|f\|_{L^p(-\pi,\pi)} \quad (1 < p < \infty).$$

The preceding relation fails for $p = 1$. In fact, it is easily verified that when $h > 0$

$$\|\tilde{f}_h\|_{L^1(-\pi,\pi)} \geq c \ln \frac{1}{h} > 0,$$

where

$$f_h(x) = \begin{cases} (2h)^{-1} & \text{if } 0 \leq |x| < h, \\ 0 & \text{if } h \leq |x| \leq \pi. \end{cases}$$

Consequently there is a function $f \in L^1(-\pi,\pi)$ for which $\tilde{f} \notin L^1(-\pi,\pi)$. Consequently Re $\mathscr{H}^1$ is a proper subspace of $L^1(-\pi,\pi)$. Later we shall give a test for a function to belong to Re $\mathscr{H}^1$.

We shall call a set $I \subset [-\pi,\pi]$ a *generalized interval* if $\theta(I) = \{z: z = e^{ix}, x \in I\}$ is an arc of the unit circumference, i.e., $I$ is either an interval in $[-\pi,\pi]$ or a set of the form

$$I = [-\pi,a) \cup (b,\pi] \qquad (-\pi < a < b < \pi). \tag{47}$$

We say that a point $t$ is the *center* of the generalized interval $I$ if $e^{it}$ is the center of the arc $\theta(I)$. Finally, it is natural to call the number

$$|I| = \begin{cases} b - a, & \text{if } I = (a,b) \subset (-\pi,\pi), \\ 2\pi + a - b, & \text{if } I = [-\pi,a) \cup (b,\pi], \end{cases}$$

the *length* of the generalized interval $I$.

DEFINITION 3. A real function $a(x) \in L^\infty(-\pi,\pi)$ is called an *atom* if there is a generalized interval $I$ such that

a) $\operatorname{supp} a(x) \subset I$;
b) $\int_I a(x)\,dx = \int_{-\pi}^\pi a(x)\,dx = 0$;
c) $\|a\|_{L^\infty(-\pi,\pi)} \leq |I|^{-1}$.
We also call $a(x) = \frac{1}{2\pi}$, $x \in [-\pi,\pi]$, an *atom*.

THEOREM 4. *A necessary and sufficient condition that $f \in \operatorname{Re}\mathscr{H}^1$ is that $f(x)$ can be represented in the form*[4]

$$f(x) = \sum_k \lambda_k a_k(x), \qquad \sum_k |\lambda_k| < \infty, \tag{48}$$

*where $a_k(x)$, $k = 1, 2, \ldots$, are atoms. Here*

$$c\|f\|_{\operatorname{Re}\mathscr{H}^1} \leq \inf \sum |\lambda_k| \leq C\|f\|_{\operatorname{Re}\mathscr{H}^1}, \tag{49}$$

*where the infimum is taken over all decompositions of $f(x)$ of the form (48), and $c$ and $C$ $(0 < c < C < \infty)$ are absolute constants.*

---

[4]By conditions a) and c) in Definition 3 we have $\|a_k\|_{L^1(-\pi,\pi)} \leq 1$, $k = 1, 2, \ldots$, and therefore the series (48) converges both in the $L^1(-\pi,\pi)$ norm and a.e.

PROOF. I) Let $f$ have a decomposition of the form (48). Let us show that $f \in \operatorname{Re} \mathscr{H}^1$ and $c\|f\|_{\operatorname{Re}\mathscr{H}^1} \leq \sum_k |\lambda_k|$. For this purpose it is enough to verify that the inequality

$$\|a\|_{\operatorname{Re}\mathscr{H}^1} \leq \frac{1}{c} \tag{50}$$

holds for every atom $a(x)$.

Let $I$ be a generalized interval such that

$$\operatorname{supp} a(x) \subset I, \quad \int_I a(x)\,dx = 0, \qquad \|a\|_{L^\infty(-\pi,\pi)} \leq |I|^{-1} \tag{51}$$

(the case $a(x) \equiv 1/2\pi$ is trivial). Since $\int_{-\pi}^\pi |a(x)|\,dx \leq 1$, it remains to prove (see (45)) that

$$\int_{-\pi}^\pi |\tilde{a}(x)|\,dx \leq C'. \tag{52}$$

For every measurable set $E \subset [-\pi, \pi]$, if we apply Cauchy's inequality and use Theorem 3 and (51), we obtain

$$\int_E |\tilde{a}(x)|\,dx \leq [m(E)]^{1/2}\left\{\int_E |\tilde{a}(x)|^2\,dx\right\}^{1/2}$$
$$\leq [m(E)]^{1/2}\|\tilde{a}\|_{L^2(-\pi,\pi)} \leq C[m(E)]^{1/2} \cdot |I|^{-1/2}, \tag{53}$$

from which (52) follows directly if $|I| \geq 2\pi/3$. Now suppose that $|I| < 2\pi/3$, and let $I^*$ denote the generalized interval with the same center as $I$. It follows from (53) that

$$\int_{I^*} |\tilde{a}(x)|\,dx \leq C|I^*|^{1/2} \cdot |I|^{-1/2} = \sqrt{3}C.$$

It remains to estimate the integral $\int_{[-\pi,\pi]\setminus I^*} |\tilde{a}(x)|\,dx$. We apply the evident inequality

$$\left|\sin\frac{x-t}{2}\right| \geq c'\rho(x,t), \qquad -\pi \leq x, t \leq \pi,$$

where $\rho(x,t)$ is the length of the smaller of the two arcs of the unit circle that join $e^{ix}$ and $e^{it}$, and $c' > 0$ is an absolute constant. We have, by (51) for $x \in [-\pi, \pi] \setminus I^*$,

$$\tilde{a}(x) = \frac{1}{\pi}\int_{-\pi}^\pi \frac{a(t)}{2\tan\frac{x-t}{2}}\,dt$$
$$= \frac{1}{2\pi}\int_{-\pi}^\pi a(t)\left[\frac{1}{\tan\frac{x-t}{2}} - \frac{1}{\tan\frac{x-\tau}{2}}\right]dt$$
$$= \frac{1}{2\pi}\int_I a(t)\frac{\sin\frac{t-\tau}{2}}{\sin\frac{x-t}{2}\sin\frac{x-\tau}{2}}\,dt,$$

where $\tau$ is the center of $I$. From the preceding relation, using the inequalities $\|a\|_{L^\infty(-\pi,\pi)} \le |I|^{-1}$ and $|\sin u| \le \rho(0, u)$, we deduce that

$$|\tilde{a}(x)| \le C \int_I \frac{|I|^{-1}\rho(\frac{t}{2}, \frac{\tau}{2})}{\rho(x,t)\rho(x,\tau)}\, dt \le \frac{C|I|}{[\rho(x,I)]^2},$$

$$x \in [-\pi,\pi] \setminus I^*, \quad \text{where } \rho(x,I) = \inf_{t\in I}\rho(x,t).$$

Consequently

$$\int_{[-\pi,\pi]\setminus I^*} |\tilde{a}(x)|\, dx \le C|I| \int_{[-\pi,\pi]\setminus I^*} [\rho(x,I)]^{-2}\, dx$$

$$\le C|I| \int_{|I|}^{\infty} y^{-2}\, dy \le C.$$

We have therefore proved (52), and consequently (50).

II) For a given function $f \in \text{Re}\,\mathscr{H}^1$, we construct a decomposition (48) for which

$$\sum_k |\lambda_k| \le C\|f\|_{\text{Re}\,\mathscr{H}^1}.$$

Let $F(z) \in H^1$ with $\text{Im}\, F(0) = 0$ satisfy (43), and let $F_\sigma^*(x)$ $(0 < \sigma < 1)$ be the nontangential maximal function for $F(z)$. Recall (see Appendix 2) that $F_\sigma^*(x)$ is defined by

$$F_\sigma^*(x) = \sup_{z\in\Omega_\sigma(x)} |F(z)|, \qquad x \in [-\pi,\pi], \tag{53'}$$

where $\Omega_\sigma(x)$ is a domain bounded by two tangents to the circle $|z| = \sigma$ from points $e^{ix}$, and by the longer arc of $|z| = \sigma$ included between the points of tangency.

Theorem 7 of Appendix 2 states that $\|F^*\|_{L^1(-\pi,\pi)} \le C_\sigma\|F\|_{H^1}$, and therefore (see also (46)) it is enough to find a decomposition of $f(x)$ into atoms (see (48)) such that

$$\sum_k |\lambda_k| \le C\|F_\sigma^*\|_{L^1(-\pi,\pi)}, \tag{54}$$

where the constants $C$ and $\sigma$ $(0 < \sigma < 1)$ are independent of $f$. In order to construct a decomposition (48) satisfying (54), we choose a number $\sigma$; for example, let $\sigma = \frac{1}{10}$. We may suppose, without loss of generality, that

$$\|F_\sigma^*\|_{L^1(-\pi,\pi)} = 1. \tag{55}$$

On the interval $[-\pi,\pi]$, consider the sets

$$E_0 = [-\pi,\pi], \qquad E_k = \{x \in [-\pi,\pi]: F_\sigma^*(x) > 2^k\}, \qquad k = 1,2,\ldots. \tag{56}$$

Since, for each $y > 0$, the set $\{e^{ix}: F_\sigma^*(x) > y\}$ of points of the unit circle is open, it is clear that for $k = 1, 2, \ldots$ the set $E_k$ (if not empty) is representable (in a unique way) as a sum of disjoint generalized intervals:

$$E_k = \bigcup_j I_{k,j}, \qquad I_{k,j} \cap I_{k,i} = \varnothing \quad \text{for } i \neq j,$$

$$i, j, k = 1, 2, \ldots, \qquad E_0 = I_{0,1} = [-\pi, \pi]. \tag{57}$$

Set $h_0(x) = \frac{1}{2\pi} \int_{-\pi}^{\pi} f(t)\, dt$, and for $k = 1, 2, \ldots,$

$$h_k(x) = \begin{cases} |I_{k,j}|^{-1} \int_{I_{k,j}} f(t)\, dt & \text{if } x \in I_{k,j}, \ j = 1, 2, \ldots; \\ f(x) & \text{if } x \notin E_k. \end{cases} \tag{58}$$

Since $F_\sigma^*(x)$ is finite for almost all $x$ (see (55)), it follows from the definition of $h_k(x)$, $k = 1, 2, \ldots,$ that $h(x) = f(x)$ for $k \geq k(x)$, for almost all $x \in [-\pi, \pi]$, and therefore

$$f(x) = h_0(x) + \sum_{k=0}^{\infty} [h_{k+1}(x) - h_k(x)]$$

for almost all $x \in [-\pi, \pi]$. Hence, recalling that $E_{k+1} \subset E_k$, and consequently that (see (58)) $h_{k+1}(x) = h_k(x) = f(x)$ for $x \notin E_k$, we find that

$$f(x) = h_0(x) + \sum_{k=0}^{\infty} \chi_{E_k}(x)[h_{k+1}(x) - h_k(x)], \tag{59}$$

where $\chi_{E_k}(x)$ is the characteristic function of $E_k$. From (59), using the representation $E_k = \bigcup_j I_{k,j}$ (see (57)), we obtain the following expansion of $f(x)$:

$$f(x) = h_0(x) + \sum_{k=0}^{\infty} \sum_j b_{k,j}(x) \quad \text{for almost all } x \in [-\pi, \pi] \tag{60}$$

where

$$b_{k,j}(x) = \chi_{I_{k,j}}(x)[h_{k+1}(x) - h_k(x)], \tag{61}$$

$$k = 0, 1, \ldots, \ j = 1, 2, \ldots.$$

By using the functions $b_{k,j}(x)$ we can now construct the decomposition of the form (48) that we need. We first observe that

$$\int_{I_{k,j}} b_{k,j}(x)\, dx = 0, \qquad \operatorname{supp} b_{k,j}(x) \subset I_{k,j} \tag{62}$$

for $k = 0, 1, \ldots; j = 1, 2, \ldots.$ Now we prove that

$$|h_k(x)| \leq A \cdot 2^k, \qquad k = 0, 1, \ldots, \tag{63}$$

for almost all $x \in [-\pi, \pi]$, where the constant $A = A(\sigma)$ depends only on $\sigma$, which was chosen previously.

Since $|f(x)| \leq F_\sigma^*(x)$ for almost all $x \in [-\pi, \pi]$ (see (43) and (53')), it follows from (55) that

$$|h_0(x)| = \frac{1}{2\pi} \left| \int_{-\pi}^{\pi} f(t)\, dt \right| \leq \frac{1}{2\pi} \int_{-\pi}^{\pi} F_\sigma^*(t)\, dt = \frac{1}{2\pi} < 1.$$

Now let $k = 1, 2, \ldots$ and let $I = I_{k,j}$ be one of the generalized intervals in (57). Then $|I| \leq 2^{-k} \leq \frac{1}{2}$ (see (55) and (56)), and if $e^{i\alpha_1}$, $e^{i\alpha_2}$ are the endpoints of the arc $\theta(I) = \{e^{ix} : x \in I\}$ $(-\pi \leq \alpha_1, \alpha_2 \leq \pi)$, then $\alpha_1, \alpha_2 \notin E_k$ (see (57)), and therefore

$$F_\sigma^*(\alpha_i) \leq 2^k, \qquad i = 1, 2. \tag{64}$$

It follows from (64), according to (53'), that

$$|F(z)| \leq 2^k \quad \text{for } z \in \Omega_\sigma(\alpha_1) \cup \Omega_\sigma(\alpha_2). \tag{65}$$

It is easily seen (since $|I| \leq \frac{1}{2}$ and $\sigma = \frac{1}{10}$) that the sets[5]

$$\partial[\Omega_\sigma(\alpha_1) \setminus \{z : |z| = \sigma\}] \quad \text{and} \quad \partial[\Omega(\alpha_2) \setminus \{z : |z| = \sigma\}]$$

intersect at a point $\rho e^{i\alpha}$ with

$$\sigma < \rho < 1, \qquad \alpha \in I. \tag{66}$$

Let $\gamma_p$, $p = 1, 2$, be line segments joining the points $e^{i\alpha_p}$ and $\rho e^{i\alpha}$. Since $\gamma_p \subset \partial[\Omega_\sigma(\alpha_p)]$, $p = 1, 2$, it follows from the continuity of $F(z)$ for $|z| < 1$] and from (65) that $|F(z)| \leq 2^k$ if $z \in \gamma_p$, $p = 1, 2$, and $|z| < 1$.

Therefore (also see (66))

$$\left| \frac{F(z)}{z} \right| \leq \frac{2^k}{\sigma}, \qquad z \in \gamma_p, p = 1, 2, |z| < 1. \tag{67}$$

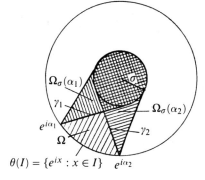

FIGURE 1

---

[5]Here $\partial(\Omega)$ denotes the boundary of $\Omega$.

Consider the domain $\Omega$ (Figure 1) bounded by the line segments $\gamma_1$ and $\gamma_2$ and the arc $\theta(I) = \{e^{ix} : x \in I\}$; also let, for $r \in (\rho, 1)$,

$$\Omega^{(r)} = \Omega \cap \{z : |z| < r\},$$

$$re^{i\alpha_p^{(r)}} = \gamma_p \cap \{z : |z| = r\}, \quad p = 1, 2.$$

By Cauchy's theorem

$$\int_{\partial(\Omega^{(r)})} \frac{F(z)}{z} \, dz = 0.$$

From this and (67), since the equation

$$\int_\Gamma \frac{F(z)}{z} \, dz = i \int_\alpha^\beta F(re^{ix}) \, dx$$

is valid for every arc $\Gamma = \{z = re^{ix} : -\pi < \alpha < x < \beta < \pi\}$, we obtain[6]

$$\left| \int_{(\alpha_1^{(r)}, \alpha_2^{(r)})} F(re^{ix}) \, dx \right| \le \int_{\gamma_1 \cup \gamma_2} \left| \frac{F(z)}{z} \right| d|z| \le \frac{2^k}{\sigma}(|\gamma_1| + |\gamma_2|).$$

But by Theorems 4 and 5 of Appendix 2,

$$\lim_{r \to 1} \int_{-\pi}^\pi |F(re^{ix}) - \Phi(x)| \, dx = 0, \qquad \Phi(x) = f(x) + i\tilde{f}(x),$$

and since $\lim_{r \to 1} \alpha_p^{(r)} = \alpha_p$, $p = 1, 2$, we obtain

$$\left| \int_I \Phi(x) \, dx \right| \le \frac{2^k}{\sigma}(|\gamma_1| + |\gamma_2|). \tag{67'}$$

It is easily seen that the ratio $(|\gamma_1| + |\gamma_2|)/|I|$ is bounded above by a number that depends only on $\sigma$, and therefore (see (67'))

$$\left| \int_I \Phi(x) \, dx \right| \le A|I|2^k, \qquad A = A(\sigma). \tag{68}$$

Since $f(x) = \text{Re}\,\Phi(x)$, it follows from (68) (also see (58)) that (63) holds for $x \in E_k$, $k = 1, 2, \dots$. For almost all $x \notin E_k$, inequality (63) follows directly from the definition of the functions $h_k(x)$ and the sets $E_k$ (also see (43)).

By using (63) and (61), we obtain $\|b_{k,j}\|_{L^\infty(-\pi,\pi)} \le 3 \cdot A \cdot 2^k$. This implies (also see (62)) that the functions

$$a_{k,j}(x) = \frac{1}{3A2^k|I_{k,j}|} b_{k,j}(x), \quad k = 0, 1, \dots, \; j = 1, 2, \dots,$$

_____

[6] Here $(\alpha_1^{(r)}, \alpha_2^{(r)})$ is a generalized interval on $I$, with endpoints $\alpha_1^{(r)}$ and $\alpha_2^{(r)}$.

are atoms. Then by transforming (60) we obtain the decomposition of $f(x)$ into atoms:

$$f(x) = \lambda_0 \cdot \frac{1}{2\pi} + \sum_{k=0}^{\infty} \sum_{j} \lambda_{k,j} a_{k,j}(x)$$

for almost all $x \in [-\pi, \pi]$, where $\lambda_0 = \int_{-\pi}^{\pi} f(t)\, dt$ and $\lambda_{k,j} = 3A2^k |I_{k,j}|$. Let us estimate the sum of the moduli of the coefficients of this decomposition. Using (55) we have

$$|\lambda_0| + \sum_{k=0}^{\infty} \sum_{j} |\lambda_{k,j}| \leq 1 + 3A \sum_{k=0}^{\infty} 2^k \sum_{j} |I_{k,j}|$$

$$= 1 + 3A \sum_{k=0}^{\infty} 2^k m(E_k)$$

$$\leq 1 + 6A \sum_{k=0}^{\infty} 2^k [m(E_k) - m(E_{k+1})]$$

$$\leq 1 + 6A \|F_\sigma^*\|_{L^1(-\pi,\pi)} = 1 + 6A.$$

This establishes (54), and hence Theorem 4.

Let us describe the space $(\mathrm{Re}\,\mathscr{H}^1)^*$ dual to the Banach space $\mathrm{Re}\,\mathscr{H}^1$. We need the following definition.

DEFINITION 4. The space BMO is the set of functions $f \in L^1(-\pi, \pi)$ that satisfy

$$\mathfrak{N}(f) \equiv \sup_{I} \left\{ |I|^{-1} \int_I |f(t) - f_I|\, dt \right\} < \infty, \qquad (69)$$

where $f_I = |I|^{-1} \int_I f(t)\, dt$, and the supremum is taken over all generalized intervals $I \subset [-\pi, \pi]$.

It is easily verified that BMO is a Banach space with norm

$$\|f\|_{\mathrm{BMO}} = \mathfrak{N}(f) + \frac{1}{2\pi} \left| \int_{-\pi}^{\pi} f(t)\, dt \right|. \qquad (70)$$

It is clear that $L^\infty(-\pi, \pi) \subset \mathrm{BMO}$. However, BMO also contains unbounded functions. For example, it is easy to verify that $\ln |x| \in \mathrm{BMO}$.

THEOREM 5. $(\mathrm{Re}\,\mathscr{H}^1)^* = \mathrm{BMO}$, i.e.,

a) if $b(x) \in \mathrm{BMO}$, and if for any $f(x) \in \mathrm{Re}\,\mathscr{H}^1$ we consider its decomposition into atoms (see Theorem 4):

$$f(x) = \sum_{k=0}^{\infty} \lambda_k a_k, \quad \sum_{k=0}^{\infty} |\lambda_k| < \infty, \quad a_k, \ k = 0, 1, \ldots \text{ are atoms}(^7), \qquad (71)$$

---

$(^7)$It is possible that $a_k(x) \equiv 0$ for $k > k_0$.

*and set*

$$L_b(f) = \langle b, f \rangle \equiv \sum_{k=0}^{\infty} \lambda_k \int_{-\pi}^{\pi} b(x) a_k(x) \, dx, \tag{72}$$

*then the sum $L_b(f)$ of the series (72) is finite, independent of the particular decomposition (71), and defines a bounded linear functional on $\operatorname{Re} \mathscr{H}^1$;*

  b) *Every bounded linear functional $L = L(f)$ on $\operatorname{Re} \mathscr{H}^1$ is representable in the form (72), where $b \in$ BMO. Furthermore,*

$$\frac{1}{C}\|L\| \le \|b\|_{\text{BMO}} \le C_1\|L\|$$

*($C$ and $C_1$ are absolute constants).*

REMARK. In general, if $f \in \operatorname{Re} \mathscr{H}^1$ and $b \in$ BMO, the function $f \cdot b$ need not be summable. In fact, let $b(x) = \ln |x| \in$ BMO and

$$f(x) = \begin{cases} \sum_{k=1}^{\infty} \dfrac{2^{k/2}}{k^2} \chi_k^{(2)}(x), & \text{if } x \in (0, 1), \\ 0, & \text{if } x \in [-\pi, \pi] \setminus (0, 1) \end{cases}$$

($\{\chi_k^{(i)}\}$ is the Haar system). Since the functions $a_k(x) = 2^{k/2}\chi_k^{(2)}(x)$, $k = 1, 2, \ldots$, are atoms, we have $f \in \operatorname{Re} \mathscr{H}^1$ by Theorem 4. Moreover, $|f(x)b(x)| > c2^k/k > 0$ for $x \in (2^{-k}, 2^{-k+1})$, $k = 2, 3, \ldots$, i.e., $fb \notin L^1(-\pi, \pi)$.

LEMMA 1. *Let $g \in L^1(-\pi, \pi)$ have the property that for every generalized interval $I$ there is a constant $C_I$ for which*

$$|I|^{-1} \int_I |g(x) - C_I| \, dx \le M,$$

*where $M$ is independent of $I$. Then $g \in$ BMO and $\mathfrak{N}(g) \le 2M$.*

PROOF. For every generalized interval $I$ we have

$$\int_I |g(x) - g_I| \, dx \le \int_I |g(x) - C_I| \, dx + |I||C_I - g_I|$$

$$\le M|I| + \int_I |C_I - g(x)| \, dx \le 2M|I|,$$

from which we obtain the conclusion of Lemma 1 by using (69).

COROLLARY 3. *If $b(x) \in$ BMO then $|b(x)| \in$ BMO and*

$$\mathfrak{N}(|b|) \le 2\mathfrak{N}(b). \tag{73}$$

Corollary 3 follows directly from Lemma 1 if we notice that

$$|I|^{-1} \int_I \left| |b(x)| - |b_I| \right| dx \le |I|^{-1} \int_I |b(x) - b_I| \, dx \le \mathfrak{N}(b)$$

for every generalized interval $I$.

PROOF OF THEOREM 5. a) Let $b \in$ BMO. Set

$$b_N(x) = \begin{cases} N & \text{if } b(x) \ge N, \\ b(x) & \text{if } -N < b(x) < N, \ N = 1, 2, \ldots. \\ -N & \text{if } b(x) \le -N, \end{cases}$$

Since $\mathfrak{N}(f + g) \le \mathfrak{N}(f) + \mathfrak{N}(g)$ (see (69)), by using the equations

$$\begin{aligned} \max(y, z) &= \tfrac{1}{2}[y + z + |y - z|], \\ \min(y, z) &= \tfrac{1}{2}[y + z - |y - z|] \qquad (y, z \in R^1), \\ b_N(x) &= \max\{\min[b(x), N], -N\}, \end{aligned}$$

we can deduce from Corollary 3 that

$$\mathfrak{N}(b_N) \le 3\mathfrak{N}(b), \qquad N = 1, 2, \ldots. \tag{74}$$

Let us suppose that $f \in L^\infty(-\pi, \pi) \subset \operatorname{Re}\mathscr{H}^1$ (see Theorem 3 and (44)). By Theorem 4 there is an expansion

$$f(x) = \sum_{k=0}^\infty \lambda_k a_k(x), \qquad \sum_{k=0}^\infty |\lambda_k| \le C\|f\|_{\operatorname{Re}\mathscr{H}^1}, \tag{75}$$

where $a_k(x)$ are atoms (see Definition 3). We have $a_0(x) \equiv \frac{1}{2\pi}$ and, for $k > 0$,

$$\operatorname{supp} a_k(x) \subset I_k, \qquad \int_{-\pi}^\pi a_k(x) \, dx = 0, \ \|a_k\|_{L^\infty(-\pi,\pi)} \le |I_k|^{-1}. \tag{76}$$

It follows from (74)–(76) that, for $N = 1, 2, \ldots,$

$$
\left| \int_{-\pi}^{\pi} f(x) b_N(x)\, dx \right| = \left| \sum_{k=0}^{\infty} \lambda_k \int_{-\pi}^{\pi} a_k(x) b_N(x)\, dx \right|
$$

$$
\leq \frac{|\lambda_0|}{2\pi} \left| \int_{-\pi}^{\pi} b_N(x)\, dx \right|
$$

$$
+ \sum_{k=1}^{\infty} |\lambda_k| \left| \int_{-\pi}^{\pi} a_k(x)[b_N(x) - (b_N)_{I_k}]\, dx \right|
$$

$$
\leq \frac{|\lambda_0|}{2\pi} \left| \int_{-\pi}^{\pi} b_N(x)\, dx \right|
$$

$$
+ \sum_{k=1}^{\infty} |\lambda_k| |I_k|^{-1} \int_{I_k} |b_N(x) - (b_N)_{I_k}|\, dx
$$

$$
\leq \frac{|\lambda_0|}{2\pi} \left| \int_{-\pi}^{\pi} b_N(x)\, dx \right| + \sum_{k=1}^{\infty} |\lambda_k| \mathfrak{N}(b_N)
$$

$$
\leq \frac{|\lambda_0|}{2\pi} \left| \int_{-\pi}^{\pi} b_N(x)\, dx \right| + 3\mathfrak{N}(b) \sum_{k=1}^{\infty} |\lambda_k|.
$$

Hence, since $f(x) b_N(x)$, $N = 1, 2, \ldots,$ has modulus not exceeding the summable function $|f(x) b(x)|$, and since $\lim_{N \to \infty} f(x) b_N(x) = f(x) b(x)$ for almost all $x \in [-\pi, \pi]$, we obtain

$$
\left| \int_{-\pi}^{\pi} f(x) b(x)\, dx \right| \leq \frac{|\lambda_0|}{2\pi} \left| \int_{-\pi}^{\pi} b(x)\, dx \right| + 3\mathfrak{N}(b) \sum_{k=1}^{\infty} |\lambda_k|
$$

$$
\leq C \|b\|_{\text{BMO}} \|f\|_{\text{Re}\,\mathscr{H}^1}.
$$

Therefore the equation

$$
L_b(f) = \int_{-\pi}^{\pi} f(x) b(x)\, dx, \qquad f \in L^{\infty}(-\pi, \pi), \tag{77}
$$

defines a bounded linear functional on an everywhere dense linear manifold in $\text{Re}\,\mathscr{H}^1$, and evidently belongs to $L^{\infty}(-\pi, \pi)$. (The density of elements of $L^{\infty}(-\pi, \pi)$ in $\text{Re}\,\mathscr{H}^1$ follows, for example, from Theorem 4, since for every function $f \in \text{Re}\,\mathscr{H}^1$, the partial sums of (48) converge to $f$ in the norm of $\text{Re}\,\mathscr{H}^1$ (see (49)).) Therefore the functional $L_b$ has a natural extension to $\text{Re}\,\mathscr{H}^1$:

$$
|L_b(f)| \leq C \|b\|_{\text{BMO}} \|f\|_{\text{Re}\,\mathscr{H}^1}, \qquad f \in \text{Re}\,\mathscr{H}^1. \tag{78}
$$

It remains to prove that, for every expansion of the form (71) of a function $f \in \text{Re}\,\mathscr{H}^1$ the series (72) converges and its sum is $L_b(f)$. This proposition follows directly from (77) and the convergence of the series (71) to $f(x)$ in the norm of Re $\mathscr{H}^1$:

$$L_b(f) = \lim_{N \to \infty} L_b \left( \sum_{k=0}^{N} \lambda_k a_k \right) = \lim_{N \to \infty} \sum_{k=0}^{N} \lambda_k L_b(a_k)$$

$$= \sum_{k=0}^{\infty} \lambda_k \int_{-\pi}^{\pi} a_k(x) b(x) \, dx.$$

b) Let $L$ be any bounded linear functional on Re $\mathscr{H}^1$. Then (see Theorem 3 and (45)), for every $f \in L^2(-\pi, \pi)$,

$$|L(f)| \leq \|L\| \, \|f\|_{\text{Re} \mathscr{H}^1} \leq C \|L\| \, \|f\|_{L^2(-\pi,\pi)}$$

($C$ is an absolute constant). This means that $L$ is a bounded linear functional on $L^2(-\pi, \pi)$. Consequently there is a function $b \in L^2(-\pi, \pi)$ such that

$$\|b\|_{L^2(-\pi,\pi)} \leq C \|L\|, \tag{79}$$

for which

$$L(f) = \int_{-\pi}^{\pi} f(x) b(x) \, dx \qquad \text{if } f \in L^2(-\pi, \pi). \tag{80}$$

In particular, (80) is satisfied if $f(x)$ is any atom. Let us show that

$$\|b\|_{\text{BMO}} \leq C_1 \|L\|. \tag{81}$$

Let $I$ be any generalized interval, and $g \in L^\infty(-\pi, \pi)$ any function satisfying $\|g\|_{L^\infty(-\pi,\pi)} \leq 1$. Then the function

$$a(x) = \tfrac{1}{2} |I|^{-1} \chi_I(x) [g(x) - g_I], \qquad x \in [-\pi, \pi],$$

is an atom, and $\|a\|_{\text{Re} \mathscr{H}^1} \leq C$ by Theorem 4. Therefore

$$C \|L\| \geq \|a\|_{\text{Re} \mathscr{H}^1} \|L\| \geq |L(a)| = \left| \int_{-\pi}^{\pi} a(x) b(x) \, dx \right|$$

$$= \left| \int_{-\pi}^{\pi} a(x) [b(x) - b_I] \, dx \right|$$

$$= \tfrac{1}{2} |I|^{-1} \left| \int_I g(x) [b(x) - b_I] \, dx \right|.$$

If in the preceding inequality we choose $g(x)$ in the best possible way, we find that

$$|I|^{-1} \int_I |b(x) - b_I| \, dx \leq 2C \|L\|,$$

for every generalized interval $I$. Taking account of the inequalities

$$\frac{1}{2\pi} \left| \int_{-\pi}^{\pi} b(x) \, dx \right| \leq \|b\|_{L^2(-\pi,\pi)} \leq C \|L\|$$

(see (79)), we establish (81).

Consequently if $f \in L^\infty(-\pi, \pi) \subset \operatorname{Re} \mathcal{H}^1$, the value of $L(f)$ equals the value of the bounded linear functional $L_b$ on the element $f$ (see (77) and proposition a) of Theorem 5 (already proved)). Since $L^\infty(-\pi, \pi)$ is dense in $\operatorname{Re} \mathcal{H}^1$, we have $L(f) = L_b(f)$ for every $f \in \operatorname{Re} \mathcal{H}^1$. This equation (see also (78) and (81)) completes the proof of Theorem 5.

## §3. The spaces $\mathcal{H}(\Delta)$ and BMO($\Delta$) (nonperiodic case)

In this section we consider nonperiodic analogs of the spaces $\operatorname{Re} \mathcal{H}^1$ and BMO. The results obtained here will be needed in the next chapter for the study of the Franklin orthonormal system.

DEFINITION 5. A real function $a(x), x \in \Delta = [0, 1]$, is a $\Delta$-atom if there is a subinterval $I \subset \Delta$ such that

a') $\operatorname{supp} a(x) \subset I$;

b') $\int_I a(x)\, dx = \int_0^1 a(x)\, dx = 0$;

c') $\|a\|_\infty \le |I|^{-1}$.

The function $a(x) = 1$, $x \in \Delta$, is also a $\Delta$-*atom*.

DEFINITION 6. The space $\mathcal{H}(\Delta)$ is the set of functions $f \in L^1(0, 1)$ representable in the form([8])

$$f(x) = \sum_{k=0}^\infty \lambda_k a_k(x), \qquad \sum_{k=0}^\infty |\lambda_k| < \infty, \tag{82}$$

where $a_k(x)$, $k = 1, 2, \ldots$, are $\Delta$-atoms.

It is easily seen that $\mathcal{H}(\Delta)$ becomes a Banach space if we set

$$\|f\|_{\mathcal{H}(\Delta)} = \inf \sum_{k=0}^\infty |\lambda_k|, \tag{83}$$

where the infimum is taken over all expansions of $f(x)$ in the form (82). Moreover, of course,

$$\|f\|_1 \le \|f\|_{\mathcal{H}(\Delta)}. \tag{84}$$

DEFINITION 7. BMO($\Delta$) is the set of functions $f \in L^1(0, 1)$ that satisfy

$$\mathfrak{N}_\Delta(f) \equiv \sup_I \left\{ |I|^{-1} \int_I |f(x) - f_I|\, dx \right\} < \infty,$$

where the supremum is taken over all subintervals $I$ of $\Delta$.

BMO($\Delta$) is a Banach space with the norm

$$\|f\|_{\mathrm{BMO}(\Delta)} = \mathfrak{N}_\Delta(f) + \left| \int_0^1 f(x)\, dx \right|.$$

---

([8])The series (82) converges in the norm of $L^1(0, 1)$, and also a.e.

Let Re $\mathscr{H}_+^1$ and BMO$_+$ denote the spaces of even elements of Re $\mathscr{H}^1$ and BMO, respectively. The following proposition and its corollaries are of an auxiliary nature. They allow results from the periodic case to be extended to the nonperiodic case, and conversely.

PROPOSITION 1. *The linear operator $Q: L^1(0,1) \to L^1(-\pi,\pi)$ defined by*

$$Q(f,x) = \begin{cases} f\left(\dfrac{x}{\pi}\right) & \text{if } x \in [0,\pi], \\ f\left(-\dfrac{x}{\pi}\right) & \text{if } x \in [-\pi,0] \end{cases} \quad f \in L^1(0,1), \qquad (85)$$

*is a one-to-one mapping of $\mathscr{H}(\Delta)$ on Re $\mathscr{H}_+^1$ and of BMO($\Delta$) on BMO$_+$. Moreover,*

1) $c\|f\|_{\mathscr{H}(\Delta)} \leq \|Q(f)\|_{\mathrm{Re}\,\mathscr{H}^1} \leq C\|f\|_{\mathscr{H}(\Delta)}$, $f \in \mathscr{H}(\Delta)$;
2) $c\|g\|_{\mathrm{BMO}(\Delta)} \leq \|Q(g)\|_{\mathrm{BMO}} \leq C\|g\|_{\mathrm{BMO}(\Delta)}$, $g \in \mathrm{BMO}(\Delta)$ (*C and* $c > 0$ *are absolute constants*).

PROOF. Let $f \in \mathscr{H}(\Delta)$, $q(x) = Q(f,x)$, and $\varepsilon > 0$. Using the definition of the norm in $\mathscr{H}(\Delta)$ (see (83)), we find the decomposition of $f(x)$ into $\Delta$-atoms:

$$f(x) = \lambda_0 + \sum_{k=1}^{\infty} \lambda_k a_k(x), \qquad \sum_{k=0}^{\infty} |\lambda_k| \leq \|f\|_{\mathscr{H}(\Delta)} + \varepsilon$$

$$\left( \int_0^1 a_k(x)\,dx = 0, k = 1,2,\dots \right).$$

Then when $x \in [-\pi,\pi]$

$$q(x) = \lambda_0 + \sum_{k=1}^{\infty} \lambda_k Q(a_k,x)$$

$$= 2\pi\lambda_0 \cdot \frac{1}{2\pi} + \pi \sum_{k=1}^{\infty} [\lambda_k b_k(x) + \lambda_k b_k(-x)],$$

where $b_k(x) = \frac{1}{\pi}\chi_{[0,\pi]}(x)Q(a_k,x)$, $k = 1,2,\dots$. It follows directly from Definitions 3 and 6 that the functions $b_k(x)$ and $b_k(-x)$, $x \in [-\pi,\pi]$, $k = 1,2,\dots$, are atoms. Consequently, by Theorem 4,

$$\|q\|_{\mathrm{Re}\,\mathscr{H}^1} \leq \frac{2\pi}{c} \sum_{k=0}^{\infty} |\lambda_k| \leq \frac{2\pi}{c}(\|f\|_{\mathscr{H}(\Delta)} + \varepsilon)$$

(*c* is the same constant as in (49)). Since $\varepsilon > 0$ can be arbitrarily small, it follows from the preceding inequality that

$$\|Q(f)\|_{\mathrm{Re}\,\mathscr{H}^1} \leq C\|f\|_{\mathscr{H}(\Delta)} \qquad \left( C = \frac{2\pi}{c} \right). \qquad (86)$$

Now let $q(x) \in \operatorname{Re} \mathscr{H}_+^1$. By using Theorem 4, we obtain the decomposition

$$q(x) = \frac{\lambda_0}{2\pi} + \sum_{k=1}^{\infty} \lambda_k u_k(x), \qquad \sum_{k=0}^{\infty} |\lambda_k| \leq C \|q\|_{\operatorname{Re}\mathscr{H}^1},$$

where $u_k(x)$, $x \in [-\pi, \pi]$, $k = 1, 2, \ldots$, are atoms $(\int_{-\pi}^{\pi} u_k(x)\,dx = 0)$.
Since $q(x)$ is an even function, we have for $x \neq 0, \pm\pi$,

$$q(x) = \frac{1}{2}[q(x) + q(-x)]$$

$$= \frac{\lambda_0}{2\pi} + \frac{1}{2}\sum_{k=1}^{\infty} \lambda_k [u_k(x) + u_k(-x)]$$

$$= \frac{\lambda_0}{2\pi} + \sum_{k=1}^{\infty}[\lambda_k b_k(x) + \lambda_k b_k(-x)],$$

$$b_k(x) = \frac{1}{2}\chi_{(0,\pi)}(x)[u_k(x) + u_k(-x)], \qquad k = 1, 2, \ldots.$$

It is easily seen that, for every atom $u_k(x)$ $(u_k \neq \frac{1}{2\pi})$, the function $\frac{1}{2}\chi_{(0,\pi)}[u_k(x) + u_k(-x)]$ is also an atom; hence the function $a_k(x) = b_k(\pi x)$, $x \in [0, 1]$, $k = 1, 2, \ldots$, is a $\Delta$-atom.
Set

$$f(x) = \frac{\lambda_0}{2\pi} + \sum_{k=1}^{\infty} \lambda_k a_k(x), \qquad x \in [0, 1].$$

Then $f \in \mathscr{H}(\Delta)$, $q(x) = Q(f, x)$ $(x \neq 0, \pm\pi)$, and

$$\|f\|_{\mathscr{H}(\Delta)} \leq \sum_{k=0}^{\infty} |\lambda_k| \leq C\|q\|_{\operatorname{Re}\mathscr{H}^1}. \tag{87}$$

Combining (86) and (87), we obtain relation 1). Relation 2) follows immediately from the fact that, for every function $b \in \operatorname{BMO}_+$,

$$\mathfrak{N}(b) \leq \mathfrak{N}(b\chi_{(0,\pi)}) + \mathfrak{N}(b\chi_{(-\pi,0)})$$

$$\leq 2 \sup_{I \subset (0,\pi)} \left\{ |I|^{-1} \int_I |b(x) - b_I|\,dx \right\},$$

where the supremum is taken over all intervals $I$ contained in $(0, \pi)$. This completes the proof of Proposition 1.

COROLLARY 4. $(\mathscr{H}(\Delta))^* = \operatorname{BMO}(\Delta)$ (here the equation is to be understood as in Theorem 5).

PROOF. From the very definition of $\mathscr{H}(\Delta)$, the bounded functions are dense in $\mathscr{H}(\Delta)$. Consequently it is enough to prove that

a) for every $b \in \text{BMO}(\Delta)$

$$\int_0^1 f(x)b(x)\,dx \leq C\|f\|_{\mathcal{H}(\Delta)}\|b\|_{\text{BMO}(\Delta)}, \qquad f \in L^\infty(0,1);$$

b) for every bounded linear functional $L = L(f)$ on $\mathcal{H}(\Delta)$ there is a function $b(x) \in \text{BMO}(\Delta)$ with $\|b\|_{\text{BMO}(\Delta)} \leq C'\|L\|$ such that

$$L(f) = \int_0^1 f(x)b(x)\,dx, \quad \text{if } f \in L^\infty(0,1)$$

($C$ and $C'$ are absolute constants).

If we use Theorem 5 and Proposition 1 (see, in particular, (77) and (85)), we have

$$\int_0^1 f(x)b(x)\,dx = \frac{1}{2\pi}\int_{-\pi}^\pi Q(f,x)Q(b,x)\,dx$$

$$\leq \frac{1}{c}\|Q(f)\|_{\text{Re}\,\mathcal{H}^1}\|Q(b)\|_{\text{BMO}}$$

$$\leq C\|f\|_{\mathcal{H}(\Delta)}\|b\|_{\text{BMO}(\Delta)},$$

from which a) follows.

We notice in addition that, when $q \in \text{Re}\,\mathcal{H}^1$ and $b \in \text{BMO}$,

$$\begin{aligned}\|q(x) + q(-x)\|_{\text{Re}\,\mathcal{H}^1} &\leq 2\|q\|_{\text{Re}\,\mathcal{H}^1}, \\ \|b(x) + b(-x)\|_{\text{BMO}} &\leq 2\|b\|_{\text{BMO}}.\end{aligned} \tag{88}$$

If $L(f)$ is a bounded linear functional on $\mathcal{H}(\Delta)$, then by Proposition 1 and (88) the functional

$$\tilde{L}(q) = L(f), \qquad q \in \text{Re}\,\mathcal{H}^1, \qquad Q(f,x) = \tfrac{1}{2}[q(x) + q(-x)]$$

is a bounded linear functional on $\text{Re}\,\mathcal{H}^1$. Consequently, by Theorem 5 (see (80) and (81)) the following equation holds for $q \in L^\infty(-\pi, \pi)$:

$$\tilde{L}(q) = \int_{-\pi}^\pi b(x)q(x)\,dx, \qquad \|b\|_{\text{BMO}} \leq C\|\tilde{L}\| \leq C_1\|L\|.$$

Therefore, for $q \in \text{Re}\,\mathcal{H}_+^1 \cap L^\infty(-\pi, \pi)$,

$$\tilde{L}(q) = \int_{-\pi}^\pi q(x)\frac{b(x) + b(-x)}{2}\,dx = \int_{-\pi}^\pi q(x)b'(x)\,dx,$$

where $b'(x) \in \text{BMO}$ and $\|b'\|_{\text{BMO}} \leq C_2\|L\|$. Finally, if we again use Proposition 1, we obtain

$$L(f) = \tilde{L}(Q(f)) = \int_{-\pi}^\pi Q(f,x)b'(x)\,dx$$

$$= \int_{-\pi}^\pi Q(f,x)Q(b'',x)\,dx = 2\pi\int_0^1 f(x)b''(x)\,dx \qquad (f \in L^\infty(0,1)),$$

where $\|b''\|_{\text{BMO}(\Delta)} \leq C'\|L\|$. This completes the proof of Corollary 4.

The next corollary is an immediate consequence of Corollary 4 and the Hahn–Banach theorem.

COROLLARY 5. *For every $f \in \mathscr{H}(\Delta)$*

$$c\|f\|_{\mathscr{H}(\Delta)} \leq \sup_{\|b\|_{\mathrm{BMO}(\Delta)} \leq 1} \langle b, f \rangle \leq C\|f\|_{\mathscr{H}(\Delta)},$$

*where $c > 0$ and $C$ are absolute constants.*

REMARK. If $f \in L^\infty(0,1)$, the value of $\langle b, f \rangle$ for every $b(x) \in \mathrm{BMO}(\Delta)$ is equal to the integral $\int_0^1 f(x)b(x)\,dx$ (see the proof of Corollary 4).

COROLLARY 6. *Let $\{f_n(x)\}_{n=1}^\infty$, $x \in [0,1]$, be an O.N.S. with $f_1 \equiv 1$, $f_n \in \mathrm{Lip}\,1$, $n = 1, 2, \ldots$.*
*If $\{f_n\}_{n=1}^\infty$ is a basis in $\mathscr{H}(\Delta)$ then*
*1) the system $\Phi = \{\varphi_1(x), \varphi_n(x), \tilde{\varphi}_n(x)\}$, $x \in [-\pi, \pi]$, with (see (85))*

$$\varphi_n(x) = \frac{1}{\sqrt{2\pi}} Q(f_n, x), \qquad x \in [-\pi, \pi], \ n = 1, 2, \ldots, \qquad (89)$$

*is a basis for $\mathrm{Re}\,\mathscr{H}^1$;*
*2) the system $\mathscr{F} = \{F_n(z)\}_{n=1}^\infty$, $|z| < 1$, with*

$$F_1(z) \equiv \frac{1}{\sqrt{2\pi}},$$

$$F_n(re^{i\theta}) = \frac{1}{2\sqrt{2\pi}} \int_{-\pi}^{\pi} [\varphi_n(t+\theta) + i\tilde{\varphi}_n(t+\theta)]P_r(t)\,dt, \qquad n = 2, 3, \ldots \quad (90)$$

*is a basis for $\mathscr{H}^1$.*

Here $P_r(t)$ is the Poisson kernel; see Appendix 2.(5).

PROOF. We first observe that the system $\{f_n(x), \tilde{f}_n(x)\}_{n=1}^\infty$ is biorthonormal (since $f_n \in L^\infty(0,1) \subset \mathrm{BMO}(\Delta)$, $n = 1, 2, \ldots$). Therefore (see Proposition 1.3) the coefficients of the Fourier series of $f \in \mathscr{H}(\Delta)$ in the basis $\{f_n\}$ are the same as its Fourier coefficients.

It is easily seen (see Corollary 1 and Remark 2 in Appendix 2) that the systems $\{\varphi_n\}_{n=1}^\infty$ and $\{\tilde{\varphi}_n\}_{n=1}^\infty$, as well as the complex system

$$\left\{ \frac{1}{\sqrt{2}}\varphi_n + \frac{i}{\sqrt{2}}\tilde{\varphi}_n \right\}_{n=1}^\infty$$

are orthonormal. Moreover, all the functions $\varphi_n(x)$, $n = 1, 2, \ldots$, are orthogonal to the functions $\tilde{\varphi}_n(x)$, $n = 2, 3, \ldots$, since the functions $\varphi_n(x)$ are even and the functions $\tilde{\varphi}(x)$ are odd (see (40)). This means that the system $\Phi$ is an O.N.S. Finally, we observe that since $\varphi_n \in \mathrm{Lip}\,1$, it follows immediately from the definition (40) that $\tilde{\varphi}_n(x) \in L^\infty(-\pi, \pi)$, $n = 2, 3, \ldots$.

Let us show that, for every function $q \in \operatorname{Re} \mathcal{H}^1$,

$$q(x) \overset{\operatorname{Re}\mathcal{H}^1}{=} a_1(q)\varphi_1(x) + \sum_{n=2}^{\infty} a_n(q)\varphi_n(x) + b_n(q)\tilde{\varphi}_n(x), \tag{91}$$

where

$$a_n(q) = \int_{-\pi}^{\pi} q(x)\varphi_n(x)\,dx, \qquad b_n(q) = \int_{-\pi}^{\pi} q(x)\tilde{\varphi}_n(x)\,dx.$$

Let $q \in \operatorname{Re}\mathcal{H}^1$ and $\int_{-\pi}^{\pi} q(x)\,dx = 0$. Set

$$q_+(x) = \tfrac{1}{2}[q(x) + q(-x)], \qquad q_-(x) = \tfrac{1}{2}[q(x) - q(-x)];$$

then

$$q_+ \in \operatorname{Re}\mathcal{H}_+^1, \qquad \tilde{q}_- \in \operatorname{Re}\mathcal{H}_+^1, \qquad q = q_+ + q_-. \tag{92}$$

Since $q_+$ and $\varphi_n$, $n = 1, 2, \ldots$, are even functions, and $q_-$ and $\tilde{\varphi}_n$, $n = 2, 3, \ldots$, are odd (also see Corollary 1 in Appendix 2), we have

$$a_n(q) = \int_{-\pi}^{\pi} q_+(x)\varphi_n(x)\,dx = a_n(q_+),$$
$$b_n(q) = \int_{-\pi}^{\pi} q_-(x)\tilde{\varphi}_n(x)\,dx = -\int_{-\pi}^{\pi} \tilde{q}_-(x)\varphi_n(x)\,dx = -a_n(q). \tag{93}$$

But it follows from Proposition 1 and the basis property of $\{f_n\}$ in $\mathcal{H}(\Delta)$ that, for every $g \in \operatorname{Re}\mathcal{H}_+^1$,

$$g(x) \overset{\operatorname{Re}\mathcal{H}^1}{=} \sum_{n=1}^{\infty} a_n(g)\varphi_n(x)$$

and, by the definition of the norm in $\operatorname{Re}\mathcal{H}^1$,

$$\tilde{g}(x) \overset{\operatorname{Re}\mathcal{H}^1}{=} \sum_{n=1}^{\infty} a_n(g)\tilde{\varphi}_n(x) = \sum_{n=2}^{\infty} a_n(g)\tilde{\varphi}_n(x).$$

Therefore

$$\sum_{n=1}^{\infty} a_n(q_+)\varphi_n(x) \overset{\operatorname{Re}\mathcal{H}^1}{=} q_+(x), \qquad \sum_{n=2}^{\infty} a_n(\tilde{q}_-)\tilde{\varphi}_n(x) = (\widetilde{\tilde{q}_-}) = -q_-(x),$$

from which (see (92) and (93)) we obtain (91) and the first conclusion of Corollary 6.

To establish the second conclusion, it is enough, by Theorems 4 and 5 of Appendix 2, to show that the following equation is valid for every $q \in \operatorname{Re}\mathcal{H}^1$ with $\int_{-\pi}^{\pi} q(x)\,dx = 0$:

$$q(x) + i\tilde{q}(x) \overset{L^1}{=} \sum_{n=2}^{\infty} c_n(q)\frac{1}{\sqrt{2}}[\varphi_n(x) + i\tilde{\varphi}_n(x)], \tag{94}$$

where

$$c_n(q) = \frac{1}{\sqrt{2}} \int_{-\pi}^{\pi} [q(x) + i\tilde{q}(x)][\varphi_n(x) - i\tilde{\varphi}_n(x)]\,dx.$$

Since

$$c_n(q) = \frac{1}{\sqrt{2}} \int_{-\pi}^{\pi} [(q\varphi_n + \tilde{q}\tilde{\varphi}_n) + i(\tilde{q}\varphi_n - q\tilde{\varphi}_n)]\,dx$$

$$= \sqrt{2} \int_{-\pi}^{\pi} (q\varphi_n - iq\tilde{\varphi}_n)\,dx = \sqrt{2}[a_n(q) - ib_n(q)],$$

then, because

$$a_n(q) = \int_{-\pi}^{\pi} q\varphi_n\,dx = \int_{-\pi}^{\pi} \tilde{q}\tilde{\varphi}_n\,dx = b_n(\tilde{q}),$$

$$b_n(q) = \int_{-\pi}^{\pi} q\tilde{\varphi}_n\,dx = -\int_{-\pi}^{\pi} \tilde{q}\varphi_n\,dx = -a_n(q),$$

we obtain

$$c_n(q)\frac{1}{\sqrt{2}}[\varphi_n + i\tilde{\varphi}_n] = [a_n(q) - ib_n(q)][\varphi_n + i\tilde{\varphi}_n]$$

$$= a_n(q)\varphi_n + b_n(q)\tilde{\varphi}_n + i[a_n(q)\tilde{\varphi}_n - b_n(q)\varphi_n]$$

$$= a_n(q)\varphi_n + b_n(q)\tilde{\varphi}_n + i[a_n(\tilde{q})\varphi_n + b_n(\tilde{q})\tilde{\varphi}_n],$$

from which (94) follows by (91). This completes the proof of Corollary 6.

The following result shows that $\mathcal{H}(\Delta)$ contains all functions whose Fourier–Haar series have a summable majorant for their partial sums.

THEOREM 6. *If* $f \in L^*(0,1)$ *(see 3.(55)), then* $f \in \mathcal{H}(\Delta)$ *and*

$$\|f\|_{\mathcal{H}(\Delta)} \leq 13\|S^*(f)\|_1,$$

*where*

$$S^*(f,x) = \sup_{1 \leq N < \infty} \left| \sum_{n=1}^{N} c_n(f)\chi_n(x) \right|,$$

$$c_n(f) = \int_0^1 f(t)\chi_n(t)\,dt, \qquad n = 1, 2, \ldots.$$

PROOF. (Notice that here we use the construction introduced in the proof of Theorem 3.11.) We may suppose without loss of generality that $\|S^*(f)\|_1 = 1$.

We define the following sets:

$$G_0 = (0,1), \qquad G_k = \{x \in (0,1): S^*(f,x) > 2^k\}, \qquad k = 1, 2, \ldots.$$

In addition, let $h_0(x) = \int_0^1 f(t)\,dt$, and for $k = 1, 2, \ldots$ let $h_k(x)$ be the function $g'(x)$ constructed in the proof of Theorem 3.11 with $t = 2^k$ (see 3.(53) and 3.(52)).

It follows from the properties of $g'(x)$ (see 3.(52)–3.(54)) that

a) there are disjoint dyadic intervals $\delta_{k,j}$, $j = 1, 2, \ldots$, such that, for almost all $x \in (0, 1)$, we have the equation

$$h_k(x) = \begin{cases} f(x) & \text{if } x \notin E_k = \bigcup_j \delta_{k,j}, \\ |\delta_{k,j}|^{-1} \int_{\delta_{k,j}} f(t)\,dt & \text{if } x \in \delta_{k,j} \end{cases} \tag{95}$$

(the set $(0, 1) \setminus E_k$ coincides, up to a countable set, with the set $E$ defined in 3.(53′); also see 3.(45) and 3.(11′));

b) $\|h_k\|_\infty \leq 2^k$;

c) $E_{k+1} \subset E_k$;

d) $m(E_k) \leq 2m(G_k)$.

Since $m(G_k) \to 0$ as $k \to \infty$ and $h_k(x) = f(x)$ for $x \in (0, 1) \setminus E_k$, it follows from c) and d) that, for almost all $x \in (0, 1)$,

$$f(x) = h_0(x) + \sum_{k=0}^\infty [h_{k+1}(x) - h_k(x)]$$

$$= h_0(x) + \sum_{k=0}^\infty \chi_{E_k}(x)[h_{k+1}(x) - h_k(x)]. \tag{96}$$

Recalling that $E_k = \bigcup_j \delta_{k,j}$, $k = 1, 2, \ldots$, we find from (96) that

$$f(x) = h_0(x) + \sum_{k=0}^\infty \sum_j b_{k,j}(x) \qquad x \in (0, 1), \tag{97}$$

where $b_{k,j}(x) = \chi_{\delta_{k,j}}(x)[h_{k+1}(x) - h_k(x)]$. By a) and b) the functions $b_{k,j}(x)$ have the following properties:

1) $\operatorname{supp} b_{k,j}(x) \subset \delta_{k,j}$;

2) $\int_0^1 b_{k,j}(x)\,dx = 0$;

3) $\|b_{k,j}\|_\infty \leq 3 \cdot 2^k$,

i.e., the functions $a_{k,j}(x) = (3 \cdot 2^k |\delta_{k,j}|)^{-1} b_{k,j}(x)$ are $\Delta$-atoms. Then the following decomposition is a consequence of (97):

$$f(x) = \lambda_0 + \sum_{k=0}^\infty \sum_j \lambda_{k,j} a_{k,j}(x),$$

$$\lambda_0 = \int_0^1 f(t)\,dt, \qquad \lambda_{k,j} = 3 \cdot 2^k |\delta_{k,j}|.$$

By a) and d),

$$\sum_{k=0}^{\infty}\sum_{j}|\lambda_{k,j}| = 3\sum_{k=0}^{\infty}2^k\sum_{j}|\delta_{k,j}| = 3\sum_{k=0}^{\infty}2^k m(E_k)$$

$$\leq 6\sum_{k=0}^{\infty}2^k m(G_k) \leq 12\sum_{k=0}^{\infty}2^k[m(G_k) - m(G_{k+1})]$$

$$\leq 12\|S^*(f)\|_1 = 12.$$

The conclusion of Theorem 6 is a consequence of the preceding inequality if we recall that $|\lambda_0| \leq 1$.

In concluding this chapter, we establish a property of elements of BMO($\Delta$). It is evident that BMO($\Delta$) $\supset L^\infty(\Delta)$. On the other hand, BMO($\Delta$) $\subset L^q(\Delta)$ for every $q < \infty$ (this follows, for example, from Theorem 5). By using the method that we applied in Chapter 3 to study the Haar system, we can obtain a more precise result.

THEOREM 7. *For every $f \in$ BMO($\Delta$) and every interval $I \subset \Delta$, we have the inequality*

$$m\{x \in I: |f(x) - f_I| > y\} \leq B|I| \exp\left\{\frac{-by}{\mathfrak{N}_\Delta(f)}\right\}, \qquad y > 0, \qquad (98)$$

*where $B > 0$ and $b > 0$ are absolute constants.*

PROOF. Let $I = (a, b) \subset \Delta$. We may suppose without loss of generality that

$$\mathfrak{N}_\Delta(f) = 1, \qquad f_I = |I|^{-1}\int_I f(t)\,dt = 0, \qquad 1 < y < \infty \qquad (99)$$

(since $\mathfrak{N}(f - C) = \mathfrak{N}(f)$, and when $\mathfrak{N}(f) = 1$ and $y \leq 1$, the inequality in Theorem 7 is satisfied with $B = e$ and $b = 1$). For $x \in I$, set

$$\tilde{M}_I(f, x) = \sup_{\delta \equiv x}|\delta|^{-1}\int_\delta |f(t)|\,dt,$$

where the supremum is taken over all intervals $\delta$ of the form

$$\delta_k^i = \left(a + \frac{i-1}{2^k}(b-a), a + \frac{i}{2^k}(b-a)\right),$$

$$k = 0, 1, \ldots, \quad i = 1, \ldots, 2^k, \qquad (100)$$

that contain $x$. If $T: (0, 1) \to (a, b)$ is a similarity transformation (i.e., $T(x) = a + (b-a)z$, $z \in (0, 1)$), it is clear that $T(\Delta_k^i) = \delta_k^i$. Consequently the intervals $\delta_k^i$ preserve properties 1) and 2) of the binary intervals $\Delta_k^i$ (see §3.1), and we have

$$\tilde{M}_I(f, x) = \tilde{M}(f(T), T^{-1}(x)), \qquad x \in I, \ T^{-1}(x) \in (0, 1),$$

where $\tilde{M}(f, x)$ was defined in Chapter 3 (see 3.(29)).

For $y > 1$, let $G(y) = \{x \in I : \tilde{M}_I(f, x) > y\}$.

If $G$ is not empty, then by using Lemma 1 from Theorem 3.7 we can find disjoint intervals $\delta_s$, $s = 1, 2, \ldots$, of the form (100), such that

a) $G(y) = \bigcup_s \delta_s$;

b) $y < |\delta_s|^{-1} \int_{\delta_s} |f(x)| \, dx \le 2y$, $s = 1, 2, \ldots$

(by (99), $|I|^{-1} \int_I |f(x)| \, dx \le \mathfrak{N}_\Delta(f) = 1 < y$, and therefore $G(y) \ne I$).

For $f(x) \in$ BMO, relation b) can be made more precise. Let us show that

c) $y < |\delta_s|^{-1} \int_{\delta_s} |f(x)| \, dx \le y + 2$, $s = 1, 2, \ldots$.

In fact, since $\tilde{M}_I(f, x)$ does not exceed $y$ at the endpoints of $\delta_s$ (see a)), then because $\delta_s$ (together with one of its endpoints) lies in an interval $\omega$ of the form (100) with $|\omega| = 2|\delta_s|$, we obtain $|\omega|^{-1} \int_\omega |f(x)| \, dx \le y$ and

$$|\delta_s|^{-1} \int_{\delta_s} |f(x)| \, dx \le |\delta_s|^{-1} \int_{\delta_s} |f(x) - f_\omega| \, dx + |f_\omega|$$

$$\le 2|\omega|^{-1} \int_\omega |f(x) - f_\omega| \, dx + |\omega|^{-1} \int_\omega |f(x)| \, dx$$

$$\le 2\mathfrak{N}(f) + y = y + 2.$$

We now establish the inequality

d) $m\{G(y + 4)\} \le \frac{1}{2} m\{G(y)\}$, $y > 1$.

We may evidently suppose that $G(y + 4) \ne \varnothing$. Then (see a))

$$G(y) = \bigcup_s \delta_s, \qquad G(y + 4) = \bigcup_j \delta_j',$$

where each interval $\delta_j'$ is contained in an interval $\delta_s$ (since $G(y + 4) \subset G(y)$ and the intervals $\delta_s$ are disjoint). Let

$$E_s = \bigcup_{j : \, \delta_j' \subset \delta_s} \delta_j'.$$

Then by b) (with $y + 4$ in place of $y$)

$$(y + 4)m(E_s) = (y + 4) \sum_{j : \, \delta_j' \subset \delta_s} |\delta_j'| \le \int_{E_s} |f(x)| \, dx.$$

Consequently (see d) and (99)), if $m(E_s) > 0$ we have

$$y + 4 \leq \{m(E_s)\}^{-1} \int_{E_s} |f(x)| \, dx$$

$$\leq \{m(E_s)\}^{-1} \int_{E_s} |f(x) - f_{\delta_s}| \, dx + |f_{\delta_s}|$$

$$\leq \frac{|\delta_s|}{m(E_s)} \cdot |\delta_s|^{-1} \int_{\delta_s} |f(x) - f_{\delta_s}| \, dx + |\delta_s|^{-1} \int_{\delta_s} |f(x)| \, dx$$

$$\leq \frac{|\delta_s|}{m(E_s)} \mathfrak{N}_\Delta(f) + y + 2 = \frac{|\delta_s|}{m(E_s)} + y + 2,$$

i.e., it is always true that $m(E_s) \leq \frac{1}{2}|\delta_s|$, $s = 1, 2, \ldots$. Summing over $s = 1, 2, \ldots$, we obtain d).

The conclusion of Theorem 7 follows easily from d). In fact, according to (99) we need only prove that

$$m\{x \in I : |f(x)| > y\} \leq B|I| \exp(-by), \qquad y > 1. \tag{101}$$

Let $y > 1$, $k = [(y - 1)/4]$, $p = 1 + 4k$. Then $1 \leq p \leq y$, and by d) we have

$$m\{G(y)\} \leq m\{G(p)\} \leq 2^{-(k-1)}|I| \leq B|I|2^{-y/4},$$

from which we obtain (101) by using the inequality $|f(x)| \leq \tilde{M}_I(f, x)$ for almost all $x \in I$ (see 3.(30)). This completes the proof of Theorem 7.

COROLLARY 7. *For every* $f \in L^1(0, 1)$,[9]

$$c\mathfrak{N}_{\Delta,2}(f) \leq \mathfrak{N}_\Delta(f) \leq \mathfrak{N}_{\Delta,2}(f), \tag{102}$$

*where* $c > 0$ *is an absolute constant and*

$$\mathfrak{N}_{\Delta,2}(f) = \sup_{I \subset (0,1)} \left\{ |I|^{-1} \int_I |f(x) - f_I|^2 \, dx \right\}^{1/2} \tag{103}$$

(in (103) the supremum is taken over all intervals $I \subset (0, 1)$).

PROOF. The right-hand inequality in (102) follows directly from Cauchy's inequality: for every interval $I$,

$$|I|^{-1} \int_I |f - f_I| \, dx \leq |I|^{-1}|I|^{1/2} \left\{ \int_I |f - f_I|^2 \, dx \right\}^{1/2} \leq \mathfrak{N}_{\Delta,2}(f),$$

---

[9]It is, of course, possible that all the numbers in (102) are equal to $+\infty$.

i.e., $\mathfrak{N}_\Delta(f) \leq \mathfrak{N}_{\Delta,2}(f)$. Now let $\mathfrak{N}_\Delta(f) = 1$, let $I$ be any interval, and let

$$\lambda(y) = m\{x \in I : |f(x) - f_I| > y\}, \qquad y > 0.$$

Then by (98) (also see equation (2) in Appendix 1)

$$\int_I |f(x) - f_I|^2 \, dx = 2 \int_0^\infty y\lambda(y) \, dy \leq 2B|I| \int_0^\infty y e^{-by} \, dy$$

$$= 2B|I|b^{-1} \int_0^\infty e^{-by} \, dy = 2Bb^{-2} \cdot |I|,$$

i.e., (see (103)), $\mathfrak{N}_{\Delta,2}(f) \leq (2B)^{1/2} \cdot b^{-1}$. This completes the proof of Corollary 7.

# CHAPTER VI

# The Faber–Schauder and Franklin Systems

In this chapter we discuss two important systems: the Faber–Schauder functions and the Franklin functions. The Faber–Schauder system was the first example of a basis in the space of continuous functions on [0, 1]; the Franklin system was the first example of an orthogonal basis in this space. Subsequently the properties of these systems have been quite thoroughly investigated, and not at all in vain. Recently the Faber–Schauder and Franklin systems have found various interesting applications in analysis.

## §1. The Faber–Schauder system

DEFINITION 1. The *Faber–Schauder system* consists of the functions $\Phi = \{\varphi_n(x)\}_{n=0}^{\infty}$, $x \in [0, 1]$, for which

$$\varphi_0(x) = 1, \qquad \varphi_1(x) = x, \ x \in [0, 1],$$

and, for $n = 2^k + i$, $k = 0, 1, \ldots$, $i = 1, \ldots, 2^k$,

$$\varphi_n(x) \equiv \varphi_k^{(i)}(x) = \begin{cases} 0 & \text{if } x \notin \left( \dfrac{i-1}{2^k}, \dfrac{i}{2^k} \right), \\[2mm] 1 & \text{if } x = \dfrac{2_{i-1}}{2^{k+1}}, \\[2mm] \text{linear and continuous on} \\[1mm] \left[ \dfrac{i-1}{2^k}, \dfrac{2i-1}{2^{k+1}} \right] \text{ and } \left[ \dfrac{2i-1}{2^{k+1}}, \dfrac{i}{2^k} \right]. \end{cases} \tag{1}$$

The Faber–Schauder system can also be defined by integrating the Haar functions. In fact, we have

$$\varphi_1(x) = \int_0^x \chi_1(t) \, dt, \qquad \varphi_n(x) = 2\|\chi_n\|_\infty \int_0^x \chi_n(t) \, dt,$$

$$n = 2, 3, \ldots. \tag{2}$$

Let us consider a Faber–Schauder series

$$\sum_{n=0}^{\infty} A_n \varphi_n(x) \equiv A_0 \varphi_0(x) + A_1 \varphi_1(x) + \sum_{k=0}^{\infty} \sum_{i=1}^{2^k} A_{k,i} \varphi_k^{(i)}(x) \tag{3}$$

and suppose that it converges at all points of $[0, 1]$ to a finite function $f(x)$. Let us show that in this case the coefficients $\{A_n\}$ are uniquely determined by $f(x)$; in fact, that

$$A_0 = A_0(f) = f(0), \qquad A_1 = A_1(f) = f(1) - f(0),$$

$$A_n = A_n(f) = A_{k,i}(f) = f\left(\frac{2i-1}{2^{k+1}}\right) - \frac{1}{2}\left[f\left(\frac{i-1}{2^k}\right) + f\left(\frac{i}{2^k}\right)\right], \quad (4)$$

where $n = 2^k + i$, $k = 0, 1, \ldots$; $i = 1, \ldots, 2^k$.

Using the equation (see (1))

$$A_0 = A_0\varphi_0(0) = \sum_{n=0}^{\infty} A_n\varphi_n(0),$$

$$A_0 + A_1 = A_0\varphi_0(1) + A_1\varphi_1(1) = \sum_{n=0}^{\infty} A_n\varphi_n(1),$$

we find that $A_0 = f(0)$, $A_1 = f(1) - f(0)$. Now if $n = 2^k + i$, $k = 0, 1, \ldots$; $i = 1, \ldots, 2^k$, then by (1) we have

$$A_n = \sum_{s=2^k+1}^{2^{k+1}} A_s\varphi_s\left(\frac{2i-1}{2^{k+1}}\right) = S_{2^{k+1}}\left(\frac{2i-1}{2^{k+1}}\right) - S_{2^k}\left(\frac{2i-1}{2^{k+1}}\right), \quad (5)$$

where $S_N(x)$ are the partial sums of (3):

$$S_N(x) = \sum_{n=0}^{N} A_n\varphi_n(x), \qquad N = 0, 1, \ldots.$$

We see from (1) that the functions $\varphi_n(x)$ are zero at the points $x = 1/2^k$, $l = 0, 1, \ldots, 2^k$, if $n > 2^k$. Consequently, since $S_{2^k}(x)$ is linear in each interval $[(i-1)/2^k, i/2^k]$, $i = 1, \ldots, 2^k$, we find from (5) that

$$A_n = S_{2^{k+1}}\left(\frac{2i-1}{2^{k+1}}\right) - \frac{1}{2}\left[S_{2^k}\left(\frac{i-1}{2^k}\right) + S_{2^k}\left(\frac{i}{2^k}\right)\right]$$

$$= f\left(\frac{2i-1}{2^{k+1}}\right) - \frac{1}{2}\left[f\left(\frac{i-1}{2^k}\right) + f\left(\frac{i}{2^k}\right)\right].$$

From (4), which we have proved, it follows in particular that when $N = 1, 2, \ldots$, the equation

$$\sum_{n=1}^{N} a_n\varphi_n(x) \equiv 0 \quad \text{on } [0, 1]$$

is satisfied only when $a_n = 0$ for $n = 0, 1, \ldots, N$, i.e., when the functions $\{\varphi_n\}_{n=0}^{N}$ are linearly independent. The following propositions follow from the definition of $\varphi_n$, $n = 0, 1, \ldots, N$ (see (1)), and from their linear independence.

(A) For $N = 1, 2, \ldots$, the space $G_N = G_N(\Phi)$ of polynomials in the Faber–Schauder system, of the form $P_N(x) = \sum_{n=0}^{N} a_n \varphi_n(x)$, has dimension $N + 1$ and is identical to the space $L_N$ defined as follows:

$$L_1 = \{f \in C(0, 1): f''(x) = 0 \quad \text{for } x \in (0, 1)\},$$

$$L_N = \left\{ f \in C(0, 1): f''(x) = 0 \right.$$

$$\left. \text{for } x \in \left( \bigcup_{s=1}^{2i} \Delta_{k+1}^s \right) \cup \left( \bigcup_{s=i+1}^{2^k} \Delta_k^s \right) \right\} \qquad (6)$$

$$(N = 2^k + i, \ k = 0, 1, \ldots, \ i = 1, \ldots, 2^k).$$

We also notice the following property of the Faber–Schauder system:

(B) For every $f(x)$ and $N \geq 1$ the sum

$$S_N(f, x) = \sum_{n=0}^{N} A_n(f) \varphi_n(x),$$

with coefficients defined by (4), coincides with $f(x)$ on the set $\pi_N$:

$$\pi_1 = \{0, 1\}, \qquad \pi_N = \left\{ \frac{s}{2^k} \right\}_{s=0}^{2^k} \cup \left\{ \frac{2s - 1}{2^{k+1}} \right\}_{s=1}^{i}, \qquad (7)$$

$$N = 2^k + i, \ k = 0, 1, \ldots, \ i = 1, \ldots, 2^k.$$

In fact, let the function $g(x) \in L_N$ have the property that $g(x) = f(x)$ for $x \in \pi_N$. Then, by (A), $g(x)$ is a polynomial in the system $\Phi$: $g(x) = \sum_{n=0}^{N} A_n \varphi_n(x)$, and, as was shown above, $A_n = A_n(g)$, $n = 0, 1, \ldots, N$. But $g(x) = f(x)$ for $x \in \pi_N$; therefore (see (4)) $A_n(g) = A_n(f)$, $n = 0, 1, \ldots$, i.e., $S_N(f, x) \equiv g(x)$ and $S_N(f, x) = f(x)$ for $x \in \pi_N$.

Propositions (A) and (B) immediately imply the uniform convergence of the series $\sum_{n=0}^{\infty} A_n(f) \varphi_n(x)$ to the arbitrary function $f(x)$. The uniqueness of the series representing $f(x)$ was verified earlier (see (3) and (4)). Hence we have established the following theorem.

THEOREM 1. *The Faber–Schauder system is a basis in $C(0, 1)$. Moreover, the coefficients of the expansion*

$$f(x) = \sum_{n=0}^{\infty} A_n \varphi_n(x), \qquad f \in C(0, 1),$$

*are given by* (4), *and the partial sums* $S_N(f, x)$ *of the expansion belong to* $L_N$ *and satisfy*

$$S_N(f, x) = f(x) \quad \text{for } x \in \pi_N, \ N = 1, 2, \ldots. \tag{8}$$

REMARK. It was shown in Chapter 1 (see Theorem 1.9) that an orthonormal basis in $C(0, 1)$ is also a basis in $L^p(0, 1)$, $1 \leq p < \infty$. The example of the Faber–Schauder system shows that the situation can be different for nonorthogonal bases. The Faber–Schauder system is not even minimal in $L^p(0, 1)$ for $1 \leq p < \infty$.

In fact, it is easy to construct, for each $\varepsilon \in (0, 1)$, a polynomial $P(x) = \sum_{n=1}^{N} a_n \varphi_n(x)$ such that $P(x) = 1$ for $x \in [\varepsilon, 1]$ and $0 \leq P(x) \leq 1$ for $x \in [0, \varepsilon]$; and this means that $\|\varphi_0 - P\|_p \leq \varepsilon^{1/p}$.

COROLLARY 1. *Let* $f \in C(0, 1)$. *We have the inequalities*

a) $|A_N(f)| \leq \omega^{(2)}(1/N, f)$, $\quad N = 1, 2, \ldots$, *where*

$$\omega^{(2)}(\delta, f) = \sup_{\substack{0 < h \leq \delta, \\ h \leq x \leq 1-h}} |f(x + h) + f(x - h) - 2f(x)|;$$

b) $\|f - S_N(f)\|_C \leq \omega^{(2)}(1/N, f)$.

PROOF. Inequality a) follows directly from (4). We also notice that the points of $\pi_N$ divide $[0, 1]$ into intervals of length $< 2/N$. Consequently b) will be established (see (8)) if we verify that, for every interval $(\alpha, \beta) \subset (0, 1)$

$$I = \max_{x \in [\alpha, \beta]} |f(x) - \gamma(x)| \leq \omega^{(2)}\left(\frac{\beta - \alpha}{2}, f\right),$$

where

$$\gamma(x) = f(\alpha) + \frac{f(\beta) - f(\alpha)}{\beta - \alpha}(x - \alpha), \qquad x \in [0, 1].$$

Let $u(x) = f(x) - \gamma(x) \in C(0, 1)$. Since $u(\alpha) = u(\beta) = 0$, we can find a point $x_0 \in (\alpha, \beta)$ such that

$$|u(x_0)| = \max_{x \in (\alpha, \beta)} |u(x)| = I.$$

We may suppose without loss of generality that $h = x_0 - \alpha \leq \beta - x_0$. Then $\alpha = x_0 - h < x_0 < x_0 + h \leq \beta$, and we have

$$|u(x_0 + h) + u(x_0 - h) - 2u(x_0)| = |u(x_0 + h) - 2u(x_0)| \geq I.$$

But $\gamma(x)$ is linear; consequently $\gamma(x_0 + h) + \gamma(x_0 - h) - 2\gamma(x_0) = 0$, and since $0 < h < \frac{1}{2}(\beta - \alpha)$, we obtain

$$I \leq |f(x_0 + h) + f(x_0 - h) - 2f(x_0)| \leq \omega^{(2)}\left(\frac{\beta - \alpha}{2}, f\right).$$

This completes the proof of Corollary 1.

It is easy to deduce from inequality a) of Corollary 1 that if $f(x)$ has

$$\omega(\delta, f) = O\left(\frac{1}{\log^{1+\varepsilon}\frac{1}{\delta}}\right)$$

as $\delta \to 0$ ($\varepsilon > 0$), then its Faber–Schauder series converges absolutely (and therefore unconditionally) in the norm of $C(0,1)$. However, the series $\sum_{n=0}^{\infty}|A_n(f)\varphi_n(x)|$ does not converge uniformly for every continuous $f(x)$. This is a consequence of a general result whose proof comes naturally into this section.

THEOREM 2. *There is no unconditional basis in* $C(0,1)$.

PROOF. Let $\{g_n(x)\}_{n=1}^{\infty}$ be a basis in $C(0,1)$ and let $\{\psi_n(x)\}$ be its dual system (the dual space of $C(0,1)$ is the space $V$ of functions of bounded variation; hence $\psi_n \in V$, $n = 1, 2, \ldots$; also see Corollary 1.1).

Choose a point $x_0 \in (0,1)$ at which all the functions $\psi_n(x)$, $n = 1, 2, \ldots$, are continuous. Let $n_0$ be a positive integer such that $n_0^{-1} < \min(x_0, 1-x_0)$. For $n = n_0, n_0 + 1, \ldots$, we obtain

$$\tau_n(x) = \begin{cases} 1 & \text{if } x = x_0, \\ 0 & \text{if } x \in \left[0, x_0 - \frac{1}{n}\right] \cup \left[x_0 + \frac{1}{n}, 1\right] \\ \text{linear and continuous on } \left[x_0 - \frac{1}{n}, x_0\right] \\ \text{and } \left[x_0, x_0 + \frac{1}{n}\right]. \end{cases} \tag{9}$$

It is easily seen (by Leibniz's alternating series theorem) that for every sequence

$$n_0 = n_1 < n_2 < \cdots \tag{10}$$

of integers, the function

$$F_N(x) = \sum_{k=1}^{N} \frac{(-1)^k}{k} \tau_{n_k}(x), \qquad N = 1, 2, \ldots,$$

has a norm not exceeding unity in $C(0,1)$:

$$\|F_N\|_C \leq 1, \qquad N = 1, 2, \ldots. \tag{11}$$

Suppose that we have found a sequence (10) such that the functions $\tau_{n_k}(x)$, $k = 1, 2, \ldots$, can be represented in the form

$$\tau_{n_k}(x) = P_k(x) + \eta_k(x), \qquad \|\eta_k\|_C \leq 2^{-k}, \ k = 1, 2, \ldots, \tag{12}$$

where $P_k(x)$, $k = 1, 2, \ldots$, are "nonintersecting" polynomials in the system $\{g_n\}$:

$$P_k(x) = \sum_{n=l_k+1}^{l_{k+1}} c_n g_n(x), \qquad t_k < t_{k+1}, \; k = 1, 2, \ldots.$$

Then, for $N = 1, 2, \ldots$, if we consider the functions

$$E_N(x) = \sum_{n=t_1+1}^{t_{N+1}} c'_n g_n(x) \equiv \sum_{k=1}^{N} \frac{(-1)^k}{k} P_k(x)$$

$$\left( c'_n = \frac{(-1)^k}{k} c_n, \text{ if } t_k < n \le t_{k+1} \right),$$

it follows from (11) and (12) that

$$\|E_N\|_C = \left\| \sum_{k=1}^{N} \frac{(-1)^k}{k} (\tau_{n_k} - \eta_k) \right\|_C \le \|F_N\|_C + \sum_{k=1}^{N} \|\eta_k\|_C \le 2. \qquad (13)$$

On the other hand, if we set $\varepsilon_n = (-1)^k$ for $t < n \le t_{k+1}$, $k = 1, \ldots, N$, we have, by (9) and (12),

$$\left\| \sum_{n=t_1+1}^{t_{N+1}} \varepsilon_n c'_n g_n(x) \right\|_C \ge \sum_{k=1}^{N} \frac{1}{k} P_k(x_0)$$

$$\ge \sum_{k=1}^{N} \frac{1}{k} \tau_{n_k}(x_0) - \sum_{k=1}^{N} \|\eta_k\|_C \ge \sum_{k=1}^{N} \frac{1}{k} - 1 \ge c \ln N,$$

$$N = 1, 2, \ldots. \qquad (14)$$

By Theorem 1.10 we find from (13) and (14) that $\{g_n\}_{n=1}^{\infty}$ is not an unconditional basis in $C(0, 1)$.

Consequently, in order to complete the proof of Theorem 2 it remains only to construct a sequence $\{n_k\}_{k=1}^{\infty}$ and functions $P_k(x)$ and $\eta_k(x)$, $k = 1, 2, \ldots$, that satisfy (10) and (12). Let us set $n_1 = n_0$ and $t_1 = 0$, and choose (remembering that $\{g_n\}$ is a basis in $C(0, 1)$) a number $t_2$ so that the polynomial

$$P_1(x) = \sum_{n=t_1+1}^{t_2} \langle \psi_n, \tau_{n_1} \rangle g_n(x)$$

satisfies the inequality

$$\|\eta_1\|_C \equiv \|\tau_{n_i} - P_1\|_C \le \tfrac{1}{2}.$$

Now suppose that the numbers $n_1, \ldots, n_k$ and functions $P_1, \ldots, P_k$ have already been constructed. Then, using the continuity of $\psi_n(x)$ at $x_0$, we choose $n_{k+1} > n_k$ so that

$$W = \left\{ \max_{1 \le n \le t_{k+1}} \text{Var}\left( \left[ x_0 - \frac{1}{n_{k+1}}, x_0 + \frac{1}{n_{k+1}} \right], \psi_n \right) \right\}$$

$$\times \sum_{n=1}^{t_{k+1}} \|g_n\|_C \le 2^{-k-2},$$

where $\text{Var}([\alpha, \beta], \psi)$ is the total variation of $\psi(x)$ on $[\alpha, \beta]$. With this choice of $n_{k+1}$ we have the inequality

$$\left\| \sum_{n=1}^{t_{k+1}} \langle \psi_n, \tau_{n_{k+1}} \rangle g_n(x) \right\|_C \le \sum_{n=1}^{t_{k+1}} \left| \int_{x_0-1/n_{k+1}}^{x_0+1/n_{k+1}} \tau_{n_{k+1}}(x) \, d\psi_n(x) \right| \cdot \|g_n\|_C$$

$$\le W \le 2^{-k-2}. \tag{15}$$

Now choose $t_{k+2} > t_{k+1}$ so that

$$\left\| \tau_{n_{k+1}}(x) - \sum_{n=1}^{t_{k+2}} \langle \psi_n, \tau_{n_{k+1}} \rangle g_n(x) \right\|_C \le 2^{-k-2}, \tag{16}$$

and set

$$P_{k+1}(x) = \sum_{n=t_{k+1}+1}^{t_{k+2}} \langle \psi_n, \tau_{n_{k+1}} \rangle g_n(x),$$

$$\eta_{k+1}(x) = \tau_{n_{k+1}}(x) - P_{k+1}(x).$$

It follows from (15) and (16) that $\|\eta_{k+1}\|_C \le 2^{-k-1}$. This completes the proof of Theorem 2.

We now return to the Faber–Schauder system. We have the following theorem.

THEOREM 3. *Let $0 < \alpha < 1$. A necessary and sufficient condition for the series (3) to be the expansion of a function $f \in \text{Lip}\,\alpha$ is that*

$$|A_n| \le Cn^{-\alpha}, \qquad n = 1, 2, \dots. \tag{17}$$

PROOF. If $f \in \text{Lip}\,\alpha$, it follows directly from inequality a) of Corollary 1 that $|A_n(f)| \le Cn^{-\alpha}$, $n = 1, 2, \dots$

Now let there be given a series of the form (3) whose coefficients satisfy (17). Then this series converges absolutely and uniformly (because

$$\sum_{n=2^k+1}^{2^{k+1}} |A_n \varphi_n(x)| \le \max_{2^k < n \le 2^{k+1}} |A_n| \le C2^{-k\alpha}$$

for $k = 0, 1, 2, \ldots$ ), and consequently is the expansion of its sum $f(x)$ in the system $\Phi$. Moreover, when $x, y \in [0, 1]$,

$$|f(x) - f(y)| \leq \sum_{n=0}^{\infty} |A_n| |\varphi_n(x) - \varphi_n(y)|$$

$$= |A_1| |x - y| + \sum_{k=0}^{\infty} \sum_{i=1}^{2^k} |A_{k,i}| |\varphi_k^{(i)}(x) - \varphi_k^{(i)}(y)|. \quad (18)$$

Notice (see (1)) that

$$|\varphi_k^{(i)}(x) - \varphi_k^{(i)}(y)| \leq \min\{1, 2^{k+1}|x - y|\}$$

for every $k = 0, 1, \ldots$, and $x, y \in [0, 1]$. Moreover, since the supports of the functions $\varphi_k^{(i)}(x)$, $i = 1, \ldots, 2^k$, are disjoint for each $k$, it follows that for all $x, y \in [0, 1]$ there are at most two nonzero terms in the sum

$$\sum_{i=1}^{2^k} |A_{k,i}| |\varphi_k^{(i)}(x) - \varphi_k^{(i)}(y)|.$$

If we use these remarks and choose $k_0$ so that $2^{-k_0-1} < |x - y| \leq 2^{-k_0}$, we find from (18) that

$$|f(x) - f(y)| \leq |A_1| |x - y| + \sum_{k=0}^{\infty} 2 \left( \max_{1 \leq i \leq 2^k} |A_{k,i}| \right) \min(1, 2^{k+1}|x - y|)$$

$$\leq |A_1| |x - y| + 2C \sum_{k=0}^{k_0} 2^{-k\alpha} 2^{k+1} |x - y| + 2C \sum_{k=k_0+1}^{\infty} 2^{-k\alpha}$$

$$\leq |A_1| |x - y| + 4C|x - y| \sum_{k=0}^{k_0} 2^{k(1-\alpha)} + C' 2^{-k_0 \alpha}$$

$$\leq K|x - y|^{\alpha}.$$

This completes the proof of Theorem 3.

REMARK. It is easily seen that

$$f(x) = \sum_{k=1}^{\infty} 2^{-k-1} \varphi_k^{(1)}(x) \notin \text{Lip } 1,$$

although evidently $|A_n(f)| \leq n^{-1}$. Therefore the conclusion of Theorem 3 fails when $\alpha = 1$.

The following result, which is useful in problems on "improving" functions by changes of variable (see, for example, Theorem 4.9), shows that for each $f \in C(0, 1)$ we can find a change of variable $\tau(x)$ such that $f \circ \tau(x) \equiv f(\tau(x))$ has a lacunary Faber–Schauder series.[1]

---

[1] That is, $f \circ \tau(x) = \sum_{k=0}^{\infty} A_{n_k} \varphi_{n_k}(x)$, $n_{k+1}/n_k \geq \gamma > 1$, $k = 1, 2, \ldots$.

PROPOSITION 1. *Let $f \in C(0,1)$, $f(0) = f(1) = 0$, and let $1 < k_0 < k_1 < \cdots$ be a sequence of positive integers. There exist integers $i_s$, $1 \le i_s \le 2^{k_s}$, $s = 0, 1, \ldots$, and a function $\tau(x)$, continuous on $[0, 1]$, satisfying*

$$0 = \tau(0) < \tau(x) < \tau(y) < \tau(1) = 1 \quad \text{for } 0 < x < y < 1 \qquad (19)$$

*and such that the expansion of the composite function $F(x) = f \circ \tau(x)$ in a Faber–Schauder series has the form*

$$F(x) = \sum_{s=0}^{\infty} A_{k_s, i_s} \varphi_{k_s}^{(i_s)}(x). \qquad (20)$$

PROOF. By virtue of formula (4), in order to prove Proposition 1 it is sufficient to construct points $\{\{a_k^i\}_{i=0}^{2^k}\}_{k=0}^{\infty}$ and a sequence $\{i_s\}_{s=0}^{\infty}$, $1 \le i_s \le 2^{k_s}$, such that

  a) $a_0^0 = 0$, $a_0^1 = 1$, $a_k^i < a_m^j$ if $i/2^k < j/2^m$, and $a_k^i = a_m^j$ if $i/2^k = j/2^m$;

  b) $\lim_{k \to \infty} \{\max_{1 \le i \le 2^k} (a_k^i - a_k^{i-1})\} = 0$;

  c) $f(a_{m+1}^{2i-1}) = \frac{1}{2}[f(a_m^{i-1}) + f(a_m^i)]$ if $(m, i) \notin \bigcup_{s=0}^{\infty}(k_s, i_s)$.

In fact, if we have constructed such points, and if we then set $\tau(i/2^k) = a_k^i$, $k = 0, 1, \ldots$; $i = 0, 1, \ldots, 2^k$, we can, by a) and b), continue $\tau(x)$ uniquely to the whole interval $[0, 1]$ as a continuous strictly monotonic function (with $\tau(0) = 0$, $\tau(1) = 1$). Moreover, it follows from c) and (4) that the series of functions $F(x)$ has the form (20).

In order to construct the points $\{a_k^i\}$, which we shall do inductively, we must also see that the following condition is satisfied:

  d) for every pair $(m, i)$, $m = 1, 2, \ldots$, $i = 1, \ldots, 2^m$, either

    (I) $f(a_m^{i-1}) \neq f(a_m^i)$, or

    (II) there is an interval $(\alpha_m^i, \beta_m^i) \subset (a_m^{i-1}, a_m^i)$ such that $f(x) \equiv \text{const} = f(a_m^{i-1}) = f(a_m^i)$ for $x \in (\alpha_m^i, \beta_m^i)$.

Set $a_0^0 = a_1^0 = 0$, $a_0^1 = a_1^2 = 1$, and select $a_1^1 \in (0, 1)$ so that $f(a_1^1) \neq 0 = f(0) = f(1)$ (if this is impossible, $f(x) \equiv 0$ on $[0, 1]$, and we may take $\tau(x) \equiv x$).

Now suppose that the points $\{\{a_k^i\}_{i=0}^{2^k}\}$ have been defined in such a way that conditions a), c), and d) are satisfied, and let $(a_m^{l_m-1}, a_m^{l_m})$ be the longest interval $(a_m^{i-1}, a_m^i)$, $i = 1, \ldots, 2^m$ (or one of the longest intervals). We are to construct the points $\{a_{m+1}^i\}_{i=0}^{2^{m+1}}$; for this purpose we begin by setting

$$a_{m+1}^{2i} = a_m^i, \qquad i = 0, 1, \ldots, 2^m. \qquad (21)$$

We divide the construction of the points $\{a_{m+1}^{2i-1}\}_{i=1}^{2^m}$ into two steps.

*Case 1.* $m \notin \{k_s\}_{s=0}^{\infty}$ or $m = k_s$ for some $s$ but $i \neq l_{k_s}$.

If the pair $(m, i)$ satisfies (I), we may, by the continuity of $f(x)$, choose $a_{m+1}^{2i-1}$ so that

$$a_m^{i-1} < a_{m+1}^{2i-1} < a_m^i, \qquad f(a_{m+1}^{2i-1}) = \frac{1}{2}[f(a_m^{i-1}) + f(a_m^i)]. \tag{22}$$

If the pair $(m, i)$ satisfies (II), we take

$$a_{m+1}^{2i-1} = \frac{1}{2}(\alpha_m^i + \beta_m^i).$$

*Case 2.* $(m, i) = (k_s, l_{k_s})$ for some $s$. If the interval

$$\left( a_m^{i-1} + \frac{1}{4}(a_m^i - a_m^{i-1}), \ a_m^i - \frac{1}{4}(a_m^i - a_m^{i-1}) \right)$$

contains a point $\xi$ such that $f(\xi) \neq f(a_m^{i-1})$, $f(\xi) \neq f(a_m^i)$, we take $a_{m+1}^{2i-1} = \xi$. In the opposite case $f(x) \equiv \text{const}$ on this interval (and $f(x) = f(a_m^{i-1})$, or $f(x) = f(a_m^i)$, and we take $a_{m+1}^{2i-1} = \frac{1}{2}(a_m^{i-1} + a_m^i))$.

In either case,

$$\max\{a_{m+1}^{2i-1} - a_{m+1}^{2i-2}, a_{m+1}^{2i} - a_{m+1}^{2i-1}\} \leq \frac{3}{4}(a_m^i - a_m^{i-1}). \tag{23}$$

We have now constructed the points $\{a_{m+1}^i\}_{i=0}^{2^{m+1}}$. Continuing the same process, we define the points $\{\{a_k^i\}_{i=0}^{2^k}\}_{k=0}^\infty$. We also set $i_s = l_{k_s}$ for $s = 0, 1, \ldots$. It follows directly from the construction that conditions a), c), and d) are satisfied (see, in particular, (21) and (22)). Moreover, since for $s = 0, 1, \ldots$ the longest of the intervals $(a_{k_s}^{i-1}, a_{k_s}^i)$, $i = 1, \ldots, 2^{k_s}$, is divided by $a_{k_s+1}^{2i_s-1}$ into two parts whose lengths are less than $\frac{3}{4} \max_{1 \leq i \leq 2^{k_s}} (a_{k_s}^i - a_{k_s}^{i-1})$ (see (23)), we have

$$\lim_{k \to \infty} \left\{ \max_{1 \leq i \leq 2^k} (a_k^i - a_k^{i-1}) \right\} = 0,$$

i.e., condition b) is also satisfied. This completes the proof of Proposition 1.

COROLLARY 2. *For every $f \in C(0, 1)$ with $f(0) = f(1)$ there is a function $\tau(x)$ of the form (19) such that the second-order modulus of continuity of the composition[2] $F(x) = f \circ \tau(x)$ satisfies the condition*

$$\omega_1^{(2)}(\delta, F) = o(\delta) \qquad \text{as } \delta \to 0. \tag{24}$$

---

[2] Here we suppose that $F(x)$ has been continued with period 1 from [0, 1] to the real line. The second-order modulus of continuity of periodic functions is defined by (6′) in Appendix 1.

PROOF. By Proposition 1, it is enough to show that (24) holds for every $F(x) \in C(0, 1)$ of the form

$$F(x) = \sum_{k=0}^{\infty} A_k \varphi_k^{(i_k)}(x) \qquad (1 \leq i_k \leq 2).$$

It is easily verified that

$$\omega_1^{(2)}(\delta, \varphi_k^{(i)}) \leq C \min\{2^k \delta^2, 2^{-k}\}$$

for $k = 0, 1, \ldots$; $i = 1, \ldots, 2^k$; $\delta > 0$ ($C$ is an absolute constant).

Consequently, if we set $k(\delta) = [\frac{1}{2}|\log_2 \delta|]$, $\delta > 0$, we obtain

$$\omega_1^{(2)}(\delta, F) \leq C \left[ \sum_{k=0}^{k(\delta)-1} |A_k| 2^k \delta^2 + \sum_{k=k(\delta)}^{2k(\delta)} |A_k| 2^k \delta^2 + \sum_{k=2k(\delta)+1}^{\infty} |A_k| 2^{-k} \right]$$

$$\leq C \left[ \max_{0 \leq k < \infty} |A_k| \delta^{3/2} + \max_{k(\delta) \leq k \leq 2k(\delta)} |A_k| \delta + \max_{k > 2k(\delta)} |A_k| \delta \right]$$

$$\leq \varepsilon(\delta) \cdot \delta,$$

where

$$\varepsilon(\delta) = C \left[ \max_{0 \leq k < \infty} |A_k| \cdot \delta^{1/2} + 2 \max_{k \geq k(\delta)} |A_k| \right] \to 0$$

as $\delta \to 0$, since $A_k \to 0$ as $k \to \infty$. This completes the proof of Corollary 2.

## §2. Systems of Faber–Schauder type

We consider a sequence of points $\{\{a_k^i\}_{i=0}^{2^k}\}_{k=0}^{\infty}$ such that

$$a_0^0 = 0, \qquad a_0^1 = 1, \qquad a_k^i < a_m^j \text{ if } i/2^k < j/2^m,$$

$$\text{and} \quad a_k^i = a_m^j \text{ if } i/2^k = j/2^m,$$

and use these points to define a *system of Faber–Schauder type* on $[0, 1]$ by setting $\tilde{\varphi}_0(x) \equiv 1$, $\tilde{\varphi}_1(x) \equiv x$, and, for $n = 2^k + i$, $k = 0, 1, \ldots$; $i = 1, \ldots, 2^k$,

$$\tilde{\varphi}_n(x) = \tilde{\varphi}_k^{(i)}(x) = \begin{cases} 0 & \text{if } x \notin (a_k^{i-1}, a_k^i), \\ 1 & \text{if } x = a_{k+1}^{2i-1} \\ \text{linear and continuous on the intervals} \\ [a_k^{i-1}, a_{k+1}^{2i-1}] \text{ and } [a_{k+1}^{2i-1}, a_k^i]. \end{cases}$$

It is clear that when $a_k^i = i/2^k$ for $k = 0, 1, \ldots$; $i = 0, 1, \ldots, 2^k$, the system $\{\tilde{\varphi}_n\}$ is just the Faber–Schauder system.

In just the same way as for the Faber–Schauder system, we can show that the functions $\{\tilde{\varphi}_n\}_{n=0}^N$ ($N = 1, 2, \ldots$) are linearly independent, and

($\tilde{A}$) the space $\tilde{G}_N$ of polynomials of the form $\tilde{P}_N(x) = \sum_{n=0}^{\infty} a_n \tilde{\varphi}_n(x)$
($N = 2^k + i;\ k = 0, 1, \dots;\ i = 1, \dots, 2^k$) is the same as the space

$$\tilde{L}_N = \left\{ f \in C(0,1) \colon f''(x) = 0 \right.$$

$$\left. \text{for } x \in \left[ \bigcup_{s=1}^{2i} (a_{k+1}^{s-1}, a_{k+1}^s) \right] \cup \left[ \bigcup_{s=i+1}^{2^k} (a_k^{s-1}, a_k^s) \right] \right\};$$

($\tilde{B}$) for every $f(x)$ and $N = 2^m + j,\ m = 0, 1, \dots;\ j = 1, \dots, 2^k$, the sum

$$\tilde{S}_N(x) = \tilde{S}_N(f, x) = \sum_{n=0}^{N} \tilde{A}_n(f) \tilde{\varphi}_n(x),$$

coincides with $f(x)$ on the set

$$\tilde{\pi}_N = \{a_m^s\}_{s=0}^{2^m} \cup \{a_{m+1}^{2s-1}\}_{s=1}^{j}.$$

if we define the coefficients $\tilde{A}_n(f)$ by

$$\tilde{A}_0(f) = f(0), \qquad \tilde{A}_1(f) = f(1) - f(0)$$

and, for $N = 2^k + i,\ k = 0, 1, \dots;\ i = 1, \dots, 2^k$, by

$$\tilde{A}_n(f) = \tilde{A}_{k,i}(f)$$
$$= f(a_{k+1}^{2i-1}) - \left[ \frac{a_k^i - a_{k+1}^{2i-1}}{a_k^i - a_k^{i-1}} f(a_k^{i-1}) + \frac{a_{k+1}^{2i-1} - a_k^{i-1}}{a_k^i - a_k^{i-1}} f(a_k^i) \right], \quad (25)$$

In the case when

$$\lambda_k = \max_{1 \le i \le 2^k} (a_k^i - a_k^{i-1}) \to 0 \quad \text{as } k \to \infty \tag{26}$$

we can derive the following result from ($\tilde{A}$) and ($\tilde{B}$):

THEOREM 4. *The system $\{\tilde{\varphi}_n\}_{n=0}^{\infty}$ of Faber–Schauder type, constructed from a sequence $\{\{a_k^i\}_{i=0}^{2^k}\}_{k=0}^{\infty}$ under condition (26), is a basis in $C(0,1)$. Moreover, the coefficients in the expansion*

$$f(x) = \sum_{n=0}^{\infty} \tilde{A}_n(f) \tilde{\varphi}_n(x), \qquad f \in C(0,1),$$

*are determined by the equations (25), and the partial sums $S_N(f, x),\ N > 1$, of this expansion belong to $L_N$ and coincide with $f(x)$ for $x \in \tilde{\pi}_N$.*

COROLLARY 3. *The coefficients and partial sums of the expansion of a continuous function $f(x)$ in the system of Faber–Schauder type satisfy the* *following inequalities for* $N = 2^k + i$; $k = 0, 1, \ldots$; $i = 1, \ldots, 2^k$ :

a) $|\tilde{A}_N(f)| \leq \omega(\lambda_k, f)$;

b) $\|f - \tilde{S}_N(f)\|_C \leq \omega(\lambda_k, f)$,

*where* $\omega(\delta, f)$ *is the modulus of continuity of* $f(x)$, *and the numbers* $\lambda_k$ *are defined in* (26).

### §3. The Franklin system.
### Definition, elementary properties

We determine a Faber–Schauder orthogonal system $\Phi = \{\varphi_n(x)\}_{n=0}^{\infty}$ by the Gram-Schmidt method: Let

$$f_0(x) = \varphi_0(x), \qquad f_n(x) = \sum_{i=0}^{n} \alpha_i^{(n)} \varphi_i(x), \qquad \alpha_n^{(n)} > 0,$$

$$(f_n, \varphi_i) = 0, \qquad (f_n, f_n) = 1, \tag{27}$$

$$n = 0, 1, \ldots, \quad i = 0, 1, \ldots, n - 1,$$

where $(f, g)$ is the scalar product in $L^2(0, 1)$: $(f, g) = \int_0^1 fg\,dx$.

DEFINITION 2. The *Franklin system* is the O.N.S. $F = \{f_n(x)\}_{n=0}^{\infty}$ defined by (27).

It follows from Theorem 1 that $F$ is complete in $C(0, 1)$, and therefore also in $L^p(0, 1)$, $1 \leq p < \infty$. Let us consider the Fourier-Franklin series of continuous functions:

$$f(x) = \sum_{n=0}^{\infty} c_n(f) f_n(x), \qquad c_n(f) = c_n(f, F) = (f, f_n), \quad n = 0, 1, \ldots. \tag{28}$$

As follows from general results (see Proposition 1.5), a partial sum of (28):

$$S_N(f, x) = \sum_{n=0}^{N} c_n(f) f_n(x), \qquad N = 1, 2, \ldots,$$

is identical with the orthogonal projection of $f(x)$ on the space $G_N = G_N(F)$ of polynomials of the form $\sum_0^N a_n f_n(x)$. By the construction of the Franklin system, $G_N(F) = G_N(\varphi) = L_N$ (see (6)). Therefore (see Theorem 1.6), in order to prove that the Franklin system is a basis in $C(0, 1)$ it is enough to verify that the operators of orthogonal projection on $L_N$, considered as operators from $C(0, 1)$ to $C(0, 1)$, have uniformly bounded norms for $N = 1, 2, \ldots$ We are going to investigate the following more general problem.

Let $\pi$ be a partition of $[0, 1]$ into $n$ intervals by the points

$$0 = t_{-1} = t_0 < t_1 < \cdots < t_n = t_{n+1} = 1, \qquad I_k = (t_{k-1}, t_k),$$
$$k = 0, \ldots, n + 1, \tag{29}$$

and let $S_\pi$ be the space of broken lines with vertices at the points $t_k$, $k = 0, 1, \ldots, n$, i.e.,

$$S_\pi = \{f \in C(0, 1): f''(x) = 0 \text{ for } x \in I_k, \ k = 1, \ldots, n\} \qquad (29')$$

(in particular, $G_N = L_N = S_{\pi_N}$ (see (6) and (7))). Also let $P_\pi$ be the operator of orthogonal projection from $L^2(0, 1)$ on $S_\pi$.

THEOREM 5. *For every partition $\pi$ of the form* (29), *and any function $f \in C(0, 1)$,*

$$\|P_\pi(f)\|_C \leq 3\|f\|_C.$$

PROOF. Take the following basis in $S_\pi$:

$$N_j(t) = \begin{cases} 0 & \text{if } t \in [0, 1] \setminus (I_j \cup I_{j+1}), \\ \dfrac{t - t_{j-1}}{\delta_j} & \text{if } t \in \overline{I}_j, \ j = 0, \ldots, n, \\ \dfrac{t_{j+1} - t}{\delta_{j+1}} & \text{if } t \in \overline{I}_{j+1}, \end{cases} \qquad (30)$$

where $\delta_j = |I_j|$, $\overline{I}_j = [t_{j-1}, t_j]$, $j = 1, \ldots, n$; $\overline{I}_0 = \overline{I}_{n+1} = \varnothing$. It is easily seen that

$$N_j(t_s) = \begin{cases} 1 & \text{if } s = j, \\ 0 & \text{if } s \neq j, \end{cases} \quad s, j = 0, 1, \ldots, n,$$

$$\sum_{j=0}^{n} N_j(t) = 1, \qquad t \in [0, 1], \qquad (31)$$

$$(N_j, 1) = \frac{1}{2}(t_{j+1} - t_{j-1}), \quad (N_j, N_j) = \frac{1}{3}(t_{j+1} - t_{j-1}), j = 0, 1, \ldots, n,$$

and that for an arbitrary function $g \in S_\pi$

$$g(t) = \sum_{j=0}^{n} g(t_j) N_j(t), \qquad \|g\|_C = \max_{0 \leq j \leq n} |g(t_j)|. \qquad (32)$$

Let $f \in C(0, 1)$; then (see (32)) for $t \in [0, 1]$

$$P_\pi(f, t) = \sum_{j=0}^{n} b_j N_j(t),$$

$$\text{where } b_j = b_j(f) = P_\pi(f, t_j), \ j = 0, 1, \ldots, n, \qquad (33)$$

and, by the definition of $P_\pi$, for $k = 0, 1, \ldots, n$,

$$(N_k, f) = (N_k, P_\pi(f)) = b_k(N_k, N_k) + \sum_{\substack{j=0 \\ j \neq k}}^{n} b_j(N_k, N_j).$$

If we take $k$ so that $|b_k| = \max_{0 \le j \le n} |b_j|$, we obtain from the preceding inequality, by using (31),

$$|b_k| \left[ (N_k, N_k) - \sum_{\substack{j=0 \\ j \ne k}}^{n} (N_k, N_j) \right] \le |(N_k, f)|,$$

and therefore (since $\sum_{j=0}^{n} N_j(t) \equiv 1$)

$$|b_k|[2(N_k, N_k) - (N_k, 1)] \le \|f\|_C (N_k, 1).$$

Consequently (see (31)–(33))

$$\|P_\pi(f)\|_C = \max_{0 \le j \le n} |b_j| = |b_k| \le \frac{(N_k, 1)}{2(N_k, N_k) - (N_k, 1)} \|f\|_C$$

$$= \frac{\frac{1}{2}(t_{k+1} - t_{k-1})}{\frac{2}{3}(t_{k+1} - t_{k-1}) - \frac{1}{2}(t_{k+1} - t_{k-1})} \|f\|_C = 3\|f\|_C$$

as was to be proved.

The following theorem is, as we noticed above, a consequence of Theorem 5.

THEOREM 6. *The Franklin system is a basis in $C(0, 1)$.*

THEOREM 7. *The Franklin system is a basis in $L^p(0, 1)$, $1 \le p < \infty$. Moreover, the partial sums $S_N(f, x)$ of the Fourier–Franklin series of $f(x) \in L^p(0, 1)$, namely*

$$\sum_{n=0}^{\infty} c_n(f) f_n(x); \quad c_n(f) = \int_0^1 f(x) f_n(x) \, dx, \qquad n = 0, 1, \dots$$

*satisfy*[3]

$$\|f - S_N(f)\|_p \le C \omega_p \left( \frac{1}{N}, f \right), \qquad N = 1, 2, \dots \tag{34}$$

*where $C$ is an absolute constant.*

PROOF. It follows from Theorems 1.9 and 6 that the Franklin system is a basis in $L^p(0, 1)$, $1 \le p < \infty$, and satisfies the inequality

$$\|S_N(f)\|_p \le M\|f\|_p, \qquad f \in L^p(0, 1), \tag{35}$$

where $M$ is an absolute constant (see the proof of Theorem 1.9).

If $f \in L^p(0, 1)$ and $g \in G_N$, then since $S_N(g, x) = g(x)$, we have

$$\|f - S_N(f)\|_p \le \|f - g\|_p + \|g - S_N(f)\|_p$$

$$= \|f - g\|_p + \|S_N(f - g)\|_p \le (M + 1)\|f - g\|_p.$$

---

[3] It is easily seen that (34) is also valid when $p = \infty$, for continuous functions $f(x)$.

Hence in order to prove (34) it is enough to verify that

$$E_N(f) = \inf_{g \in G_N} \|f - g\|_p \leq B\omega_p\left(\frac{1}{N}, f\right). \tag{36}$$

Let us choose a function $f \in L^p(0,1)$ and a number $k$ $(k = 0, 1, 2, \dots)$. It follows from the results of Chapter 3 (see Theorem 3.3 and equation 3.(8)) that

$$\|\theta(x) - f(x)\|_p \leq C\omega_p(2^{-k}, f); \qquad \theta(x) = \theta_i = 2^k \int_{\Delta_k^i} f(t)\, dt$$

for

$$x \in \Delta_k^i = \left(\frac{i-1}{2^k}, \frac{i}{2^k}\right), \qquad i = 1, \dots, 2^k. \tag{37}$$

Let us consider (for $k \geq 1$) the function $g(x) \in G_{2^k} = L_{2^k}$ (see (6)) which is uniquely determined by the conditions

$$g\left(\frac{i}{2^k}\right) = \theta_i \quad \text{for } i = 1, \dots, 2^k;\ g(0) = \theta_1.$$

Then $g(x) = \theta_1$ for $x \in \Delta_k^1$ and

$$g(x) = 2^k(\theta_i - \theta_{i-1})\left(x - \frac{i-1}{2^k}\right) + \theta_{i-1}$$

for $x \in \Delta_k^i$, $i = 2, 3, \dots, 2^k$. Consequently

$$\|g - \theta\|_p^p = \sum_{i=1}^{2^k} \int_{\Delta_k^i} |g(x) - \theta_i|^p\, dx$$

$$= \sum_{i=2}^{2^k} 2^{kp}|\theta_i - \theta_{i-1}|^p \int_{\Delta_k^i} \left|x - \frac{i}{2^k}\right|^p dx$$

$$= \sum_{i=2}^{2^k} 2^{kp}|\theta_i - \theta_{i-1}|^p \frac{1}{p+1} 2^{-k(p+1)}$$

$$= \frac{2^{-k}}{p+1} \sum_{i=2}^{2^k} \left|2^k \cdot \int_{\Delta_k^i} f(t)\, dt - 2^k \cdot \int_{\Delta_k^{i-1}} f(t)\, dt\right|^p$$

$$\leq \frac{1}{p+1} 2^{k(p-1)} \sum_{i=2}^{2^k} \left[\int_{\Delta_k^{i-1}} |f(t + 2^{-k}) - f(t)|\, dt\right]^p.$$

If we apply Hölder's inequality to the last inequality, we obtain (with $1/p + 1/q = 1$)

$$\|g - \theta\|_p^p \leq \frac{1}{p+1} 2^{k(p-1)} \sum_{i=2}^{2^k} 2^{-kp/q} \int_{\Delta_k^{i-1}} |f(t + 2^{-h}) - f(t)|^p \, dt$$

$$\leq \frac{1}{p+1} \int_0^{1-2^{-k}} |f(t + 2^{-k}) - f(t)|^p \, dt \leq \frac{1}{p+1} \omega_p^p(2^{-k}, f).$$

From this and (37), inequality (36) follows for $N = 2^k$, $k = 0, 1, \ldots$. Finally, if $2^k < N < 2^{k+1}$, then by using the inequality $\omega_p(2\delta, f) \leq 2\omega_p(\delta, f)$, $\delta > 0$ (see Appendix 1), we obtain

$$E_N(f) \leq E_{2^k}(f) \leq B'\omega_p(2^{-k}, f) \leq 2B'\omega_p(2^{-k-1}, f) \leq B\omega_p\left(\frac{1}{N}, f\right).$$

This completes the proof of Theorem 7.

## §4. The exponential inequality for the Franklin functions

This section consists of two parts. The first part is auxiliary and is devoted to the study of the basis $\{N_j(t)\}_{j=0}^N$ in the space $S_\pi$ of broken lines corresponding to an arbitrary partition $\pi$ of the form (29) (see (29'), (30)). The second part is devoted to the fundamental Theorem 9, which exhibits the nature of the behavior of the functions in the Franklin system on the interval $[0, 1]$.

**1°.** Let $\pi$ be an arbitrary partition of the form (29) and let $B = \{b_{i,j}\}_{i,j=0}^n$ be the Gram matrix of the system of functions $\{N_j\}_{j=0}^n$, i.e., the matrix in which the element in the $i$th row and $j$th column is defined as follows (see (30) and (31)):

$$b_{i,j} = (N_i, N_j) = \begin{cases} 0 & \text{if } |i - j| \geq 2, \\ \frac{1}{3}(\delta_j + \delta_{j+1}) & \text{if } i = j, \\ \frac{1}{6}\delta_{i+1} & \text{if } i = j - 1, \\ \frac{1}{6}\delta_i & \text{if } i = j + 1. \end{cases} \tag{38}$$

Let $A = \{a_{i,j}\}_{i,j=0}^n$ be the inverse of $B$,[4] and

$$N_j^*(t) = \sum_{k=0}^n a_{k,j} N_k(t), \qquad j = 0, 1, \ldots, n. \tag{39}$$

---

[4] $\det B \neq 0$, since the system of functions $\{N_j\}_{j=0}^n$ is linearly independent (see §3 of Appendix 1).

Then, by definition, for $i, j = 0, 1, \ldots, n$, we have

$$(N_i, N_j^*) = \sum_{k=0}^{n} a_{k,j} b_{i,k} = \begin{cases} 1 & \text{if } i = j, \\ 0 & \text{if } i \neq j. \end{cases} \tag{40}$$

It follows from (40) that $\{N_j, N_j^*\}_{j=0}^{n}$ is a biorthogonal system, and therefore $\{N_j^*\}_{j=0}^{n}$ is a basis in $S_\pi$; and every $g \in S_\pi$ satisfies the equation

$$g(t) = \sum_{j=0}^{n} (g, N_j^*) N_j(t) = \sum_{j=0}^{n} (g, N_j) N_j^*(t). \tag{41}$$

Similarly, if for every function $f \in C(0, 1)$ we use the equation

$$(P_\pi(f), N_j^*) = (f, N_j^*), \qquad (P_\pi(f), N_j) = (f, N_j), \qquad j = 0, 1, \ldots, n,$$

we obtain (see (41))

$$P_\pi(f, t) = \sum_{j=0}^{n} (f, N_j^*) N_j(t) = \sum_{j=0}^{n} (f, N_j) N_j^*(t). \tag{42}$$

In what follows we shall also need another theorem:

THEOREM 8. *For every partition $\pi$ of the form (29), the elements of the matrix $A$ satisfy*

$$|a_{i,j}| \leq \frac{4 \cdot 2^{-|i-j|}}{\max(t_{i+1} - t_{i-1}, t_{j+1} - t_{j-1})}, \qquad i, j = 0, 1, \ldots, n. \tag{43}$$

The conclusion of Theorem 8 follows immediately from two lemmas, which we now prove.

LEMMA 1. *The elements of $A$ satisfy*
1) $a_{i,j} = (-1)^{i+j} |a_{i,j}|, \ i, j = 0, 1, \ldots, n$;
2) $|a_{i,j}| \leq 2^{-|i-j|} \min(a_{i,i}, a_{j,j}), \ i, j = 0, 1, \ldots, n$.

PROOF. We first observe that $a_{i,i} = (N_i^*, N_i^*) > 0, \ i = 0, 1, \ldots, n$ (see (39)). For given $j = 1, \ldots, n$, we have, by (38) and (40),

$$\frac{\delta_1}{3} a_{0,j} + \frac{\delta_1}{6} a_{1,j} = 0, \tag{44}$$

and if $0 < k < j$, then

$$\frac{\delta_k}{6} a_{k-1,j} + \frac{\delta_k + \delta_{k+1}}{3} a_{k,j} + \frac{\delta_{k+1}}{6} a_{k+1,j} = 0. \tag{45}$$

We are going to prove, by induction, the relations

$$(I_k): \begin{cases} \operatorname{sgn} a_{0,j} = -\operatorname{sgn} a_{1,j} = \cdots = (-1)^k \operatorname{sgn} a_{k,j}, \\ |a_{i-1,j}| \leq \tfrac{1}{2} |a_{i,j}|, \qquad i = 1, \ldots, k. \end{cases}$$

The validity of $(I_1)$ follows from (44). Suppose now that $(I_k)$ holds for $1 \leq k \leq j$; we prove $(I_{k+1})$. According to $(I_k)$ we have

$$\frac{\delta_k + \delta_{+1}}{3}|a_{k,j}| \geq \frac{\delta_k + \delta_{k+1}}{3}2|a_{k-1,j}| > \frac{\delta_k}{6}|a_{k-1,j}|,$$

and consequently $\mathrm{sgn}(\delta_k a_{k-1,j}/6 + (\delta_k + \delta_{k+1})a_{k,j}/3) = \mathrm{sgn}\, a_{k,j}$. Then, from (45), we obtain $\mathrm{sgn}\, a_{k,j} = -\mathrm{sgn}\, a_{k+1,j}$. Moreover, if we use $(I_k)$ and the equation $\mathrm{sgn}\, a_{k,j} = -\mathrm{sgn}\, a_{k+1,j}$, we obtain from (45)

$$\begin{aligned}
\frac{\delta_{k+1}}{6}|a_{k+1,j}| &= \frac{\delta_k + \delta_{k+1}}{3}|a_{k,j}| - \frac{\delta_k}{6}|a_{k-1,j}| \\
&\geq \frac{\delta_k + \delta_{k+1}}{3}|a_{k,j}| - \frac{\delta_k}{6} \cdot \frac{1}{2}|a_{k,j}| \\
&= \left(\frac{1}{4}\delta_k + \frac{1}{3}\delta_{k+1}\right)|a_{k,j}| > \frac{1}{3}\delta_{k+1}|a_{k,j}|,
\end{aligned}$$

i.e., $|a_{k,j}| < \frac{1}{2}|a_{k+1,j}|$. Therefore $(I_{k+1})$, and hence $(I_j)$, are established.

Similarly we can verify that, for $j = 0, 1, \ldots, n-1$,

$$(I_j'): \quad \begin{cases} \mathrm{sgn}\, a_{n,j} = -\mathrm{sgn}\, a_{n-1,j} = \cdots = (-1)^{n-j}\,\mathrm{sgn}\, a_{j,j}, \\ |a_{i+1,j}| \leq \frac{1}{2}|a_{i,j}|, \qquad i = j, \ldots, n-1. \end{cases}$$

Conclusion 1) of Lemma 1, and the inequality

$$|a_{i,j}| \leq 2^{-|i-j|}|a_{j,j}|, \qquad i, j = 0, 1, \ldots, n. \tag{46}$$

follow from $(I_j)$, $j = 1, \ldots, n$, and $(I_j')$, $j = 0, 1, \ldots, n-1$. Since the matrix $A$ is symmetric (as the inverse of the symmetric matrix $B$), we also find from (46) that $|a_{i,j}| = |a_{j,i}| \leq 2^{-|i-j|}|a_{i,i}|$. This completes the proof of Lemma 1.

LEMMA 2. *The diagonal elements of A satisfy*

$$0 < a_{j,j} < \frac{4}{\delta_j + \delta_{j+1}}, \qquad j = 0, 1, \ldots, n. \tag{47}$$

PROOF. The positivity of the numbers $a_{j,j}$, $j = 0, 1, \ldots, n$, was noticed in the proof of Lemma 1. In addition, according to (38) and (40) we have, for $j = 0, 1, \ldots, n$,

$$\frac{\delta_j}{6}a_{j-1,j} + \frac{\delta_j + \delta_{j+1}}{3}a_{j,j} + \frac{\delta_{j+1}}{6}a_{j+1,j} = 1$$

(where $a_{-1,0} = a_{n+1,n} = 0$). But by Lemma 1 we have $a_{j-1,j} \leq 0$, $a_{j+1,j} \leq 0$, and $\max(|a_{j-1,j}|, |a_{j+1,j}|) \leq \frac{1}{2}a_{j,j}$. Consequently

$$-\frac{\delta_j}{12}a_{j,j} + \frac{\delta_j + \delta_{j+1}}{3}a_{j,j} - \frac{\delta_{j+1}}{12}a_{j,j} \leq 1, \qquad j = 0, 1, \ldots, n,$$

from which (47) follows. This establishes Lemma 2, and with it also Theorem 8.

2°.  Now let $\pi_n$, $n = 1, 2, \ldots$, be the partition determined by (7), i.e., $\pi_1 = \{0, 1\}$, and for $n = 2^k + i$, $k = 0, 1, \ldots$; $i = \ldots, 2^k$,

$$\pi_n = \{t_s\}_{s=0}^n, \quad \text{where } t_s = t_s(n) = \frac{s}{2^{k+1}} \quad \text{if } s = 0, 1, \ldots, 2i,$$

$$\text{and} \quad t_s = t_s(n) = \frac{s - i}{2^k} \quad \text{if } s = 2i + 1, \ldots, n. \tag{48}$$

Recall that we denote the space $S_{\pi_n}$ of broken lines by $L_n$ (see (29′) and (6)). In addition, let $L_0$ be the space of constant functions on $[0, 1]$, and let $P_n$, $n = 0, 1, \ldots$, be the operator of orthogonal projection on $L_n$ (i.e., $P_n = P_{\pi_n}$).

As we remarked at the beginning of §3, the space of polynomials of the form $\sum_{k=0}^n a_k f_k(x)$ in the Franklin system coincides with $L_n$. Therefore

$$f_n \in L_n, \qquad (f_n, f_n) = 1, \qquad n = 0, 1, \ldots, \qquad (f_n, g) = 0,$$
$$\text{if } g \in L_{n-1}, \ n = 1, 2, \ldots. \tag{49}$$

Now we are ready to prove the fundamental result of this section.

THEOREM 9.  *For $n = 1, 2, \ldots$, and $t \in [0, 1]$, we have the inequality*

$$|f_n(t)| \le 3 \cdot 2^8 n^{1/2} 2^{-(n/2)|t - z_n|}, \tag{50}$$

*where $z_n = (2i - 1)/2^{k+1}$ if $n = 2^k + i$; $k = 0, 1, \ldots$; $i = 1, \ldots, 2^k$; $z_1 = 1$.*

REMARK.  The conclusion of Theorem 9 is known as the *exponential inequality* for the Franklin functions because the right-hand side of (50) contains the factor $2^{-n|t - x_n|/2}$, which decreases exponentially for points $t$ far from $\Delta_n$.

PROOF OF THEOREM 9.  Inequality (50) is evident for $n = 1$, since $f_1(x) = 2\sqrt{3}(x - \frac{1}{2})$ (see (27)), and $\|f_1\|_C < 2$.

Let $n = 2^k + i$, $k = 0, 1, \ldots$; $i = 1, \ldots, 2^k$. Consider the functions $\{N_j(t)\}_{j=0}^n$ [$N_j(t) = N_j(t, \pi_n)$], defined by (30) and forming a basis in $L_n = S_{\pi_n}$. Notice that $\pi_n$ is obtained from $\pi_{n-1}$ by adjoining the point $t_{2i-1} = (2i - 1)/2^{k+1}$. From this it follows easily (see the definition of $N_j(t)$) that $N_j(t) = N_j(t, \pi_n) \in L_{n-1}$ if $|j - (2i - 1)| > 1$, $j = 0, 1, \ldots, n$. Consequently (also see (41) and (49))

$$f_n(t) = \sum_{j=0}^n (f_n, N_j) N_j^*(t) = \sum_{j; |j - (2i-1)| \le 1} (f_n, N_j) N_j^*(t). \tag{51}$$

Since $0 \le N_j(t) \le 1$ and $m\{t \in (0, 1): N_j(t) \ne 0\} \le 2^{-k+1}$, $j = 0, 1, \ldots, n$ (see (48) and (30)), we have

$$|(f_n, N_j)| \le \|f_n\|_2 \|N_j\|_2 \le 2^{-(k-1)/2} \le 2n^{-1/2},$$

and we deduce from (51) that

$$|f_n(t)| \le 3 \cdot 2n^{-1/2} \max_{j:\ |j-(2i-1)| \le 1} |N_j^*(t)| \tag{52}$$

when $t \in [0, 1]$.

Let us choose a point $t \in I_s = (t_{s-1}, t_s)$, $0 < s \le n$. By the definition of $N_j(t)$ (see (30)) it follows that $N_l(t) = 0$ if $|l - s| > 1$. Therefore (also see (39))

$$N_j^*(t) = \sum_{l=0}^{n} a_{l,j} N_l(t) = \sum_{l=s-1}^{s+1} a_{l,j} N_l(t).$$

From the last equation and (52), we obtain, using $\sum_{l=0}^{n} N_l(t) = 1$, $t \in [0, 1]$ (see (31)),

$$|f_n(t)| \le 6n^{-1/2} \max_{\substack{j:\ |j-(2i-1)| \le 1 \\ l:\ |l-s| \le 1}} |a_{l,j}|, \qquad t \in I_s. \tag{53}$$

But by Theorem 8

$$\max_{\substack{j:\ |j-(2i-1)| \le 1 \\ l:\ |l-s| \le 1}} |a_{l,j}| \le 4 \cdot 2^{k+1} \max_{\substack{j:\ |j-(2i-1)| \le 1 \\ l:\ |l-s| \le 1}} 2^{-|l-j|} \le 2^3 \cdot n \cdot 2^{-|s-(2i-1)|+2}. \tag{54}$$

In addition, since $z_n = (2i - 1)/2^{k+1} = t_{2i-1}$, we have

$$|t - z_n| \le (|s - (2i - 1)| + 2)2^{-k} \le (|s - (2i - 1)| + 2)2n^{-1},$$

for $t \in I_s$, i.e., (see (53) and (54)) for every $t \in [0, 1]$,

$$|f_n(t)| \le 6n^{-1/2} \cdot 2^3 n 2^{-(n/2)|t-z_n|+4} \le 3 \cdot 2^8 n^{1/2} 2^{-(n/2)|t-z_n|}.$$

This completes the proof of Theorem 9.

COROLLARY 4. *There are absolute constants $C > 0$ and $q$, $0 < q < 1$, such that the following inequalities hold for $n = 1, 2, \dots$ and $t \in [0, 1]$:*
1) $|f_n(t)| \le Cn^{1/2} q^{n|t-z_n|}$;
2) $|f_n'(t)| \le Cn^{3/2} q^{n|t-z_n|}$;
3) $|\int_0^t f_n(x)\, dx| \le Cn^{-1/2} q^{n|t-z_n|}$;
4) $|f_n(t) - f_n(x)| \le Cn^{3/2}(t - x)q^{nd([x,t],z_n)}$ *for $x < t$, where $d(I, I') = \min_{y \in I, y' \in I'} |y - y'|$ if $I$, $I' < R^1$.*

PROOF. Inequality 1) was established in Theorem 9, and 2) follows from 1) because, for $t \in (t_{s-1}, t_s)$, $1 \le s \le n$ (see (48)),

$$|f_n'(t)| = \left| \frac{f(t_s) - f(t_{s-1})}{t_s - t_{s-1}} \right| \le 2n|f(t_s) - f(t_{s-1})|.$$

Inequality 3) also follows from 1) (for $t > z_n$ it is necessary to recall that $\int_0^1 f_n(x)\,dx = 0$, i.e., $\int_0^t f_n(x)\,dx = -\int_t^1 f_n(x)\,dx$). Finally, 4) follows immediately from 2).

## §5. Unconditional convergence of Fourier-Franklin series in the spaces $\mathscr{H}(\Delta)$ and $L^p(0,1)$

In this section we prove the following theorem.

THEOREM 10. *The Franklin system* $F = \{f_n\}$ *is an unconditional basis in* $\mathscr{H}(\Delta)$.

In Chapter 5 we showed (see §5.3) that, starting from any basis in $\mathscr{H}(\Delta)$, we can construct bases in the space $\operatorname{Re}\mathscr{H}^1$ and $H^1$. In particular, if we start from the Franklin system and define, according to Corollary 5.6 (see equation (90) of Chapter 5), the system $\mathscr{F} = \{F_n(z)\}$, $|z| < 1$, we obtain the following corollary of Theorem 10 and Corollary 5.6.

COROLLARY 5. *The system* $F = \{F_n\}$ *is an unconditional basis in* $H^1$.

Therefore, in contrast to the space $L^1(0,1)$ (see Theorem 2.13), there are unconditional bases in $\mathscr{H}(\Delta)$, $\operatorname{Re}\mathscr{H}^1$, and $H^1$.

It is interesting to notice that the properties of the Franklin O.N.S. turn out to be useful in an apparently unrelated problem—the construction of an unconditional basis in the space $H^1$ of analytic functions.

We shall deduce Theorem 10 from the following result.

THEOREM 11. *There are constants* $c > 0$ *and* $C > 0$ *such that for every* $g \in L^2(0,1)$

$$c\|g\|_{\mathrm{BMO}(\Delta)} \le \sup_{1 \le n < \infty} \left[ n \sum_{m:\,\Delta_m \subset \Delta_n} c_m^2(g) \right]^{1/2} \le C\|g\|_{\mathrm{BMO}(\Delta)}, \qquad (55)$$

*where the* $c_m(g)$, $m = 0, 1, \ldots$, *are the Fourier–Franklin coefficients of* $g(x)$, *and* $\Delta_m$, $m = 0, 1, \ldots$, *are the usual double integrals (see, for example,* 3.(1)).

We shall need three lemmas for the proof of Theorem 11.

LEMMA 1. *There is a constant* $C > 0$ *such that*

$$\|f_n\|_{\mathscr{H}(\Delta)} \le Cn^{-1/2}, \qquad n = 1, 2, \ldots. \qquad (56)$$

PROOF. We may, of course, restrict ourselves to the case $n > 1$, since $f_1(x) = 2\sqrt{3}(x - \frac{1}{2})$ and (see the definition of the norm in $\mathscr{H}(\Delta)$) we have

$\|f_1\|_{\mathcal{H}(\Delta)} \leq \sqrt{3}$. By Corollary 5.5, it is enough to prove that, for every $g \in \text{BMO}(\Delta)$ with $\|g\|_{\text{BMO}(\Delta)} \leq 1$,

$$\left| \int_0^1 f_n(x) g(x)\, dx \right| \leq C n^{-1/2}, \qquad n = 2, 3, \ldots. \tag{57}$$

Let $n = 2^k + i(k = 0, 1, \ldots; i = 1, \ldots, 2^k)$ be given, and let $\|g\|_{\text{BMO}(\Delta)} \leq 1$.

We first prove that

$$\int_{\Delta_k^j} |g(x) - g_{\Delta_k^i}|\, dx \leq 2^{-k+1}(|i - j| + 1), \qquad j = 1, \ldots, 2^k, \tag{58}$$

where, as usual, $g_I = |I|^{-1} \int_I g(t)\, dt$ and $I = (\alpha, \beta) < (0, 1)$. Let $I$ be the shortest interval that contains $\Delta_k^i$ and $\Delta_k^j$ (it is clear that $|I| = (|i - j| + 1) 2^{-k}$). Then

$$\int_{\Delta_k^j} |g(x) - g_{\Delta_k^i}|\, dx \leq \int_{\Delta_k^j} |g(x) - g_I|\, dx + |\Delta_k^j||g_{\Delta_k^i} - g_I|$$

$$\leq \int_I |g(x) - g_I|\, dx + \int_{\Delta_k^i} |g(x) - g_I|\, dx$$

$$\leq 2|I| \|g\|_{\text{BMO}(\Delta)} \leq 2^{-k+1}(|i - j| + 1).$$

We now prove (57). We use the equations $\int_0^1 f_n(x)\, dx = 0$, $n = 1, 2, \ldots$, and the following inequality which is a direct consequence of Theorem 9:

$$|f_n(x)| \leq C 2^{k/2} p^{|i-j|}, p = 2^{-1/2}, \ x \in \Delta_k^j, \ j = 1, \ldots, 2^k.$$

Then we obtain

$$\left| \int_0^1 f_n(x) g(x)\, dx \right| = \left| \int_0^1 f_n(x)[g(x) - g_{\Delta_k^i}]\, dx \right|$$

$$\leq \sum_{j=1}^{2^k} \int_{\Delta_k^j} |f_n(x)| |g(x) - g_{\Delta_k^i}|\, dx$$

$$\leq C 2^{k/2} \sum_{j=1}^{2^k} p^{|i-j|} \int_{\Delta_k^j} |g(x) - g_{\Delta_k^i}|\, dx$$

$$\leq 2C 2^{-k/2} \sum_{j=1}^{2^k} (|i - j| + 1) p^{|i-j|} \leq C' 2^{-k/2} \leq C n^{-1/2}.$$

This completes the proof of Lemma 1.

LEMMA 2. *Let (as in Theorem 9) $z_m$ be the center of the interval $\Delta_m$, $m = 1, 2, \ldots$, and let $q$, $0 < q < 1$, be a given number. We have the following inequalities:*

1) *for every $x \in (0, 1)$ and $\rho \in (0, 1]$*

$$S(x, \rho) = \sum_{\substack{m:\ m > \rho^{-1} \\ |x - z_m| > \rho}} q^{m|x - z_m|} \leq C_q;$$

2) *for every interval $I \subset (0, 1)$([5])*

$$\sum_{1 \leq m \leq 2|I|^{-1}} m q^{md(I, z_m)} \leq C_q |I|^{-1},$$

3) $\sum_{m:\ \Delta_m \subset \Delta_n} m^{-1/2} q^{md([0,1] \setminus \Delta_n, \Delta_m)} \leq C_q n^{-1/2}$, $n = 1, 2, \ldots$;

4) $\sum_{m \geq n,\ \Delta_m \not\subset \Delta_n} m^{-1/2} q^{md(\Delta_n, \Delta_m)} \leq C_q n^{-1/2}$, $n = 1, 2, \ldots$.

We shall prove only inequality 1). The proofs of the other inequalities are equally elementary and of the same nature.

It is clear that

$$S(x, \rho) \leq 1 + \sum_{k:\ 2^k \geq \frac{1}{2}\rho^{-1}} S_k(x, \rho),$$

where

$$S_k(x, \rho) = \sum_{\substack{i:\ 1 \leq i \leq 2^k, \\ |x - (2i-1)/2^{k+1}| > \rho}} q^{2^k |x - (2i-1)/2^{k+1}|}.$$

Let $k$ be given and let $i$ be a number such that

$$x \in \left[ \frac{i_0 - 1}{2^k}, \frac{i_0}{2^k} \right].$$

Then

$$\left\| x - \frac{2i - 1}{2^{k+1}} \right\| - \left\| \frac{i}{2^k} - \frac{i_0}{2^k} \right\| < \frac{2}{2^k},$$

and therefore

$$S_k(x, \rho) \leq \sum_{i:\ |i/2^k - i_0/2^k| > \rho - 1/2^{k-1}} q^{2^k(|i_0/2^k - i/2^k| - 1/2^{k-1})}$$

$$\leq q^{-2} \sum_{s:\ |s| > (\rho - 1/2^{k-1})2^k} q^{|s|} \leq C_q' q^{\rho 2^k},$$

$$S(x, \rho) \leq 1 + C_q' \sum_{k:\ 2^k > \frac{1}{2}\rho^{-1}} q^{2^k \rho} \leq C_q' \sum_{k=0}^{\infty} q^{k/2} \leq C_q.$$

This completes the proof of inequality 1).

---

([5])$d(I, I')$ is defined as in Corollary 4.

LEMMA 3. *For* $g \in L^2(0,1)$ *and* $I = (\alpha, \beta) \subset (0,1)$ *let*

$$\theta(g, I) = \left[ \int_I (g(x) - g_1)^2 \, dx \right]^{1/2}. \tag{59}$$

*Then, for* $g$, *and* $g' \in L^2(0,1)$ *and* $\lambda \in R^1$,

$$\theta(\lambda g, I) = |\lambda| \theta(g, I), \qquad \theta(g + g', I) \leq \theta(g, I) + \theta(g', I),$$

$$\theta(g, I) \leq \left[ \int_I g^2(x) \, dx \right]^{1/2}.$$

In verifying this we need only the following inequality. Let $u(c) = \int_I [g(x(-c]^2 \, dx$. Then the minimum, $\min_{-\infty < c < \infty} u(c)$, is always attained for $c = g_I$. In fact, $u'(c) = -2 \int_I g(x) \, dx + 2c|I|$, i.e., $u'(c) = 0$ only for $c = g_I$. Consequently $u(g_I) \leq u(0)$. This completes the proof of Lemma 3.

We now turn to the proof of Theorem 11.

a) *An upper bound for* $\|g\|_{\mathrm{BMO}(\Delta)}$. Let there be given a sequence $\{c_m\}_{m=0}^\infty$ that satisfies the condition

$$\sum_{m:\, \Delta_m \subset \Delta_n} c_m^2 \leq n^{-1}, \qquad n = 1, 2, \ldots. \tag{60}$$

Let us show that then the function

$$g(x) \overset{L^2}{=} \sum_{m=0}^\infty c_m f_m(x) \in \mathrm{BMO}(\Delta) \quad \text{and} \quad \|g\|_{\mathrm{BMO}(\Delta)} \leq A$$

($A$ is an absolute constant). For this purpose we take an arbitrary interval $I = (\alpha, \beta) \subset (0,1)$ and estimate

$$\left[ |I|^{-1} \int_I |g(x) - g_I|^2 \, dx \right]^{1/2} = |I|^{-1/2} \cdot \theta(g, I).$$

It is easily seen that we can find binary intervals $\Delta_{n_1}$ and $\Delta_{n_2}$ (possibly coincident) such that

$$\bar{\delta} \equiv \overline{\Delta_{n_1} \cup \Delta_{n_2}} \supset (\alpha - |I|, \beta + |I|) \cap (0,1),$$

$$|I| \leq |\Delta_{n_1}| = |\Delta_{n_2}| \leq 6|I| \tag{61}$$

(in (61), $\bar{\delta}$ is the closure of the set $\delta = \Delta_{n_1} \cup \Delta_{n_2}$; also let $|\delta| = m(\delta)$).

Then we consider the decomposition

$$g(x) = \sum_{m=0}^\infty c_m f_m(x)$$

$$= \sum_{m:\, \Delta_m \subset \delta} + \sum_{\substack{m:\, \Delta_m \cap \delta = \varnothing, \\ |\Delta_m| < |I|}} + \sum_{\substack{m:\, |\Delta_m| \geq |I|, \\ \Delta_m \not\subset \delta}}$$

$$= G_1(x) + G_2(x) + G_3(x).$$

(recall that, by the general properties of binary intervals, if $|\Delta_m| < |\delta|$ then either $\Delta_m \subset \delta$ or $\Delta_m \cap \delta = \varnothing$).

From (60) and (61), by using the inequality $|\Delta_n| \geq n^{-1}$, $n = 1, 2, \ldots,$ we obtain

$$
\int_I G_1^2(x)\,dx \leq \int_0^1 G_1^2(x)\,dx
$$

$$
\leq \sum_{m:\,\Delta_m \subset \Delta_{n_1}} c_m^2 + \sum_{m:\,\Delta_m \subset \Delta_{n_2}} c_m^2
$$

$$
\leq n_1^{-1} + n_2^{-1} \leq 12|I|.
$$

Consequently (see Lemma 3)

$$
\theta(G_1, I) \leq \left[ \int_I G_1^2(x)\,dx \right]^{1/2} \leq (12|I|)^{1/2}. \tag{62}
$$

To estimate $G_2(x)$, we use the inequality $c_m^2 \leq m^{-1}$, $m = 1, 2, \ldots$ (see (60)) and conclusion 1) of Corollary 4:

$$
|G_2(x)| \leq \sum_{\substack{m:\,\Delta_m \cap \delta = \varnothing \\ |\Delta_m| < |I|}} |c_m|\,|f_m(x)|
$$

$$
\leq \sum_{\substack{m:\,\Delta_m \cap \delta = \varnothing \\ m > |I|^{-1}}} m^{-1/2}|f_m(x)| \leq C \sum_{\substack{m:\,\Delta_m \cap \delta = \varnothing \\ m > |I|^{-1}}} q^{m|x - z_m|}
$$

$$
(q < 1).
$$

According to (61), if $\Delta_m \cap \delta = \varnothing$ and $x \in I$, we have $|x - z_m| > |I|$; consequently, if we use inequality 1) from Lemma 2 we obtain

$$
|G_2(x)| \leq C \sum_{\substack{m:\,m > |I|^{-1} \\ |x - z_m| > |I|}} q^{m|x - z_m|} \leq A' \tag{63}
$$

for $x \in I$. To estimate $G_3(x)$ we use conclusion 4) of Corollary 4 and inequality 2) of Lemma 2. Then, for $x, t \in I$ and $x < t$, we obtain

$$
|G_3(t) - G_3(x)| \leq \sum_{m:\,|\Delta_m| \geq |I|} c_m|f_m(t) - f_m(x)|
$$

$$
\leq \sum_{1 \leq m \leq 2|I|^{-1}} m^{-1/2} \cdot m^{3/2}|I|q^{md([x,t],z_m)}
$$

$$
\leq C|I| \sum_{1 \leq m \leq 2|I|^{-1}} m q^{md(I, z_m)} \leq A''. \tag{64}
$$

It follows from (62)–(64) (also see Lemma 3) that

$$\left[ |I|^{-1} \int_I |g(x) - g_I|^2 \, dx \right]^{1/2} = |I|^{-1/2} \theta(g, I)$$

$$\leq |I|^{-1/2} \sum_{i=1}^{3} \theta(G_i, I)$$

$$\leq |I|^{-1/2} A''' |I|^{1/2} = A'''$$

for every interval $I \subset (0, 1)$; i.e., (see (103) of Chapter 5) $\mathfrak{N}_{\Delta,2}(g) \leq A'''$, and a fortiori (see Corollary 5.7) $\mathfrak{N}_{\Delta}(g) \leq A'''$.

Moreover, it follows from (60) for $n = 1$ that $|\int_0^1 g(x) \, dx| = |c_0| = 1$. We have therefore proved that

$$g \in \mathrm{BMO}(\Delta) \quad \text{and} \quad \|g\|_{\mathrm{BMO}(\Delta)} = \left| \int_0^1 g(x) \, dx \right| + \mathfrak{N}_{\Delta}(g) \leq A.$$

b) Now let $g \in \mathrm{BMO}(\Delta)$ and $\|g\|_{\mathrm{BMO}(\Delta)} \leq 1$. Let us show then that

$$\sum_{m: \, \Delta_m \subset \Delta_n} c_m^2 \leq B n^{-1}, \qquad c_m = c_m(g), \quad n, m = 1, 2, \ldots \qquad (65)$$

($B$ is an absolute constant). We consider the case $n = 1$ separately.

By Corollary 5.7 and Parseval's equation,

$$\sum_{m=1}^{\infty} c_m^2 = \int_0^1 |g(x) - c_0|^2 \, dx = \int_0^1 |g(x) - g_{(0,1)}|^2 \, dx$$

$$\leq \mathfrak{N}_{\Delta,2}(g) \leq C \mathfrak{N}_{\Delta}(g) \leq C.$$

Moreover,

$$|c_0|^2 = \left[ \int_0^1 g(x) \, dx \right]^2 \leq \|g\|_{\mathrm{BMO}(\Delta)}^2 \leq 1,$$

and we obtain (65) for $n = 1$.

Before discussing the case $n > 1$, we observe that, by Lemma 1 and Corollary 5.5 (also see the remark after Corollary 5.5)

$$|c_m(g)| = \left| \int_0^1 g f_m \, dx \right| \leq C \|g\|_{\mathrm{BMO}(\Delta)} \|f_m\|_{\mathscr{H}(\Delta)} \leq C m^{-1/2},$$

$$m = 1, 2, \ldots. \qquad (66)$$

Now let $n > 1$ be given. Consider the decomposition

$$g(x) \overset{L_2}{=} \sum_{m=0}^{\infty} c_m f_m(x) = \sum_{m: \, \Delta_m \subset \Delta_n} + \sum_{m \geq n: \, \Delta_m \not\subset \Delta_n} + \sum_{m < n}$$

$$= E_1(x) + E_2(x) + E_3(x).$$

Let us show that (see (59))

$$\theta(E_i, \Delta_n) = Cn^{-1/2}, \qquad i = 2, 3. \tag{67}$$

We need the inequality

$$\int_I f_m^2(x)\,dx \le Cq^{2md(\Delta_m, I)}, \tag{68}$$

$$I = (\alpha, \beta) \subset (0, 1), \qquad m = 1, 2, \ldots,$$

where $q < 1$ is the same as in Corollary 4.

If $\Delta_m \cap I \ne \varnothing$, inequality (68) is evident, since $\|f_m\|_2 = 1$. On the other hand, if $\Delta_m \cap I = \varnothing$ we use conclusion 1) of Corollary 4 to obtain

$$\int_I f_m^2(x)\,dx \le Cm \int_I q^{2m|x - z_m|}\,dx$$

$$\le Cm \left| \frac{1}{m \ln q} q^{2m|x - z_m|} \right|_\alpha^\beta \le Cq^{2md(\Delta_m, I)},$$

which establishes (68). We can estimate the number $\theta(E_2, \Delta_n)$ by using (68) and inequality 4) from Lemma 2 (also see Lemma 3). We have (also see (66))

$$\theta(E_2, \Delta_n) \le \sum_{m \ge n:\ \Delta_m \not\subset \Delta_n} |c_m| \theta(f_m, \Delta_n)$$

$$\le C' \sum_{m \ge n:\ \Delta_m \not\subset \Delta_n} m^{-1/2} q^{1/2 md(\Delta_m, \Delta_n)} \le Cn^{-1/2}.$$

For $i = 3$, inequality (67) follows from the inequality

$$|E_3(t) - E_3(x)| \le C \quad \text{for } t, x \in \Delta_n,$$

which can be verified in the same way as (64). Since (see Corollary 5.7)

$$\theta(g, \Delta_n) \le |\Delta_n|^{1/2} \mathfrak{N}_{\Delta, 2}(g) \le C|\Delta_n|^{1/2} \|g\|_{\mathrm{BMO}(\Delta)} \le Cn^{-1/2},$$

it follows from (67) that

$$\theta(E_1, \Delta_n) \le \theta(g, \Delta_n) + \theta(E_2, \Delta_n) + \theta(E_3, \Delta_n)$$
$$\le Cn^{-1/2},$$

and therefore

$$\left\{ \int_{\Delta_n} E_1^2(x)\,dx \right\}^{1/2} \le \theta(E_1, \Delta_n) + |\Delta_n|^{-1/2} \left| \int_{\Delta_n} E_1(x)\,dx \right|$$

$$\le Cx^{-1/2} + |\Delta_n|^{-1/2} \left| \int_{\Delta_n} E_1(x)\,dx \right|. \tag{69}$$

By the equation (Parseval's equation)

$$\sum_{m:\,\Delta_m\subset\Delta_n} c_m^2 = \int_0^1 E_1^2(x)\,dx = \int_{\Delta_n} E_1^2(x)\,dx + \int_{[0,1]\backslash\Delta_n} E_1^2(x)\,dx$$

and (69), in order to complete the proof of (65) it remains only to prove that

1) $\left|\int_{\Delta_n} E_1(x)\,dx\right| \le Cn^{-1}$;

2) $\int_{[0,1]\backslash\Delta_n} E_1^2(x)\,dx \le Cn^{-1}$.

It follows from conclusion 1) of Corollary 4 that (with $\Delta_n = (a,b)$)

$$\int_{[0,1]\backslash\Delta_n} |f_m(x)|\,dx \le C\left[ m^{1/2}\int_0^a q^{m|x-z_m|}\,dx + m^{1/2}\int_b^1 q^{m|x-z_m|}\,dx \right]$$

$$\le C m^{1/2}\max_{x\in[0,1]\backslash\Delta_n}\left|\frac{1}{m\ln q} q^{m|x-z_m|}\right|$$

$$\le C m^{-1/2} q^{md([0,1]\backslash\Delta_n,\Delta_m)}.$$

From this and (66), if we recall that $\int_0^1 f_m(x)\,dx = 0$ for $m \ge 1$, and apply inequality 3) of Lemma 2, we obtain

$$\left|\int_{\Delta_n} E_1(x)\,dx\right| \le \sum_{m:\,\Delta_m\subset\Delta_n} |c_m|\left|\int_{\Delta_n} f_m(x)\,dx\right|$$

$$\le C\sum_{m:\,\Delta_m\subset\Delta_n} m^{-1/2}\left|\int_{[0,1]\backslash\Delta_n} f_m(x)\,dx\right|$$

$$\le Cn^{-1/2}\sum_{m:\,\Delta_m\subset\Delta_n} m^{-1/2} q^{md([0,1]\backslash\Delta_n,\Delta_m)} \le Cn^{-1}.$$

Finally, again using (66), (68), and inequality 3) of Lemma 2, we obtain

$$\left[\int_{[0,1]\backslash\Delta_n} E_1^2(x)\,dx\right]^{1/2} \le \sum_{m:\,\Delta_m\subset\Delta_n} \|c_m f_m\|_{L^2([0,1]\backslash\Delta_n)}$$

$$\le \sum_{m:\,\Delta_n\subset\Delta_n} Cm^{-1/2}\left[\int_{[0,1]\backslash\Delta_n} f_m^2(x)\,dx\right]^{1/2}$$

$$\le C\sum_{m:\,\Delta_m\subset\Delta_n} m^{-1/2} q^{md([0,1]\backslash\Delta_n,\Delta_m)} \le Cn^{-1/2}.$$

Inequalities 1) and 2), and hence (65), are now established. Hence we have completed the proof of Theorem 11.

PROOF OF THEOREM 10. The completeness of $F$ in $\mathscr{H}(\Delta)$ follows from the density of the bounded functions in $\mathscr{H}(\Delta)$ (see Definition 5.6) and the inequality $\|f\|_{\mathscr{H}(\Delta)} \le C\|f\|_2$ (see, for example, Theorems 5.6 and 3.4), according to which every function in $L^\infty(0,1)$ can be approximated arbitrarily closely in the $\mathscr{H}(\Delta)$ norm by polynomials in the system $F$. It follows from Theorem 1.2 ($F$ is self-dual) that $F$ is minimal. Therefore, by Theorem 1.10 it is enough to show that, for every $f \in \mathscr{H}(\Delta)$ and every sequence $\varepsilon_n = \pm 1$, $n = 0, 1, \ldots$,

$$\|T_{\varepsilon,N}(f)\|_{\mathscr{H}(\Delta)} \le C\|f\|_{\mathscr{H}(\Delta)}, \qquad T_{\varepsilon,N}(f,x) = \sum_{n=0}^{N} \varepsilon_n c_n(f) f_n(x), \qquad (70)$$

where $C$ is an absolute constant, and $c_n(f) = \int_0^1 f(x) f_n(x)\, dx$, $n = 0, 1, \ldots$, are the Fourier–Franklin coefficients of $f(x)$ (note that $\langle f, f_n \rangle = \int_0^1 f(x) f_n(x)\, dx$, since $f_n \in L^\infty(0,1)$).

By Corollary 5.5,

$$\|T_{\varepsilon,N}(f)\|_{\mathscr{H}(\Delta)} \le C \sup_{\|g\|_{\mathrm{BMO}(\Delta)} \le 1} \left| \int_0^1 T_{\varepsilon,N}(f,x) g(x)\, dx \right|.$$

But, by the orthonormality of the Franklin system and Corollary 5.4,

$$\left| \int_0^1 T_{\varepsilon,N}(f,x) g(x)\, dx \right| = \left| \sum_{n=0}^{N} \varepsilon_n c_n(f) c_n(g) \right|$$

$$= \left| \int_0^1 f(x) T_{\varepsilon,N}(g,x)\, dx \right|$$

$$\le C\|f\|_{\mathscr{H}(\Delta)} \|T_{\varepsilon,N}(g)\|_{\mathrm{BMO}(\Delta)},$$

and it now remains only to verify that,

$$\|T_{\varepsilon,N}(g)\|_{\mathrm{BMO}(\Delta)} \le C\|g\|_{\mathrm{BMO}(\Delta)}, \qquad N = 1, 2, \ldots$$

for every function $g \in \mathrm{BMO}(\Delta)$ and every sequence $\varepsilon_n = \pm 1$, $n = 0, 1, \ldots$. The last inequality follows directly from Theorem 11 (see (55), where there is given a number equivalent to $\|g\|_{\mathrm{BMO}(\Delta)}$ but not depending on the signs of the numbers $c_n(g)$; we also note that the hypothesis of Theorem 11 that $g \in L^2(0,1)$ is always satisfied if $g \in \mathrm{BMO}(\Delta)$ (see, for example, Corollary 5.7)). This completes the proof of Theorem 10.

The following theorem can be deduced from Theorem 10 by means of the interpolation Theorem 3.12.

THEOREM 12. *The Franklin system is an unconditional basis in* $L^p(0,1)$, $1 < p < \infty$.

PROOF. By Corollary 1.3, it is enough to consider the case when $1 < p < 2$. According to Theorem 1.10 (also see Theorem 7) it is sufficient to establish the existence of a constant $C = C_p$ such that

$$\|T_{\varepsilon,N}(f)\|_p \leq C_p \|f\|_p \tag{71}$$

for every function $f \in L^p(0,1)$ and every operator $T_{\varepsilon,N}(f)$ of the form (70).

By 5.(84) and Theorem 5.6, it follows from (70) that when $f \in L^1(0,1)$, $y > 0$,

$$\begin{aligned}
m\{x \in (0,1): |T_{\varepsilon,N}(f,x)| > y\} &\leq \frac{1}{y}\|T_{\varepsilon,N}(f)\|_1 \\
&\leq \frac{1}{y}\|T_{\varepsilon,N}(f)\|_{\mathscr{H}(\Delta)} \\
&\leq \frac{C}{y}\|f\|_{\mathscr{H}(\Delta)} \\
&\leq \frac{C}{y}\|S^*(f)\|_1. \tag{72}
\end{aligned}$$

On the other hand, since the Franklin system is orthonormal,

$$m\{x \in (0,1): |T_{\varepsilon,N}(f,x)| > y\} \leq \frac{1}{y^2}\|T_{\varepsilon,N}(f)\|_2^2 \leq \frac{1}{y^2}\|f\|_2^2. \tag{73}$$

We obtain (71) from (72) and (73) by using Theorem 3.12. This completes the proof of Theorem 12.

REMARK. Theorem 12 can be considerably sharpened: it turns out that the Franklin and Haar systems are equivalent bases in $L^p(0,1)$, $1 < p < \infty$, i.e., for every polynomial $P(x) = \sum_{n=0}^N a_n f_n(x)$ in the Franklin system,

$$c_p \left\|\sum_{n=0}^N a_n \chi_{n+1}\right\|_p \leq \|P\|_p \leq C_p \left\|\sum_{n=0}^N a_n \chi_{n+1}\right\|_p.$$

We omit the proof.

In concluding this chapter we present a result that shows that we can give a test, in terms of the behavior of the Fourier–Franklin series of $f(x)$, for $f(x)$ to belong to $\mathscr{H}(\Delta)$:

THEOREM 13. *A necessary and sufficient condition for a series*

$$\sum_{n=0}^{\infty} c_n f_n(x) \tag{74}$$

*in the Franklin system to converge unconditionally in $L^1(0,1)$ is that its sum $f(x)$ belongs to $\mathscr{H}(\Delta)$.*

REMARK. Theorem 13 can be thought of as an analog of Corollary 3.4.

**LEMMA 1.** *There is a positive integer $s_0$ with the property that, for $n = 2, 3, \ldots,$*

$$\int_{H_n} f_n^2(x)\, dx \geq \frac{1}{2}, \text{ where } H_n = \bigcup_{\substack{1 \leq j \leq 2^k \\ |i-j| \leq s_0}} \Delta_k^j,$$

$$n = 2^k + i, \quad k = 0, 1, \ldots, \quad i = 1, \ldots, 2^k. \tag{75}$$

In fact, by (68)

$$\int_{(0,1)\backslash H_n} f_n^2(x)\, dx \leq 2C q^{n s_0 2^{-k}} \leq 2C q^{s_0} < \frac{1}{2},$$

if $s_0$ is sufficiently large. Hence, since $\int_0^1 f_n^2(x)\, dx = 1$, we obtain (75).

It will be convenient to set $H_0 = H_1 = (0, 1)$. Taking $s_0$ to be fixed from now on, we notice that $H_n$, $n = 0, 1, \ldots$ are, except for a finite number of points, subintervals of $(0, 1)$, and

$$|\Delta_n| \leq m(H_n) \leq (2s_0 + 1)|\Delta_n| \equiv C_0|\Delta_n|, \quad n = 0, 1, \ldots. \tag{76}$$

**PROOF OF THEOREM 13.** According to Theorem 10, if $f(x) \in \mathscr{H}(\Delta)$ its Fourier series converges unconditionally in $\mathscr{H}(\Delta)$ and therefore also in $L^1(0, 1)$ (see 5.(84)).

Now let the series (74) converge unconditionally in $L^1(0, 1)$ to $f(x)$. Then, according to Theorem 1.1, the series $\sum_{n=0}^{\infty} r_n(t)c_n f_n(x)$ ($\{r_n(t)\}$ is the Rademacher system) converges for every $t \in (0, 1)$, and consequently (see Theorem 2.12)

$$P(f, x) = \left\{ \sum_{n=0}^{\infty} c_n^2(f) f_n^2(x) \right\}^{1/2} \in L^1(0, 1). \tag{77}$$

Hence to complete the proof of Theorem 13 it is enough to show that

$$\|f\|_{\mathscr{H}(\Delta)} \leq C \|P(f)\|_1. \tag{78}$$

Moreover, we may suppose without loss of generality that $f$ is a polynomial in the Franklin system—this follows from

$$\lim_{N \to \infty} \left\| P(f) - P\left( \sum_{n=0}^{N} c_n(f) f_n \right) \right\|_1 = 0 \Bigg)$$

—and $\|P(f)\|_1 = 1$.

We choose a function (a polynomial) $f(x)$ with $\|P(f)\|_1 = 1$ and set $\Omega_{-1} = E_{-1} = (0, 1)$, and for $r = 0, 1, \ldots$ (see (76))

$$\Omega_r = \{x \in (0, 1): P(f, x) > 2^r\},$$

$$E_r = \left\{ x \in (0, 1): M(\chi_{\Omega_r}, x) > \frac{1}{4C_0} \right\},$$

where, as usual, $\chi_\Omega$ is the characteristic function of $\Omega$, and $M(g,x)$ is the Hardy–Littlewood maximal function of $g$ (see Appendix 1). By the inequality of weak type for the maximal function (see Theorem 2 of Appendix 1)

$$m(E_r) \leq C\|\chi_{\Omega_r}\|_1 = Cm(\Omega_r), \qquad r = 0, 1, \ldots \tag{79}$$

On the other hand, since $M(g,x) \geq |g(x)|$ for almost all $x$, we have

$$m(\Omega_r \setminus E_r) = 0, \qquad r = 0, 1, \ldots \tag{80}$$

In addition, for $r = 0, 1, \ldots$, let

$$\Lambda_r = \left\{ n : m(H_n \cap \Omega_{r-1}) > \frac{|\Delta_n|}{4}, m(H_n \cap \Omega_r) \leq \frac{|\Delta_n|}{4} \right\}.$$

The set $\Lambda_r$ may be empty; however, it is clear that every $n \geq 0$ belongs to some $\Lambda_r$.

Let us show that the inequality (78) that we need can be deduced from the following two inequalities:

a) $\sum_{n \in \Lambda_r} c_n^2(f) \leq C \int_{E_{r-1} \setminus \Omega_r} P^2(f, x) \, dx, r = 0, 1, \ldots$;

b) every function $b \in \mathrm{BMO}(\Delta)$ with $b(x) = \sum_0^\infty b_n f_n(x)$ satisfies

$$\sum_{n \in \Lambda_r} b_n^2 \leq Cm(E_{r-1})\|b\|^2_{\mathrm{BMO}(\Delta)}, \qquad r = 0, 1, \ldots.$$

($C$ is an absolute constant).

In fact, by using a), b), and (79), we find that, for every

$$b(x) = \sum_{n=0}^\infty b_n f_n(x) \in \mathrm{BMO}(\Delta)$$

with $\|b\|_{BMO(\Delta)} \le 1$, we have[6]

$$
\begin{aligned}
\left| \int_0^1 f(x)b(x)\, dx \right| &= \left| \sum_{n=0}^\infty c_n(f) b_n \right| \le \sum_{r=0}^\infty \sum_{n \in \Lambda_r} |c_n(f) b_n| \\
&\le \sum_{r=0}^\infty \left\{ \sum_{n \in \Lambda_r} c_n^2(f) \right\}^{1/2} \left\{ \sum_{n \in \Lambda_r} b_n^2 \right\}^{1/2} \\
&\le C \sum_{r=0}^\infty \left\{ \int_{E_{r-1} \setminus \Omega_r} P^2(f,x)\, dx \right\}^{1/2} \{ m(E_{r-1}) \}^{1/2} \\
&\le C \sum_{r=0}^\infty 2^r m(E_{r-1}) \le C' + C' \sum_{r=1}^\infty 2^r m(\Omega_{r-1}) \\
&\le C' + 2C' \sum_{r=1}^\infty 2^r [m(\Omega_{r-1}) - m(\Omega_r)] \\
&= C' + 2C' \sum_{r=1}^\infty 2^r m\{ x \in (0,1) : 2^{r-1} < P(f,x) \le 2^r \} \\
&\le 3C' = C = C\|P(f)\|_1,
\end{aligned}
$$

from which we obtain (78) by using Corollary 5.5.

We still have to prove inequalities a) and b). Let us prove a). By the definition of $P(f,x)$,

$$
\int_{E_{r-1} \setminus \Omega_r} P^2(f,x)\, dx \ge \sum_{n \in \Lambda_r} c_n^2(f) \int_{E_{r-1} \setminus \Omega_r} f_n^2(x)\, dx, \qquad r = 0, 1, \ldots . \quad (81)
$$

Moreover (see (76)), if $n \in \Lambda_r$,

$$
|H_n|^{-1} \int_{H_n} \chi_{\Omega_{r-1}}(x)\, dx > [C_0 |\Delta_n|]^{-1} \frac{1}{4} |\Delta_n| = \frac{1}{4C_0}.
$$

Therefore $M(\chi_{\Omega_{r-1}}, x) > 1/4C_0$ if $x \in H_n$ and

$$
H_n \subset E_{r-1}, \qquad n \in \Lambda_r, \ r = 0, 1, \ldots . \quad (82)
$$

Consequently inequality a) will follow from (81) if we show that, for $n \in \Lambda_r$, $r = 0, 1, \ldots$,

$$
\int_{H_n \setminus \Omega_r} f_n^2(x)\, dx \ge \alpha > 0, \quad (83)
$$

where $\alpha$ is an absolute constant.

---

[6] Notice that $c_n(f) = 0$ for $n > n_0$, since $f$ is a polynomial.

Inequality (83) is evident for $n = 0$ and $n = 1$, since $m(H_n \cap \Omega_r) \leq \frac{1}{4}$ when $n \in \Lambda_r$. Now let $n = 2^k + i$, $k = 0, 1, \ldots$; $i = 1, \ldots, 2^k$. Then

$$\int_{H_n \setminus \Omega_r} f_n^2(x)\, dx = \sum_{j:\, \Delta_k^j \subset H_n} \int_{\Delta_k^j \setminus \Omega_r} f_n^2(x)\, dx$$

$$= \sum_{s:\, \Delta_{k+1}^s \subset H_n} \int_{\Delta_{k+1}^s \setminus \Omega_r} f_n^2(x)\, dx. \tag{84}$$

By the construction of the sets $\Lambda_r$, the inequality $m(\Delta_k^j \setminus \Omega_r) > \frac{3}{4}|\Delta_k^j|$ is valid for $n \in \Lambda_r$ and every interval $\Delta_k^j \subset H_n$. Consequently

$$m(\Delta_{k+1}^s \setminus \Omega_r) \geq \frac{1}{2}|\Delta_{k+1}^s|, \quad \text{if } \Delta_{k+1}^s \subset H_n. \tag{85}$$

But since $f_n(x)$ is linear in each interval $\Delta_{k+1}^s$, $s = 1, \ldots, 2^{k+1}$ (see (49) and (6)), it is easily seen that, for every set $G \subset \Delta_{k+1}^s$ with $m(G) \geq \delta|\Delta_{k+1}^s|$ $(\delta > 0)$, we have

$$\int_G f_n^2(x)\, dx \geq C_\delta \int_{\Delta_{k+1}^s} f_n^2(x)\, dx.$$

Hence we obtain, by using (85),

$$\int_{\Delta_{k+1}^s \setminus \Omega_r} f_n^2(x)\, dx \geq c \int_{\Delta_{k+1}^s} f_n^2(x)\, dx, \qquad \Delta_{k+1}^s \subset H_n,$$

and (see (84))

$$\int_{H_n \setminus \Omega_r} f_n^2(x)\, dx \geq c \int_{H_n} f_n^2(x)\, dx, \qquad n \in \Lambda_r,\ r = 0, 1, \ldots.$$

To obtain (83) we have only to apply Lemma 1. Inequality a) is now established. Let us prove b).

If $\Lambda_r \neq 0$, then by using property 2) of binary intervals (see §3.1) it is easy to find disjoint binary intervals $\Delta_{n_q}$, $n_q \in \Lambda_r$, $q = 1, \ldots$, such that for every $n \in \Lambda_r$ the interval $\Delta_n$ is contained in one of the intervals $\Delta_{n_q}$, $q = 1, \ldots$. Notice also that $\Delta_n \subset H_n \subset E_{r-1}$ for $n \in \Lambda_r$, by (82). Then, if we apply Theorem 11 (see (55)), we find that, for every function $b \in \mathrm{BMO}(\Delta)$, with $b(x) = \sum_{n=0}^\infty b_n f_n(x)$,

$$\sum_{n \in \Lambda_r} b_n^2 \leq \sum_q \sum_{m:\, \Delta_m \subset \Delta_{n_q}} b_m^2 \leq C\|b\|_{\mathrm{BMO}(\Delta)} \sum_q \frac{1}{n_q}$$

$$\leq C\|b\|_{\mathrm{BMO}(\Delta)} \sum_q |\Delta_{n_q}| \leq C\|b\|_{\mathrm{BMO}(\Delta)} m(E_{r-1}).$$

This completes the proof of inequality b), and hence of Theorem 13.

# CHAPTER VII

# Orthogonalization and Factorization Theorems

In this chapter we discuss two fundamental topics from the theory of orthogonal series. Many of the results of the chapter are unified by showing that, in a number of cases, objects more general than orthogonal systems reduce in essence to orthogonal systems.

For example, we show in §3 that if a system $\Phi = \{\varphi_n(x)\}_{n=1}^{\infty}$, $x \in (0, 1)$, has the property that if the series $\sum_{n=1}^{\infty} a_n \varphi_n(x)$ converges in measure on $(0, 1)$ for every sequence $\{a_n\}_{n=1}^{\infty}$ with $\sum_{n=1}^{\infty} a_n^2 < \infty$, then for every $\varepsilon > 0$ there are a set $E_\varepsilon \subset (0, 1)$ with $m(E) > 1 - \varepsilon$, and a system of functions $\Psi = \{\psi_n\}_{n=1}^{\infty}$, orthonormal on $(0, 1)$, such that $\varphi_n(x) = C_\varepsilon \psi_n(x)$, $n = 1, 2, \ldots$, $x \in E\varepsilon$.

In many problems about series of functions it becomes necessary to extend a system $\Phi$ of functions defined on $(0, 1)$ to a larger set $G$ $(G \supset (0, 1)$, $m(G) > 1)$ so that $\Phi$ is, on $G$, either an orthogonal system or a complete orthogonal system. Moreover, it is sometimes necessary for $\Phi$ to have some additional properties, for example to be uniformly bounded on $G \backslash (0, 1)$. In this chapter, §1 is devoted to problems of this kind.

Beginning with §2, we consider what are known as factorization theorems, and present their applications to the study of the properties of the majorant operator for the partial sums $S_\Phi^*$ generated by an orthogonal convergence system $\Phi$ (see Definition 1.2).

The term "factorization theorem" makes sense in the context of the history of the first theorem of this kind:

Every bounded linear operator $T: L^2(0, 1) \to L^1(0, 1)$ can be represented as a product, $T = T_f \cdot T'$, where $T': L^2(0, 1) \to L^2(0, 1)$ is a bounded linear operator, and $T_f$ is the operator of multiplication by the function $f \in L^2(0, 1): T_f(g) = f(x)g(x)$.

Factorization theorems for operators that act on a Banach space have important applications, and have been intensively studied in functional analysis. For the theory of functions, there are especially interesting results for operators that act on the space $L^0(0, 1)$ (of functions that are measurable and finite almost everywhere). We shall state factorization theorems for such operators (see Theorems 5 and 6 in §2) in an equivalent form as statements about series of functions.

### §1. Orthogonalization of a system of functions by means of extension to a larger set

We begin with the following classical result.

THEOREM 1. *On the interval* $(0, 1)$ *let there be given a set of functions* $\Phi = \{\varphi_n(x)\}_{n=1}^\infty$, $\varphi_n \in L^2(0, 1)$, $n = 1, 2, \ldots$, *and also a set* $E$ *with* $(0, 1) \subset E \subset R^1$ *and* $m(E) > 1$. *A necessary and sufficient condition for the existence of a system of functions* $\{\psi_n(x)\}_{n=1}^\infty$, *orthonormal on* $E$, *and such that* $\psi_n(x) = \varphi_n(x)$ *for* $n = 1, 2, \ldots$ *and* $x \in (0, 1)$, *is that the following inequalities are satisfied:*

$$\sup_{\{a_n\}:\sum_{n=1}^N a_n^2 \leq 1} \left\| \sum_{n=1}^N a_n \varphi_n \right\|_{L^2(0,1)} \leq 1, \quad N = 1, 2, \ldots. \tag{1}$$

PROOF. The necessity of condition (1) is evident, since if $\Phi$ can be extended to an orthonormal system we have

$$\left\| \sum_{n=1}^N a_n \varphi_n \right\|_{L^2(0,1)} \leq \left\| \sum_{n=1}^N a_n \psi_n \right\|_{L^2(E)} \leq \left( \sum_{n=1}^N a_n^2 \right)^{1/2}.$$

In proving the sufficiency, we observe that (1) is equivalent to having, for $N = 1, 2, \ldots,$[1]

$$\sup_{\{a_n\}:\sum_{n=1}^N a_n^2 \leq 1} \sum_{n,m=1}^N a_n a_m (\varphi_n, \varphi_m) \leq 1. \tag{2}$$

Let $G_N$ be the Gram matrix of $\{\varphi_n\}_{n=1}^N$, that is, $G_N = \{(\varphi_n, \varphi_m)\}_{n,m=1}^N$, and let $I_N$ be the unit matrix of order $N$. If $A$ is a matrix, we denote its elements by $(A)_{n,m}$.

It follows from (2) that the quadratic form defined by the matrix $G'_N = I_N - G_N$ is nonnegative definite, and consequently (as is shown in linear algebra) $G'_N$ can be represented in the form $G'_N = A_N A_N^*$, or equivalently there are vectors $\{z_n\}_{n=1}^N \subset R^N$ such that

$$(z_n, z_m) = (G'_N)_{n,m}, \quad 1 \leq n, m \leq N. \tag{3}$$

---

[1] Here $(\varphi, \psi) = \int_0^1 \varphi(x)\psi(x)\, dx.$

It follows from (3) and the isometry of systems of vectors with the same Gram matrix (see Appendix 1, Proposition 4) that

a) for every $N \geq 1$ there is a set of functions $\{\psi_n^N(x)\}_{n=1}^N \subset L^2(E\backslash(0,1))$ such that

$$\int_{E\backslash(0,1)} \psi_n^N(x)\psi_m^N(x)\,dx = (G_N')_{n,m}, \qquad 1 \leq n, \quad m \leq N; \qquad (4)$$

b) if the set $\{\psi_n^{N-1}\}_{n=1}^{N-1}$ has already been constructed, we can choose the set $\{\psi_n^N\}_{n=1}^N$ satisfying (4) so that

$$\psi_n^{N-1}(x) = \psi_n^N(x), \qquad 1 \leq n \leq N - 1. \qquad (5)$$

We choose $\psi_1^1(x)$ so that

$$\int_0^1 \varphi_1^2(x)\,dx + \int_{E\backslash(0,1)} (\psi_1^1(x))^2\,dx = 1,$$

and then for $N = 2, 3, \ldots$, we choose $\{\psi_n^N(x)\}_{n=1}^N$ so that (4) and (5) are satisfied.

For $n = 1, 2, \ldots$, set

$$\psi_n(x) = \begin{cases} \varphi_n(x) & \text{if } x \in (0,1), \\ \psi_n^n(x) & \text{if } x \in E\backslash(0,1). \end{cases}$$

The functions $\{\psi_n(x)\}$ satisfy the requirements of the theorem, since we have, for $n \leq m$, by (4) and (5),

$$\int_E \psi_n(x)\psi_m(x)\,dx = \int_0^1 \varphi_n(x)\varphi_m(x)\,dx + \int_{E\backslash(0,1)} \psi_n^n(x)\psi_m^m(x)\,dx$$

$$= \int_0^1 \varphi_n(x)\varphi_m(x)\,dx + \int_{E\backslash(0,1)} \psi_n^m(x)\psi_m^m(x)\,dx$$

$$= (G_m)_{n,m} + (G_m')_{n,m} = (I)_{n,m}.$$

This completes the proof of Theorem 1.

REMARK. Since all $N$-dimensional Euclidean spaces are isomorphic, it is clear that when a finite set of functions $\{\varphi_n(x)\}_{n=1}^N$, $x \in (0,1)$, is orthogonalized, the functions $\{\psi_n(x)\}_{n=1}^N$, $x \in E\backslash(0,1)$ (see the statement of Theorem 1), can be taken to be in any $N$-dimensional subspace of $L^2(E\backslash(0,1))$. In particular, if $E = (0,b)$, $b > 1$, the functions $\psi_n(x)$, $x \in (1,b)$, can be taken to be in $\mathscr{D}_N(1,b)$.

Some problems require an extension of the functions $\{\varphi_n(x)\}_{n=1}^N$ such that the $\varphi_n(x)$ are pairwise orthogonal on $E \supset (0,1)$ and in addition are uniformly bounded (by a constant independent of $N$) on $E\backslash(0,1)$. Such an extension is easily constructed directly if $\{\varphi_n\}$ satisfies the condition, more restrictive than (1),

$$\left| \int_0^1 \varphi_n(x)\varphi_m(x)\,dx \right| \leq \frac{C}{N^2} \quad \text{for } n \neq m.$$

Somewhat more careful considerations lead to the following result.

THEOREM 2. *Let* $\{\varphi_n\}_{n=1}^N \subset L^2(0,1)$ *satisfy*

1) $\left| \int_0^1 \varphi_n(x)\varphi_m(x)\,dx \right| \leq \gamma_{|n-m|}, \qquad 1 \leq n, m \leq N;$
2) $\sum_{p=1}^{N-1} \gamma_p < M.$

*Then the functions* $\varphi_n(x)$, $1 \leq n \leq N$, *can be defined on the interval* $[1, 2M+1]$ *so that*

a) $\left| \int_0^{2M+1} \varphi_n(x)\varphi_m(x)\,dx \right| = 0$ *for* $n \neq m$; $1 \leq n, m \leq N$;
b) $|\varphi_n(x)| = 1$ *for* $1 \leq x \leq 2M+1$, $1 \leq n \leq N$;
c) *the functions* $\varphi_n(x)$, $1 \leq n \leq N$, *are piecewise constant on* $[1, 2M+1]$ *with a finite number of intervals of constancy.*

PROOF. We choose a partition of the interval $[1, 2M+1]$ by points $1 = t_0 < \cdots < t_{N-1} = 2M+1$, such that the length of the interval $\delta_p = (t_{p-1}, t_p)$ is $|\delta_p| > 2\gamma_p$, $1 \leq p \leq N-1$, and show that

(∗) for each $p$, $1 \leq p \leq N-1$, we can define a piecewise constant function $\varphi_n(x)$, $1 \leq n \leq N$, such that

1) $|\varphi_n(x)| = 1$, $x \in \delta_p$;
2) $\int_{\delta_p} \varphi_n\varphi_m\,dx = -\int_0^1 \varphi_n\varphi_m\,dx$ if $|m - n| = p$, $1 \leq m, n \leq N$;
3) $\int_{\delta_p} \varphi_n\varphi_m\,dx = 0$ if $|m - n| \neq p$.

It is easily seen then that the functions $\varphi_n(x)$, $n = 1, 2, \ldots$, $x \in \bigcup_{p=1}^{N-1} \delta_p$, yield the extension of the set $\{\varphi_n(x)\}$, $x \in (0,1)$, that is called for in Theorem 2.

To prove (∗), we choose a number $p$, $1 \leq p \leq N-1$, and divide the block of numbers $\{1, \ldots, N\}$ into two parts $P$ and $P'$, where $P = \{n \in [1, N]: 2sp + 1 \leq n \leq (2s+1)p, s = 0, 1, \ldots\}$ and $P' = \{1, \ldots, N\}\backslash P$. In addition, let $\delta_p^-$ and $\delta_p^+$ be the left-hand and right-hand halves of $\delta_p$:

$$\delta_p^- = \left( \frac{t_{p-1} + t_p}{2}, t_p \right), \qquad \delta_p^+ = \left( t_{p-1}, \frac{t_{p-1} + t_p}{2} \right).$$

On $\delta_p$ we define piecewise constant functions $\varphi_n(x)$ with $|\varphi_n(x)| \equiv 1$ for $n = 1, \ldots, N$, in such a way that

$$\int_{\delta_p^+} \varphi_n\varphi_m\,dx = \begin{cases} -\int_0^1 \varphi_n\varphi_m\,dx & \text{if } m - n = p \text{ and } n \in P, \\ 0 & \text{if } m - n \neq p \text{ or } n \notin P, \end{cases} \tag{6}$$

$$\int_{\delta_p^-} \varphi_n \varphi_m \, dx = \begin{cases} -\int_0^1 \varphi_n \varphi_m \, dx & \text{if } m - n = p \text{ and } n \in P', \\ 0 & \text{if } m - n \neq p \text{ or } n \notin P'. \end{cases} \tag{7}$$

It is clear that $(*)$ follows from (6) and (7). To begin with, for $x \in \delta_p^+$ and $n \in P$, we set

$$\widetilde{\varphi}_n(x) = 1,$$

$$\widetilde{\varphi}_{n+p} = \begin{cases} 1 & \text{if } t_{p-1} < x \le \alpha_n, \\ -1 & \text{if } \alpha_n < x < \dfrac{t_{p-1} + t_p}{2}, \end{cases}$$

where

$$\alpha_n = -\frac{1}{2} \int_0^1 \varphi_n \varphi_{n+p} \, dx + t_{p-1} + \frac{t_p - t_{p-1}}{4}$$

(since $\frac{1}{2}(t_p - t_{p-1}) > \gamma_p \ge |\int_0^1 \varphi_n \varphi_{n+p} \, dx|$, we have $\alpha_n \in \delta_p'$). Then, when $n \in P$,

$$\int_{\delta_p^+} \widetilde{\varphi}_n \widetilde{\varphi}_{n+p} \, dx = (\alpha_n - t_{p-1}) - \left( \frac{t_{p-1} + t_p}{2} - \alpha_n \right)$$

$$= -\int_0^1 \varphi_n \varphi_{n+p} \, dx. \tag{8}$$

The numbers $\alpha_n$, $n \in P$, determine a partition of $\delta_p^+$ into intervals $\omega_s$, $1 \le s \le s_0$. We select a system of piecewise constant functions $\psi_n(x)$, $n \in P$, $x \in \delta_p^+$, such that

1) $|\psi_n(x)| \equiv 1$;
2) $\int_{\omega_s} \psi_n \psi_{n'} \, dx = 0$ for $n \neq n'$, $n, n' \in P$, $1 \le s \le s_0$

(such a system is easily constructed by transforming appropriate Rademacher functions to each interval $\omega_2$).

For $x \in \delta_p^+$, $n \in P$, we set

$$\varphi_n(x) = \widetilde{\varphi}_n(x) \psi_n(x), \qquad \varphi_{n+p}(x) = \widetilde{\varphi}_{n+p}(x) \psi_n(x)$$
$$(\text{for } n + p \le N). \tag{9}$$

By means of (9), the functions $\varphi_n(x)$ are defined on $\delta_p^+$ for all $n$, $1 \le n \le N$. Moreover (by (8) and the properties of $\psi_n(x)$), equation (6) is satisfied. By analogy with the construction on $\delta_p^+$, we define the functions $\varphi_n(x)$, $1 \le n \le N$, on $\delta_p^-$ so that (7) is satisfied. This completes the proof of Theorem 2.

REMARK. Since the transformation $f(x) \to \sqrt{b} f(x/b)$ is an isometry between $L^2(0, 1)$ and $L^2(0, b)$, it is clear that if we have an extension of $\{\varphi_n(x)\}_{n=1}^N$ from $(0, 1)$ to an O.N.S. on $(0, M)$, $M > 1$, with

$\sup_{1 \le x \le M, 1 \le n \le N} |\varphi_n(x)| \le K$, then for every $M' > 1$ there is an extension of the same system to an O.N.S. on $(0, M')$ such that

$$\sup_{\substack{1 \le x \le M' \\ 1 \le n \le N}} |\varphi_n(x)| \le K \left( \frac{M-1}{M'-1} \right)^{1/2}.$$

Condition (1) in Theorem 1 is equivalent to the condition that the norm of the Gram matrix $G_\Phi = \{(\varphi_n, \varphi_m)\}_{n,m=1}^\infty$ of the system $\Phi$ does not exceed 1—recall that

$$\|G_\Phi\| = \sup_{\sum_{n=1}^\infty a_n^2 = \sum_{n=1}^\infty b_n^2 = 1} \sum_{n,m=1}^\infty a_n a_m (G_\Phi)_{n,m}.$$

We can also give, in terms of the Gram matrix of $\Phi$, a criterion for $\Phi$ to be extendable to a complete O.N.S.

THEOREM 3. *Let there be given on $(0,1)$ a set of functions $\Phi = \{\varphi_n(x)\}_{n=1}^\infty$ with $\varphi_n \in L^2(0,1)$, $n = 1, 2 \ldots$; and also a set $E$ with $(0,1) \subset E \subset R^1$, $m(E) > 1$. A necessary and sufficient condition for the existence of an orthonormal system $\{\psi_n(x)\}_{n=1}^\infty$ which is complete in $L^2(E)$ and satisfies $\psi_n(x) = \varphi_n(x)$, $n = 1, 2, \ldots$; $x \in (0,1)$, is that all of the following three conditions are satisfied:*

1) *$\{\varphi_n\}$ is complete in $L^2(0,1)$;*
2) *$G_\Phi^2 = G_\Phi$;*
3) *the matrix $G = I - G_\Phi$ has infinite rank.*

PROOF. I) Let there exist a system $\{\psi_n\}$ as specified in Theorem 3. It is evident that condition 1) is satisfied. Now set $\psi_n'(x) = \psi_n(x)\chi_{(0,1)}(x)$, $x \in E$, $n = 1, 2, \ldots$ (where $\chi_U(x)$ is the characteristic function of $U$). Then by Parseval's equation (see 1.(23')) we obtain

$$
\begin{aligned}
(G_\Phi)_{m,n} &= \int_0^1 \varphi_m \varphi_n \, dx = \int_E \psi_m' \psi_n' \, dx \\
&= \sum_{s=1}^\infty \int_E \psi_m' \psi_s \, dx \int_E \psi_n' \psi_s \, dx \\
&= \sum_{s=1}^\infty \int_0^1 \varphi_m \varphi_s \, dx \int_0^1 \varphi_n \varphi_s \, dx \\
&= \sum_{s=1}^\infty (G_\Phi)_{m,s} (G_\Phi)_{n,s}.
\end{aligned}
\tag{10}
$$

Since $(G_\Phi)_{n,s} = (G_\Phi)_{s,n}$, equation (2) follows from (10) (notice that the series on the right of (10) is absolutely convergent, since

$$\sum_{s=1}^{\infty} (G_\Phi)_{m,s}^2 = \|\psi_m'\|_{L^2(E)}^2 \leq 1$$

for every $m$). Finally, since the system $\widetilde{\Phi} = \{\psi_n(x)\chi_{E\backslash(0,1)}(x)\}_{n=1}^{\infty}$ is complete in $L^2(E\backslash(0,1))$, it contains an infinite linearly independent subsystem, and therefore the rank of $G_{\widetilde{\Phi}}$ is infinite. Since $G_{\widetilde{\Phi}} = I - G_\Phi$, condition 3) is established.

II). Let the system $\{\varphi_n(x)\}_{n=1}^{\infty}$, $x \in (0,1)$, satisfy conditions 1)–3). Select any O.N.S. $\{f_s\}_{s=1}^{\infty}$ that is complete in $L^2(E\backslash(0,1))$, and set

$$p_n(x) = \sum_{s=1}^{\infty} (\widetilde{G})_{n,s} f_s(x), \qquad x \in E\backslash(0,1) \tag{11}$$

(the series in (11) converges in $L^2(E\backslash(0,1))$, since $\sum_{s=1}^{\infty} (\widetilde{G})_{n,s}^2 < \infty$ for $n = 1, 2, \ldots$, by condition 2)). Since $(\widetilde{G})^2 = (I - G_\Phi)^2 = I - 2G_\Phi + G_\Phi^2 = I - G_\Phi$ (see 2)), we obtain by Parseval's equation, for $1 \leq n, k < \infty$,

$$\int_{E\backslash(0,1)} p_n p_k \, dx = \sum_{s=1}^{\infty} (\widetilde{G})_{n,s} (\widetilde{G})_{k,s} = \sum_{s=1}^{\infty} (\widetilde{G})_{n,s} (\widetilde{G})_{s,k} = (\widetilde{G})_{n,k}. \tag{12}$$

Let $P$ be the closed linear manifold of $L^2(E\backslash(0,1))$ spanned by the functions $p_n(x)$ (by condition 3), $P$ is infinite-dimensional), and let $U: P \to L^2(E\backslash(0,1))$ be the isometric operator that maps $P$ onto $L^2(E\backslash(0,1))$. For $n = 1, 2, \ldots$, set

$$\psi_n(x) = \begin{cases} \varphi_n(x) & \text{if } x \in (0,1), \\ U(p_n, x) & \text{if } x \in E\backslash(0,1). \end{cases}$$

Then (see (12))

$$\int_E \psi_n \psi_k \, dx = \int_0^1 \varphi_n \varphi_k \, dx + \int_{E\backslash(0,1)} U(p_n) U(p_k) \, dx$$

$$= (G_\Phi)_{n,k} + \int_{E\backslash(0,1)} p_n p_k \, dx = (I)_{n,k},$$

i.e., $\{\psi_n(x)\}$, $x \in E$, is an O.N.S. To establish the completeness of $\{\psi_n\}$, it is sufficient to verify Parseval's equation for the $\{\psi_n\}$-expansion of every element of a system $\Omega$ that is complete in $L^2(E)$. To do this, we take $\Omega$ to be the system

$$\Omega = \{\psi_n \chi_{(0,1)}, \ \psi_n \chi_{E\backslash(0,1)}, \ n = 1, 2, \ldots\}$$

(it is easily verified that $\Omega$ is complete in $L^2(E)$). Using condition 2) for $n = 1, 2, \ldots$, we obtain

$$\sum_{s=1}^{\infty} \left( \int_E \psi_n \chi_{(0,1)} \psi_s \, dx \right)^2 = \sum_{s=1}^{\infty} \left( \int_0^1 \varphi_n \varphi_s \, dx \right)^2$$

$$= \sum_{s=1}^{\infty} (G_\Phi)_{n,s}^2 = \sum_{s=1}^{\infty} (G_\Phi)_{n,s} (G_\Phi)_{s,n}$$

$$= (G_\Phi)_{n,n} \int_0^1 \varphi_n^2 \, dx = \int_E (\psi_n \chi_{(0,1)})^2 \, dx.$$

Similarly, by (12),

$$\sum_{s=1}^{\infty} \left( \int_{E \setminus (0,1)} \psi_n \psi_s \, dx \right)^2 = \sum_{s=1}^{\infty} \left( \int_{E \setminus (0,1)} p_n p_s \, dx \right)^2 = \sum_{s=1}^{\infty} (\tilde{G})_{n,s}^2 = (\tilde{G}_{n,n})$$

$$= \int_{E \setminus (0,1)} p_n^2(x) \, dx = \int_{E \setminus (0,1)} \psi_n^2(x) \, dx.$$

We have therefore established the completeness of $\{\psi_n\}$, and hence Theorem 3 is established.

The following proposition is often useful in constructing various orthogonal systems.

THEOREM 4. *Let* $\Phi = \{\varphi_n(x)\}_{n=1}^N$ *and* $\Phi' = \{\varphi_n(x)\}_{n=N+1}^M$ $[M > N, x \in (0, 1)]$ *be two sets of orthogonal functions. Then for* $x \in (0, 1)$ *there is a function* $\varepsilon(x)$*, with* $|\varepsilon(x)| = 1$*, such that the functions* $\{\psi_n(x)\}_{n=1}^M$ *defined by the equations*

$$\psi_n(x) = \begin{cases} \psi_n(x) & \text{if } 1 \leq n \leq N, \\ \varepsilon(x) \varphi_n(x) & \text{if } N < n \leq M, \end{cases} \qquad x \in (0, 1),$$

*form an O.N.S.*

LEMMA 1. *Let* $f_1, f_2 \in L^1(0, 1)$ *and let* $m\{x \in (0, 1): f_1(x) = 0\} = 0$*. Then if, for some* $\alpha_0 > 0$*,*

$$\min_{\alpha: |\alpha| < \alpha_0} \|f_1 + \alpha f_2\|_1 \geq \|f_1\|_1, \tag{13}$$

*we have*

$$\int_0^1 f_2(x) \, \text{sgn} \, f_1(x) \, dx = 0. \tag{14}$$

PROOF. Since $f_1(x) \neq 0$ a.e. on $(0, 1)$, we have

$$\|f_1 + \alpha f_2\|_1 = \|f_1 \, \text{sgn} \, f_1 + \alpha f_2 \, \text{sgn} \, f_1\|_1.$$

Consequently it is enough to consider the case $f_1(x) \geq 0$, $x \in (0, 1)$, and show that in this case (13) implies that

$$\int_0^1 f_2(x)\, dx = 0. \tag{15}$$

In this case $f_1(x) > 0$ a.e. on $(0, 1)$ and

$$I_\alpha = \int_0^1 |f_1 + \alpha f_2|\, dx = \int_0^1 \left|1 + \alpha \frac{f_2}{f_1}\right| f_1\, dx = \int_0^1 |1 + \alpha g|\, d\mu, \tag{16}$$

where $\mu = f_1\, dx$ is an absolutely continuous measure and $g(x) = f_2(x)/f_1(x) \in L^1(\mu)$. It follows from (16) with $\alpha > 0$ that

$$
\begin{aligned}
I_\alpha &= \int_{\{x:1+\alpha g \geq 0\}} [1 + \alpha g]\, d\mu - \int_{\{x:1+\alpha g < 0\}} [1 + \alpha g]\, d\mu \\
&= \int_0^1 [1 + \alpha g]\, d\mu - 2\int_{\{x:1+\alpha g < 0\}} [1 + \alpha g]\, d\mu \\
&= \int_0^1 f_1(x)\, dx + \alpha \int_0^1 g\, d\mu - 2\mu \left\{ x: g(x) < -\frac{1}{\alpha} \right\} \\
&\quad - 2\alpha \int_{\{x:g<-1/\alpha\}} g\, d\mu.
\end{aligned} \tag{17}
$$

Similarly, when $\alpha < 0$,

$$
\begin{aligned}
I_\alpha &= \int_0^1 f_1\, dx + \alpha \int_0^1 g\, d\mu - 2\mu \left\{ x: g > -\frac{1}{u} \right\} \\
&\quad - 2\alpha \int_{\{x:g>-1/\alpha\}} g\, d\mu.
\end{aligned} \tag{18}
$$

But since $g \in L^1(\mu)$ and $\mu$ is absolutely continuous, we have

$$\mu\{x: |g(x)| > t\} \leq \frac{1}{t} \int_{\{x:|g(x)|>t\}} |g(x)|\, d\mu = o(1/t) \quad \text{as } t \to \infty.$$

Consequently we find from (17) and (18) that when $\alpha \to 0$

$$I_\alpha = \int_0^1 f_1(x)\, dx + \alpha \int_0^1 g(x)\, d\mu + o(\alpha),$$

and therefore (see (13))

$$\int_0^1 g(x)\, d\mu = \int_0^1 f^2(x)\, dx = 0.$$

This establishes (15) and therefore Lemma 1.

LEMMA 2. *Let $g \in L^1(0,1)$, $f_i \in L^1(0,1)$ for $i = 1,\dots,m$, and let $f_0(x) = \sum_{i=1}^{M} \alpha_1^0 f_i(x)$ have the property that*

$$\|g - f_0\|_1 = \inf_{\{\alpha_i\}} \left\| g - \sum_{i=1}^{m} \alpha_i f_i \right\|_1. \tag{19}$$

*Then if*

$$m\left\{ x \in (0,1): g(x) - \sum_{i=1}^{m} \alpha_i f_i(x) = 0 \right\} = 0, \tag{20}$$

*for all $\alpha_i$, $i = 1,\dots,m$, we have the equations*

$$\int_0^1 f_i(x) \operatorname{sgn}[g(x) - f_0(x)] \, dx = 0, \qquad i = 1,\dots,m.$$

In fact, if $\psi(x) = g(x) - f_0(x)$, then $m\{x \in (0,1): \psi(x) = 0\} = 0$ by (20), and (see (19)) for any $\alpha$ we have $\|\psi - \alpha f_i\| > \|\psi\|_1$ for $i = 1, 2, \dots$. Therefore, by Lemma 1, we have $\int_0^1 f_i(x) \operatorname{sgn} \psi(x) \, dx = 0$, $i = 1,\dots,m$. This completes the proof of Lemma 2.

LEMMA 3. *Let there be given a set of functions $\{f_i\}_{i=1}^{m} \subset L^1(0,1)$. Then there is a function $g \in L^1(0,1)$ such that (20) is valid for arbitrary numbers $\alpha_i$, $i = 1,\dots,m$.*

PROOF. We shall show that $g(x)$ may be taken to be the function $e^{\lambda x}$ for some $\lambda \in (0,1)$. Suppose, on the contrary, that for every $\lambda \in (0,1)$ there is a set $\{\alpha_i\}_{i=1}^{m}$ of coefficients such that

$$m(E_\lambda) \equiv m\left\{ x \in (0,1): e^{\lambda x} = \sum_{i=1}^{m} \alpha_i f_i(x) \right\} > 0.$$

Then it is clear that we can find a number $\delta \in (0,1)$ and a sequence of pairwise different numbers $\{\lambda_j\}_{j=1}^{\infty} \subset (0,1)$ such that

$$m(E_{\lambda_j}) > \delta, \qquad j = 1, 2, \dots.$$

Then, for $R = 1, 2, \dots$,

$$R\delta < \int_0^1 \sum_{j=1}^{R} \chi_{E_{\lambda_j}}(x) \, dx \le Rm\left\{ \sum_{j=1}^{R} \chi_{E_{\lambda_j}}(x) > \frac{R\delta}{2} \right\} + \frac{R\delta}{2}$$

and consequently

$$m\left\{ x: \sum_{j=1}^{R} \chi_{E_{\lambda_j}}(x) > \frac{R\delta}{2} \right\} > \frac{R\delta}{2}. \tag{21}$$

It follows from (21) that

$$m\left\{x:\sum_{j=1}^{R_0}\chi_{E_{\lambda_j}}(x) > m+1\right\} > 0 \quad \text{if } R_0 > \frac{2}{\delta}(m+1),$$

and therefore among the numbers $\lambda_1,\ldots,\lambda_R$ there are numbers $\lambda_{j_s}$, $s = 1,\ldots,m+1$, such that

$$m(G) > 0 \quad \text{where } G = \bigcap_{s=1}^{m+1} E_{\lambda_{j_s}}. \tag{22}$$

Inequality (22) implies that

$$\exp(\lambda_{j_s}x) = \sum_{i=1}^{m}\alpha_i^{(s)}f_i(x), \quad s = 1,\ldots,m+1, \ x \in G,$$

i.e., the functions $\exp(\lambda_{j_s}x)$, $x \in G$, lie in an $m$-dimensional linear space, and consequently are linearly dependent:

$$\sum_{s=1}^{m+1}\beta_s\exp(\lambda_{j_s}x) = 0, \quad x \in G, \ m(G) > 0, \ \sum_{s=1}^{m+1}\beta_s^2 > 0.$$

Since the function $F(x) = \sum_{s=1}^{m+1}\beta_s\exp(\lambda_{j_s}z)$ is entire, it follows from the preceding equation and the uniqueness theorem for analytic functions that $F(x) = 0$, $-\infty < x < \infty$. Thus we have arrived at a contradiction, since it is evident that, as $x \to +\infty$,

$$F(x) \sim \beta_\nu\exp(\lambda_{j_\nu}x),$$

where $\lambda_{j_\nu} = \max\{\lambda_{j_s}: \beta_s \neq 0, \ s = 1,\ldots,m+1\}$.

This completes the proof of Lemma 3.

COROLLARY 1. *For every set* $\{f_i(x)\}_{i=1}^{m} \subset L^1(0,1)$ *there is a function* $\varepsilon(x)$ *with* $|\varepsilon(x)| = 1$ *for almost every* $x \in (0,1)$, *such that*

$$\int_0^1 f_i(x)\varepsilon(x)\,dx = 0, \quad i = 1,\ldots,m.$$

PROOF. Using Lemma 3, we can find a function $g(x)$ for which (20) is satisfied for every $\alpha_i$, $i = 1,\ldots,m$. Furthermore, if $L \subset L^1(0,1)$ is the subspace spanned by $f_i$, $i = 1,\ldots,m$, we can choose $f_0 \in L$ so that

$$\|f_0 - g\|_1 = \inf_{f\in L}\|f - g\|_1.$$

Then, by Lemma 2, the function $\varepsilon(x) = \operatorname{sgn}[g(x) - f_0(x)]$ satisfies the requirements of Corollary 1.

The conclusion of Theorem 4 follows at once from Corollary 1. In fact, if we consider the set

$$\varphi_n(x)\varphi_m(x), \qquad 1 \le n \le N < m \le M,$$

then if we apply Corollary 1 to this set, we obtain a function $\varepsilon(x)$ $[|\varepsilon(x)| = 1, x \in (0, 1)]$ for which

$$\int_0^1 \varphi_n(x)[\varepsilon(x)\varphi_m(x)]\,dx = 0, \qquad 1 \le n \le N < m \le M.$$

It follows from the preceding equation that $\{\psi_n\}$ is orthonormal. This completes the proof of Theorem 4.

## §2. Two theorems on sequences of functions

In this section we prove two theorems, which can be used to obtain many factorization theorems. Some applications of the results obtained here will be given in later sections. Before stating the theorems, we require two definitions.

DEFINITION 1. A system

$$\{f_n(x)\}_{n=1}^\infty, \qquad x \in (0, 1), \tag{23}$$

of measurable functions is called an *A-system* if, for every sequence $\{\xi_n\}_{n=1}^\infty$ with $\sum_{n=1}^\infty |\xi_n| < \infty$,

$$\lim_{n \to \infty} \xi_n f_n(x) = 0 \quad \text{for almost every } x \in (0, 1).$$

DEFINITION 2. A system (23) of measurable functions is said to be an *absolute-convergence system* if, for every sequence $\{\xi_n\}_{n=1}^\infty$, $\sum_{n=1}^\infty |\xi_n| < \infty$,

$$\sum_{n=1}^\infty |\xi_n f_n(x)| < \infty \quad \text{for almost every } x \in (0, 1).$$

REMARK. It is clear that (23) is an *A*-system or an absolute-convergence system at the same time as $\{|f_n(x)|\}_{n=1}^\infty$.

THEOREM 5. *A necessary and sufficient condition for a system* (23) *to be an A-system is that for each $\varepsilon > 0$ there are a set $E_\varepsilon \subset (0, 1)$, $m(E_\varepsilon) > 1 - \varepsilon$, and a constant $C_\varepsilon$ such that for every $y > 0$*

$$m\{x \in E_e : |f_n(x)| > y\} \le \frac{C_\varepsilon}{y}, \qquad n = 1, 2, \dots. \tag{24}$$

THEOREM 6. *A necessary and sufficient condition for a system* (23) *to be an absolute convergence system is that for each $\varepsilon > 0$ there are a set $E_\varepsilon \subset (0, 1)$, $m(E_\varepsilon) > 1 - \varepsilon$, and a constant $C_\varepsilon$ such that*

$$\sup_{1 \le n < \infty} \int_{E_\varepsilon} |f_n(x)|\,dx \le C_\varepsilon. \tag{25}$$

In the proofs of both theorems we shall need the following simple proposition.

LEMMA 1. *Let the sequence $\{g_k\}_{k=1}^\infty$ of functions converge weakly in $L^p(0,1)$ $(1 \le p < \infty)$ to $g(x)$, and for $k = 1, 2, \ldots$, let*
1) $0 \le g_k(x) \le 1, \qquad x \in (0,1)$;
2) $\int_0^1 g_k(x)\,dx \ge \delta > 0$.
*Then $0 \le g(x) \le 1$ for almost every $x \in (0,1)$, and there is a number $\varepsilon > 0$ such that*

$$m\{x \in (0,1): g(x) > \varepsilon\} \ge \delta. \tag{26}$$

PROOF. By the weak convergence of $g_k(x)$ to $g(x)$ we have, for every set $E \subset (0,1)$ with $m(E) > 0$,

$$\int_E g(x)\,dx = \int_0^1 \chi_E(x) g(x)\,dx = \lim_{k\to\infty} \int_0^1 \chi_E(x) g_k(x)\,dx \ge 0,$$

from which it follows that $g(x) \le 0$ for almost every $x \in (0,1)$. Similarly, the relation

$$\int_E [1 - g(x)]\,dx = \lim_{k\to\infty} \int_E [1 - g_k(x)]\,dx \ge 0, \qquad E \subset (0,1),$$

implies that $g(x) \le 1$ for almost all $x \in (0,1)$. Then

$$m(G) \ge \int_0^1 g(x)\,dx = \lim_{k\to\infty} \int_0^1 g_k(x)\,dx \ge \delta.$$

If $m(G) > \delta$, then, since $\lim_{\varepsilon\to0} m\{x: 0 < g(x) < \varepsilon\} = 0$, the inequality (26) is satisfied for sufficiently small $\varepsilon$. It is possible to have $m(G) = \delta$ only if $g(x) = 1$ for almost all $x \in G$, and then (26) again follows. This completes the proof of Lemma 1.

PROOF OF THEOREM 5. *Sufficiency of* (24). Choose $\varepsilon > 0$; we shall show that when (24) is satisfied, the sequence $\xi f_n(x)$ $(\sum_{m=1}^\infty |\xi_n| < \infty)$ converges to zero for almost all $x \in E_\varepsilon$. Hence, since $\varepsilon > 0$ is arbitrary and $m(E_\varepsilon) > 1 - \varepsilon$, it follows that the sequence $\xi_n f_n(x)$ converges to zero for almost all $x \in (0,1)$. By (24),

$$\sum_{n=1}^\infty m\{x \in E_\varepsilon: |\xi_n f_n(x)| > \delta\} \le C_\varepsilon \delta^{-1} \sum_{n=1}^\infty |\xi_n| < \infty,$$

for all $\delta > 0$ and $\{\xi_n\}$ with $\sum_{n=1}^\infty |\xi_n| < \infty$, and consequently

$$m\left\{ \varlimsup_{n\to\infty} \{x \in E_\varepsilon: |\xi_n f_n(x)| > \delta\} \right\} = 0. \tag{27}$$

Since the number $\delta > 0$ in (27) can be arbitrarily small, we find that $\lim_{n\to\infty} |\xi_n f_n(x)| = 0$ for almost all $x \in E_\varepsilon$.

*Necessity of* (24). First we obtain that it is enough to consider the case when

$$0 \le f_n(x) \le M_n < \infty, \qquad n = 1, \ldots, x \in (0,1). \tag{28}$$

We have already noticed that we may restrict our attention to nonnegative functions (see the remark following Definition 2). If $f'_n > 0$, $n = 1, 2, \ldots$, is an $A$-system, for each $\varepsilon > 0$ we can find $M_n$ such that

$$\sum_{n=1}^{\infty} m(G_n) \le \frac{\varepsilon}{2}, \qquad G_n \equiv \{x \in (0,1): f'_n(x) > M_n\}.$$

Then it is clear that the functions $f_n(x)$, $n = 1, 2, \ldots$, with

$$f_n(x) = \begin{cases} f'_n(x) & \text{if } x \notin G_n, \\ 0 & \text{if } x \in G_n, \end{cases} \qquad x \in (0,1),$$

also form an $A$-system and satisfy (28). Supposing that the necessity of (24) has already been established for the system $\{f_n\}$, we find a set $E_{\varepsilon/2} \subset (0,1)$ with $m(E_{\varepsilon/2}) > 1 - \varepsilon/2$, and a constant $C_\varepsilon$, for which (24) is satisfied. But then, setting $C'_\varepsilon = C_{\varepsilon/2}$ and $E'_\varepsilon = E_{\varepsilon/2} \setminus \bigcup_{n=1}^{\infty} G_n$, we obtain

$$m\{x \in E'_\varepsilon: f'_n(x) > y\} \le \frac{C'_\varepsilon}{y}, \qquad n = 1, 2, \ldots, \qquad m(E'_\varepsilon) > 1 - \varepsilon,$$

i.e., (24) also holds for $\{f'_n\}$ and every $\varepsilon > 0$.

Now let $\{f_n\}$ be an $A$-system satisfying (28). For convenience, we break the proof of the necessity of condition (28) into several steps.

a) We need to show that for every $\varepsilon > 0$ there are a set $E_\varepsilon \subset (0,1)$ and a constant $C$ such that

$$\int_{E_\varepsilon} \chi_{n,z}(x)\, dx \le C_\varepsilon z, \qquad n = 1, \ldots, z > 0, \tag{29}$$

where $\chi_{n,z}(x)$ is the characteristic function of the set $E_{n,z} = \{x \in \alpha(0,1): f_n(x) > 1/z\}$.

There are two possibilities for the sequence $\{f_n\}$:

I) For each $\varepsilon > 0$ and $z_0 > 0$ there are a set $F_{\varepsilon,z_0}$ with $m(F_{\varepsilon,z_0}) \ge 1 - \varepsilon$, and a constant $A_\varepsilon$ (depending only on $\varepsilon$), such that

$$\int_{F_{\varepsilon,z_0}} \chi_{n,z}(x)\, dx \le A_\varepsilon z, \qquad n = 1, 2, \ldots, z \ge z_0.$$

II) There is an $\varepsilon_0 > 0$ such that for every $A$ there is a number $z_0 = z_0(A)$ such that, for every set $F \subset (0,1)$ with $m(F) > 1 - \varepsilon_0$,

$$m\left\{ x \in F: f_{n'}(x) > \frac{1}{z'} \right\} > Az' \tag{30}$$

for some $n' \ge 1$ and $z' > z_0$.

Let us show that if case I) occurs for $\{f_n\}$ then (29) is satisfied, and consequently also the required inequality (24). We fix $\varepsilon > 0$, take the sequence $z_j = 1/j$, $j = 1, 2, \ldots$, and let $\Phi_j = F_{\varepsilon, z_j}$, $j = 1, 2, \ldots$, be the set such that $m(\Phi_j) \geq 1 - \varepsilon$ for $j = 1, 2, \ldots$, and

$$\int_{\Phi_j} \chi_{n,z}(x)\,dx \leq A_\varepsilon z \quad \text{for } n = 1, 2, \ldots; \ z \geq \frac{1}{j}. \tag{31}$$

By the weak compactness of the unit ball in $L^2(0, 1)$, we can select, from the sequence $\{\chi_{\Phi_j}(x)\}_{j=1}^\infty$ of characteristic functions of the sets $\Phi_j$, a subsequence $\chi_{\Phi_{j_k}}(x)$, $k = 1, 2, \ldots$, that converges weakly in $L^2(0, 1)$ to a function $\chi^0(x)$. Then by (31)

$$\int_0^1 \chi^0(x)\chi_{n,z}(x)\,dx \leq A_\varepsilon z \quad \text{for } z > 0, \ n = 1, 2, \ldots. \tag{32}$$

Here (see Lemma 1) $m\{x \in (0, 1) : \chi^0(x) \geq \eta\} \geq 1 - \varepsilon$ for some number $\eta = \eta(\varepsilon) > 0$. Let $E_\varepsilon = \{x \in (0, 1) : \chi^0(x) > \eta\}$. Then (see (32))

$$\int_{E_\varepsilon} \chi_{n,z}(x)\,dx \leq \frac{1}{\eta}\int_0^1 \chi^0(x)\chi_{n,z}(x)\,dx \leq \frac{A_\varepsilon}{\eta} z$$

for $n = 1, 2, \ldots$, and $z > 0$, i.e., we have the required inequality (29).

b) Let us show that case II) is impossible for an $A$-system $\{f_n\}_{n=1}^\infty$, and therefore that the only possibility is case I), which has already been discussed.

Suppose the contrary, i.e, that the $A$-system $\{f_n\}$ is in case II).

The next step is to construct (using (30)), corresponding to each constant $A > 0$, numbers $z_1, \ldots, z_N$ and pairwise different indices $n_1, \ldots, n_N$, such that

1) $z_i > 0$;
2) $\sum_{i=1}^N z_i \leq \frac{1}{A}$;
3) $z_i f_{n_i}(x) \geq 1$ for $x \in E_i$, $i = 1, \ldots, N$, $\qquad$ (33)

where the sets $E_i$ are pairwise disjoint and $m(\bigcup_{i=1}^n E_i) \geq \varepsilon_0 > 0$. For this purpose we construct successively (for $s = 1, 2, \ldots$) collections $\{\eta_i^s\}_{i=1}^{i_s}$ of positive numbers, $\{n_i^s\}_{i=1}^{i_s}$ of pairwise different positive integers, and $\{E_i^s\}_{i=1}^{i_s}$ of disjoint sets, such that the following relations hold:

$$\begin{aligned}
&1) && 0 < \eta_i^s \leq \frac{m(E_i^s)}{A}; \\
&2) && \eta_i^s f_{n_i^s}(x) \geq 1 \quad \text{for } x \in E_i^s; \\
&3) && \sum_{i=1}^{i_s} m(E_i^s) \geq \min\left\{\varepsilon_0,\, Az_0 + \sum_{i=1}^{i_{s-1}} m(E_i^{s-1})\right\} \\
& && (z_0 = z_0(A); \text{ see } (30)).
\end{aligned} \tag{34}$$

It follows at once from inequality 3) that (33) will be satisfied for some $s_0$ ($s_0 < \varepsilon_0/Az_0$) if we set $N = i_{s_0}$, $z_i = \eta_i^{s_0}$, $n_i = n_i^{s_0}$, $E_i = E_i^{s_0}$, $i = 1, \ldots, i_{s_0}$ (and stop the process of construction at the step with index $s_0$).

Let $i_1 = 1$ for $s = 1$, and choose the numbers $n_1^1$, $\eta_1^1$ and the set $E_1^1$ so that

$$E_1^1 = \{x \in (0,1): \eta_1^1 f_{n_1^1}(x) > 1\}, \qquad m(E_1^1) > A\eta_1^1 \geq Az_0$$

(it follows from (30) that this choice is possible). Then it is clear that relations 1), 2), and 3) in (34) are satisfied when $s = 1$ (we take $i_0 = 0$, $\sum_1^0 = 0$).

Suppose that the sets $\{\eta_i^s\}$, $\{n_i^s\}$, and $\{E_i^s\}$ have already been constructed and that $\sum_{i=1}^{i_s} m(F_i^s) < \varepsilon_0$. Let us construct $\{\eta_i^{s+1}\}$, $\{n_i^{s+1}\}$, and $\{E_i^{s+1}\}$. If we set $\mathscr{F}_0 = (0,1) \setminus \bigcup_{i=1}^{i_s} E_i^s$, then $m(\mathscr{F}_0) < 1 - \varepsilon_0$ and, by (30), there are numbers $\tilde{n}$ and $\tilde{\eta} > z_0$ for which

$$m(\tilde{E}) > A\tilde{\eta} > Az_0, \quad \text{where } \tilde{E} = \left\{x \in \mathscr{F}_0: f_{\tilde{n}}(x) > \frac{1}{\tilde{\eta}}\right\}. \tag{35}$$

Two cases are possible.

i) $\tilde{n} \notin \{n_i^s\}_{i=1}^{i_s}$. Then we set $i_{s+1} = i_s + 1$, and take $n_i^{s+1} = n_i^s$, $\eta_i^{s+1} = \eta_i^s$, $E_i^{s+1} = E_i^s$ if $1 \leq i \leq i_s$; $n_{i_{s+1}}^{s+1} = \tilde{n}$, $\eta_{i_{s+1}}^{s+1} = \tilde{\eta}$, $E_{i_{s+1}}^{s+1} = \tilde{E}$. It is easily seen (by using (35)) that (34) remains valid for the sets defined in this way.

ii) $\tilde{n} \in \{n_i^s\}_{i=1}^{i_s}$, i.e., $\tilde{n} = n_{i_0}^s$ for some $i_0$, $1 \leq i_0 \leq i_s$. Then we have

$$\begin{aligned}
&\text{a)} \quad \tilde{\eta} \leq \frac{m(E)}{A}; \\
&\text{b)} \quad \eta_{i_0}^s \leq \frac{m(E_{i_0}^s)}{A}; \\
&\text{c)} \quad \tilde{\eta} f_{n_{i_0}^s}(x) \geq 1 \quad \text{for } x \in \tilde{E}; \\
&\text{d)} \quad \eta_{i_0}^s f_{n_{i_0}^s}(x) \geq 1 \quad \text{for } x \in E_{i_0}^s.
\end{aligned} \tag{36}$$

Set $i_{s+1} = i_s$, and let, for $i \neq i_0$ and $1 \leq i \leq i_{s+1}$, $n_i^{s+1} = n_i^s$, $\eta_i^{s+1} = \eta_i^s$, $E_i^{s+1} = E_i^s$. Also let

$$n_i^{s+1} = n_i^s, \qquad \eta_i^{s+1} = \max\{\tilde{\eta}, \eta_{i_0}^s\}, \qquad E^{s+1} = \tilde{E} \cup E_{i_0}^s. \tag{37}$$

Then it follows directly from (36) that 1) and 2) of (34) are valid, and inequality 3) of (34) follows from the definition of the set $E_{i_s}^{s+1}$ and (35).

Having completed the induction, we have established the existence, for $s = 1, 2, \ldots$, of sets with the properties (34), and consequently the existence for every constant $A > 0$ of sets $\{z_i\}$, $\{n_i\}$, and $\{E_i\}$ that satisfy (33). If we now take the number $A$ in (33) successively equal to 1, $2^2$, $3^2, \ldots$, then we have, for $k = 1, 2, \ldots$, sets $\{z_i^{(k)}\}_{i=1}^{N_k}$, $\{n_i^{(k)}\}_{i=1}^{N_k}$, such that

$$z_i^{(k)} > 0, \qquad 1 \le i \le N_k, \qquad \sum_{i=1}^{N_k} z_i^{(k)} \le \frac{1}{k^2}, \qquad k = 1, 2, \ldots,$$

$$\max_{1 \le i \le N_k} z_i^{(k)} f_{n_i^{(k)}}(x) \ge 1 \quad \text{for } x \in Q_k, \ m(Q_k) \ge \varepsilon_0 > 0.$$

(38)

Moreover, as is easily seen by using the boundedness of $f_n(x)$ (see (28)), the sets $\{n_i^{(k)}\}$ can be chosen so that

$$\max_{1 \le i \le N_k} n_i^{(k)} < \min_{1 \le i \le N_{k+1}} n_i^{(k+1)}, \qquad k = 1, 2, \ldots.$$

Let us define a sequence of numbers by setting, for $n = 1, 2, \ldots$,

$$\xi_n = \begin{cases} z_i^{(k)} & \text{if } n = n_i^{(k)} \text{ for some } k = 1, 2, \ldots \text{ and} \\ & 1 \le i \le N_k, \\ 0 & \text{if } n \notin \{n_i^{(k)}\}, \end{cases}$$

and let $Q = \overline{\lim}_{k \to \infty} Q_k$. Then it follows from (38) that

$$\sum_{n=1}^{\infty} \xi_n \le \sum_{k=1}^{\infty} \frac{1}{k^2} < \infty, \qquad m(Q) \ge \varepsilon_0 > 0$$

and $\overline{\lim}_{n \to \infty} \xi_n f_n(x) \ge 1$ for every $x \in Q$, which contradicts the assumption that $\{f_n\}$ is an $A$-system. We have therefore shown that case II) is impossible for $A$-systems. This completes the proof of Theorem 5.

PROOF OF THEOREM 6. If inequalities (25) are satisfied, we can find a sequence $\{E_k\}_{k=1}^{\infty}$ of sets such that $m(E_k) \ge 1 - 1/k$ and

$$\sup_{1 \le n < \infty} \int_{E_k} |f_n(x)| \, dx = C_k < \infty.$$

Then for every sequence $\xi_n$ with $\sum_{n=1}^{\infty} |\xi_n| < \infty$ we have

$$\sum_{n=1}^{\infty} \int_{E_k} |\xi_n f_n(x)| \, dx \le C_k \sum_{n=1}^{\infty} |\xi_n| < \infty,$$

whence it follows by Beppo Levi's theorem that the series $\sum_{n=1}^{\infty} \xi_n f_n(x)$ converges almost everywhere on $E_k$ for $k = 1, 2, \ldots$, and therefore almost everywhere on $(0, 1)$.

The necessity of (25) is an essentially deeper result. For its proof we require the following proposition   (von Neumann's minimax theorem; for a proof see Appendix 1, §3):

Let $A = \{a_{ij}\}$, $a_{ij} \geq 0$, $j = 1, \ldots, n$; $i = 1, \ldots, m$, be a matrix with nonnegative elements, and let

$$\sigma_m = \left\{ x \in R^m \colon x = (x_1, \ldots, x_m), x_i \geq 0, \sum_{i=1}^{m} x_i = 1 \right\}.$$

Then the quadratic form

$$F(x, y) = \sum_{\substack{1 \leq i \leq m \\ 1 \leq j \leq n}} a_{i,j} x_i y_j \qquad (x = \{x_i\}, y = \{y_i\})$$

satisfies the equation

$$\min_{y \in \sigma_n} \max_{x \in \sigma_m} F(x, y) = \max_{x \in \sigma_m} \min_{y \in \sigma_n} F(x, y). \tag{39}$$

We prove the following lemma by using (39).

LEMMA 2. *Let there be given a set $\{f_n(x)\}_{n=1}^{N}$, $x \in (0, 1)$, of nonnegative bounded functions, and numbers $R > 0$ and $\delta \in (0, 1)$. In addition, let there be, for every set $\overline{\xi} = \{\xi_n\}_{n=1}^{N} \in \sigma_N$, a set $E_{\overline{\xi}}$ for which $m(E_{\overline{\xi}}) \geq \delta$ and*

$$\int_{E_{\overline{\xi}}} \sum_{n=1}^{N} \xi_n f_n(x) \leq R.$$

*Then there is a function $\chi(x)$, $x \in (0, 1)$, for which*

1) $0 \leq \chi(x) \leq 1$, $x \in (0, 1)$;
2) $\int_0^1 \chi(x) \, dx \geq \delta$; $\qquad\qquad\qquad\qquad$ (40)
3) $\int_0^1 \chi(x) \sum_{n=1}^{N} \xi_n f_n(x) \, dx \leq 2R$

*for every set $\overline{\xi} = \{\xi_n\} \in \sigma_N$.*

PROOF OF THE LEMMA. Let $\Omega = \{\overline{\xi}^{(1)}, \ldots, \overline{\xi}^{(k)}\} \subset \sigma_N$ be a collection of sets such that for each set $\overline{\xi} = \{\xi_n\}_{n=1}^{N} \in \sigma_N$ there is an element $\overline{\xi}^{(i)} = \{\xi_n^{(i)}\}_{n=1}^{N} \in \Omega$ such that

$$\sum_{n=1}^{N} \int_0^1 |\xi_n - \xi_n^{(i)}| f_n(x) \, dx \leq R \tag{41}$$

(the existence of the collection $\Omega$ follows at once from the compactness of $\sigma_N$). By the hypothesis of Lemma 2, there is, for each $\overline{\xi}^{(i)} \in \Omega$, a set $E_i$, $m(E_i) \geq \delta$, for which

$$\int_0^1 \chi_{E_i}(x) \sum_{n=1}^{N} \xi_n^{(i)} f_n(x) \, dx \leq R, \tag{42}$$

where $\chi_{E_i}(x)$ is the characteristic function of $E_i$. For $\bar{\xi} = \{\xi_n\}_{n=1}^N \in \sigma_N$ and $\bar{\eta} = \{\eta_i\}_{i=1}^k \in \sigma_k$, we set

$$F(\bar{\xi}, \bar{\eta}) = \sum_{n=1}^N \sum_{i=1}^k \xi_n \eta_i \int_0^1 \chi_{E_i}(x) f_n(x) \, dx$$

and show that for each $\bar{\xi} \in \sigma_N$

$$\min_{\bar{\eta} \in \sigma_k} F(\bar{\xi}, \bar{\eta}) \le 2R. \tag{43}$$

In fact, let $\bar{\xi}^{(i)} \in \Omega$ be a set for which (41) is satisfied. Then (see (42))

$$\int_0^1 \sum_{n=1}^N \xi_n \chi_{E_i}(x) f_n(x) \, dx \le \int_0^1 \sum_{n=1}^N \xi_n^{(i)} \chi_{E_i}(x) f_n(x) \, dx$$

$$+ \int_0^1 \sum_{n=1}^N |\xi_n - \xi_n^{(i)}| f_n(x) \, dx \le 2R,$$

and this means that $F(\bar{\xi}, \bar{\eta}) \le 2R$ for $\bar{\eta} = (0, \ldots, 0, \underset{i}{1}, 0, \ldots, 0)$. This establishes (43). We deduce from (43), by using von Neumann's theorem (see (39)), that

$$\min_{\bar{\eta} \in \sigma_k} \max_{\bar{\xi} \in \sigma_N} F(\bar{\xi}, \bar{\eta}) = \max_{\bar{\xi} \in \sigma_N} \min_{\bar{\eta} \in \sigma_k} F(\bar{\xi}, \bar{\eta}) \le 2R. \tag{44}$$

Let $\bar{\eta}^0 = \{\eta_i^0\}_{i=1}^k$ be a set for which

$$\max_{\bar{\xi} \in \sigma_N} F(\bar{\xi}, \bar{\eta}^0) = \min_{\bar{\eta} \in \sigma_k} \max_{\bar{\xi} \in \sigma_N} F(\bar{\xi}, \bar{\eta})$$

and

$$\chi(x) = \sum_{i=1}^k \eta_i^0 \chi_{E_i}(x).$$

Then:

1) $0 \le \chi(x) \le \sum_{i=1}^k \eta_i^0 = 1, x \in (0, 1)$;
2) $\int_0^1 \chi(x) \, dx = \sum_{i=1}^k \eta_i^0 \int_0^1 \chi_{E_i}(x) \, dx \ge \delta, \sum_{i=1}^k \eta_i^0 = \delta$;
3) by (44) and the definition of the set $\bar{\eta}^0$ we have, for every $\bar{\xi} = \{\xi_n\} \in \sigma_N$,

$$\int_0^1 \chi(x) \sum_{n=1}^N \xi_n f_n(x) \, dx = \int_0^1 \sum_{n=1}^N \xi_n \sum_{i=1}^k \eta_i^0 \chi_{E_i}(x) f_n(x) \, dx$$

$$= F(\bar{\xi}, \bar{\eta}) \le 2R.$$

The function $\chi(x)$ has the required properties. This completes the proof of Lemma 2.

We now turn directly to the proof of the necessity of inequality (25) of Theorem 6. We may assume without loss of generality that the sequence $\{f_n\}$ consists of nonnegative and individually bounded functions. This is established just as in Theorem 5 (see the discussion after (28)).

Assume the contrary, i.e., that there is an absolute-convergence system $\{f_n\}_{n=1}^{\infty}$ $(0 < f_n(x) \le M_n, \; x \in (0,1), \; n = 1,2,\dots)$ such that for some $\varepsilon_0 > 0$

$$\sup_{1 \le n < \infty} \int_E f_n(x)\, dx = \infty \tag{45}$$

for every $E \subset (0,1)$, $m(E) \ge \delta_0 = 1 - \varepsilon_0$.

a) Let us show that (45) implies the existence for each $k = 1,2,\dots$, of a sequence $\overline{\xi}^{(k)} = \{\xi_n^{(k)}\}_{n=1}^{\infty}$ with $\xi_n^{(k)} \ge 0$, $\sum_{n=1}^{\infty} \xi_n^{(k)} = 1$, such that

$$\int_E \sum_{n=1}^{\infty} \xi_n^{(k)} f_n(x)\, dx \ge k^2 \quad \text{for every } E \subset (0,1), \; m(E) \ge \delta_0. \tag{46}$$

In fact, in the contrary case there would be a number $k_0$ such that for every sequence $\overline{\xi} = \{\xi_n\}_{n=1}^{\infty}$ $(\sum_{n=1}^{\infty} \xi_n = 1, \; \xi_n \ge 0)$ there would be a set $E_{\overline{\xi}}$, with $m(E_{\overline{\xi}}) \ge \delta_0$, for which

$$\int_{E_{\overline{\xi}}} \sum_{n=1}^{\infty} \xi_n f_n(x)\, dx \le k_0^2. \tag{47}$$

But then, fixing $N$, we could find for each set $\overline{\xi} = \{\xi\} \in \sigma_N$ a set $E_{\overline{\xi}}$ with $m(E_{\overline{\xi}}) \ge \delta_0$ for which (47) holds. Then we could use Lemma 2 to obtain a function $\chi_N(x)$ $(0 \le \chi_N(x) \le 1, \int_0^1 \chi_N(x)\, dx \ge \delta_0)$ for which

$$\sup_{\overline{\xi} \in \sigma_N} \sum_{n=1}^{N} \xi_n \int_0^1 \chi_N(x) f_N(x)\, dx \le 2k_0^2.$$

Consequently (by the equation $\sup_{\overline{\xi} \in \sigma_N} \sum_{n=1}^{N} a_n \xi_n = \max_{1 \le n \le N} a_n$, $a_n \ge 0$, $1 \le n \le N$),

$$\max_{1 \le n \le N} \int_0^1 \chi_N(x) f_n(x)\, dx \le 2k_0^2, \qquad N = 1,2,\dots. \tag{48}$$

By using the weak compactness of the unit ball in $L^2(0,1)$, and then Lemma 1, we could extract from the sequence $\{\chi_N(x)\}_{N=1}^{\infty}$ a subsequence $\chi_{N_s}(x)$, $s = 1,2,\dots$, that converges weakly in $L^2(0,1)$ to a nonnegative function $\chi(x)$ for which, for some $\gamma > 0$, we would have

$$\sup_{1 \le n < \infty} \int_E f_n(x)\, dx \le \frac{1}{\gamma} \sup_{1 \le n < \infty} \int_0^1 \chi(x) f_n(x)\, dx \le \frac{2k_0^2}{\gamma} < \infty,$$

$$m(E) \ge \delta_0.$$

This contradicts the assumption that (45) holds. Hence we have shown that (45) implies the existence of a sequence $\bar{\xi}^{(k)}$, $k = 1, 2, \ldots$, for which (46) is valid.

b) We define a sequence $\{\xi_n^0\}_{n=1}^\infty$ by setting

$$\xi_n^0 = \sum_{k=1}^\infty \frac{1}{k^2} \xi_n^{(k)}, \qquad n = 1, 2, \ldots,$$

where $\{\xi_n^{(k)}\}$ satisfies (46), and consider the series

$$\sum_{n=1}^\infty \xi_n^0 f_n(x). \tag{49}$$

By (46), we have, for every set $E \subset (0, 1)$ with $m(E) \geq \delta_0$,

$$\sum_{n=1}^\infty \int_E \xi_n^0 f_n(x)\, dx = \sum_{n=1}^\infty \sum_{k=1}^\infty \frac{\xi_n^{(k)}}{k^2} \int_E f_n(x)\, dx$$

$$= \sum_{k=1}^\infty \frac{1}{k^2} \sum_{n=1}^\infty \xi_n^{(k)} \int_E f_n(x)\, dx \geq \sum_{k=1}^\infty 1 = \infty. \tag{50}$$

It follows from (50) that (49) diverges on a set of positive measure[2] although $\sum_{n=1}^\infty \xi_n^0 \leq \sum_{k=1}^\infty 1/k^2 < \infty$, and hence we have a contradiction of the assumption that $\{f_n\}$ is an absolute-convergence system. This contradiction refutes (45) and completes the proof of Theorem 6.

## §3. Structure of systems with convergence in measure for $l^2$

DEFINITION 3. A system of measurable functions $\Phi = \{\varphi_n(x)\}_{n=1}^\infty$, $x \in (0, 1)$, is said to be a *system with convergence in measure for $l^2$* if every series of the form

$$\sum_{n=1}^\infty a_n \varphi_n(x), \qquad \sum_{n=1}^\infty a_n^2 < \infty, \tag{51}$$

converges in measure on $(0, 1)$.

THEOREM 7. *A necessary and sufficient condition for a system* $\Phi = \{\varphi_n(x)\}_{n=1}^\infty$ *to be a system with convergence in measure for $l^2$ is that for each $\varepsilon > 0$ there are a set $E_\varepsilon \subset (0, 1)$ with $m(E_\varepsilon) > 1 - \varepsilon$, and a constant $C_\varepsilon$, such that every series of the form (51) converges in $L^2(E_\varepsilon)$ and the*

---

[2]Since for every series of positive terms that converges almost everywhere there is a set $E_0$, $m(E_0) \geq \delta_0$, on which the sum of the series is uniformly bounded; and then (50) is impossible for $E = E_0$.

*following inequality is satisfied:*

$$\sup_{\{a_n\}:\sum_{n=1}^{\infty} a_n^2 = 1} \left\| \sum_{n=1}^{\infty} a_n \varphi_n \right\|_{L^2(E_\varepsilon)} \leq C_\varepsilon. \tag{52}$$

The following description of systems with convergence in measure for $l^2$ follows immediately from Theorems 1 and 7.

COROLLARY 2. *A necessary and sufficient condition for a system* $\Phi = \{\varphi_n\}$ *to be a system with convergence in measure for* $l^2$ *is that for each* $\varepsilon > 0$ *there are a set* $E_\varepsilon \subset (0, 1)$ *with* $m(E) > 1 - \varepsilon$, *a constant* $C_\varepsilon$, *and a system of functions* $\Psi_\varepsilon = \{\psi_n(x)\}_{n=1}^{\infty}$ *which is orthonormal on* $(0, 1)$, *such that*

$$\varphi_n(x) = C_\varepsilon \psi_n(x), \qquad n = 1, 2, \ldots, \; x \in E_\varepsilon.$$

PROOF OF THEOREM 7. a) Suppose that for the system $\Phi = \{\varphi_n\}$ there is, for $\varepsilon = \frac{1}{2}, \frac{1}{3}, \ldots$, a set $E_\varepsilon$ with $m(E_\varepsilon) > 1 - \varepsilon$ such that every series of the form (51) converges in $L^2(E_\varepsilon)$. Then it is evident that this series converges in measure on $E_\varepsilon$, $\varepsilon = \frac{1}{2}, \frac{1}{3}, \ldots$, and consequently converges in measure on $(0, 1)$, i.e., $\Phi$ is a system with convergence in measure for $l^2$.

b) Let $\Phi = \{\varphi_n\}$ be a system with convergence in measure for $l^2$. Consider the sequence $\{f_k\}_{k=1}^{\infty}$ of functions which is composed of the squares of polynomials in the system $\Phi$ with rational coefficients (enumerated in arbitrary order), on the unit ball of $l^2$, i.e.,

$$f_k(x) = \left\{ \sum_{n=1}^{N_k} a_n^k \varphi_n(x) \right\}^2, \qquad \sum_{n=1}^{N_k} (a_n^k)^2 \leq 1,$$

where $a_n^k$ are rational numbers.

We shall use the notation

$$a^k = \{a_n^k\}_{n=1}^{\infty}, \qquad P_k(x) = \sum_{n=1}^{N_k} a_n^k \varphi_n(x), \qquad k = 1, 2, \ldots. \tag{53}$$

In order to complete the proof of Theorem 7 it is enough to show that for every $\varepsilon > 0$ there are a closed set $E_\varepsilon \subset (0, 1)$ with $m(E_\varepsilon) > 1 - \varepsilon$, and a constant $C_\varepsilon$, such that

$$\sup_{1 \leq k < \infty} \int_{E_\varepsilon} f_k(x)\, dx \leq C_\varepsilon. \tag{54}$$

In fact, if (54) has been proved, then for every polynomial $P(x) = \sum_{n=1}^{N} a_n \varphi_n(x)$ with $\sum_{n=1}^{N} a_n^2 \leq 1$ we find, by continuity, that

$$\int_{E_\varepsilon} P^2(x)\, dx \leq C_\varepsilon.$$

It then follows easily that every series (50) converges in $L^2(E_\varepsilon)$, and

$$\left\| \sum_{n=1}^{\infty} a_n \varphi_n \right\|_{L^2(E_\varepsilon)} \leq C_\varepsilon \left( \sum_{n=1}^{\infty} a_n^2 \right)^{1/2}.$$

By Theorem 6, inequality (54) (and consequently also Theorem 7) will be established if we verify the following lemma.

LEMMA 1. *The system $\{f_k\}$ is an absolute-convergence system.*

In proving Lemma 1, we shall need two auxiliary propositions.

LEMMA 2. *Let $\Phi = \{\varphi_n\}_{n=1}^{\infty}$ be a system with convergence in measure for $l^2$, and let $\{b^k\}_{k=1}^{\infty}$ be a sequence of elements of $l^2$ such that $\lim_{k \to \infty} \|b^k - b^0\|_{l^2} = 0$ $(b^k = \{b_n^k\}_{n=1}^{\infty}, k = 0, 1, \ldots)$. Then the sequence $g_k(x) = \sum_{n=1}^{\infty} b_n^k \varphi_n(x)$, $k = 1, 2, \ldots$, converges in measure to a function $g_0(x) = \sum_{n=1}^{\infty} b_n^0 \varphi_n(x)$.*

PROOF OF LEMMA 2. We may suppose without loss of generality that $b^0 \equiv 0$. Suppose, on the contrary, that $\{g_k\}$ does not converge in measure to the zero function, although $\lim_{k \to \infty} \|b^k\|_{l^2} = 0$. This means that there are numbers $\varepsilon > 0$ and $\delta > 0$, and a sequence $\{k_s\}_{s=1}^{\infty}$ of indices, such that, for $s = 1, 2, \ldots$,

$$1) \quad m\{x \in (0,1): |g_{k_s}(x)| > \varepsilon\} > \delta; \qquad 2) \quad \|b^{k_s}\|_{l^2} \leq \frac{1}{s^2}. \qquad (55)$$

By inequalities (55) and 1), and the convergence in measure of the series $\sum_{n=1}^{\infty} b_n^k \varphi_n(x)$, we can find numbers $N_s$, $s = 1, 2, \ldots$, such that

$$m \left\{ x \in (0,1): \left| \sum_{n=1}^{N_s} b_n^{k_s} \varphi_n(x) \right| > \frac{\varepsilon}{2} \right\} > \frac{\delta}{2}, \qquad s = 1, 2, \ldots. \qquad (56)$$

Moreover, it follows from (55) and 2) that $\sum_{n=1}^{N} b_n^{k_s} \varphi_n(x) \to 0$ in measure for each given $N$, as $s \to \infty$. Therefore there is a sequence $N_s'$, $s = 1, 2, \ldots$; $N_s' \leq N_s$; $\lim_{s \to \infty} N_s' = \infty$, such that

$$m \left\{ x \in (0,1): \left| \sum_{n=N_s'}^{N_s} b_n^{k_s} \varphi_n(x) \right| > \frac{\varepsilon}{3} \right\} > \frac{\delta}{3}, \qquad s = 1, 2, \ldots. \qquad (57)$$

Choose a sequence of integers $\{s_j\}_{j=1}^{\infty}$ which increases so fast that $N_{s_j} < N_{s_{j+1}}'$ for $j = 1, 2, \ldots$, and set

$$\beta_n = \begin{cases} b_n & \text{if } N_{s_j}' \leq N_{s_j}, \ j = 1, 2, \ldots, \\ 0 & \text{for all other values of } n. \end{cases}$$

It then follows from (57) that the series $\sum_{n=1}^{\infty} \beta_n \varphi_n(x)$ does not converge in measure on $(0, 1)$, although $\sum_{n=1}^{\infty} \beta_n^2 \leq \sum_{s=1}^{n} \|b^{k_s}\|_{l^2}^2 < \infty$, i.e., $\Phi$ is not a system with convergence in measure for $l^2$. The resulting contradiction establishes Lemma 2.

LEMMA 3. *Let* $\{a^k\}_{k=1}^{\infty} \subset l^2$ *and* $\sum_{k=1}^{\infty} \|a^k\|_{l^2}^2 < \infty$. *Then the series*

$$\sum_{k=1}^{\infty} r_k(t) a^k,$$

*converges in* $l^2$ *for almost all* $t \in (0, 1)$, *where* $\{r_k(t)\}$ *is the Rademacher system.*

Lemma 3 follows immediately from the isomorphism between $l^2$ and and $L^2$, and Corollary 2.5.

PROOF OF LEMMA 1. Let there be given any sequence $\{\xi_k\}_{k=1}^{\infty}$ with $\sum_{k=1}^{\infty} |\xi_k| < \infty$. Consider the series

$$\sum_{k=1}^{\infty} r_k(t) |\xi_k|^{1/2} a^k \tag{58}$$

in $l^2$, where $a^k$, $k = 1, 2, \ldots$, are defined in (53). Since $\|a^k\|_{l^2} \leq 1$ for $k = 1, 2, \ldots$, we have $\sum_{k=1}^{\infty} \||\xi_k|^{1/2} a^k\|_{l^2}^2 < \infty$. Then by Lemma 3 the series (58) converges in $l^2$ for almost all $t \in (0, 1)$. But then, by Lemma 2 (also see (53)), the series

$$\sum_{k=1}^{\infty} r_k(t) |\xi_k|^{1/2} P_k(x),$$

converges in measure for almost all $t \in (0, 1)$. Hence by Theorem 2.11 we deduce that

$$\sum_{n=1}^{\infty} [|\xi_k|^{1/2} P_k(x)]^2 = \sum_{k=1}^{\infty} |\xi_k f_k(x)| < \infty$$

for almost all $x \in (0, 1)$. This completes the proof of Lemma 1 and hence of Theorem 7.

## §4. Properties of the majorant operator for the partial sums

Recall that an O.N.S. $\Phi = \{\varphi_n(x)\}_{n=1}^{\infty}$, $x \in (0, 1)$, is a *convergence system* if every series of the form

$$\sum_{n=1}^{\infty} a_n \varphi_n(x), \qquad \sum_{n=1}^{\infty} a_n^2 < \infty \tag{59}$$

converges a.e. on $(0, 1)$ (see Definition 1.2). Examples of convergence systems are the Rademacher, Haar, Walsh, and trigonometric systems.

Each convergence system $\Phi$ generates the majorant operator $S_\Phi^*: l^2 \to L^0(0,1)$ for the partial sums (see 1.(5)):

$$S_\Phi^*(a) = S_\Phi^*(a,x) = \sup_{1 \le N < \infty} \left| \sum_{n=1}^N a_n \varphi_n(x) \right|, \qquad a = \{a_n\} \in l^2. \tag{60}$$

The problem of a.e. convergence of series is one of the most interesting topics in the theory of orthogonal series, and the majority of the results that have been obtained on this topic can be stated in terms of properties of $S_\Phi^*$. Hence we shall investigate the properties of $S_\Phi^*$ quite extensively, both for particular and for general systems $\Phi$.

In this section we describe the additional properties of $S_\Phi^*$ that are generated by an arbitrary convergence system.

In Chapter 8 (see Corollary 8.1) we shall construct an O.N.S. which is not a convergence system. This construction is based on the following fact discovered by D. E. Men′shov (see Lemma 1 for Theorem 8.2):

*For each $N = 1, 2, \ldots$ there is an O.N.S. $\Phi_0(N) = \{\psi_n^N(x)\}_{n=1}^N$, $x \in (0,1)$, such that*

$$m\left\{ x \in (0,1): \max_{1 \le j \le N} \left| \sum_{n=1}^j \psi_n^N(x) \right| > c_0 N^{1/2} \ln N \right\} \ge \frac{1}{4}, \tag{61}$$

*where $c_0 > 0$ is an absolute constant.*

Postponing the proof of (61) to Chapter 8, we shall use it here to obtain the following result.

**THEOREM 8.** *There is an orthonormal convergence system $\Phi = \{\varphi_n(x)\}_{n=1}^\infty$, $x \in (0,1)$, such that for some sequence $a = \{a_n\}_{n=1}^\infty \in l^2$*

$$S_\Phi^*(a) \notin \bigcup_{p>0} L^p(0,1).$$

**PROOF.** For $k = 0, 1, \ldots$, we denote by $\omega_k$ the interval

$$\left( \frac{1}{k+2}, \frac{1}{k+1} \right)$$

and by $M_k$ the number $2^{2^{k+1}} - 2^{2^k}$, $k = 0, 1, \ldots$. We define the system $\Phi$ by setting, for $n \in [2^{2^k}, 2^{2^{k+1}} - 1]$, $k = 0, 1, \ldots$,

$$\varphi_{n-1}(x) = \begin{cases} [(k+1)(k+2)]^{1/2} \psi_{n+1-2^{2^k}}^{M_k} \left[ (k+1)(k+2)\left( x - \frac{1}{k+2} \right) \right] \\ \qquad\qquad\qquad\qquad\qquad\qquad\qquad \text{for } x \in \omega_k, \\ 0 \quad \text{for } x \notin \omega_k. \end{cases}$$

In other words, the functions $\varphi(x)$ with indices from $2^{2^k} - 1$ to $2^{2^{k+1}} - 2$ form a set $\Phi_0(M_k)$ (see (61)) which is transformed from $(0, 1)$ to a similar set (with preservation of the $L^2(0, 1)$ norms) on $\omega_k$.

It is clear that, for every $x \in (0, 1)$, we have $\varphi_n(x) = 0$ for $n > n(x)$, and consequently $\Phi$ is an absolute-convergence system.

We define the sequence $a = \{a_n\}_{n=1}^\infty \in l^2$ by taking

$$a_n = \frac{1}{k+1} M_k^{-1/2} \quad \text{for } 2^{2^k} - 1 \leq n \leq 2^{2^{k+1}} - 2, \ k = 1, 2, \ldots.$$

Then

$$\sum_{n=1}^\infty a_n^2 \leq \sum_{k=1}^\infty \frac{1}{k^2} < \infty.$$

At the same time, we find, using (61), that for every $p > 0$ $(^3)$

$$\int_0^1 [S_\Phi^*(a, x)]^p \, dx \geq \sup_{k \geq 0} \int_{\omega_k} [S_\Phi^*(a, x)]^p \, dx$$

$$\geq \sup_{k \geq 0} (C 2^k)^p \frac{1}{4(k+1)(k+2)} = \infty.$$

This completes the proof of Theorem 8.

It follows from Theorem 8 that in general $S_\Phi^*$ is not a bounded operator on $L^0(0, 1)$. The following result in the opposite direction is based on Theorem 5. It shows that, for every convergence system, $S_\Phi^*$ is a bounded operator from $l^2$ to $L^p(E_\varepsilon)$ for every $p < 2$ and some set $E_\varepsilon(0, 1)$ with $m(E_\varepsilon) \geq 1 - \varepsilon$.

**THEOREM 9.** *Let the O.N.S. $\Phi = \{\varphi_n(x)\}_{n=1}^\infty$, $x \in (0, 1)$, be a convergence system. Then for each $\varepsilon > 0$ there are a set $E_\varepsilon \subset (0, 1)$ with $m(E_\varepsilon) \geq 1 - \varepsilon$ and a constant $C_\varepsilon$ such that*

$$m\{x \in E_\varepsilon : S_\Phi^*(a, x) > y\} \leq C_\varepsilon \frac{\sum_{n=1}^\infty a_n^2}{y^2}$$

*for every sequence $a = \{a_n\} \in l^2$ and every $y > 0$.*

**LEMMA 1.** *Let the O.N.S. $\Phi = \{\varphi_n\}$ be a convergence system, and let $\{a^k\}_{k=1}^\infty$ ($a^k = \{a_n^k\}_{n=1}^\infty$) be a sequence of elements of $l^2$ such that*

$$\sum_{k=1}^\infty \|a^k\|_{l^2}^2 < \infty.$$

*Then for almost all $x \in (0, 1)$*

$$\lim_{k \to \infty} S_\Phi^*(a^k, x) = 0.$$

___

$(^3)$Here $C > 0$ is an absolute constant.

PROOF. Let us show that, under the hypotheses of the lemma,

$$\overline{\lim_{k\to\infty}} S_\Phi^*(a^k, x) < \infty \quad \text{for almost all } x \in (0, 1). \tag{62}$$

If we then apply (62) to the sequence $\tilde{a}^k = \lambda_k a^k$, $k = 1, 2, \ldots$, where the numbers $\lambda_k > 0$ have the property that $\lim_{k\to\infty} \lambda_k = \infty$ but $\sum_{k=1}^\infty \|\tilde{a}^k\|_{l^2}^2 < \infty$, and use the property $S_\Phi^*(\tilde{a}^k) = \lambda_k S_\Phi^*(a^k)$, we obtain the conclusion of Lemma 1.

Let $N_k(x)$, $k = 1, 2, \ldots$, be a measurable function which is integral-valued and bounded [$N_k(x) < M_k$ for $x(0, 1)$], such that

$$\left| \sum_{n=1}^{N_k(x)} a_n^k \varphi_n(x) \right| > \frac{1}{2} S_\Phi^*(a^k, x) \quad \text{for } x \in (0, 1) \backslash E_k, \ m(e_k) < 2^{-k}. \tag{63}$$

Now, for $k, m = 1, 2, \ldots$, we set

$$S_{k,m}(x) = \sum_{n=1}^{N_k(x)} a_n^m \varphi_n(x).$$

It follows from (63) that for each $x \in G = (0, 1) \backslash \overline{\lim}_{k\to\infty} E_k$ we can find a number $k(x)$ such that

$$|S_{k,k}(x)| > \tfrac{1}{2} G_\Phi^*(a^k, x) \quad \text{for } k > k(x),$$

and since $m(G) = 1$ (see (63)), then (62), and consequently Lemma 1, will be established if we verify that

$$\overline{\lim_{k\to\infty}} |S_{k,k}(x)| < \infty \quad \text{for almost every } x \in (0, 1). \tag{64}$$

Consider the series

$$\sum_{m,n=1}^\infty a_n^m r_m(t) \varphi_n(x) = \sum_{n=1}^\infty \left[ \sum_{m=1}^\infty a_n^m r_m(t) \right] \varphi_n(x) \equiv \sum_{n=1}^\infty c_n(t) \varphi_n(x), \tag{65}$$

where $r_m(t)$ are the Rademacher functions. By the hypothesis

$$\sum_{m,n=1}^\infty (a_n^m)^2 < \infty,$$

the series on the left-hand side of (65) converges (in any order) in $L^2((0, 1) \times (0, 1))$, and the series $\sum_{m=1}^\infty a_n^m r_m(t)$, $n = 1, 2, \ldots$, converge in $L^2(0, 1)$.

Since the Rademacher system is orthonormal,

$$\sum_{n=1}^\infty \int_0^1 c_n^2(t)\, dt = \sum_{n=1}^\infty \sum_{m=1}^\infty (a_n^m)^2 < \infty,$$

whence we deduce by Beppo Levi's theorem that

$$\sum_{n=1}^{\infty} \int_0^1 c_n^2(t) < \infty \quad \text{for almost all } t \in (0, 1). \tag{66}$$

Since $\Phi$ is a convergence system, it follows from (66) that

$$A(t, x) \equiv \sup_{1 \leq N < \infty} \left| \sum_{n=1}^{N} c_n(t)\varphi_n(x) \right| < \infty \tag{67}$$

$$\text{a.e. on } (0, 1) \times (0, 1).$$

For $k = 1, 2, \ldots$, we consider the sums

$$
\begin{aligned}
F_k(t, x) &\equiv \sum_{n=1}^{N_k(x)} c_n(t)\varphi_n(x) = \sum_{n=1}^{N_k(x)} \left[ \sum_{m=1}^{\infty} a_n^m r_m(t) \right] \varphi_n(x) \\
&= \sum_{m=1}^{\infty} r_m(t) \sum_{n=1}^{N_k(x)} a_n^m \varphi_n(x) = \sum_{m=1}^{\infty} r_m(t) S_{k,m}(x).
\end{aligned}
\tag{68}
$$

(Equation (68) is valid for almost all points of the square $(0, 1) \times (0, 1)$. The justification of the change of order of summation in (68) presents no difficulty if we use the boundedness of the function $N_k(x)$, and the convergence of the series $\sum_{m=1}^{\infty} a_n^m r_m(t)$, $n = 1, 2, \ldots$, for almost all $t \in (0, 1)$, which follows from Theorem 2.9.)

From (68) and (67) we find that, for $k = 1, 2, \ldots$,

$$\left| \sum_{m=1}^{\infty} r_m(t) S_{k,m}(x) \right| \leq A(t, x) < \infty \quad \text{a.e. on } (0, 1) \times (0, 1) \tag{69}$$

We choose an arbitrary number $\alpha$, $0 < \alpha < 1$, and find a constant $N_\alpha$ such that the planar measure

$$m_2(G_\alpha) \equiv m_2\{(t, x) \in (0, 1) \times (0, 1) : A(t, x) < N_\alpha\} > 1 - \alpha^2.$$

Then it is easily seen that $m(P_\alpha) \geq 1 - \alpha$, where

$$P_\alpha = \{x \in (0, 1) : m\{t \in (0, 1) : (t, x) \in G_\alpha\} > 1 - \alpha\}. \tag{70}$$

By (69), for each $x \in P_\alpha$, for $k = 1, 2, \ldots$,

$$m\left\{ t \in (0, 1) : \left| \sum_{m=1}^{\infty} r_m(t) S_{k,m}(x) \right| \geq N_\alpha \right\} \leq \alpha.$$

Hence, by Theorem 2.8 for $\alpha < c_1$ (where $c_1$ is the constant in Theorem 2.8), we obtain

$$\sum_{m=1}^{\infty} S_{k,m}^2(x) \leq (4N_\alpha)^2,$$

for $x \in P_\alpha$, and $k = 1, 2, \ldots$ . Therefore we also have

$$S^2_{k,k}(x) \le (4N_\alpha)^2, \quad \alpha < c_1, \ x \in P_\alpha, \ m(P_\alpha) \ge 1 - \alpha, \ k = 1, 2, \ldots. \quad (71)$$

Since the number $\alpha$ in (71) can be arbitrarily small, the required inequality (64) follows from (71). This completes the proof of Lemma 1.

PROOF OF THEOREM 9. Let $\{a^k\}_{k=1}^\infty$ be a sequence of elements of $l^2$ which is dense in the unit ball of this space ($a^k = \{a^k_n\}_{n=1}^\infty$, $\|a^k\|_{l^2} \le 1$).

In addition, let $\{\xi_k\}$, $\sum_{k=1}^\infty |\xi_k| < \infty$ be any sequence in $l^1$. Then, setting $b^k = |\xi_k|^{1/2} a^k$, $k = 1, 2, \ldots$, we will have

$$\sum_{k=1}^\infty \|b^k\|_{l^2}^2 = \sum_{k=1}^\infty |\xi_k| \, \|a^k\|_{l^2}^2 < \infty,$$

from which it follows by Lemma 1 that for almost all $x \in (0, 1)$

$$\lim_{k \to \infty} |\xi_k|^{1/2} S^*_\Phi(a^k, x) = \lim_{k \to \infty} S^*_\Phi(b^k, x) = 0. \quad (72)$$

Squaring the right-hand and left-hand sides of (72), we find that, for every sequence $\{\xi_k\}$ with $\sum_{k=1}^\infty |\xi_k| < \infty$,

$$\lim_{k \to \infty} |\xi_k| [S^*_\Phi(a^k, x)]^2 = 0 \quad \text{for almost every } x \in (0, 1),$$

i.e., $[S^*(a^k, x)]^2$ is an $A$-system.

Now, by using Theorem 5, we can find, for every $\varepsilon > 0$, a set $E_\varepsilon \subset (0, 1)$ with $m(E_\varepsilon) \ge 1 - \varepsilon$, and a constant $C_\varepsilon$ such that

$$m\{x \in E_\varepsilon : [S^*_\Phi(a^k, x)]^2 > y\} \le \frac{C_\varepsilon}{y},$$

for $k = 1, 2, \ldots$, and $y > 0$, or, what amounts to the same thing,

$$m\{x \in E_\varepsilon : S^*_\Phi(a^k, x) > y\} \le \frac{C_\varepsilon}{y^2}, \quad y > 0. \quad (73)$$

But then it follows by considerations of continuity that, for every sequence $a = \{a_n\} \in l^2$ with $\|a\|_{l^2} \le 1$ and every bounded integral-valued measurable function $N(x)$, we have the inequality

$$m\left\{ x \in E_\varepsilon : \left| \sum_{n=1}^{N(x)} a_n \varphi_n(x) \right| > y \right\} \le \frac{C_\varepsilon}{y^2}, \quad y > 0 \quad (74)$$

(in fact, it is easily seen that if $\lim_{s \to \infty} \|a^{k_s} - a\|_{l^2} = 0$, then

$$\lim_{s \to \infty} \sum_{n=1}^{N(x)} a_n^{k_s} \varphi_n(x) = \sum_{n=1}^{N(x)} a_n \varphi_n(x)$$

in measure).

From (74) we at once obtain

$$m\{x \in E_\varepsilon : S_\Phi^*(a, x) > y\} \le \frac{C_\varepsilon}{y}, \qquad y > 0,$$

$$\|a\|_{l^2} \le 1, \ m(E_\varepsilon) \ge 1 - \varepsilon. \quad (75)$$

Finally, recalling that $S_\Phi^*(a) = \|a\|_{l^2} S_\Phi^*(a/\|a\|_{l^2})$, we can deduce the conclusion of Theorem 9 from (75).

If we compare Theorems 7 and 9, it is natural to raise the question of the possibility of strengthening Theorem 9, i.e., of whether we can always find a set $E_\varepsilon \subset (0, 1)$ with $m(E_\varepsilon) > 1 - \varepsilon$, for which the majorant $S_\Phi^*(a) \in L^2(E_\varepsilon)$ for all $a \in l^2$. The answer to this question is negative; more precisely, we have the following result:

*There is an orthonormal convergence system* $\Phi_0 = \{\varphi_n(x)\}_{n=1}^\infty$, $x \in (0, 1)$, such that for every set $E \subset (0, 1)$ with $m(E) > 0$ there is a sequence $a = \{a_n\}_{n=1}^\infty \in l^2$ for which $S_\Phi^*(a) \notin L^2(E)$.

The proof of this proposition (which is based on (61)) will be given in Chapter 10 (see Theorem 10.3).

We note that in Chapter 10 we shall clarify the significance, for problems on the representation of measurable functions by orthogonal series in the system $\Phi$, of the question of whether or not one can obtain inequalities for $S_\Phi^*(a)$ in the $L^2$ norm for $\Phi$.

# CHAPTER VIII

# Theorems on the Convergence
# of General Orthogonal Series

The existence of significant general "positive" propositions on the convergence of orthogonal series is one of the main arguments for making a systematic investigation of the properties of orthogonal systems, and every such property is of considerable interest. In this chapter we present a number of now classical theorems on the convergence almost everywhere of general orthogonal series (see, for example, Theorems 1, 4, and 5); in some cases we establish the necessity of their hypotheses. Moreover, we discuss questions on the properties of integrated O.N.S. and on the possibility of extracting from a given O.N.S. a subsystem with "nice" properties.

## §1. Convergence of orthogonal series almost everywhere

THEOREM 1. *Let* $\Phi = \{\varphi_n(x)\}_{n=1}^{\infty}$, $x \in (0,1)$, *be an arbitrary O.N.S. Then every series*

$$\sum_{n=1}^{\infty} a_n \varphi_n(x), \tag{1}$$

*whose coefficients satisfy the condition*

$$L = \sum_{n=1}^{\infty} a_n^2 \log_2^2(n+1) < \infty \tag{2}$$

*converges for almost all* $x \in (0,1)$. *Moreover, if* $S_{\Phi}^*(a,x)$ $(a = \{a_n\}_{n=1}^{\infty})$ *is the majorant of the partial sums of* (1) *(see Definition 1.1), then under condition* (2) *we have the inequality*

$$\|S_{\Phi}^*(a,x)\|_2 \le CL^{1/2}, \tag{3}$$

*where C is an absolute constant.*

The proof of Theorem 1 is based on the following lemma.

**LEMMA 1.** *Let* $\Phi = \{\varphi_n\}$ *be an O.N.S. The following inequality is satisfied for every set of numbers* $\{a_n\}_{n=1}^N$:

$$\int_0^1 \delta^2(x)\,dx \le K \left(\sum_{n=1}^N a_n^2\right) \log_2^2(N+1),$$

$$\delta(x) = \max_{1 \le j \le N} \left|\sum_{n=1}^j a_n \varphi_n(x)\right|,$$

*where* $K$ *is an absolute constant.*

PROOF. We may suppose without loss of generality that $N = 2^r$, $r = 1, 2, \ldots$ (by setting $a_n = 0$ for $N < n \le 2^r$ if $2^{r-1} < N < 2^r$).

For each $j$, $j = 1, 2, \ldots, 2^r$, we write $j$ in the binary system:

$$j = \sum_{k=0}^r \varepsilon_k 2^{r-k}, \qquad \varepsilon_k = \varepsilon_k(j) = 0 \text{ or } 1, \; k = 0, 1, \ldots, r.$$

It is clear then that every sum of the form $\sum_1^j b_n$ can be represented as follows:

$$\sum_{n=1}^j b_n = \sum_{k:\varepsilon_k \ne 0} \; \sum_{\sum_{s=0}^{k-1} \varepsilon_s 2^{r-s} < n \le \sum_{s=0}^k \varepsilon_s 2^{r-s}} b_n. \tag{4}$$

From this we obtain, by using Cauchy's inequality,

$$\left|\sum_{n=1}^l b_n\right|^2 \le (r+1) \sum_{k:\varepsilon_k \ne 0} \left(\sum_{\sum_{s=0}^{k-1} \varepsilon_s 2^{r-s} < n \le \sum_{s=0}^k \varepsilon_s 2^{r-s}} b_n\right)^2$$

$$\le (r+1) \sum_{k=0}^r \sum_{p=0}^{2^k-1} \left(\sum_{n=p2^{r-k}+1}^{(p+1)\cdot 2^{r-k}} b_n\right)^2.$$

We apply this inequality in order to estimate the sums $|\sum_{n=1}^{j(x)} a_n \varphi_n(x)|$, where $j(x)$ is chosen so that $|\sum_{n=1}^\infty a_n \varphi_n(x)| = \delta(x)$, $x \in (0, 1)$:

$$\delta^2(x) \le (r+1) \sum_{k=0}^r \sum_{p=0}^{2^k-1} \left[\sum_{n=p2^{r-k}+1}^{(p+1)2^{r-k}} a_n \phi_n(x)\right]^2. \tag{5}$$

If we integrate (5) and apply the orthonormality of $\{\varphi_n\}_{n=1}^N$, we obtain

$$\int_0^1 \delta^2(x)\,dx \le (r+1)\sum_{k=0}^r \sum_{n=1}^N a_n^2 = (r+1)^2 \sum_{n=1}^N a_n^2$$

$$= (1 + \log_2 N)^2 \sum_{n=1}^N a_n^2.$$

This completes the proof of Lemma 1.

PROOF OF THEOREM 1. We first suppose that the sequence

$$S_{2^k}(a,x) = \sum_{n=1}^{2^k} a_n \varphi_n(x), \qquad k = 1, 2, \ldots,$$

converges for almost all $x \in (0,1)$, and estimate the $L^2(0,1)$ norm of $S'(a,x) = \sup_{0<k<\infty} |S_{2^k}(a,x)|$. Let

$$\Psi_k(x) = \sum_{n=2^k}^{2^{k+1}-1} a_n \varphi_n(x), \qquad k = 0, 1, \ldots.$$

Then $\|\Psi_k\|_2^2 = \sum_{n=2}^{2^{k+1}-1} a_n^2$ and, by (2), $\sum_{k=0}^\infty \|\Psi_k\|_2^2 (k+1)^2 \le 2L$. Consequently

$$\sum_{k=0}^\infty \|\Psi_k\|_1 \le \sum_{k=0}^\infty \|\Psi_k\|_2 = \sum_{k=0}^\infty \|\Psi_k\|_2 (k+1)(k+1)^{-1}$$

$$\le \left\{ \sum_{k=0}^\infty \|\Psi_k\|_2^2 (k+1)^2 \right\}^{1/2} \left\{ \sum_{k=0}^\infty (k+1)^{-2} \right\}^{1/2} \le 2L^{1/2}. \quad (6)$$

It follows from (6) and Beppo Levi's theorem that the series $\sum_{k=0}^\infty |\Psi_k(x)|$ converges for almost all $x \in (0,1)$, and consequently so does $S_{2^k}(a,x)$, $k = 0, 1, \ldots$. In addition, $S'(a,x) \le \sum_{k=0}^\infty |\Psi_k(x)|$, $x \in (0,1)$, whence (see (6)) we obtain

$$\|S'(a,x)\|_2 \le \left\| \sum_{k=0}^\infty |\Psi_k(x)| \right\|_2 \le \sum_{k=0}^\infty \|\Psi_k\|_2 \le 2L^{1/2}. \quad (7)$$

Now consider the function

$$S''(a,x) = \sup_{0<k<\infty} \delta_k(x),$$

$$\delta_k(x) = \max_{2^k \le j < 2^{k+1}} \left| \sum_{n=2^k}^j a_n \varphi_n(x) \right|, \qquad k = 1, 2, \ldots.$$

By Lemma 1,

$$\sum_{k=1}^{\infty} \int_0^1 \delta_k^2(x)\,dx \le K \sum_{k=1}^{\infty}(k+1)^2 \sum_{n=2^k}^{2^{k+1}-1} a_n^2 \le 2KL. \tag{8}$$

From this it follows that $\lim_{k\to\infty}\delta_k(x) = 0$ for almost all $x \in (0,1)$; and this, together with the convergence of $S_{2^k}(a,x)$, $k = 0,1,\ldots$ (proved above), guarantees the convergence a.e. on $(0,1)$ of the series $(1)$. Moreover, since

$$S_\Phi^*(a,x) \le S'(a,x) + S''(a,x), \qquad x \in (0,1),$$

and $[S''(a,x)]^2 \le \sum_{k=1}^{\infty}\delta_k^2(x)$, then we obtain, by using $(7)$ and $(8)$,

$$\|S_\Phi^*(a,x)\|_2 \le \|S'(a,x)\|_2 + \|S''(a,x)\|_2$$
$$\le 2L^{1/2} + (2KL)^{1/2} = CL^{1/2}.$$

This completes the proof of Theorem 1.

DEFINITION 1. A sequence $\{\omega_n\}_{n=1}^{\infty}$, $1 = \omega_1 \le \omega_2 \le \cdots$, is called a *Weyl set for the convergence a.e. of the O.N.S.* $\Phi = \{\varphi_n\}_{n=1}^{\infty}$ if the convergence of the series $\sum_{n=1}^{\infty}a_n^2\omega_n$ guarantees that the series $(1)$ converges a.e.

In terms of the notion of a Weyl set, the result obtained in Theorem 1 can be stated in the following form: the sequence $\omega_n = \log_2^2(n+1)$, $n = 1,2,\ldots$, is a Weyl set for convergence almost everywhere of every O.N.S. As the following proposition shows, Theorem 1 is sharp.

THEOREM 2. *There exists an O.N.S.* $\Phi = \{\varphi_n(x)\}_{n=1}^{\infty}$, $x \in (0,1)$, *such that corresponding to each sequence* $\{\omega_n\}_{n=1}^{\infty}$ *with* $1 = \omega_1 \le \omega_2 \le \cdots$ *and* $\omega_n = o(\log_2^2 n)$ *as* $n \to \infty$, *there is a series*

$$\sum_{n=1}^{\infty}a_n\psi_n(x),$$

*that diverges a.e. and whose coefficients satisfy*

$$\sum_{n=1}^{\infty}a_n^2\omega_n < \infty. \tag{9}$$

COROLLARY 1. *There exists an almost everywhere divergent orthogonal series with coefficients in* $l^2$.

The proof of Theorem 2 is based on the following lemma.

LEMMA 1. *There is a constant $c_0 > 0$ such that corresponding to every $N = 1, 2, \ldots$, there is a set of orthonormal functions $\Psi(N) = \{\psi_k^N\}_{k=1}^N$, $x \in (0, 1)$, for which*

$$m \left\{ x \in (0, 1) : \max_{1 \le j \le N} \left| \sum_{k=1}^{j} \psi_k^N(x) \right| > c_0 N^{1/2} \log_2 N \right\} \ge \frac{1}{4} \qquad (10)$$

*and in addition*

$$\psi_k^N(x) \in \mathscr{D}_{4N} \quad and \quad \int_0^1 \psi_k^N(x)\, dx = 0, \qquad k = 1, \ldots, N. \qquad (11)$$

PROOF. Choose a number $N > 1$ and let $\chi(x)$ denote the function $\chi(x) = \operatorname{sgn} \sin 4N\pi x$, $x \in (0, 1)$. We first define the set $\Psi(N)$ on the interval $(0, \frac{1}{2})$. To do this, we divide $(0, \frac{1}{2})$ into $N$ equal parts $I_s = (\frac{s-1}{2N}, \frac{s}{2N})$, $s = 1, \ldots, N$, and set

$$\psi_k^N(x) = \begin{cases} \chi(x) \dfrac{\gamma N^{1/2}}{s - k} & \text{for } x \in I_s,\ s \ne k, \\ 0 & \text{for } x \in I_k, \end{cases} \quad 1 \le s, k \le N, \qquad (12)$$

where the constant $\gamma$, independent of $N$, will be specified later.

Since $|\chi(x)| = 1$ for almost all $x \in (0, \frac{1}{2})$, we obtain([1])

$$
\begin{aligned}
I_N &\equiv \sup_{\{a_k\} \in B_2^N} \left\| \sum_{k=1}^{N} a_k \psi_k^N \right\|_{L^2(0, 1/2)} \\
&= \sup_{\{a_k\} \in B_2^N} \left\| \sum_{k=1}^{N} a_k \chi(x) \psi_k^N \right\|_{L^2(0, 1/2)} \\
&= \gamma N^{1/2} \sup_{\{b_s\} \in B_2^N} \sup_{\{a_k\} \in B_2^N} (2N)^{-1/2} \sum_{s=1}^{N} b_s \sum_{\substack{k=1 \\ k \ne s}}^{N} \frac{a_k}{s - k} \\
&= \frac{\gamma}{2} \sup_{\{b_s\} \in B_2^N} \sup_{\{a_k\} \in B_2^N} \sum_{\substack{s,k=1 \\ s \ne k}}^{N} \frac{a_k b_s}{s - k}.
\end{aligned}
$$

Using the estimate for the norm of the Hilbert matrix obtained in Chapter 5 (see 5.(24)), we choose $\gamma > 0$ so that $I_N < 1$ for $N = 2, 3, \ldots$. Then, applying the orthogonalization theorem (see Theorem 7.1 and the remark after that theorem), we can choose $\psi_k^N(x)$, $x \in (\frac{1}{2}, 1)$, $k = 1, 2, \ldots$, so that the set $\Psi(N) = \{\psi_k^N(x)\}_{k=1}^N$ is orthonormal on $(0, 1)$ and, in addition, the functions $\psi_k^N(x)$, $x \in (\frac{1}{2}, 1)$, have the form $\psi_k^N(x) = \chi(x) \tilde{\psi}_k^N(x)$,

---

([1])Recall that $B_2^N = \{ \{a_k\}_{k=1}^N : \sum_{k=1}^N a_k^2 \le 1 \}$.

where $\tilde{\psi}_k^N(x)$, $k = 1, \ldots, N$, are constant on each interval $(\frac{s-1}{2N}, \frac{s}{2N})$, $s = N + 1, \ldots, 2N$.

Then $\psi_k^N(x) \in \mathscr{D}_{4N}$ for $k = 1, \ldots, N$ (when appropriately defined at the points $s/4N$, $s = 0, 1, \ldots, 4N$, a definition which is not significant for us) and $\int_0^1 \psi_k^N(x) \, dx = 0$, since for $s = 1, \ldots, 2N$ and $k = 1, \ldots, N$ we have

$$\int_{(s-1)/2N}^{s/2N} \psi_k^N(x) \, dx = C \int_{(s-1)/2N}^{s/2N} \chi(x) \, dx = 0.$$

Let us establish (10). For this purpose we set $j(x) = s$ for $x \in I_s$, $s = 1, \ldots, N$. Then for $x \in I_s$, $s = 1, \ldots, N$,

$$\left| \sum_{k=1}^{j(x)} \psi_k^N(x) \right| = \gamma N^{1/2} \sum_{k=1}^{s-1} \frac{1}{s-k} = \gamma N^{1/2} \sum_{p=1}^{s-1} \frac{1}{p} \geq c N^{1/2} \ln s. \tag{13}$$

It follows from (13) that

$$\left| \sum_{k=1}^{j(x)} \psi_k^N(x) \right| > c_0 N^{1/2} \log_2 N$$

for almost every $x \in (\frac{1}{4}, \frac{1}{2})$. This completes the proof of Lemma 1.

PROOF OF THEOREM 2. We continue the functions $\psi_k^N(x)$, $k = 1, \ldots, N$, $N = 2, 3, \ldots$, constructed in Lemma 1, From $[0, 1]$ to the whole real line. We define a sequence of numbers $R_q$, $q = 0, 1, \ldots$, by

$$R_0 = 1, \qquad R_{q+1} = 4^{q+1} R_q, \qquad q = 0, 1, \ldots, \tag{14}$$

and set $\psi_1(x) = 1$, $x \in (0, 1)$; and if $n = 2^q + m$, $0 \leq m \leq 2^q - 1$, $q = 1, 2, \ldots$, then

$$\psi_n(x) = \psi_{m+1}^{2^q}(R_q x), \qquad x \in (0, 1). \tag{15}$$

Let us show that the system $\Psi = \{\psi_n(x)\}_{n=1}^{\infty}$ satisfies the requirements of Theorem 2. We first verify that $\Psi$ is an O.N.S. If $2^q < n, n' \leq 2^{q+1}$, $q = 1, 2, \ldots$, it follows immediately from Lemma 1 and (15) that

$$\int_0^1 \psi_n(x) \psi_{n'}(x) \, dx = \begin{cases} 1 & \text{for } n = n', \\ 0 & \text{for } n \neq n'. \end{cases}$$

If, however, $n' < 2^q \leq n < 2^{q+1}$, $q = 1, 2, \ldots$, then, since $\psi_n(x)$ has period $R_q^{-1}$ and $\int_0^1 \psi_n(x) \, dx = 0$ (see (11)), and $\psi_{n'}(x)$ is constant on each interval of the form $((i-1)/N_{q-1}, i/N_{q-1})$, where $N_{q-1} = 4 \cdot 2^{q-1} R_{q-1} = 2^{-(q-1)} R_q$ (see (11), (14), and (15)), we find that

$$\int_0^1 \psi_{n'}(x) \psi_n(x) \, dx = \sum_{i=1}^{N_{q-1}} c_i \int_{(i-1)/N_{q-1}}^{i/N_{q-1}} \psi_n(x) \, dx = 0.$$

Let there be given a sequence $\{\omega_n\}$ with $\omega_n = o(\log_2^2 n)$ as $n \to \infty$ $(1 \le \omega_1 \le \omega_2 \le \cdots)$. Then we can find an increasing sequence of integers $q_j$, $j = 1, 2, \ldots$, such that

$$1 \le \omega_{2^{q_j+1}}^{1/2} < j^{-3} q_j, \qquad j = 1, 2, \ldots \qquad (16)$$

Now we define the sequence of coefficients $\{a_n\}_{n=1}^\infty$:

$$a_n = \begin{cases} [2^{q_j} \omega_{2^{q_j+1}}]^{-1/2} j^{-1} & \text{if } 2^{q_j} \le n < 2^{q_j+1},\ j = 1, 2, \ldots, \\ 0 & \text{for other values of } n. \end{cases}$$

for $j = 1, 2, \ldots$, we have by (10) and (15)

$$m\left\{ x \in (0, 1) : \max_{2^{q_j} \le l < 2^{q_j+1}} \left| \sum_{n=2^{q_j}}^{l} a_n \psi_n(x) \right| > c_0 a_{2^{q_j}} 2^{(1/2)q_j} q_j \right\} \ge \frac{1}{4},$$

from which we obtain, for $j = 1, 2, \ldots$, by using (16) and (17),

$$m(E_j) \equiv m\left\{ x \in (0, 1) : \max_{2^{q_j} \le l < 2^{q_j+1}} \left| \sum_{n=2^{q_j}}^{l} a_n \psi_n(x) \right| > c_0 j^2 \right\} \ge \frac{1}{4}. \qquad (18)$$

It follows at once from (18) that the series $\sum_{n=1}^\infty a_n \psi_n(x)$ with the coefficients (17) diverges on a set of positive measure. Moreover, it is easily seen that the sets $E_j$ are independent (see Definition 2.2), and hence by the Borel–Cantelli lemma (see §2.2) and (18) it follows that

$$m(\overline{\lim_{j \to \infty}} E_j) = 1.$$

This implies that the series $\sum_{n=1}^\infty a_n \psi_n(x)$ diverges a.e. on $(0, 1)$.

It remains only to verify that the sequence (17) of coefficients satisfies (9). In fact,

$$\sum_{n=1}^\infty a_n^2 \omega_n = \sum_{j=1}^\infty \sum_{n=2^{q_j}}^{2^{q_j+1}-1} a_n^2 \omega_n \le \sum_{j=1}^\infty a_{2^{q_j}}^2 \omega_{2^{q_j+1}} 2^{q_j} \le \sum_{j=1}^\infty \frac{1}{j^2} < \infty.$$

This completes the proof of Theorem 2.

It should be remarked that the constructions of all known examples of sets $\{\psi_k^N(x)\}_{k=1}^N$ of orthogonal functions for which (10) holds depend, in one way or another, on the properties of the Hilbert matrix $\{1/i - j\}$. At the same time, at first glance such examples can be very different from the set constructed in the proof of Theorem 2. For example, we have the following theorem.

THEOREM 3. *There is an O.N.S.* $\Psi = \{\psi_n(x)\}_{n=1}^\infty$, $x \in (0,1)$, *with* $|\psi_n(x)| = 1$ *for almost all* $x \in (0,1)$ *and* $n = 1,2,\ldots$, *which satisfies all the requirements of Theorem 2.*

LEMMA 1. *There are constants* $A_1$, $A_2 > 0$ *such that, for* $p = 1,2,\ldots$, *there is an O.N.S.* $\Psi(p) = \{\psi_n(x)\}_{n=1}^{2p^2}$, $x \in (0,1)$, *for which* $|\psi_n(x)| = 1$ *for almost all* $x \in (0,1)$, $n = 1,\ldots,2p^2$, *and*

$$m\left\{x \in (0,1): \max_{1 \le j \le 2p^2}\left|\sum_{n=1}^{j}\psi_n(x)\right| > A_1 p \log_2 p\right\} \ge A_2.$$

*Moreover, the functions* $\psi_n(x)$ *are piecewise constant, with a finite number of intervals of constancy.*

We prove Lemma 1 below. We omit the deduction of Theorem 3 from Lemma 1, since it is completely analogous to the proof of Theorem 2.

PROOF OF LEMMA 1. We first define an auxiliary system of functions $f_n(x)$, $n = 1,\ldots,2p$, on the interval $(0,4)$:

$$f_n(x) = \frac{1}{2(s-p-n-1/2)} \quad \text{for } x \in \left(\frac{s-1}{p}, \frac{s}{p}\right),$$
$$s = 1,\ldots,4p. \quad (19)$$

For $1 \le n, 1 \le 2p$, let

$$\alpha_{n,l} = \int_0^4 f_n(x)f_l(x)\,dx.$$

Then

$$\alpha_{n,n} \le \frac{C}{p}\sum_{s=-\infty}^{\infty}\frac{1}{(s-1/2)^2} \le \frac{C_1}{p}, \qquad 1 \le n \le 2p. \quad (20)$$

If $n > l$, we have

$$\alpha_{n,l} = \frac{1}{4p}\sum_{s=1}^{4p}\frac{1}{(s-p-n-1/2)(s-p-l-1/2)}$$

$$= \frac{1}{4p(n-l)}\sum_{s=1}^{4p}\left\{\frac{1}{s-p-n-1/2} - \frac{1}{s-p-l-1/2}\right\}$$

$$= \frac{1}{4p(n-l)}\left\{\sum_{s=1-p-n}^{3p-n}\frac{1}{s-1/2} - \sum_{s=1-p-l}^{3p-l}\frac{1}{s-1/2}\right\}$$

$$= \frac{1}{4p(n-l)}\left\{\sum_{s=1-p-n}^{-p-l}\frac{1}{s-1/2} - \sum_{s=3p-n+1}^{3p-l}\frac{1}{s-1/2}\right\},$$

from which it follows at once that

$$\alpha_{n,l} \leq \frac{C_2}{p^2}, \qquad n \neq l. \tag{21}$$

It is easily verified (just as in the proof of Lemma 1 and Theorem 2: see (13)) that

$$\max_{1 \leq j \leq 2p} \sum_{n=1}^{j} f_n(x) \geq c_3 \log_2 p \quad \text{for almost all } x \in (2, 3). \tag{22}$$

Set

$$g_{r+(s-1)p}(x) = f_s(x), \qquad x \in (0, 4), \; r = 1, \ldots, p, \; s = 1, \ldots, 2p. \tag{23}$$

Then it follows from (20) and (21) that

$$\int_0^4 g_n(x) g_m(x) \, dx \leq \gamma_{|n-m|}, \qquad n, m = 1, \ldots, 2p^2,$$

where

$$\gamma_i = \begin{cases} C_1 p^{-1} & \text{if } 0 \leq i \leq p - 1, \\ C_2 p^{-2} & \text{if } p \leq i \leq 2p^2 - 1. \end{cases}$$

Since $\sum_{i=1}^{2p^2-1} \gamma_i \leq C_4$, by using Theorem 7.2 we can define the functions $g_n(x)$, $n = 1, \ldots, 2p^2$, on the interval $(4, M)$, $M = 5 + C_4$, in such a way that

1) $\{g\}_{n=1}^{2p^2}$ is an orthonormal system on $(0, M)$;

2) $|g_n(x)| = 1$ for almost all $x \in (4, M)$ and $|g_n(x)| \leq 1$ for $x \in (0, 4)$ (see (19));

3) the functions $g_n(x)$ are piecewise constant on $(4, M)$, and consequently (see (19)) on $(0, M)$.

Therefore the system

$$h_n(x) = g_n(Mx), \qquad x \in (0, 1), \; 1 \leq n \leq 2p^2,$$

is an orthogonal (but not normal) system of piecewise constant functions, for which $h_n(x) \leq 1$ for $x \in (0, 1)$ and $n = 1, \ldots, 2p^2$. By (22) and (23),

$$m \left\{ x \in (0, 1) : \max_{1 \leq j \leq 2p^2} \left| \sum_{n=1}^{j} h_n(x) \right| \geq c_3 p \log_2 p \right\} \geq c_5 > 0. \tag{24}$$

Let us divide the interval $(0, 1)$ into pairwise nonintersecting intervals $I_r$, $r = 1, \ldots, R$, on each of which the functions $h_n(x)$ are constant:

$$h_n(x) = \rho_n(r),$$
$$x \in I_r = (a_r, b_r), \qquad 1 \leq r \leq R, \; 1 \leq n \leq 2p^2. \tag{25}$$

For $r = 1, \ldots, R$, let $\{\chi_n^r(x)\}_{n=1}^{2p^2}$ be a set of independent piecewise constant functions, defined on $I_r$ and such that $\int_{I_r} \chi_n^r(x)\,dx = 0$ and $\chi_n^r(x)$ has the two values $1 - \rho_n(r)$ and $-1 - \rho_n(r)$ (see Theorem 2.1).

We can define the required set $\{\psi_n(x)\}_{n=1}^{2p^2}$ by using the functions $\chi_n^r(x)$. For $x \in I_r$, $r = 1, \ldots, R$, set $\psi_n(x) = h_n(x) + \chi_n^r(x)$, $n = 1, \ldots, 2p^2$; or, what amounts to the same thing,$(^2)$

$$\psi_n(x) = h_n(x) + \psi_n'(x),$$
$$\text{where } \psi_n'(x) = \sum_{r=1}^{R} \chi_n^r(x), \quad x \in (0, 1). \tag{26}$$

Then, by construction, $\psi_n(x)$ is piecewise constant and $|\psi_n(x)| = 1$ for almost all $x \in (0, 1)$ and $n = 1, \ldots, 2p^2$. Let us verify that the system $\{\psi_n\}_{n=1}^{2p^2}$ is orthogonal. By (26), for $k \neq l$,

$$\int_0^1 \psi_k(x)\psi_l(x)\,dx = \int_0^1 h_k(x)h_l(x)\,dx + \int_0^1 h_k(x)\psi_l'(x)\,dx$$
$$+ \int_0^1 h_l(x)\psi_k'(x)\,dx + \int_0^1 \psi_k'(x)\psi_l'(x)\,dx = 0,$$

since the system $\{h_n\}$ is orthogonal and, for all $k$ and $l \neq k$ (see (25)),

$$\int_0^1 h_k(x)\chi_l^r(x)\,dx = \rho_k(r) \int_{I_r} \chi_l^r(x)\,dx = 0,$$
$$\int_{I_r} \chi_k^r(x)\chi_l^r(x)\,dx = 0, \qquad r = 1, \ldots, R.$$

Let $I_r$ be any interval for which

$$\left| \sum_{n=1}^{j(r)} h_n(x) \right| \geq c_3 p \log_2 p, \qquad x \in I_r \tag{27}$$

for some $j = j(r)$, $1 \leq j \leq 2p^2$.

By (24), the sum of the measures of such intervals is at least $c_5$. Therefore, to complete the proof of Lemma 1 it is enough to show that when $p \geq p_0$

$$m\left\{ x \in I_r : \left| \sum_{n=1}^{j(r)} \psi_n(x) \right| \geq \frac{c_3}{2} p \log_2 p \right\} \geq \frac{1}{2}|I_r| \tag{28}$$

for each interval $I_r$ that satisfies (27).

---

$(^2)$We suppose that the functions are defined and equal to zero outside $I_r$.

Using the orthogonality of the functions $\chi_n^r$ and the inequality $\|\chi_n^r\|_\infty \leq 2$, $n = 1, \ldots, p^2$ (see the definition of $\chi_n^r(x)$), we obtain

$$\int_{I_r} \left[ \sum_{n=1}^{j(r)} \chi_n^r(x) \right]^2 dx = \sum_{n=1}^{j(r)} \int_{I_r} [\chi_n^r(x)]^2 \, dx \leq 2p^2 4|I_r| = 8p^2|I_r|. \qquad (29)$$

Applying Chebyshev's inequality, we deduce from (29) that

$$m \left\{ x \in I_r : \left| \sum_{n=1}^{j(r)} \chi_n^r(x) \right| > \frac{c_3}{2} p \log_2 p \right\} \leq \frac{c_6|I_r|}{\log_2^2 p} \leq \frac{1}{2}|I_r|, \qquad (30)$$

if $p$ is sufficiently large ($c_6/\log_2^2 p < \frac{1}{2}$). It follows from (27) and (30) that

$$m \left\{ x \in I_r : \left| \sum_{n=1}^{j(r)} [h_n(x) + \chi_n^r(x)] \right| \geq \frac{c_3}{2} p \log_2 p \right\} \geq \frac{1}{2}|I_r|.$$

Since $h_n(x) + \chi_n^r(x) = \psi_n(x)$ for $x \in I_r$, we have established (28), and therefore Lemma 1.

Up to now there are not any general conditions on the system $\{\varphi_n\}$ that ensure a strengthening of Theorem 1. However, there are some systems for which a stronger form can be found by using the following result.

THEOREM 4. *Let* $\Phi = \{\varphi_n(x)\}_{n=1}^\infty$. $x \in (0, 1)$, *be an O.N.S., and let* $L_n(t)$, $t \in (0, 1)$, $n = 1, 2, \ldots$, *be its Lebesgue functions (see Definition 1.11). If the relation*

$$L_n(t) \leq C\lambda_n, \qquad n = 1, 2, \ldots, \qquad \text{where } \lambda_1 \leq \lambda_2 \leq \cdots, \qquad (31)$$

*holds for almost all $t$ from some set $E \subset (0, 1)$, then the finiteness of the sums*

$$\sum_{n=1}^\infty a_n^2 \lambda_n < \infty \qquad (32)$$

*implies that* (1) *converges almost everywhere on $E$.*

If we apply Theorem 4 to the trigonometric system, and take account of 4.(11), we obtain the conclusion of Theorem 4.6. There is a similar result for Walsh series.

The proof of Theorem 4 depends on the following lemma, which is also applicable to the study of the convergence of sequenes of partial sums of orthogonal series.

**LEMMA 1.** *For some sequence of indices $\{\nu_k\}_{k=1}^{\infty}$ ($\nu_1 < \nu_2 < \cdots$) let the Lebesgue functions $L_{\nu_k}(t)$ of the O.N.S. $\Phi = \{\varphi_n\}$ have the property that*

$$L_{\nu_k}(t) \le C\lambda_{\nu_k}, \qquad k = 1, 2, \ldots, x \in E, \ 0 < \lambda_{\nu_1} \le \lambda_{\nu_2} \le \cdots, \tag{33}$$

*on some set $E \subset (0,1)$ of positive measure. Then the partial sums of every series (1) for which*

$$\sum_{n=1}^{\infty} a_n^2 < \infty, \tag{34}$$

*satisfy the inequality*

$$|S_{\nu_k}(x)| \le C_x \lambda_{\nu_k}^{1/2}, \qquad k = 1, 2, \ldots, \qquad \text{for almost all } \ x \in E,$$

*where $C_x$ is a constant that depends on $x$ but not on $k$.*

**PROOF.** Let $k_n(x)$ be the smallest index among $\nu_1, \ldots, \nu_n$ for which $\max S_{\nu_k}(x)\lambda_{\nu_k}^{-1/2}$ is attained, i.e.,

$$\delta_n^+(x) = \max_{1 \le k \le n} S_{\nu_k}(x)\lambda_{\nu_k}^{-1/2} = S_{k_n(x)}(x)\lambda_{k_n(x)}^{-1/2}, \qquad x \in (0,1). \tag{35}$$

It is evident that

$$\delta_1^+(x) \le \delta_2^+(x) \le \cdots \quad \text{and} \quad \int_0^1 |\delta_1^+(x)|\, dx < \infty. \tag{36}$$

Let us show that

$$\lim_{n \to \infty} \delta_n^+(x) < \infty \quad \text{for a.e. } x \in E. \tag{37}$$

For this purpose, it is enough, by (36), to verify that

$$\lim_{n \to \infty} \int_E \delta_n^+(x)\, dx < \infty. \tag{38}$$

Using the equation (see 1.(14))

$$S_N(x) = \int_0^1 S_\nu(t) K_N(t, x)\, dx,$$

$$N \le \nu, \qquad K_N(t, x) = \sum_{n=1}^{N} \varphi_n(t)\varphi_n(x)$$

and Cauchy's inequality, we obtain

$$
\int_E \delta_n^+(x)\,dx = \int_E \lambda_{k_n(x)}^{-1/2} \int_0^1 S_{\nu_n}(t) K_{k_n(x)}(t,x)\,dt\,dx
$$
$$
= \int_0^1 S_{\nu_n}(t) \int_E K_{k_n(x)}(t,x)\lambda_{k_n(x)}^{-1/2}\,dx\,dt
$$
$$
\leq \left\{ \int_0^1 S_{\nu_n}^2(t)\,dt \right\}^{1/2} \left\{ \int_0^1 \left[ \int_E K_{k_n(x)}(t,x)\lambda_{k_n(x)}^{-1/2}\,dx \right]^2 dt \right\}^{1/2}.
$$
$$
\tag{39}
$$

Since the series (1) converges in $L^2(0,1)$ (see (34)),

$$
\left\{ \int_0^1 S_{\nu_n}^2(t)\,dt \right\}^{1/2} \leq \left\{ \sum_{p=1}^{\infty} a_p^2 \right\}^{1/2} = C < \infty.
$$

If we use this and write the square of the integral in (39) as a product of integrals, we deduce from (39) that

$$
\int_E \delta_n^+(x)\,dx \leq C \left\{ \int_0^1 \int_E \int_E K_{k_n(x)}(t,x) K_{k_n(y)}(t,y) \right.
$$
$$
\left. \times \lambda_{k_n(x)}^{-1/2} \lambda_{k_n(y)}^{-1/2}\,dx\,dy\,dt \right\}^{1/2}. \tag{40}
$$

If in (40) we integrate first with respect to $t$ and use both the monotonicity of $\{\lambda_{\nu_k}\}$ and the equation (which follows from the orthonormality of $\Phi$)

$$
\int_0^1 K_{k_n(x)}(t,x) K_{k_n(y)}(t,y)\,dt = K_{k_n(x,y)}(x,y),
$$

where $k_n(x,y) = \min(k_n(x), k_n(y))$, we find that

$$
\int_E \delta_n^+(x)\,dx \leq C \left\{ \int_E \int_E |K_{k_n(x,y)}(x,y)\lambda_{k_n(x,y)}^{-1}|\,dx\,dy \right\}^{1/2}
$$
$$
\leq C \left\{ \int_E \int_E |K_{k_n(x)}(x,y)|\lambda_{k_n(x)}^{-1}\,dx\,dy \right.
$$
$$
\left. + \int_E \int_E |K_{k_n(y)}(x,y)|\lambda_{k_n(y)}^{-1}\,dx\,dy \right\}^{1/2}
$$
$$
\leq C \left\{ \int_E L_{k_n(x)}(x)\lambda_{k_n(x)}^{-1}\,dx + \int_E L_{k_n(y)}(y)\lambda_{k_n(y)}^{-1}\,dy \right\}^{1/2}.
$$

Then (38), and therefore (37), follow at once from the preceding inequality by using (33).

If we consider the series $\sum_{n=1}^{\infty} -a_n\varphi_n(x)$ instead of (1), we find from (37) that

$$\lim_{n\to\infty} \delta_n^-(x) > -\infty$$

for almost all $x \in E$, $\quad \delta_n^-(x) = \min_{1\le k\le n} S_{\nu_k}(x)\lambda_{\nu_k}^{-1/2}.$ $\qquad$ (41)

Relations (37) and (41) show that

$$f^-(x) = \lim_{n\to\infty} \delta_n^-(x) \le S_{\nu_k}(x)\lambda_{\nu_k}^{-1/2} \le \lim_{n\to\infty} \delta_n^+(x) = f^+(x),$$

$$k = 1, 2, \ldots,$$

a.e. on $E$, where $f^-(x)$ and $f^+(x)$ are finite a.e. on $E$. Consequently

$$|S_{\nu_k}(x)| \le \{|f^-(x)| + |f^+(x)|\}\lambda_{\nu_k}^{1/2}.$$

This completes the proof of Lemma 1.

PROOF OF THEOREM 4. We find a sequence $\{\mu_n\}$, $1 \le \mu_1 \le \mu_2 \le \cdots$, for which (see (32))

$$\lim_{n\to\infty} \mu_n = \infty, \qquad \sum_{n=1}^{\infty} a_n^2 \lambda_n \mu_n < \infty, \qquad (42)$$

and let $\sigma_N(x)$ denote the $N$th partial sum of the series

$$\sum_{n=1}^{\infty} a_n(\lambda_n\mu_n)^{1/2}\varphi_n(x). \qquad (43)$$

In order to estimate the partial sums $S_N(x)$ of the series (1), we consider the evident equation $(1 \le p, q < \infty)$

$$S_{q+p}(x) - S_q(x) = \sum_{n=q+1}^{q+p} \frac{1}{(\lambda_n\mu_n)^{1/2}} a_n(\lambda_n\mu_n)^{1/2}\varphi_n(x) \qquad (44)$$

and apply Abel's transformation to the sum. We obtain[3]

$$|S_{q+p}(x) - S_q(x)| \le \sum_{n=q+1}^{q+p-1} [(\lambda_n\mu_n)^{1/2} - (\lambda_{n+1}\mu_{n+1})^{-1/2}]|\sigma_n(x)|$$

$$+ (\lambda_{q+1}\mu_{q+1})^{-1/2}|\sigma_q(x)|$$

$$+ (\lambda_{q+p}\mu_{q+p})^{-1/2}|\sigma_{q+p}(x)|. \qquad (45)$$

Using Lemma 1 (with $\nu_k = k$) for an estimate of the partial sums of (43), we obtain, for almost all $x \in E$, the inequality

$$|\sigma_{q+p}(x)| \le C_x\lambda_{q+p}^{1/2}|\sigma_q(x)| \le C_x\lambda_q^{1/2}, \qquad 1 \le q, \quad p < \infty.$$

---

[3] When $p = 1$, we take $\sum_{n=q+1}^{q} = 0.$

Hence it follows from (45) that

$$|S_{q+p}(x) - S_q(x)| \le \sum_{n=q+1}^{q+p-1} [(\lambda_n\mu_n)^{-1/2} - (\lambda_{n+1}\mu_{n+1})^{-1/2}]|\sigma_n(x)| + o_x(1),$$

as $q \to \infty$, for almost all $x \in E$, where $o_x(1)$ depends on $x$ and $q$, and tends to zero as $q \to \infty$. To prove Theorem 4, it remains only to show that, for almost all $x \in E$,

$$\sup_{1 \le p < \infty} \sum_{n=q+1}^{q+p-1} [(\lambda_n\mu_n)^{-1/2} - (\lambda_{n+1}\mu_{n+1})^{-1/2}]|\sigma_n(x)| = o_x(1)$$

as $q \to \infty$. For this purpose it is enough, by Beppo Levi's theorem, to verify that

$$S = \sum_{n=1}^{\infty} \int_0^1 [(\lambda_n\mu_n)^{-1/2} - (\lambda_{n+1}\mu_{n+1})^{-1/2}]|\sigma_n(x)|\, dx < \infty. \qquad (46)$$

Using the monotonicity of the sequence $\lambda_n\mu_n$, $n = 1, 2, \ldots$, and (42), we obtain

$$S \le \sum_{n=1}^{\infty} [(\lambda_n\mu_n)^{-1/2} - (\lambda_{n+1}\mu_{n+1})^{-1/2}]\|\sigma_n\|_2$$

$$\le \left\{ \sum_{n=1}^{\infty} a_n^2 \lambda_n\mu_n \right\}^{1/2} \sum_{n=1}^{\infty} [(\lambda_n\mu_n)^{-1/2} - (\lambda_{n+1}\mu_{n+1})^{-1/2}] < \infty.$$

This establishes (46), and consequently also Theorem 4.

## §2. Unconditional convergence almost everywhere

In this section we investigate the unconditional convergence a.e. (see Definition 1.4) of series of the form

$$\sum_{n=1}^{\infty} a_n\varphi_n(x), \quad \text{where } \{\varphi_n(x)\}_{n=1}^{\infty}, \ x \in (0, 1), \text{ is an O.N.S.} \qquad (47)$$

Already, in connection with series in the Haar system, we have shown (see Corollary 3.7) that, in contrast to unconditional convergence in a metric space, unconditional convergence a.e. of a series $\sum_{n=1}^{\infty} f_n(x)$ does not in general imply the convergence a.e. of all series of the form

$$\sum_{n=1}^{\infty} \varepsilon_n f_n(x), \quad \varepsilon_n = \pm 1, \ n = 1, 2, \ldots.$$

Hence in order to solve problems on the unconditional convergence a.e. of orthogonal series, we often have to overcome difficulties connected with the investigation of all possible rearrangements of a given O.N.S. As for conditions on the coefficients of (47) which will ensure, for any O.N.S. $\Phi = \{\varphi_n\}$, the unconditional convergence of this series a.e., there are known to be (best possible) conditions, of which we present one.

THEOREM 5. *If the coefficients of the series* (47) *satisfy the inequality*

$$\sum_{k=0}^{\infty} \left[ \sum_{n=\nu_k+1}^{\nu_{k+1}} a_n^2 \log_2^2 n \right]^{1/2} < \infty, \qquad \nu_k = 2^{2^k}, \ k = 0, 1, \ldots, \qquad (48)$$

*then* (47) *converges unconditionally a.e.*

PROOF. We may evidently suppose that $a_1 = a_2 = 0$. For $k = 0, 1, 2, \ldots$, we denote the block of integers $\{\nu_k + 1, \ldots, \nu_{k+1}\}$ by $H_k$. In addition, let

$$\sum_{n=1}^{\infty} a_{\sigma(n)} \varphi_{\sigma(n)}(x) \qquad (49)$$

be any rearrangement of (47). For $k = 0, 1, \ldots$, we define a sequence $\{\varepsilon_n^k\}_{n=1}^{\infty}$ of zeros and ones by setting

$$\varepsilon_n^k = \begin{cases} 1 & \text{if } \sigma(n) \in H_k, \\ 0 & \text{if } \sigma(n) \notin H_k. \end{cases}$$

Then it is clear that, for all $p, q$ $(p \leq q)$ and $x \in (0, 1)$,

$$\sum_{n=p}^{q} a_{\sigma(n)} \varphi_{\sigma(n)}(x) = \sum_{k=0}^{\infty} \sum_{n=p}^{q} \varepsilon_n^k a_{\sigma(n)} \varphi_{\sigma(n)}(x). \qquad (50)$$

(The series on the right of (50) converges for all $x \in (0, 1)$, since all its terms are 0 from some index onward.) For $k = 0, 1, \ldots$, set

$$\delta_k(x) = \sup_{1 \leq p < q < \infty} \left| \sum_{n=p}^{q} \varepsilon_n^k a_{\sigma(n)} \varphi_{\sigma(n)}(x) \right|. \qquad (51)$$

By Lemma 1 for Theorem 1 and the inequality

$$\delta_k(x) \leq 2 \sup_{1 \leq q < \infty} \left| \sum_{n=1}^{q} \varepsilon_n^k a_{\sigma(n)} \phi_{\sigma(n)}(x) \right|,$$

we have

$$\int_0^1 \delta_k(x) \, dx \leq \left\{ \int_0^1 \delta_k^2(x) \, dx \right\}^{1/2} \leq K 2^k \left\{ \sum_{n \in H_k} a_n^2 \right\}^{1/2}.$$

It follows from this (see (48)) that $\sum_{k=0}^{\infty} \int_0^1 \delta_k(x)\,dx < \infty$ and, by Beppo Levi's theorem,

$$\sum_{k=0}^{\infty} \delta_k(x) < \infty, \qquad x \in G \subset (0,1), \quad m(G) = 1. \tag{52}$$

For each $x \in G$ and each $\varepsilon > 0$ there is a number $N = N(x,\varepsilon)$ such that

$$\sum_{k=N}^{\infty} \delta_k(x) < \varepsilon.$$

But then, if $p = p(x,\varepsilon)$ is so large that the sum $\sum_{n=1}^{p-1} a_{\sigma(n)}\varphi_{\sigma(n)}$ contains all the $\varphi_n(x)$ with indices in $H_k$, $0 < k \le N(x,\varepsilon)$, then for each $q \ge p$ we have, by (50),

$$\left| \sum_{n=p}^{q} a_{\sigma(n)}\varphi_{\sigma(n)}(x) \right| \le \sum_{k=N}^{\infty} \delta_k(x) < \varepsilon. \tag{53}$$

Since the number $\varepsilon > 0$ in (53) can be arbitrarily small, we have proved the convergence of (49) at each $x \in G$, $m(G) = 1$. This completes the proof of Theorem 5.

Theorem 5 immediately implies the following corollary.

COROLLARY 2. *If for some $\varepsilon > 0$, the coefficients of (47) satisfy*

$$\sum_{n=2}^{\infty} a_n^2 \log_2^2 n (\log_2 \log_2 n)^{1+\varepsilon} < \infty,$$

*then (47) converges unconditionally a.e.*

COROLLARY 3. *If*

$$\sum_{n=1}^{\infty} |a_n|^{2-\varepsilon} < \infty,$$

*for some $\varepsilon > 0$, the series (47) converges unconditionally a.e.*

In fact, let us consider a rearrangement

$$\sum_{n=1}^{\infty} a_{\sigma(n)}\varphi_{\sigma(n)}(x), \tag{54}$$

of the series (47) such that the numbers $|a_{\sigma(n)}|$ form a nonincreasing sequence

$$|a_{\sigma(1)}| \ge |a_{\sigma(2)}| \ge \cdots$$

Then[4] $|a_{\sigma(n)}|^{2-\varepsilon} \le C n^{-1}$, $n = 1, 2, \ldots$, and therefore

$$|a_{\sigma(n)}| \le C n^{-(1/2+\delta)}, \qquad n = 1, 2, \ldots,$$

_____

[4] Here we used the elementary proposition that if $\sum_{n=1}^{\infty} b_n < \infty$ and $b_1 \ge b_2 \ge \cdots$, then $b_n = O(n^{-1})$ as $n \to \infty$.

for some $\delta > 0$. Then Theorem 5 implies the unconditional convergence a.e. of (54); or, what is equivalent, the unconditional convergence of (47) a.e. This completes the proof of Corollary 3.

The following result establishes the definitive character of Theorem 5.

**THEOREM 6.** *There is an O.N.S.* $\Phi = \{\varphi_n(x)\}_{n=1}^{\infty}$, $x \in (0, 1)$, *such that, for every sequence* $\{b_k\}_{k=1}^{\infty}$ *with*

$$\sum_{k=0}^{\infty} b_k = \infty \qquad (b_k > 0) \tag{55}$$

*the series (with* $\nu_k = 2^{2^k}$ *)*

$$\sum_{k=0}^{\infty} \frac{b_k}{2^k \nu_k} \sum_{n=\nu_k+1}^{\nu_{k+1}} \varphi_n(x) \equiv \sum_{n=3}^{\infty} a_n \varphi_n(x) \tag{56}$$

*diverges a.e. on* $(0, 1)$ *after rearrangement of its terms.*

REMARK. Hypothesis (55) of Theorem 6 is necessary in the sense that if $\sum_{k=0}^{\infty} |b_k| < \infty$, then the coefficients in (56) satisfy

$$\sum_{k=0}^{\infty} \left[ \sum_{n=\nu_k+1}^{\nu_{k+1}} a_n^2 \log_2^2 n \right]^{1/2} \leq \sum_{k=0}^{\infty} |b_k| < \infty.$$

Then the series (56) converges unconditionally a.e. by Theorem 5. As to the condition $b_k \geq 0$, it is not essential, and was introduced only for convenience (see also Theorem 2.14).

Taking $b_k = 1/k \log_2 k$, $k = 2, 3, \ldots$, in Theorem 6, we obtain the following corollary.

COROLLARY 4. *There is an orthogonal series* (47) *which diverges a.e. after a rearrangement of its terms and has the property that*

$$\sum_{n=2}^{\infty} a_n^2 \log_2^2 n \log_2 \log_2 n < \infty.$$

We shall denote by $\mathscr{D}_N^0(I)$, $I = (a, b) \subset (0, 1)$, the $(N - 1)$-dimensional subspace of $\mathscr{D}_N(I)$ consisting of the functions $f(x)$ for which $\int_I f(x)\,dx = 0$.

The proof of Theorem 6 depends on the following lemma.

LEMMA 1. *For every* $N \geq 100$ *and for every interval* $I = (a, b) \subset (0, 1)$ *there is an orthonormal basis*

$$\Psi_{2N-1}(I) = \{\psi_n^{2N-1}(x, I)\}_{n=1}^{2N-1} = \{\psi_n(x)\}_{n=1}^{2N-1}$$

*in $\mathscr{D}_{2N}^0(I)$ whose elements satisfy the conditions:*

1) *for $n = 1, \ldots, N - 1$, the function $\psi_n(x)$ changes sign at least once on the interval $(\frac{a+b}{2}, b)$ i.e., for some point $\xi_n \in [\frac{a+b}{2}, b]$ we have $\psi_n(x) \leq 0$ if $\frac{a+b}{2} < x < \xi_n$, and $\psi_n(x) \geq 0$ if $\xi_n < x < b$, with*

$$\frac{a+b}{2} \leq \xi_1 \leq \cdots \leq \xi_{N-1} \leq b \tag{57}$$

*(the points $\xi_n$ are said to be points of essential change of sign for $\psi_n(x)$);*

2) *the condition*

$$\sum_{n=1}^{m(x)} \psi_n(x) \geq c \frac{N^{1/2}}{|I|^{1/2}} \ln N, \tag{58}$$

*is satisfied for almost all $x \in (a + \frac{5}{8}|I|, a + \frac{7}{8}|I|)$; here $c > 0$ is an absolute constant.*

Before proving Lemma 1, let us recall that it was shown in Chapter 5 (see 5.(31)) that there is an absolute constant $C > 0$ such that, for $A = 1, 2, \ldots,$ and $N = 1, 2, \ldots,$ we have

$$\|H_N(A)\| \leq C, \qquad H_N(A) = \{h_{s,k}\}_{s,k=1}^N, \tag{59}$$

where

$$h_{s,k} = \begin{cases} \dfrac{1}{s-k} & \text{if } 0 < |s - k| < A, \\ 0 & \text{if } s = k \text{ or } |s - k| > A, \\ s, k = 1, \ldots, N. \end{cases}$$

PROOF OF LEMMA 1. It is easily seen that it is enough to consider the case when $I = (0, 1)$, and then to apply a similarity transformation (preserving the $L^2(0, 1)$ norm) to carry functions in the system $\Psi_{2N-1}((0, 1))$ to $I$; we denote the resulting system by $\Psi_{2N-1}(I)$.

The set $\Psi_{2N-1}((0, 1))$ will be a modification of the O.N.S. constructed in Theorem 2 (see (10)). To begin with, for $N^{1/2} < n < N - N^{1/2}$ and $x \in (\frac{1}{2}, 1)$, we set

$$\psi_n(x) = \begin{cases} \dfrac{\gamma N^{1/2}}{s - n} & \text{if } x \in \left(\dfrac{1}{2} + \dfrac{s-1}{2N}, \dfrac{1}{2} + \dfrac{s}{2N}\right), \\ & \qquad\qquad 0 < |s - n| \leq N^{1/2}, \tag{60} \\ 0 & \text{for other } x \in \left(\dfrac{1}{2}, 1\right), \end{cases}$$

where $\gamma > 0$ will be chosen later, independently of $N$.

In addition, for $x \in (\frac{1}{2}, 1)$ we set

$$\psi_n(x) = 0 \quad \text{for } 1 \leq n \leq N^{1/2} \text{ and } N - N^{1/2} \leq n \leq N - 1. \tag{60'}$$

Then if we take

$$
\xi_n = \begin{cases}
\dfrac{1}{2} & \text{for } 1 \le n \le N^{1/2}, \\[2mm]
\dfrac{n-1}{2N} + \dfrac{1}{2} & \text{for } N^{1/2} < N < N - N^{1/2}, \\[2mm]
1 & \text{for } N - N^{1/2} \le N \le N - 1,
\end{cases}
$$

it is easily verified that conditions 1) and 2) are satisfied for the system of functions $\{\psi_n(x)\}_{n=1}^{N-1}$, $x \in (\frac{1}{2}, 1)$, $N \ge 100$, defined by (60) and (61). Moreover, we can deduce from (59) (just as in the proof of Lemma 1 for Theorem 2) that, for a sufficiently small constant $\gamma > 0$,

$$
\sup_{\{a_n\} \in B_2^{N-1}} \left\| \sum_{n=1}^{N-1} a_n \psi_n(x) \right\|_{L^2(\frac{1}{2},1)} < 1,
$$

Then (see Theorem 7.1 and the remark after that theorem) the functions $\psi_n(x)$, $n = 1, \ldots, N - 1$, can be extended to the interval $(0, \frac{1}{2})$ in such a way that

a)    $\{\psi_n(x)\}_{n=1}^{N-1}$, $x \in (0, 1)$,   is an O.N.S.,    and

b)    $\psi_n \in \mathscr{D}_N^0\left(0, \dfrac{1}{2}\right)$,     $n = 1, \ldots, N - 1$. 

$\qquad\qquad\qquad\qquad\qquad\qquad\qquad\qquad\qquad\qquad$ (61)

Since $\int_{1/2}^1 \psi_n(x)\,dx = 0$, $n = 1, \ldots, N - 1$ (see (60) and (60′)), it follows from (61) that $\psi_n \in \mathscr{D}_{2N}^0((0, 1))$ for $n = 1, \ldots, N - 1$ (after being changed at the finite set of points $s/2N$, a change which is not significant for us). Finally, we can complete the system $\{\psi_n\}_{n=1}^{N-1}$ arbitrarily to obtain an orthonormal basis in $\mathscr{D}_{2N}^0((0, 1))$ and so obtain the required system $\Psi_{2N-1}((0, 1))$.

PROOF OF THEOREM 6. We shall construct the system $\Phi = \{\varphi_n\}_{n=1}^{\infty}$ so that for each $k$, $k = 3, 4, \ldots$, the block of functions $\{\varphi_n\}_{n=1}^{\nu_k}$ ($\nu_k = 2^{2^k}$) forms an orthonormal basis in $\mathscr{D}_{\nu_k} \equiv \mathscr{D}_{\nu_k}((0, 1))$.

We choose the functions $\varphi_n(x)$, $n = 1, \ldots, \nu_s$, arbitrarily as long as they form an orthonormal basis in $\mathscr{D}_{\nu_s}$. The functions $\varphi_n(x)$, $n = \nu_k + 1, \ldots, \nu_{k+1}$, for $k = 3, 4, \ldots$, are defined as follows: if

$$
n = \nu_k + (p-1)(\nu_k - 1)j,
$$
$$
1 \le p \le \nu_k, \quad 1 \le j = j(n) \le \nu_k - 1, \quad (62)
$$

then $\varphi_n(x)$ is supported on the interval $\Delta_{2^k}^p = ((p-1)/\nu_k), p/\nu_k)$ and equals

$$
\varphi_n(x) = \psi_j^{\nu_k - 1}(x, \Delta_{2^k}^p), \qquad\qquad (63)
$$

where $\psi_j^{2N-1}(x, I)$ where defined in Lemma 1.

It is easily seen then that $\Phi = \{\varphi_n(x)\}_{n=1}^{\infty}$, $x \in (0,1)$, is an O.N.S., and that the functions $\{\varphi_n\}_{n=1}^{\nu_k}$ form a basis in $\mathscr{D}_{\nu_k}$ $(k = 3, 4, \dots)$.

Let us show that $\Phi$ satisfies the requirements of Theorem 6. For each number $n > \nu_3$ with (see (62))

$$1 \leq j(n) \leq \tfrac{1}{2}\nu_k - 1 \tag{64}$$

(we shall call such numbers "numbers of the form (64)," for short) we denote by $\xi_n$ a point of essential change of sign of $\psi_j^{\nu_k-1}(x, \Delta_{2^k}^p)$ (the restriction (64) is needed because points of essential change of sign are defined for $\psi_j^{2N-1}(x, I)$ only when $1 \leq j \leq N - 1$; see Lemma 1).

If $\delta_{2^k}^p$ denotes the interval

$$\left( \frac{p-1}{\nu_k} + \frac{5}{8\nu_k}, \frac{p-1}{\nu_k} + \frac{7}{8\nu_k} \right),$$

$k = 3, 4, \dots$; $p = 1, \dots, \nu_k$, then it follows immediately from condition 2) of Lemma 1 and (63) that

$$\sum_{n=\nu_k+(p-1)(\nu_k-1)+1}^{m_{k,p}(x)} \varphi_n(x) \geq c\nu_k 2^k \quad \text{for almost all } x \in \delta_{2^k}^p, \tag{65}$$

where

$$m_{k,p}(x) = \max\{m : \xi_m < x, m = \nu_k + (p-1)(\nu_k-1)+1, \dots, \nu_k + p(\nu_k-1)\}.$$

Take a sequence $\{b_k\}_{k=0}^{\infty}$, $b_k \geq 0$, $\sum_{k=0}^{\infty} b_k = \infty$, and construct an almost everywhere divergent series (56). We first observe that if $\chi(x,k)$, $k = 3, 4, \dots$, are the characteristic functions of the sets $\bigcup_{p=1}^{\nu_k} \delta_{2^k}^p$, it is easily seen that the functions $\chi(x,k)$ are independent and $\int_0^1 \chi(x,k)\,dx = \tfrac{1}{4}$, $k = 3, 4, \dots$ Let us show that

$$\sum_{k=3}^{\infty} b_k \chi(x,k) = \infty \quad \text{for almost all } x \in (0,1). \tag{66}$$

Let the sequence $\{c_k\}_{k=3}^{\infty}$ have the properties that

$$0 < c_k \leq b_k, \qquad \sum_{k=3}^{\infty} c_k = \infty, \qquad \sum_{k=3}^{\infty} c_k^2 < \infty.$$

Then $\sum_{k=3}^{\infty} \tfrac{1}{4} c_k = \infty$, and by Theorem 2.9 the series $\sum_{k=3}^{\infty} c_k[\chi(x,k) - \tfrac{1}{4}]$ converges a.e. on $(0,1)$ (since

$$\left\{ \frac{4}{\sqrt{3}} \left[ \chi^{(x,k)} - \frac{1}{4} \right] \right\}_{k=3}^{\infty}, \qquad x \in (0,1),$$

is an O.N.S. of independent functions). This means that the series $\sum_{k=1}^{\infty} c_k \chi(x,k)$ diverges a.e. on $(0,1)$, whence we obtain (65) by the condition $0 \le c_k \chi(x,k) \le b_k \chi(x,k)$.

Using (65), we can find a sequence $k_s, s = 1,2,\ldots$ $(3 < k_1 < k_2 < \cdots)$ such that, for $s = 1,2,\ldots,$

$$m \left\{ x \in (0,1) : \sum_{k=k_s}^{k_{s+1}-1} b_k \chi(x,k) > s \right\} > 1 - \frac{1}{s}. \tag{67}$$

The permutation $\sigma = \{\sigma(n)\}_{n=1}^{\infty}$ will be such that

$$\{\sigma(n)\}_{n=\nu_{k_s}+1}^{\nu_{k_{s+1}}} = \{n\}_{n=\nu_{k_s}+1}^{\nu_{k_{s+1}}}, \qquad s = 1,2,\ldots,$$

and for a given $x$ the functions $\varphi_n(x)$, $n = \nu_{k_s}+1,\ldots,\nu_{k_{s+1}}$, are rearranged as follows:

a) the functions $a_n \varphi_n(x)$ with indices of the form (64) are arranged in order of increase of points of essential change of sign, i.e., so that if $\xi_{\sigma(n_1)} < \xi_{\sigma(n_2)}$ then $n_1 < n_2$;

b) the functions $a_n \varphi_n(x)$ for which the number $j(n)$ in (69) does not satisfy the inequalities (64) appear in arbitrary order after all the functions with indices of the form (64).

Let us consider the sum

$$G_s(x) = \sum_{n=\nu_{k_s}+1}^{M_s(x)} a_{\sigma(n)} \varphi_{\sigma(n)}(x), \tag{68}$$

where $M_s(x) = \max\{m : \xi_{\sigma(n)} < x, m = \nu_{k_s}+1,\ldots,\nu_{k_{s+1}}\}$. Since $\varphi_n(x) \ge 0$ for $x \in (\xi_n, 1)$ (see condition 1) in Lemma 1, and (63)), it follows that, by the construction of $\sigma$, all the terms in (68) are non-negative. Hence, by using (65), we deduce that

$$G_s(x) \ge c \sum_{k=k_s}^{k_{s+1}-1} b_k \chi(x,k).$$

for almost all $x \in (0,1)$. It follows from the preceding inequality and (67) that

$$m\{x \in (0,1) : G_s(x) > cs\} > 1 - \frac{1}{s}, \qquad s = 1,2,\ldots,$$

i.e., the series $\sum_{n=1}^{\infty} a_{\sigma(n)} \varphi_{\sigma(n)}(x)$ diverges a.e. This completes the proof of Theorem 6.

## §3. Subsequences with convergence almost everywhere

It was shown in §1 that orthogonal series with coefficients in $l^2$ can diverge a.e. on $(0,1)$. It is also true that the convergence in measure of every series

$$\sum_{n=1}^{\infty} a_n \varphi_n(x), \qquad \sum_{n=1}^{\infty} a_n^2 < \infty, \qquad \Phi = \{\varphi_n\}, \quad \text{an O.N.S.}, \qquad (69)$$

implies the convergence a.e. of some subsequence $S_{n_k}$, $k = 1, 2, \ldots$, of its partial sums. However, it is not evident that for a given system $\Phi$ the indices $N_k$ can be chosen independently of the coefficients of (69). As the following proposition shows, such a choice is possible.

THEOREM 7. *For every O.N.S.* $\Phi = \{\varphi_n(x)\}_{n=1}^{\infty}$, $x \in (0, 1)$, *there is an increasing sequence* $\{N_k\}_{k=1}^{\infty}$ *of integers such that for every series of the form* (69) *the sequence* $S_{N_k}(x) = \sum_{n=1}^{N_k} a_n \varphi_n(x)$, $k = 1, 2, \ldots$, *converges a.e. on* (0, 1) *and satisfies the inequality*([5])

$$\left\| \sup_{1 \leq q < \infty} |S_{N_q}(x)| \right\|_2 \leq C \left( \sum_{n=1}^{\infty} a_n^2 \right)^{1/2}. \qquad (70)$$

PROOF. We expand the functions in $\Phi$ in Haar series:

$$\varphi_n(x) = \sum_{j=1}^{\infty} b_{n,j} \chi_j(x),$$

$$b_{n,j} = \int_0^1 \chi_j(x) \varphi_n(x) \, dx, \qquad n, j = 1, 2, \ldots.$$

Then, by Parseval's theorem (see Theorem 1.4) and the completeness of the Haar system,

$$\sum_{j=1}^{\infty} b_{n,j}^2 = \|\varphi_n\|_2^2 = 1, \qquad n = 1, 2, \ldots, \qquad (71)$$

and by Bessel's inequality

$$\sum_{n=1}^{\infty} b_{n,j}^2 \leq \|\chi_j\|_2^2 = 1, \qquad j = 1, 2, \ldots. \qquad (72)$$

We simultaneously define the required sequence $\{N_k\}_{k=0}^{\infty}$ and an auxiliary sequence $\{w_k\}_{k=0}^{\infty}$: $N_0 = w_0 = 0$, $N_1 = w_1 = 1$; and if $N_k$ and $w_k$ have already been constructed for all positive integers $k < k_0$, then we choose

---

([5])The absolute constant $C$ in (70) is independent of the system $\Phi$.

$N_{k_0}$ and $w_{k_0}$ so that $N_{k_0} > N_{k_0-1}$, $w_{k_0} > w_{k_0-1}$, and

$$1) \quad \sum_{j=1}^{w_{k_0}-1} \sum_{n=N_{k_0}}^{\infty} b_{n,j}^2 < 4^{-k_0},$$

$$2) \quad \sum_{n=1}^{N_{k_0}-1} \sum_{j=w_{k_0}}^{\infty} b_{n,j}^2 < 4^{-k_0}. \tag{73}$$

It is easily seen (see (71) and (72)) that if $N_{k_0}$ and $w_{k_0}$ are taken sufficiently large, then (73) is satisfied.

Consider the series

$$\sum_{k=0}^{\infty} \left( \sum_{n=N_k+1}^{N_{k+1}} a_n \varphi_n(x) \right), \quad \sum_{n=1}^{\infty} a_n^2 \le 1, \tag{74}$$

whose convergence a.e. is to be established; we study its general term in detail. We have the equation

$$\sum_{n=N_k+1}^{N_{k+1}} a_n \varphi_n(x) = \sum_{n=N_k+1}^{N_{k+1}} a_n \sum_{j=1}^{\infty} b_{n,j} \chi_j(x)$$

$$= \sum_{n=N_k+1}^{N_{k+1}} a_n \left[ \sum_{j=1}^{w_{k-1}} b_{n,j} \chi_j(x) + \sum_{j=w_{k-1}+1}^{w_{k+2}} b_{n,j} \chi_j(x) \right.$$

$$\left. + \sum_{j=w_{k+2}+1}^{\infty} b_{n,j} \chi_j(x) \right]$$

$$= I_k^{(1)}(x) + I_k^{(2)}(x) + I_k^{(3)}(x),$$

where (see (73) and (74))

$$a) \quad \|I_k^{(1)}(x)\|_2 \equiv \left\| \sum_{j=1}^{w_{k-1}} \chi_j(x) \sum_{n=N_k+1}^{N_{k+1}} a_n b_{n,j} \right\|_2$$

$$= \left\{ \sum_{j=1}^{w_{k-1}} \left( \sum_{n=N_k+1}^{N_{k+1}} a_n b_{n,j} \right)^2 \right\}^{1/2}$$

$$\le \left\{ \sum_{j=1}^{w_{k-1}} \left( \sum_{n=N_k+1}^{N_{k+1}} a_n^2 \right) \times \left( \sum_{n=N_k+1}^{N_{k+1}} b_{n,j}^2 \right) \right\}^{1/2}$$

$$\le \left\{ \sum_{j=1}^{w_{k-1}} \sum_{n=N_k+1}^{\infty} b_{n,j}^2 \right\}^{1/2} < 2^{-k};$$

b)  $\|I_k^{(3)}(x)\|_2 \equiv \left\| \sum\limits_{n=N_k+1}^{N_{k+1}} a_n \sum\limits_{j=w_{k+2}+1}^{\infty} b_{n,j}\chi_j(x) \right\|_2$

$\le \sum\limits_{n=N_k+1}^{N_{k+1}} |a_n| \left\| \sum\limits_{j=w_{k+2}+1}^{\infty} b_{n,j}\chi_j(x) \right\|_2$

$\le \left\{ \sum\limits_{n=N_k+1}^{N_{k+1}} a_n^2 \right\}^{1/2} \left\{ \sum\limits_{n=N_k+1}^{N_{k+1}} \left\| \sum\limits_{j=w_{k+2}+1}^{\infty} b_{n,j}\chi_j(x) \right\|_2^2 \right\}^{1/2}$

$\le \left\{ \sum\limits_{n=N_k+1}^{N_{k+1}} \left( \sum\limits_{j=w_{k+2}+1}^{\infty} b_{n,j}^2 \right) \right\}^{1/2} < 2^{-k}.$

Therefore when $k > 1$

$$\sum_{n=N_k+1}^{N_{k+1}} a_n\varphi_n(x) = \sum_{j=w_{k-1}+1}^{w_{k+2}} \chi_j(x) \left( \sum_{n=N_k+1}^{N_{k+1}} a_n b_{n,j} \right) + \delta_k(x)$$

$$= \sum_{j=w_{k-1}+1}^{w_{k+2}} \alpha_j^{(k)}\chi_j(x) + \delta_k(x) \equiv f_k(x) + \delta_k(x),$$

$$(75)$$

where

$$\|\delta_k\|_2 \le 2 \cdot 2^{-k} \quad and \quad \alpha_j^{(k)} = \sum_{n=N_k+1}^{N_{k+1}} a_n b_{n,j}. \qquad (76)$$

By (75) and (76),

$$\sum_{j=w_{k-1}+1}^{w_{k+2}} (\alpha_j^{(k)})^2 = \left\| \sum_{j=w_{k-1}+1}^{w_{k+2}} \alpha_j^{(k)}\chi_j(x) \right\|_2^2$$

$$\le \left\{ \left\| \sum_{n=N_k+1}^{N_{k+1}} a_n\varphi_n(x) \right\|_2 + 2^{-k+1} \right\}^2$$

$$\le 2 \sum_{n=N_k+1}^{N_{k+1}} a_n^2 + 2 \cdot 2^{-2k+2},$$

from which it follows that

$$\sum_{k=2}^{\infty} \sum_{j=w_{k-1}+1}^{w_{k+2}} (\alpha_j^{(k)})^2 \le 3. \qquad (77)$$

It follows from (76), by Beppo Levi's theorem, that the series $\sum_{k=2}^{\infty} |\delta_k(x)|$ converges a.e. In addition (see (76)),

$$\left\| \sup_{2 \leq q < \infty} \left| \sum_{k=2}^{q} \delta_k(x) \right| \right\|_2 \leq \left\| \sum_{k=2}^{\infty} |\delta_k(x)| \right\|_2 \leq \sum_{k=2}^{\infty} \|\delta_k(x)\|_2 \leq 1,$$

and therefore (see (75) and (74))

$$\left\| \sup_{1 \leq q < \infty} |S_{N_q}(x)| \right\|_2 = \left\| \sup_{1 \leq q < \infty} \left| \sum_{k=0}^{q} \sum_{n=N_k+1}^{N_{k+1}} a_n \varphi_n(x) \right| \right\|_2$$

$$\leq \|a_1 \varphi_1\|_2 + \left\| \sum_{n=2}^{N_2} a_n \varphi_n \right\|_2$$

$$+ \left\| \sup_{2 \leq q < \infty} \left| \sum_{k=2}^{q} \delta_k(x) \right| \right\|_2 + \left\| \sup_{2 \leq q < \infty} \left| \sum_{k=2}^{q} f_k(x) \right| \right\|_2$$

$$\leq 3 + \left\| \sup_{2 \leq q < \infty} \left| \sum_{k=2}^{q} f_k(x) \right| \right\|_2.$$

Consequently, to prove Theorem 7 it is enough to show that the series

$$\sum_{k=2}^{\infty} f_k(x), \tag{78}$$

converges for almost all $x \in (0, 1)$, and to estimate the $L^2(0, 1)$ norm of the majorant of the partial sums of this series. It was proved in Chapter 3 (see Theorem 3.4) that every series of the form

$$\sum_{j=1}^{\infty} \alpha_j \chi_j(x), \qquad \sum_{j=1}^{\infty} \alpha_j^2 < \infty, \tag{79}$$

converges a.e. on $(0, 1)$, and that the majorant of its partial sums is subject to the inequality

$$\left\| \sup_{1 \leq M < \infty} \left| \sum_{j=1}^{M} \alpha_j \chi_j(x) \right| \right\|_2 \leq K \left( \sum_{j=1}^{\infty} \alpha_j^2 \right)^{1/2}, \tag{80}$$

where $K$ is an absolute constant. We shall use this result. Observe that the series (78) can be written as the sum of three series:

$$\sum_{p=1}^{\infty} f_{3p-1}(x) + \sum_{p=1}^{\infty} f_{3p}(x) + \sum_{p=1}^{\infty} f_{3p+1}(x), \tag{81}$$

where, by the definition of the functions $f_k$ (see (75) and also (77)), each series in (81) is a grouped series in the Haar system, of the form (79). Therefore the series in (81), and consequently the series (78), converge a.e. on $(0, 1)$ and satisfy the inequality (see (80) and (77))

$$\left\| \sup_{2 \le q < \infty} \left| \sum_{k=2}^{q} f_k(x) \right| \right\|_2 \le \sum_{i=-1}^{1} \left\| \sup_{1 \le M < \infty} \left| \sum_{p=1}^{M} f_{3p+i}(x) \right| \right\|_2 \le 3\sqrt{3}K.$$

This completes the proof of Theorem 7.

COROLLARY 5. *From every O.N.S.* $\Phi = \{\varphi_n(x)\}_{n=1}^{\infty}$, $x \in (0, 1)$, *we can extract a subsystem* $\Phi_1 = \{\varphi_{n_k}(x)\}_{k=1}^{\infty}$, $n_1 < n_2 < \cdots$, *which is a convergence system (see Definition 1.2), and for which, moreover,*

$$\|S_{\Phi_1}^*(\{a_k\}, x)\|_2 \le C\|\{a_k\}\|_{l^2},$$

*where* $S_{\Phi_1}^*$ *is the majorant operator for the partial sums and* $\{a_k\}$ *is an arbitrary sequence in* $l^2$.

In fact, if $\{N_k\}_{k=1}^{\infty}$ is the sequence of numbers that was constructed for $\Phi$ in Theorem 7, then if we choose numbers $n_k$ such that $N_k < n_k \le N_{k+1}$, $k = 1, 2, \ldots$, we obtain from Theorem 7 that $\{\varphi_{n_k}\}_{k=1}^{\infty}$ satisfies the requirements of Corollary 5.

## §4. Lacunary systems

DEFINITION 2. *An O.N.S.* $\Phi = \{\varphi_n(x)\}_{n=1}^{\infty}$, $x \in (0, 1)$, *is an* $S_p$-*system* $(2 < p < \infty)$ *if there is a constant* $K$ *such that for every polynomial* $P(x) = \sum_{n=1}^{N} a_n \varphi_n(x)$ *in the system* $\Phi$ *the following inequality is satisfied:*

$$\|P(x)\|_p \le K\|P(x)\|_2. \tag{82}$$

We observe that if $\{\varphi_n\}$ is an $S_p$-system and $L$ is the subspace of $L^2(0, 1)$ consisting of the functions of the form

$$f(x) \overset{L^2}{=} \sum_{n=1}^{\infty} a_n \phi_n(x), \qquad \sum_{n=1}^{\infty} a_n^2 < \infty,$$

then it follows immediately from Definition 2 that every orthonormal basis in $L$ is also an $S_p$-system. Therefore it is often natural to speak of $S_p$-subspaces of $L^2(0, 1)$, i.e., of subspaces $L$ such that

$$\|f\|_2 \le \|f\|_p \le K\|f\|_2, \qquad f \in L \ (2 < p < \infty).$$

Examples of $S_p$-systems (for every $p$, $2 < p < \infty$) are orthonormal systems of independent uniformly bounded functions (this follows from Theorem 2.6). However, the class of $S_p$-systems is substantially wider than the class of systems of independent functions. We shall try to clarify this in the following way.

If an O.N.S. $\Phi = \{\varphi_n(x)\}_{n=1}^N$ is a basis in $\mathscr{D}_N$, it is easily verified that $\Phi$ cannot contain a subsystem of independent functions with more than $1 + \log_2 N$ elements. On the other hand, one can show (by a method similar to that used in the proof of Theorem 9 of this section) that every O.N.S. $\Phi = \{\varphi_n(x)\}_{n=1}^N$ with $\|\varphi_n\|_\infty \le B$, $1 \le n \le N$, contains a subsystem $\{\varphi_{n_i}\}_{i=1}^{i_0}$ with more than $i_0 \ge N^\gamma$ elements, $\gamma > 0$, such that every polynomial $P(x) = \sum_{i=1}^{i_0} a_{n_i} \varphi_{n_i}(x)$ satisfies the inequality $\|P\|_4 \le C_B \|P\|_2$.

Let us show that every $S_p$-system is a convergence system.

THEOREM 8. *If* $\Phi = \{\varphi_n(x)\}_{n=1}^\infty$, $x \in (0, 1)$, *is an* $S_p$-*system (for some* $p > 2$), *then every series*

$$\sum_{n=1}^\infty a_n \varphi_n(x), \qquad \sum_{n=1}^\infty a_n^2 < \infty,$$

*converges a.e. on* $(0, 1)$, *and moreover (see* 1.(5))

$$\|S_\Phi^*(\{a_n\})\|_p \le C_\Phi \left( \sum_{n=1}^\infty a_n^2 \right)^{1/2}. \tag{83}$$

REMARK. The constant $C_\Phi$ in (83) depends only on $p$ and the constant $K = K(\Phi)$ in Definition 2 (see (82)).

Using the remark at the beginning of this section, we at once obtain the following corollary of Theorem 8.

COROLLARY 6. *Every* $S_p$-*system* $(2 < p < \infty)$ *is an unconditional convergence system (see Definition* 1.5).

To illustrate the underlying idea of the proof, we first consider the special case when $M = 2^r$, $a_1 = a_2 = \cdots = a_M = 1$. Let

$$\delta(x) = \sup_{1 \le N \le 2^r} \left| \sum_{n=1}^N a_n \varphi_n(x) \right|.$$

Using the same partition of the sum $\sum_{n=1}^N a_n \varphi_n(x)$ into "binary blocks" that was used in the proof of Theorem 1 (see equation (4)), we find that

$$\delta(x) \le \sum_{k=0}^r \max_{0 \le q \le 2^k - 1} \left| \sum_{n \in \Omega_q^k} a_n \varphi_n(x) \right|, \tag{84}$$

where

$$\Omega_q^k = \{q \cdot 2^{r-k} + 1, \ q 2^{r-k} + 2, \dots, (q+1) 2^{r-k}\}.$$

It follows from (84) that

$$\delta(x) \leq \sum_{k=0}^{r} \left\{ \sum_{q=0}^{2^k-1} \left| \sum_{n \in \Omega_q^k} a_n \varphi_n(x) \right|^p \right\}^{1/p}.$$

Hence by using the facts that $\Phi$ is an $S_p$-system and $a_n = 1$ for $n = 1, \ldots, M$, we obtain

$$\|\delta(x)\|_p \leq \sum_{k=0}^{r} \left\{ \int_0^1 \sum_{q=0}^{2^k-1} \left| \sum_{n \in \Omega_q^k} \varphi_n(x) \right|^p dx \right\}^{1/p}$$

$$\leq \sum_{k=0}^{r} \left\{ \sum_{q=0}^{2^k-1} 2^{(r-k)p/2} \right\}^{1/p} = K 2^{r/2} \sum_{k=0}^{r} 2^{(k/p)(1-p/2)}$$

$$\leq M^{1/2} K (1 - 2^{1/p-1/2})^{-1},$$

which proves (83) for this particular system.

The proof of (83) in the general case, to which we now turn, is carried out similarly ot the discussion presented above. However, here the partition of a section of the integers into equal blocks $\Omega_q^k$, $q = 0, 1, \ldots, 2^k - 1$, is replaced by a partition into blocks $\tilde{\Omega}_q^k$ such that the numbers $\sum_{n \in \tilde{\Omega}_q^k} a_n^2$ are nearly the same for each $q$, $q = 0, 1, \ldots, 2^k - 1$.

LEMMA 1. *Let a vector* $a = \{(a)_n\}_{n=1}^N$ *be given. We can determine a number* 1, $1 \leq l \leq N$, *and two vectors* $a'$ *and* $a''$ *of the following form*:

$$a' = \{(a)_1, \ldots, (a)_{l-1}, (a')_l, 0, \ldots, 0\}, \tag{85}$$
$$a'' = \{0, \ldots, 0, (a'')_l, (a)_{l+1}, \ldots, (a)_N\}$$

*so that the following relations are satisfied*:
1) $a' + a'' = a$;
2) $\sum_{n=1}^N (a')_n^2 \leq \frac{1}{2} A$, $\sum_{n=1}^N (a'')_n^2 \leq \frac{1}{2} A$ *where* $A = \sum_{n=1}^N (a)_n^2$;
3) $|(a')_l| \leq |(a)_l|$, $|(a'')_l| \leq |(a)_l|$.

In spite of its complicated statement, Lemma 1 is quite simple. If there is a number $q$, $q = 1, 2, \ldots, N - 1$, such that $\sum_{n=1}^q (a)_n^2 = \sum_{n=q+1}^N (a)_n^2$, it is enough to set $l = q$, $(a')_l = 0$, $(a'')_l = (a)_l$, and Lemma 1 is established. If there is no such number $q$, we take $l$ to be a number for which

$$\sum_{n=1}^{l-1} (a)_n^2 < \frac{A}{2} < \sum_{n=1}^{l} (a)_n^2$$

(we suppose, as usual, that $\sum_{n=1}^{0} \equiv 0$). Let $\delta = A/2 - \sum_{n=1}^{l-1}(a)_n^2$. Then:

a) $$0 < \delta < (a)_l^2;$$

b) $$\sum_{n=l+1}^{N}(a)_n^2 = A - \sum_{n=1}^{l}(a)_n^2 = A - \left(\frac{A}{2} - \delta\right) - (a)_l^2. \qquad (86)$$

Set

$$(a')_l = \frac{\delta}{(a)_l}, \qquad (a'')_l = (a)_l - (a')_l.$$

Hence we have actually defined vectors $a'$ and $a''$, with $a' + a'' = 0$. Moreover, by (86),a) $|(a')_l| = |\delta/(a)_l| < |(a)_l|$, and since $(a)_l$ and $(a')_l$ have the same sign, we have $|(a'')_l| < |(a)_l|$, i.e., condition 3) is satisfied. Finally, using (86) we find

$$\sum_{n=1}^{N}(a')_n^2 = \sum_{n=1}^{l-1}(a)_n^2 + (a')_l^2 = \frac{A}{2} - \delta + \left(\frac{\delta}{(a)_l}\right)^2 < \frac{A}{2}$$

and

$$\sum_{n=1}^{N}(a'')_n^2 = (a'')_l^2 + \sum_{n=l+1}^{N}(a)_n^2 = (a'')_l^2 + \frac{A}{2} + \delta - (a)_l^2$$

$$= \left[(a)_l - \frac{\delta}{(a)_l}\right]^2 + \frac{A}{2} + \delta - (a)_l^2$$

$$= \frac{A}{2} - \delta + \left(\frac{\delta}{(a)_l}\right)^2 < \frac{A}{2}.$$

This completes the proof of Lemma 1.

Let there be given an arbitrary set $a = \{a_n\}_{n=1}^{M}$ of numbers, which we regard as a vector in $M$-space and for which we wish to establish (83). It will be convenient to denote the components of the vector $a$ by $(a)_n$, $n = 1, \ldots, M$. By considerations of homogeneity and continuity we may suppose that $\sum_{n=1}^{M}(a)_n^2 = 1$ and $(a)_n \neq 0$ for $n = 1, \ldots, M$.

We define a decomposition of the vector $a$ into a sum of vectors:

$$a = \sum_{\nu=0}^{2^s-1} r_\nu^s, \qquad s = 0, 1, \ldots, s_0, \qquad (87)$$

where $r_0^0 = a$, and the vectors $r_\nu^s$ are constructed inductively: if $r_\nu^{s-1}$ has already been constructed, then by applying Lemma 1 (with $a = r_\nu^{s-1}$) we represent it as $r_\nu^{s-1} = r_{2\nu}^s + r_{2\nu+1}^s$, where the components of $r_{2\nu}^s$ and $r_{2\nu+1}^s$ are defined by (85). Then the vectors $r_\nu^s$, $\nu = 0, 1, \ldots, 2^s - 1$, will have the form

$$r_\nu^s = (0,\ldots,0,a'_{l_\nu^s},(a)_{l_\nu^s+1},\ldots,(a)_{l_{\nu+1}^s-1},a''_{l_{\nu+1}^s},0,\ldots,0),$$
$$\text{where } 1 = l_0^s \le l_1^s \le \cdots \le l_{2^s}^s = M. \quad (88)$$

Moreover, we have by construction (see Lemma 1)

$$\max\{\|r_{2\nu}^s\|_{l^2}^2, \|r_{2\nu+1}^s\|_{l^2}^2\} \le \tfrac{1}{2}\|r_\nu^{s-1}\|_{l^2}^2,$$
$$\max\{\|r_{2\nu}^s\|_{l^\infty}, \|r_{2\nu+1}^s\|_{l^\infty}\} \le \|r_\nu^{s-1}\|_{l^\infty}. \quad (89)$$

We take $s_0$ (see (87)) so large that the number of nonzero components of each vector $r_\nu^{s_0}$, $\nu = 0, 1, \ldots, 2^{s_0}-1$, is at most two (this is possible by the first inequality in (89)). It is also convenient to set $r_{2^s}^s = 0$, $s = 1, \ldots, s_0$.
It is clear from (87) and (88) that, for $s = 1, \ldots, s_0$; $j = 1, \ldots, 2^s$,

$$\left(\sum_{\nu=0}^{j-1} r_\nu^s\right)_n = \begin{cases} \left(\displaystyle\sum_{\nu=0}^{2^s-1} r_\nu^s\right)_n = (a)_n & \text{if } n = 1, 2, \ldots, l_j^s - 1, \\ 0 & \text{if } n = l_j^s + 1, \ldots, M, \end{cases}$$

i.e., (also see (89))

$$\sum_{\nu=0}^{j-1} r_\nu^s = ((a)_1, \ldots, (a)_{l_j^s-1}, a''_{l_j^s}, 0, \ldots, 0), |a_{l_j^s}| \le \|a\|_{l^\infty}. \quad (90)$$

If we use (90), and, for a given $N$, $N = 1, \ldots, M - 1$, choose $j = j(N)$ so that $l_{j-1}^{s_0} \le N < l_j^{s_0}$, we obtain the following expansion for the vector $a(N) = ((a)_1, \ldots, (a)_N, 0, \ldots, 0)$ :

$$a(N) = \sum_{\nu=0}^{j-1} r_\nu^{s_0} + b, |(b)_n| \le \|a\|_{l^\infty}, \qquad 1 \le n \le M, \quad (91)$$

where the vector $b$ has at most two nonzero components. Since $r_{2\nu}^s + r_{2\nu+1}^s = r_\nu^{s-1}$, we find from (91) that

$$a(N) = \sum_{s=1}^{s_0} r_{\nu(s,N)}^s + b \qquad (0 \le \nu(s,N) \le 2^s \text{ for } s = 1, \ldots, s_0), \quad (92)$$

and consequently, for arbitrary numbers $\{y_n\}_{n=1}^M$, the sum $\sum_{n=1}^N a_n y_n$ $(1 \le N \le M)$ can be represented in the form

$$\sum_{n=1}^N a_n y_n = \sum_{s=1}^{s_0} \sum_{n=1}^M (r_{\nu(s,N)}^s)_n y_n + \Delta, |\Delta| \le 2 \max_{1 \le n \le M} |a_n y_n|. \quad (93)$$

If we apply (93) in the case when $y_n = \varphi_n(x)$, and $N = N(x)$ has the property that

$$\left|\sum_{n=1}^{N} a_n \varphi_n(x)\right| = \delta(x) \equiv \max_{1 \le N' \le M} \left|\sum_{n=1}^{N'} a_n \varphi_n(x)\right|,$$

we obtain

$$\delta(x) \le \sum_{s=1}^{s_0} \left|\sum_{n=1}^{M} (r^s_{\nu(s,N(x))})_n \varphi_n(x)\right| + 2 \max_{1 \le n \le M} |a_n \varphi_n(x)|$$

$$\le \sum_{s=1}^{s_0} \left\{\sum_{\nu=0}^{2^s-1} \left|\sum_{n=1}^{M} (r^s_\nu)_n \varphi_n(x)\right|^p\right\}^{1/p} + 2 \left\{\sum_{n=1}^{M} |a_n \varphi_n(x)|^p\right\}^{1/p}. \tag{94}$$

Since $\|r^s_\nu\|^2_{l^1} \le 2^{-s}$ (see (89)) and since $\Phi$ is an $S_p$-system (see (82)), we find from (94) that

$$\|\delta(x)\|_p \le \sum_{s=1}^{s_0} \left\|\left\{\sum_{\nu=0}^{2^s-1} \left|\sum_{n=1}^{M} (r^s_\nu)_n \varphi_n(x)\right|^p\right\}^{1/p}\right\|_p + 2 \left\|\left\{\sum_{n=1}^{M} |a_n \varphi_n(x)|^p\right\}^{1/p}\right\|_p$$

$$\le \sum_{s=1}^{s_0} \left\{\sum_{\nu=0}^{2^s-1} \int_0^1 \left|\sum_{n=1}^{M} (r^s_\nu)_n \varphi_n(x)\right|^p dx\right\}^{1/p}$$

$$+ 2 \left\{\sum_{n=1}^{M} \int_0^1 |a_n \varphi_n(x)|^p dx\right\}^{1/p}$$

$$\le K \sum_{s=1}^{s_0} \left\{\sum_{\nu=0}^{2^s-1} \|r^s_\nu\|^p_{l^2}\right\}^{1/p} + 2K \left\{\sum_{n=1}^{M} |a_n|^p\right\}^{1/p}$$

$$\le K \sum_{s=1}^{s_0} 2^{s(1-p/2)1/p} + 2K \le C_\Phi < \infty.$$

This completes the proof of (83) and hence of Theorem 8.

**DEFINITION 3.** An O.N.S. $\Phi = \{\varphi_n(x)\}^\infty_{n=1}$, $x \in (0,1)$, is a *Sidon system* provided that the series

$$\sum_{n=1}^{\infty} a_n \varphi_n(x) \tag{95}$$

converges in $L^\infty(0,1)$ if and only if the sum

$$\sum_{n=1}^{\infty} |a_n| < \infty. \tag{96}$$

is finite.

It is easily verified that a necessary and sufficient condition for a system $\Phi = \{\varphi_n\}$ to be a Sidon system is that every polynomial $P(x) = \sum_{n=1}^{N} a_n \varphi_n(x)$ satisfies the inequalities

$$c \sum_{n=1}^{N} |a_n| \leq \|P\|_\infty \leq C \sum_{n=1}^{N} |a_n|, \tag{97}$$

where the constants $c > 0$ and $C$ are independent of $P(x)$.

It follows from (97) that a Sidon system must be a bounded O.N.S., i.e., a system for which

$$\|\varphi_n\|_\infty \leq K, \qquad n = 1, 2, \ldots. \tag{98}$$

However, the necessary condition (98) is not sufficient for $\Phi$ to be a Sidon system. For example, it is easily shown by using the inequality 4.(72) that if the system $\{\sqrt{2}\cos 2\pi n_k x\}_{k=1}^{\infty}$, $x \in (0, 1)$, is a Sidon system, the sequence $\{n_k\}$ must be so sparse that

$$\lambda(N) = \sum_{k : n_k < N} 1 \leq C \ln N, \qquad N = 2, 3, \ldots. \tag{99}$$

Examples of Sidon systems are the Rademacher system and systems of the form $\{\sqrt{2}\cos 2\pi n_k x\}_{k=1}^{\infty}$, where the sequence of integers $n_k$, $k = 1, 2, \ldots$, satisfies the condition $n_{k+1} > \gamma n_k$, $k = 1, 2, \ldots; \gamma > 1$.

It should be noted that if a uniformly bounded O.N.S. $\Phi = \{\varphi_n(x)\}_{n=1}^{\infty}$, $x \in (0, 1)$, has the property that if for some set $E \subset (0, 1)$ with $m(E) > 0$ (for example, $E = (\frac{1}{2}, 1)$) the convergence of the series (95) in $L^\infty(E)$ implies (96), then $\Phi$ is a Sidon system independently of the values of $\varphi_n(x)$ for $x \in (0, 1)\backslash E$. In other words, the property that $\Phi = \{\varphi_n(x)\}$, $x \in (0, 1)$, is a Sidon system does not, in general, provide any information (beyond (98)) about the behavior of $\varphi_n(x)$ on the interval $(0, \frac{1}{2})$.

One of the reasons for discussing Sidon systems is that when we study these systems it is natural to introduce Riesz products, which are a basic tool of the theory of orthogonal series, and one that we need to discuss.

It can be shown by using Riesz products that from every uniformly bounded O.N.S. $\Phi = \{\varphi_n\}$ (see (98)) we can extract a Sidon subsystem $\{\varphi_{n_k}(x)\}_{k=1}^{\infty}$. Here we obtain a quantitative analog of this result.

THEOREM 9. *From every bounded orthonormal set $\{\varphi_n(x)\}_{n=1}^{\infty}$ with $\|\varphi_n\|_\infty \leq M$ for $n = 1, \ldots, N$, we can select functions $\{\varphi_{n_k}(x)\}_{k=1}^{s}$, $1 \leq n_1 < \cdots \leq N$, with $s \geq \max\{[\frac{1}{6}\log_2 N], 1\}$ such that the inequality*

$$\frac{1}{M} \left\| \sum_{k=1}^{s} a_k \varphi_{n_k} \right\|_\infty \leq \sum_{k=1}^{s} |a_k| \leq 4M \left\| \sum_{k=1}^{s} a_k \varphi_{n_k} \right\|_\infty \tag{100}$$

*holds with arbitrary numbers $a_k$, $k = 1, \ldots, s$.*

REMARK. As the example of the trigonometric system (see (99)) shows, the order of the inequality $s \geq [\frac{1}{6} \log_2 N]$ cannot be improved.

PROOF OF THEOREM 9. We first notice that the left-hand inequality in (100) is valid for every polynomial in the system $\{\varphi\}_{n=1}^N$. Moreover, since

$$\left\| \sum_{k=1}^{s} a_k \varphi_{n_k} \right\|_\infty \geq \left\| \sum_{k=1}^{s} a_k \varphi_{n_k} \right\|_2 = \left( \sum_{k=1}^{s} a_k^2 \right)^{1/2} \geq \frac{1}{\sqrt{s}} \sum_{k=1}^{s} |a_k|,$$

the right-hand inequality in (100) is also evident for any subsystem $\{\varphi_{n_k}\}_{k=1}^s$ with $\sqrt{s} < 4M$. Consequently in order to prove the right-hand inequality (100) we may suppose without loss of generality that

$$s \geq (4M)^2 \geq 16. \tag{101}$$

(In (101) we saw that we always have $M \geq \|\varphi_1\|_\infty \geq \|\phi_1\|_2 = 1$.)

Let us show that the right-hand inequality (100) is satisfied for every polynomial in the system $\{\varphi_{n_k}\}_{k=1}^s$ provided that we have the inequality

$$I(\{n_k\}) \equiv \sum_{\{\varepsilon_k\} \in \Omega_s} \left\{ \int_0^1 \prod_{k=1}^{s} \left[ \frac{\varphi_{n_k}(x)}{M} \right]^{\varepsilon_k} dx \right\}^2 \leq 10^{-s}, \tag{102}$$

where in (102) $\Omega_s$ denotes the collection of all sets $\{\varepsilon_k\}_{k=1}^s$ such that

$$\text{a) } \varepsilon_k = 0, 1 \text{ or } 2; \quad \text{b) } \sum_{k:\varepsilon_k=2} 1 \leq 1; \quad \text{c) } \sum_{k:\varepsilon_k=1} 1 \geq 1. \tag{103}$$

In fact, let (102) be valid and let any polynomial $P(x) = \sum_{k=1}^{s} a_k \varphi_{n_k}(x)$ be given. In order to estimate the norm $\|P(x)\|_\infty$ we introduce the product (a Riesz product)

$$\prod_{k=1}^{s} \left[ 1 + \frac{\text{sgn } a_k}{M} \varphi_{n_k}(x) \right].$$

Using the nonnegativity of the product (which is immediate, since $\|\varphi_n\|_\infty \leq M$), we obtain

$$\sigma \equiv \int_0^1 P(x) \prod_{k=1}^{s} \left[ 1 + \frac{\text{sgn } a_k}{M} \varphi_{n_k}(x) \right] dx$$

$$\leq \|P(x)\|_\infty \int_0^1 \prod_{k=1}^{s} \left[ 1 + \frac{\text{sgn } a_k}{M} \varphi_{n_k}(x) \right] dx$$

$$\leq \|P(x)\|_\infty + \|P(x)\|_\infty \sum_{\{\varepsilon_k\} \in \Omega_s'} \left| \int_0^1 \prod_{k=1}^{s} \left[ \frac{\varphi_{n_k}(x)}{M} \right]^{\varepsilon_k} dx \right|, \tag{104}$$

where $\Omega'_s$ denotes the collection of all sets $\{\varepsilon_k\}^s_{k=1}$ of zeros and ones with $\sum^s_{k=1}\varepsilon_k \geq 1$. If we use (102) and apply Cauchy's inequality, we find from (104) that

$$\sigma \leq \|P(x)\|_\infty[1 + 2^{s/2}I^{1/2}(\{n_k\})] \leq 2\|P(x)\|_\infty. \qquad (105)$$

On the other hand,

$$\sigma = \int^1_0 \left[\sum^s_{p=1} a_p\varphi_{n_p}(x)\right]\prod^s_{k=1}\left[1 + \frac{\operatorname{sgn} a_k}{M}\varphi_{n_k}(x)\right]dx$$

$$= \sum^s_{p=1} a_p\int^1_0 \varphi_{n_p}(x)\,dx \qquad (106)$$

$$+ \sum^s_{p=1} a_p\int^1_0 \varphi_{n_p}(x)\sum_{\{\varepsilon_k\}\in\Omega'_s}\prod^s_{k=1}\left[\frac{\operatorname{sgn} a_n}{M}\varphi_{n_k}(x)\right]^{\varepsilon_k}dx.$$

If we pick out of the right-hand side of (106) the terms of the form

$$a_p\int^1_0 \varphi_{n_p}(x)\frac{\operatorname{sgn} a_p}{M}\varphi_{n_p}(x)\,dx = |a_p|\cdot\frac{1}{M}, \qquad p = 1,\dots,s,$$

we obtain

$$\sigma \geq \frac{1}{M}\sum^s_{k=1}|a_k| - \left(\sum^s_{k=1}|a_k|\right)M\sum_{\{\varepsilon_k\}\in\Omega_s}\left|\int^1_0 \prod^s_{k=1}\left(\frac{\varphi_{n_k}(x)}{M}\right)^{\varepsilon_k}dx\right|, \qquad (107)$$

where the summation in (107) is over sets $\{\varepsilon_k\}^s_{k=1}$ that satisfy (103).

If we apply Cauchy's inequality and again use (102), we find from (107) that

$$\sigma \geq \frac{1}{M}\sum^s_{k=1}|a_k| - \left(\sum^s_{k=1}|a_k|\right)M(s2^s)^{1/2}I^{1/2}(\{n_k\})$$

$$\geq \frac{1}{M}\sum^s_{k=1}|a_k| - \left(\sum^s_{k=1}|a_k|\right)M\left(\frac{s2^s}{10^s}\right)^{1/2} \qquad (108)$$

$$\geq \frac{1}{2M}\sum^s_{k=1}|a_k|,$$

since, by (101), $(s/5^s)^{1/2} \leq \frac{8}{5} \leq 1/2M^2$.

Combining (105) and (108), we find that if (102) is satisfied for the set $\{\varphi_{n_k}\}^s_{k=1}$, then the norm of every polynomial $P(x) = \sum^s_{k=1} a_k\varphi_{n_k}(x)$ can be estimated by

$$\|P(x)\|_\infty \geq \frac{1}{4M}\sum^s_{k=1}|a_k|,$$

i.e., Theorem 9 will be established if we prove the following proposition.

**LEMMA 1.** *From every set $\{\varphi_n\}_{n=1}^N$ of functions (with $\frac{1}{6}\log_2 N \geq 1$) that satisfy the hypotheses of Theorem 9, we can select functions $\{\varphi_{n_k}\}_{k=1}^s$, $1 \leq n_1 < \cdots < n_s \leq N$, with $s \geq [\frac{1}{6}\log_2 N]$, for which (102) is satisfied.*

PROOF. For a given $s$, we estimate the sum

$$G = \sum_{\{n_k\}\in E_N^s} I(\{n_k\})$$

$$= \sum_{\{n_k\}\in E_N^s} \sum_{\{\varepsilon_k\}\in\Omega_s} \left[\int_0^1 \prod_{k=1}^s \left(\frac{\varphi_{n_k}(x)}{M}\right)^{\varepsilon_k} dx\right]^2, \tag{109}$$

where $E_N^s$ denotes the collection of all sets $\{n_k\}_{k=1}^s$ of positive integers with $1 \leq n_1 < \cdots < n_s \leq N$. The sum $G$ consists of terms of the form (see (103))

$$\left[\frac{1}{M^j}\int_0^1 \prod_{k=1}^j \varphi_{n_k}(x)\,dx\right]^2, \qquad 1 \leq j \leq s, \quad \{n_k\} \in E_N^j,$$

and of the form

$$\left[\frac{1}{M^{j+1}}\int_0^1 \varphi_{n_q}(x)\prod_{k=1}^j \varphi_{n_k}(x)\,dx\right]^2,$$

$$1 < j \leq s, \quad 1 \leq q \leq j, \quad \{n_k\} \in E_N^j;$$

moreover, it is easily verified that when $s > 1$ each of these sums occurs in the sum (109)

$$C_{N-j}^{s-j} \text{ times} \tag{110}$$

($C_m^l$ is the binomial coefficient).

Let us estimate the sums

$$G_j' = \sum_{\{n_k\}\in E_N^j} \left[\int_0^1 \prod_{k=1}^j \varphi_{n_k}(x)\,dx\right]^2, \qquad j = 1,\ldots,s, \tag{111}$$

$$G_j'' = \sum_{\{n_k\}\in E_N^j} \sum_{q=1}^j \left[\int_0^1 \varphi_{n_q}(x)\prod_{k=1}^j \varphi_{n_k}(x)\,dx\right]^2, \qquad j = 1,\ldots,s. \tag{112}$$

By Bessel's inequality

$$G_1' = \sum_{n=1}^N \left[\int_0^1 \varphi_n(x)\,dx\right]^2 \leq \|1\|_2^2 = 1. \tag{113}$$

Moreover, when $j > 1$,

$$G_j' \leq \sum_{\{p_k\} \in E_N^{j-1}} \sum_{\{n_k\} \in E_N^j} \delta(\{p_k\}, \{n_k\}) \left[ \int_0^1 \prod_{k=1}^j \varphi_{n_k}(x)\, dx \right]^2, \qquad (114)$$

where

$$\delta(\{p_k\}, \{n_k\}) = \begin{cases} 1 & \text{if } \{p_k\} \subset \{n_k\}, \\ 0 & \text{in the opposite case.} \end{cases}$$

Since, by Bessel's inequality, for each set $\{p_k\} \subset E_N^{j-1}$,

$$\sum_{\{n_k\} \in E_N^j : \{n_k\} \supset \{p_k\}} \left[ \int_0^1 \prod_{k=1}^j \varphi_{n_k}(x)\, dx \right]^2$$

$$\leq \sum_{n=1}^N \left\{ \int_0^1 \left[ \prod_{k=1}^{j-1} \varphi_{p_k}(x) \right] \varphi_n(x)\, dx \right\}^2 \leq \left\| \prod_{k=1}^{j-1} \varphi_{p_k}(x) \right\|_2^2 \leq M^{2(j-1)},$$

it follows from (114) that

$$G_j' \leq \sum_{\{p_k\} \in E_N^{j-1}} M^{2(j-1)} = C_N^{j-1} M^{2(j-1)}, \qquad j = 2, 3, \ldots, s. \qquad (115)$$

To estimate $G_j''$ (see (112)) we write

$$G_j'' = \sum_{p=1}^N G_j''(p),$$

where

$$G_j''(p) = \sum_{\{n_k\} \in E_N^{j-1} : p \notin \{n_k\}} \left[ \int_0^1 \varphi_p^2(x) \prod_{k=1}^{j-1} \varphi_{n_k}(x)\, dx \right]^2.$$

If we estimate $G_j''(p)$ by Bessel's inequality, just as for the sum $G_j'$, we find that

$$G_j''(p) \leq C_{N-1}^{j-2} M^{2j}, \qquad j = 2, 3, \ldots, s,$$

and consequently

$$G_j'' \leq N C_{N-1}^{j-2} M^{2j}, \qquad j = 2, 3, \ldots, s. \qquad (116)$$

Using (110), we obtain the following estimate for (109) from (115) and (116):

$$G \leq \sum_{j=1}^s \frac{1}{M^{2j}} C_{N-1}^{s-j} (G_j' + G_j'') \leq \sum_{j=1}^s C_{N-1}^{s-j} [C_N^{j-1} + N C_{N-1}^{j-2}] \qquad (117)$$

$(G_1'' = C_{N-1}^{-1} = 0$; we have also used the inequality $M \geq 1$). The number of elements of $E_N^s$ is $C_N^s$; hence it follows from (117) that

$$\min_{\{n_k\} \in E_N^s} I(\{n_k\}) \leq \frac{G}{C_N^s} \leq \sum_{j=1}^{s} \frac{C_{N-j}^{s-j}}{C_N^s}[C_N^{j-1} + NC_{N-1}^{j-2}]. \qquad (118)$$

Since $jC_N^{j-1} \geq NC_{N-1}^{j-2}$, the right-hand side of (118) does not exceed

$$2(1+s) \sum_{j=1}^{s} \frac{C_{N-j}^{s-j}C_N^{j-1}}{C_N^s} = 2(1+s) \sum_{j=1}^{s} \frac{1}{N-j+1} \cdot \frac{s!}{(s-j)!(j-1)!}$$

$$\leq \frac{2(1+s)}{N-s} \sum_{j=1}^{s} C_s^j \leq \frac{(1+s)2^{s+1}}{N-s} \leq 10^{-s},$$

if $N \geq (64)^s$, i.e., $s \leq \frac{1}{6}\log_2 N$. This establishes Lemma 1, and therefore Theorem 9.

## §5. Properties of integrated orthonormal systems

Problems on the uniform convergence of orthogonal series, although they arise naturally in the study of the trigonometric and other classical systems, tend to become uninteresting if we consider series in general orthogonal systems. The point is that the elements of an O.N.S. can be quite badly behaved, for example unbounded. However, if we consider a system

$$F_n(y) = \int_0^y \varphi_n(x)\,dx, \qquad \{\varphi_n(x)\}_{n=1}^{\infty}, \text{ an O.N.S.}$$
$$(x, y \in (0, 1)), \qquad (119)$$

it turns out that we can obtain nontrivial general propositions on the uniform convergence of series in such a system. In particular, for every system of the form (119) and for almost all $t \in (0, 1)$, the following series converge uniformly (in $y$):

$$1) \; \sum_{n=1}^{\infty} r_n(t)F_n(y); \qquad 2) \; \sum_{n=1}^{\infty} \xi_n(t)F_n(y), \qquad (120)$$

where $\{r_n(t)\}_{n=1}^{\infty}$ is the Rademacher system, and $\{\xi_n(t)\}_{n=1}^{\infty}$ is an O.N.S. of independent normally distributed functions. By Theorems 2.6 and 2.10, these statements are special cases of the following result.

THEOREM 10. *Let the O.N.S.* $\{\eta_n(t)\}_{n=1}^{\infty}$, $t \in (0, 1)$, *be an $S_p$-system, with some $p > 2$. Also let $\{\varphi_n(x)\}_{n=1}^{\infty}$ be any O.N.S. Then, for almost every $t \in (0, 1)$, the series*

$$\sum_{n=1}^{\infty} \eta_n(t) \int_0^y \varphi_n(x)\,dx \qquad (121)$$

*converges uniformly (in $y$) on $[0, 1]$.*

PROOF. We use Theorem 8, according to which, for arbitrary $a_n$, $n = 1, 2, \ldots$, with $\sum_{n=1}^{\infty} a_n^2 < \infty$, the series $\sum_{n=1}^{\infty} a_n \eta_n(t)$ converges a.e., and

$$\left\| \sup_{1 \le N < \infty} \left| \sum_{n=1}^{N} a_n \eta_n(t) \right| \right\|_p \le C \left( \sum_{n=1}^{\infty} a_n^2 \right)^{1/2}. \tag{122}$$

By Bessel's inequality

$$\sum_{n=1}^{\infty} \left\{ \int_{\alpha}^{\beta} \varphi_n(x) \, dx \right\}^2 \le \int_{\alpha}^{\beta} 1^2 \, dx = \beta - \alpha, \qquad 0 \le \alpha < \beta \le 1, \tag{123}$$

and in particular

$$\sum_{n=1}^{\infty} \left\{ \int_{0}^{y} \varphi_n(x) \, dx \right\}^2 \le 1, \qquad 0 \le y \le 1. \tag{124}$$

It follows from (124) (also see (122)) that, for each given $y$, the series (121) converges for almost all $t \in (0, 1)$, and consequently

(∗) *There is a set $E_0 \subset (0, 1)$, with $m(E_0) = 1$, such that* (121) *converges for every pair* $(t, y)$ *for which* $t \in E_0$, $y \in R_2 = \{\{i/2^k\}_{i=0}^{2^k}\}_{k=0}^{\infty}$.

Let

$$S_N(t, y) = \sum_{n=1}^{N} \eta_n(t) \int_{0}^{y} \varphi_n(x) \, dx. \tag{125}$$

Also, for $m = 1, 2, \ldots$, let $\pi_m$ be the set of points $\pi_m = \{(2s - 1)/2^{m+1}\}_{s=1}^{2^m}$. Finally, we set, for each $y = (2s - 1)/2^{m+1} \in \pi_m$,

$$\rho_m(t, y) = \sup_{1 \le N < \infty} \left| S_N(t, y) - S_N \left( t, \frac{s-1}{2^m} \right) \right| \tag{126}$$

and

$$R_m(t) = \max_{y \in \pi_m} \rho_m(t, y). \tag{127}$$

Since the function $\int_{0}^{y} (x) \, dx$ is continuous, and since $S_N(t, 0) = 0$ for $t \in (0, 1)$ and $N = 1, 2, \ldots$, we find from the definition of $R_m(t)$ that for each given $t \in (0, 1)$

$$\|S_N(t, y)\|_C \le \sum_{m=1}^{\infty} R_m(t), \qquad N = 1, 2, \ldots, \tag{128}$$

$$\left| S_N(t, y) - S_N \left( t, \frac{i-1}{2^k} \right) \right| \le \sum_{m=k}^{\infty} R_m(t), \qquad y \in \left( \frac{i-1}{2^k}, \frac{i}{2^k} \right),$$

$$i = 1, \ldots, 2^k, \quad N = 1, 2, \ldots \tag{129}$$

It follows from (129) that the series (121) converges uniformly in $y$ for every $t \in E_0$ (see $(*)$) for which

$$\sum_{m=1}^{\infty} R_m(t) < \infty. \tag{130}$$

In fact, if (130) holds for $t \in E_0$, then, given any $\varepsilon > 0$, we can first find $k = 1, 2, \ldots$, such that $\sum_{m=k}^{\infty} R_m(t) < \varepsilon/3$. Then, using $(*)$, we can find a number $N_0$ such that, for all $N > N_0$ and $N' > N_0$, the inequality

$$\left| S_{N'}\left(t, \frac{i}{2^k}\right) - S_N\left(t, \frac{i}{2^k}\right) \right| < \frac{\varepsilon}{3}$$

is satisfied for all points of the form $i/2^k$, $i = 0, 1, \ldots, 2^k$. Then it follows from (129) that, for $N' > N_0$ and $N > N_0$,

$$|S_{N'}(t, y) - S_N(t, y)| < \varepsilon \quad \text{for all } y \in [0, 1].$$

Therefore in order to prove Theorem 10 it is enough to show that (130) is valied for almost every $t \in (0, 1)$.

First choose any point $y = (2s - 1)/2^{m+1} \in \pi_m$. Then we have the following equation for the function $\rho_m(t, y)$ defined in (126):

$$\rho_m(t, y) = \sup_{1 \leq N < \infty} \left| \sum_{n=1}^{N} \eta_n(t) \int_{(s-1)/2^m}^{y} \varphi_n(x) \, dx \right|,$$

from which it follows by (122) and (123) that

$$\left\{ \int_0^1 [\rho_m(t, y)]^p \, dt \right\}^{1/p} \leq C \left\{ \sum_{n=1}^{\infty} \left[ \int_{(s-1)/2^m}^{y} \varphi_n(x) \, dx \right]^2 \right\}^{1/2} \leq C \cdot 2^{-m/2}. \tag{131}$$

From (131), by applying the inequality

$$m\{t \in (0, 1) : |f(t)| > z\} \leq (\|f\|_p/z)^p,$$

we deduce that for each $y \in \pi_m$

$$m\{t \in (0, 1) : \rho_m(t, y) > 2^{(m/2p)(1-p/2)}\} \leq C^p 2^{-mp/2} 2^{(m/2)(-1+p/2)},$$

and therefore

$$m\{t \in (0, 1) : R_m(t) > 2^{(m/2p)(1-p/2)}\} \leq 2^m [C^p 2^{-mp/2} 2^{(m/2)(-1+p/2)}]$$
$$= c^p 2^{m(1/2-(1/4)p)} = C' 2^{-\gamma m},$$

where $\gamma = \gamma(p) > 0$. It follows from the preceding inequality that

$$\sum_{m=1}^{\infty} m\{t \in (0, 1) : R_m(t) < 2^{(m/2p)(1-p/2)}\} < \infty,$$

and therefore, for almost all $t \in (0, 1)$, we have $R_m(t) \leq 2^{(m/2p)(1-p/2)}$ for $m > m(t)$, and hence for almost all $t \in (0, 1)$ the sum of the series $\sum_{m=1}^{\infty} R_m(t)$ is finite. As we noted above, the conclusion of Theorem 10 follows from this.

In concluding this chapter, we consider properties of the sum $\sum_{n=1}^{\infty} F_n^2(y)$, where $F_n(y)$ is defined in (119).

If the system $\{\varphi_n\}_{n=1}^{\infty}$ is complete, then by Parseval's theorem

$$\sum_{n=1}^{\infty} F_n^2(y) = \sum_{n=1}^{\infty} \left[ \int_0^y \varphi_n(x)\, dx \right]^2 = y.$$

For an arbitrary O.N.S. we have the following result.

PROPOSITION 1. *For every O.N.S.* $\{\varphi_n(x)\}_{n=1}^{\infty}$, $x \in (0, 1)$, *the sum of the series*

$$S(y) \equiv \sum_{n=1}^{\infty} F_n^2(y), \qquad y \in [0, 1],$$

*is continuous and satisfies a Lipschitz condition of order* $\frac{1}{2}$.

In fact, we find by using Bessel's inequality that

$$|S(y) - S(y')| = \left| \sum_{n=1}^{\infty} [F_n^2(y) - F_n^2(y')] \right|$$

$$\leq \sum_{n=1}^{\infty} |F_n(y) - F_n(y')| \cdot |F_n(y) + F_n(y')|$$

$$\leq \left\{ \sum_{n=1}^{\infty} |F_n(y) - F_n(y')|^2 \right\}^{1/2} \left\{ \sum_{n=1}^{\infty} F_n^2(y) \right\}^{1/2}$$

$$+ \left\{ \sum_{n=1}^{\infty} |F_n(y) - F_n(y')|^2 \right\}^{1/2} \left\{ \sum_{n=1}^{\infty} F_n^2(y') \right\}^{1/2}$$

$$\leq 2|y - y'|^{1/2},$$

as required.

As is shown by the example of the O.N.S. $\{\varphi_n\}_{n=1}^{\infty}$ in which $\varphi_1(x) = c(1-x)^{(\varepsilon-1)/2}$, $0 < x < 1$ $(0 < \varepsilon < 1)$, with $\varphi_n(x)$, $n = 2, 3, \ldots$, zero for $x > \frac{1}{2}$, for every $\varepsilon > 0$ there is an O.N.S. $\{\varphi_n\}$ for which

$$\sum_{n=1}^{\infty} \left\{ \int_0^y \varphi_n(x)\,dx \right\}^2 \notin \mathrm{Lip}(\tfrac{1}{2} + \varepsilon),$$

i.e., the order of smoothness in Proposition 1 is exact.

# CHAPTER IX

# General Theorems on the Divergence
# of Orthogonal Series

The significant results of much work of the 1960's and 1970's on general orthogonal systems can be explained by the discovery that general theorems on the divergence of Fourier series or of series of Fourier coefficients in the trigonometric system are of a general nature and can be extended to a broad class of O.N.S. This extension required the development of new methods, which turned out also to be useful in a wider class of problems, and at the same time clarify the behavior of the examples of divergence (in one sense or another) of trigonometric series. This chapter is devoted to an exposition of general results on the divergence of orthogonal series.

## §1. Divergence almost everywhere
## of rearrangements of $L^2$ Fourier series

THEOREM 1. *There are no complete orthonormal unconditional convergence systems (see Definition 1.5).*

For the proof of Theorem 1, it is enough to show that corresponding to any C.O.N.S. $\Phi = \{\varphi_n(x)\}_{n=1}^{\infty}$, $x \in (0, 1)$, there is a function $g \in L^2(0, 1)$ whose Fourier series diverges a.e. after some rearrangement. We shall establish somewhat more, specifically that there is a function $g \in L^{\infty}(0, 1)$ for which the series

$$\sum_{n=1}^{\infty} c_n \varphi_n(x), \qquad c_n = c_n(g, \Phi) = \int_0^1 g(x)\varphi_n(x)\,dx, \quad n = 1, 2, \dots,$$

diverges unboundedly a.e. after a permutation $\sigma = \{\sigma(n)\}_{n=1}^{\infty}$ of its terms, i.e.,

$$\sup_{1 \le N < \infty} \left| \sum_{n=1}^{N} c_{\sigma(n)} \varphi_{\sigma(n)}(x) \right| = +\infty \quad \text{for almost for all } x \in (0, 1). \quad (1)$$

We shall deduce this proposition from a similar result on systems of Haar type (see §3.6 and in particular Corollary 3.9). We need the following lemma.

LEMMA 1. *For every C.O.N.S.* $\Phi = \{\varphi_n(x)\}_{n=1}^{\infty}$, $x \in (0,1)$, *there are a system* $\{\chi_n(x,\mathscr{E})\}_{n=1}^{\infty}$ *of Haar type and a system of polynomials in the system* $\Phi$:

$$P_n(x) = \sum_{s=m_{n+1}}^{m_{n+1}} a_s \varphi_s(x), \qquad 0 = m_1 < m_2 < \cdots, \ n = 1,2,\ldots, \qquad (2)$$

*such that*

a) $\chi_n(x,\mathscr{E}) = P_n(x) + \eta_n(x)$, $\quad x \in (0,1), \|\eta_n\|_2 \le 2^{-n}$, $n = 1,2,\ldots$;

b) $\displaystyle \sum_{k=1}^{n-1} \left\{ \sum_{m=m_{n+1}+1}^{\infty} [c_m(\eta_k,\Phi)]^2 \right\}^{1/2} \le 2^{-n}$, $\qquad n = 2,3,\ldots$.

PROOF. A system $\{\chi_n(x,\mathscr{E})\}$ of Haar type, or equivalently a family of sets $\mathscr{E} = \{E_k^i\}$ with properties I)–III) (see §3.6 and, in particular, 3.(112) and 3.(116)) can be constructed by induction while also satisfying the condition:

IV) every set $E_k^i \in \mathscr{E}$ can be represented as a finite union of binary intervals of equal length.

Set $E_0^1 = (0,1)$ and $\chi_1(x,\mathscr{E}) = 1$ for $x \in (0,1)$. Using the completeness of $\Phi$, we find a number $m_2 > m_1 = 0$ such that the polynomial

$$P_1(x) = \sum_{s=1}^{m_2} a_s \varphi_s(x)$$

with $a_s = c_s(\chi_1(x,\mathscr{E}),\Phi)$, $s = 1,\ldots,m_2$, satisfies condition a) (for $n = 1$).

Now suppose that, for some $k = 0,1,\ldots$, we have constructed functions $\{\chi_n(x,\mathscr{E})\}_{n=1}^{2^k}$, polynomials $\{P_n(x)\}_{n=1}^{2^k}$ and sets $\{E_k^i\}_{i=1}^{2^k}$, and that conditions a) and b) are satisfied for $n \le 2^k$. We shall construct successively for $n = 2^k + 1, 2^k + 2, \ldots, 2^{k+1}$ the functions $\chi_n(x,\mathscr{E})$, the polynomials $P_n(x)$, and the sets $\{E_{k+1}^i\}_{i=1}^{2^{k+1}}$.

We fix $n = 2^k + i$, $1 \le i \le 2^k$, and suppose that $\chi_s(x,\mathscr{E})$ and $P_s(x)$, with $s < n$, have already been constructed and that therefore the number $m_n$ is defined (see (2)), and we consider the functions

$$\psi_j(x) = 2^{k/2} r_j(x) \chi_{E_k^i}(x), \qquad x \in (0,1), \ j = 1,2,\ldots, \qquad (3)$$

where $\{r_j(x)\}$ is the Rademacher system and $\chi_E(x)$ is the characteristic function of $E$. Since $m(E_k^i) = 2^{-k}$, we have $\|\psi_j\|_2 = 1$, $j = 1,2,\ldots$. Moreover, it follows from condition IV) and the definition of the Rademacher system that when $j'$ is sufficiently large the functions $\psi_j(x)$, $j \geq j'$, are pairwise orthogonal. Consequently, for $m = 1, 2, \ldots$, the Fourier coefficients

$$c_m(\psi_j, \Phi) = \int_0^1 \psi_j(x)\varphi_m(x)\,dx \to 0 \quad \text{as } j \to \infty,$$

and therefore there is a number $j_0 > j'$ such that

$$\left\| \sum_{m=1}^{m_n} c_m(\psi_{j_0}, \Phi)\varphi_m(x) \right\|_2 = \left\{ \sum_{m=1}^{m_n} c_m^2(\psi_{j_0}, \Phi) \right\}^{1/2} \leq 2^{-n-1}. \tag{4}$$

Furthermore, if we suppose that $\Phi$ is a C.O.N.S. we can choose a number $m_{n+1} > m_n$ so large that inequality b) and the inequality

$$\left\| \sum_{m=1}^{m_{n+1}} c_m(\psi_j, \Phi)\varphi_m(x) - \psi_{j_0}(x) \right\|_2 \leq 2^{-n-1} \tag{5}$$

are both satisfied. Set

$$\chi_n(x, \mathscr{E}) = \chi_k^i(x, \mathscr{E}) = \psi_{j_0}(x),$$

$$P_n(x) = \sum_{m=m_n+1}^{m_{n+1}} c_m(\psi_{j_0}, \Phi)\varphi_m(x), \qquad x \in (0,1), \tag{6}$$

$$E_{k+1}^{2i-1} = \{x \in (0,1)\colon \psi_{j_0}(x) = 2^{k/2}\},$$

$$E_{k+1}^{2i} = \{x \in (0,1)\colon \psi_{j_0}(x) = -2^{k/2}\}.$$

Then it follows immediately from (4) and (5) that

$$\|\chi_n(x, \mathscr{E}) - P_n(x)\|_2 \leq 2^{-n},$$

i.e., relation a) is satisfied.

It is clear from the construction (see (3) and (6)) that the set $E_{k+1}^i$ $(1 \leq i \leq 2^{k+1})$ consists of binary intervals of equal length and

$$m(E_{k+1}^i) = 2^{-k-1}, E_{k+1}^i \cap E_{k+1}^j = \varnothing,$$

$$\text{if } i \neq j, i, j = 1, \ldots, 2^{k+1},$$

$$E_k^i = E_{k+1}^{2i-1} \cup E_{k+1}^{2i}, \qquad i = 1, \ldots, 2^k.$$

Carrying out the construction for all $k \geq 0$, we obtain a family $\mathscr{E} = \{E_k^i\}$ for which conditions I)—III) (see §3.6) are satisfied, and consequently $\{\chi_n(x, \mathscr{E})\}_{n=1}^\infty$ is a system of Haar type. The validity of a) and b) for $n = 1, 2, \ldots$, has already been established. This completes the proof of Lemma 1.

PROOF OF THEOREM 1. Let $\Phi = \{\varphi_n(x)\}$ be a C.O.N.S. Using Lemma 1, we find a system $\chi_{\mathscr{E}} = \{\chi_n(x, \mathscr{E})\}_{n=1}^\infty$ of Haar type and polynomials $P_n(x)$,

$n = 1, 2, \ldots$, of the form (2), satisfying relations a) and b) of Lemma 1. By Corollary 3.9 there is a function $g \in L^\infty(0, 1)$ whose Fourier series in the system $\chi_\mathcal{E}$,

$$\sum_{n=1}^{\infty} b_n \chi_n(x, \mathcal{E}), \qquad b_n = \int_0^1 g(t) \chi_n(t, \mathcal{E}) \, dt \qquad n = 1, 2, \ldots, \quad (7)$$

after a rearrangement $\sigma' = \{\sigma'(n)\}_{n=1}^{\infty}$ of its terms, diverges unboundedly on $(0, 1)$, i.e.,

$$\sup_{1 \le N < \infty} \left| \sum_{n=1}^{N} b_{\sigma'(n)} \chi_{\sigma'(n)}(x, \mathcal{E}) \right| = \infty \quad \text{for almost all } x \in (0, 1). \quad (8)$$

Here, of course, the series (7) converges in $L^2$ to $g(x)$. Let us show that the Fourier series of $g(x)$ in the system $\Phi$ diverges unboundedly after a rearrangement. For this purpose it is enough to verify that

$$\sup_{1 \le N < \infty} \left| \sum_{n=1}^{N} Q_{\sigma'(n)}(x) \right| = \infty \quad \text{for almost all } x \in (0, 1), \quad (9)$$

where

$$Q_n(x) = \sum_{m=m_n+1}^{m_{n+1}} c_m \varphi_m(x), \qquad c_m = c_m(g, \Phi), \; n, m = 1, 2, \ldots.$$

For $n = 1, 2, \ldots$, set

$$\beta_n(x) = Q_n(x) - b_n \chi_n(x, \mathcal{E}); \quad (10)$$

we shall show that the series

$$\sum_{n=1}^{\infty} \|\beta_n\|_2 < \infty \quad (11)$$

converges. Hence (see (8) and (10)) we obtain (9) at once, since, by Beppo Levi's theorem, it follows from (11) that $\sum_{n=1}^{\infty} |\beta_n(x)| < \infty$ for almost all $x \in (0, 1)$.

For $n = 1, 2, \ldots$ and $m_n < m \le m_{n+1}$ we have (see Lemma 1)

$$c_m = c_m(g, \Phi) = \int_0^1 \left[ \sum_{\nu=1}^\infty b_\nu \chi_\nu(x, \mathscr{E}) \right] \varphi_m(x)\, dx$$

$$= \sum_{\nu=1}^\infty b_\nu \left[ \int_0^1 P_\nu(x) \varphi_m(x)\, dx + \int_0^1 \eta_\nu(x) \varphi_m(x)\, dx \right]$$

$$= \sum_{\nu=1}^\infty b_\nu \int_0^1 \sum_{s=m_\nu+1}^{m_{\nu+1}} a_s \varphi_s(x) \varphi_m(x)\, dx$$

$$+ \sum_{\nu=1}^\infty b_\nu c_m(\eta_\nu, \Phi) = b_n a_m + \sum_{\nu=1}^\infty b_\nu c_m(\eta_\nu, \Phi)$$

(the series on the right-hand side of the preceding series converges, since, by Lemma 1, we have $|c_m(\eta_\nu, \Phi)| \le \|\eta_\nu\|_2 \le 2^{-\nu}$, $\nu = 1, 2, \ldots$).

Consequently (see (10) and Lemma 1),

$$\beta_n(x) \equiv \sum_{m=m_n+1}^{m_{n+1}} c_m \varphi_m(x) - b_n \chi_n(x, \mathscr{E})$$

$$= b_n P_n(x) + \sum_{m=m_n+1}^{m_{n+1}} \left| \sum_{\nu=1}^\infty b_\nu c_m(\eta_\nu, \Phi) \right| \varphi_m(x) - b_n \chi_n(x, \mathscr{E})$$

$$= b_n \eta_n(x) + \sum_{\nu=1}^\infty b_\nu \sum_{m=m_n+1}^{m_{n+1}} c_m(\eta_\nu, \Phi) \varphi_m(x).$$

Hence if we set $b = \sup_{1 \le n < \infty} |b_n|$ and apply Bessel's inequality and relations a) and b) of Lemma 1, we obtain

$$\|\beta_n\|_2 \le b \cdot 2^{-n} + b \sum_{\nu=1}^\infty \left\{ \sum_{m=m_n+1}^{m_{n+1}} c_m^2(\eta_\nu, \Phi) \right\}^{1/2}$$

$$\le b \cdot 2^{-n} + b \sum_{\nu=1}^{n-2} \left\{ \sum_{m=m_n+1}^{m_{n+1}} c_m^2(\eta_\nu, \Phi) \right\}^{1/2} + b \sum_{\nu=n-1}^\infty \|\eta_\nu\|_2$$

$$\le b \cdot 2^{-n} + b \cdot 2^{-n+1} + b \cdot 2^{-n+2} = 7b \cdot 2^{-n}, \qquad n = 3, 4, \ldots.$$

This establishes the convergence of the series (11), and Theorem 1 is established.

COROLLARY 1. *Every C.O.N.S.* $\Phi = \{\varphi_n(x)\}_{n=1}^\infty$, $x \in (0, 1)$, *satisfies*

$$[\varphi_n^+(x)]^2 = \infty \quad \text{for almost all } x \in (0, 1),$$
$$[\varphi_n^-(x)]^2 = \infty \quad \text{for almost all } x \in (0, 1),$$

(12)

*where* $\varphi_n^+(x) = \max[0, \varphi_n(x)\}$, $\varphi_n^-(x) = \varphi_n(x) - \varphi_n^+(x)$.

PROOF. It is clear that it is enough to establish (12). Suppose the contrary: there are a set $E \subset (0, 1)$, $m(E) > 0$, and a constant $M > 0$ such that

$$\sum_{n=1}^{\infty} [\varphi_n^+(x)]^2 \leq M^2 \quad \text{for } x \in E.$$

Let us show that then, for every sequence $\{c_n\}_{n=1}^{\infty} \in l^2$,

$$\sum_{n=1}^{\infty} |c_n \varphi_n(x)| < \infty \quad \text{for almost all } x \in E. \tag{13}$$

This will lead to a contradiction, since in the proof of Theorem 1 we constructed a function $g \in L^{\infty}(0, 1) \subset L^2(0, 1)$ whose Fourier series satisfies (1).

We may suppose, without loss of generality, that $c_n \geq 0$, $n = 1, 2, \ldots$, and $\sum_{n=1}^{\infty} c_n^2 = 1$. For $x \in E$ we have

$$\sum_{n=1}^{\infty} c_n \varphi_n^+(x) \leq \left\{ \sum_{n=1}^{\infty} c_n^2 \right\}^{1/2} \cdot \left\{ \sum_{n=1}^{\infty} [\varphi_n^+(x)]^2 \right\}^{1/2} \leq M < \infty, \tag{14}$$

and consequently the series $\sum_{n=1}^{\infty} c_n \varphi_n^+(x)$ converges in $L^2(E)$ (we have used the fact that $c_n \varphi_n^+(x) \geq 0$). But then, using the convergence of the series $\sum_{n=1}^{\infty} c_n \varphi_n(x)$ in $L^2(E)$ and the equation $\varphi_n^-(x) = \varphi_n(x) - \varphi_n^+(x)$, we find that the series

$$\sum_{n=1}^{\infty} c_n \varphi_n^-(x) \tag{15}$$

also converges in $L^2(E)$. Since $c_n \varphi_n^-(x) \leq 0$ for $n = 1, 2, \ldots$, and $x \in E$, the convergence of the series (15) in $L^2(E)$ implies its convergence a.e.:

$$\sum_{n=1}^{\infty} c_n \varphi_n^-(x) = M'(x) > -\infty \quad \text{a.e. on } E.$$

From this and (14), relation (13) follows by the equation $|\varphi_n(x)| = \varphi_n^+(x) - \varphi_n^-(x)$. This completes the proof of Corollary 1.

Theorem 1 can be strengthened:

THEOREM 2. *For every C.O.N.S.* $\Phi = \{\varphi_n(x)\}_{n=1}^{\infty}$, $x \in (0, 1)$, *there is a function* $f \in C(0, 1)$ *whose Fourier series in the system* $\Phi$ *diverges unboundedly a.e. on* $(0, 1)$ *after rearrangement of its terms.*

In the proof of Theorem 1, we constructed, for every C.O.N.S. $\{\varphi_n(x)\}_{n=1}^{\infty}$, a function $g \in L^{\infty}(0, 1)$ whose Fourier series diverges unboundedly a.e. after rearrangement of its terms (see (1)). Hence to prove Theorem 2 it is enough to verify the following lemma.

LEMMA 1. *If for an O.N.S.* $\Phi = \{\varphi_n(x)\}_{n=1}^{\infty}$ *there is a function* $g \in L^{\infty}(0,1)$ *for which the Fourier series in the system* $\Phi$:

$$\sum_{n=1}^{\infty} c_n(g)\varphi_n(x), \qquad c_n(g) = c_n(g,\Phi) = \int_0^1 g(x)\varphi_n(x)\,dx \qquad (16)$$

*diverges unboundedly a.e. on* $(0,1)$, *then there also is a function* $f(x)$, *continuous on* $(0,1)$, *which has the same property.*

LEMMA 2. *Let* $f_1(x)$ *and* $f_2(x)$ *be measurable functions on* $(0,1)$, *and* $n$ *a positive integer. Then there is an integer* $j$, $1 \le j \le n$, *such that*

$$m\left\{x \in (0,1): |f_1(x) + jf_2(x)| \ge \frac{1}{2}|f_2(x)|\right\} \ge 1 - \frac{1}{n}. \qquad (17)$$

PROOF. Let $G = \{x \in (0,1): f_2(x) \ne 0\}$, and for $j = 1,\ldots,n$, let

$$E_j = \left\{x \in G: f_1(x) \in \left(-jf_2(x) - \frac{1}{2}|f_2(x)|, -jf_2(x) + \frac{1}{2}|f_2(x)|\right)\right\}.$$

Since the sets $E_j \subset (0,1)$, $j = 1,\ldots,n$, are disjoint, it is clear that $m(E_j) \le \frac{1}{n}$ for some $j = 1,\ldots,n$. Hence we obtain (17) from the equation

$$m\{x \in (0,1): |f_1(x) + jf_2(x)| < \tfrac{1}{2}|f_2(x)|\} = m(E_j).$$

PROOF OF LEMMA 1. We may suppose, without loss of generality, that $\|g\|_{\infty} = 1$. Since the series (16) diverges unboundedly a.e. on $(0,1)$, then for each $k = 1,2,\ldots$, we can find a number $M_k > 0$ and a measurable integral-valued function $N_k(x)$, $1 \le N_k(x) \le M_k$, such that

$$m\left\{x \in (0,1): \left|\sum_{n=1}^{N_k(x)} c_n(g)\varphi_n(x)\right| > 2^k + 1\right\} \ge 1 - 2^{-k-1}. \qquad (18)$$

Since $|c_n(g) - c_n(h)| \le \|g - h\|_2$, $n = 1,2,\ldots$, when $h \in L^2(0,1)$, we can state that if the function

$$f_k(x) \in C(0,1), \qquad \|f_k\|_C \le 1, \qquad (19)$$

is taken sufficiently close to $g(x)$ in the $L^2(0,1)$ norm, then we have the inequality

$$m\{x \in (0,1): |F_k(x)| > 2^k\} \ge 1 - 2^{-k}, \qquad (20)$$

where

$$F_k(x) = \sum_{n=1}^{N_k(x)} c_n(f_k)\varphi_n(x).$$

Let us show that, for a sufficiently sparse sequence of integers $\{k_s\}_{s=1}^\infty$ and some numbers $b_s$, $1 \le b_s \le k_s^2$, $s = 1, 2, \ldots$, the function

$$f(x) = \sum_{s=1}^\infty \frac{1}{k_s^4} b_s f_{k_s}(x), \qquad x \in (0, 1), \tag{21}$$

satisfies the requirements of Lemma 1, i.e., $f \in C(0, 1)$ and the series

$$\sum_{n=1}^\infty c_n(f) \varphi_n(x)$$

diverges unboundedly a.e. on $(0, 1)$. We shall construct the numbers $k_s$ and $b_s$, $s = 1, 2, \ldots$, inductively for $s = 1, 2, \ldots$. Here we use the notation

$$Q_j(x) = \sum_{s=1}^j \frac{1}{k_s^4} b_s f_{k_s}(x), \qquad j = 1, 2, \ldots.$$

The sequence $\{k_s\}_{s=1}^\infty$ will be chosen so that it satisfies

$$k_1 = 1, \qquad \sum_{s=j+1}^\infty \frac{1}{k_s^2} \le (M_{kj} \cdot 2^j)^{-1}, \qquad j = 1, 2, \ldots. \tag{22}$$

Furthermore, we set $b_1 = 1$, and if $b_s$, $s = 1, 2, \ldots, s_0$, have already been defined, then, using Lemma 2, we choose $b_{s_0+1}$, $1 \le b_{s_0+1} \le k_{s_0+1}^2$, so that the following inequality is satisfied (see (20)):

$$m \left\{ x \in (0, 1) : \left| \sum_{n=1}^{N_{k_{s_0+1}}(x)} c_n(Q_{s_0}) \varphi_n(x) + b_{s_0+1} k_{s_0+1}^{-4} F_{k_{s_0+1}}(x) \right| \ge \frac{1}{2} \frac{|F_{k_{s_0+1}}(x)|}{k_{s_0+1}^4} \right\}$$

$$\ge 1 - \frac{1}{k_{s_0+1}^2}.$$

Then we obtain, by (20),

$$m \left\{ x \in (0, 1) : \left| \sum_{n=1}^{N_{k_{s_0+1}}(x)} c_n(Q_{s_0+1}) \varphi_n(x) \right| \ge \frac{2^{k_{s_0+1}}}{2 k_{s_0+1}^4} \right\}$$

$$\ge 1 - 2^{-k_{s_0+1}} - k_{s_0+1}^{-2}. \tag{23}$$

We have now constructed the numbers $k_s$, $b_s$ for $s = 1, 2, \ldots$, and therefore also the function $f(x)$. Then the uniform convergence of the series (21), and hence the continuity of $f(x)$, follow from (19) and the inequalities $b_s \le k_s^2$, $s = 1, 2, \ldots$. Let us show that the Fourier series of $f(x)$ diverges unboundedly a.e. on $(0, 1)$. Let $j > 1$ and

$$\Delta_j(x) = \sum_{s=j+1}^\infty \frac{1}{k_s^4} b_s f_{k_s}(x).$$

Then

$$\sum_{n=1}^{N_{k_j}(x)} c_n(f)\varphi_n(x) = \sum_{n=1}^{N_{k_j}(x)} c_n(Q_j)\varphi_n(x) + \sum_{n=1}^{N_{k_j}(x)} c_n(\Delta_j)\varphi_n(x)$$
$$= \psi_j^{(1)}(x) + \psi_j^{(2)}(x). \tag{24}$$

According to (23),

$$m\{x \in (0,1): |\psi_j^{(1)}(x)| > A_j\} \geq 1 - \varepsilon_j, \tag{25}$$

where $A_j \to \infty$ and $\varepsilon_j \to 0$ as $j \to \infty$. On the other hand (see (19) and (22)),

$$\|\Delta_j\|_c \leq \sum_{s=j+1}^{\infty} \frac{1}{k_s^2} \leq (M_{k_j} \cdot 2^j)^{-1}.$$

Consequently

$$\|\psi_j^{(2)}\|_2 \leq \sum_{n=1}^{M_{k_j}} |c_n(\Delta_j)| \leq \sum_{n=1}^{M_{k_j}} \|\Delta_j\|_2 \leq 2^{-j},$$

and therefore

$$m\{x \in (0,1): |\psi_j^{(2)}(x)| > 1\} \leq \|\psi_j^{(2)}\|_2^2 \leq 2^{-2j}, \qquad j = 1, 2, \ldots.$$

From this and from (25) (see also (24)) we obtain

$$m\left\{x \in (0,1): \left|\sum_{n=1}^{N_{k_j}(x)} c_n(f)\varphi_n(x)\right| > A_j - 1\right\} > 1 - \varepsilon_j - 2^{-2j},$$

$$j = 1, 2, \ldots.$$

This completes the proof of Lemma 1.

## §2. Fourier coefficients of continuous functions

Theorem 2 and Corollary 8.3 immediately imply the following theorem.

THEOREM 3. *For every C.O.N.S.* $\Phi = \{\varphi_n(x)\}_{n=1}^{\infty}$, $x \in (0,1)$, *there is a function* $f(x)$, *continuous on* $[0,1]$, *such that*

$$\sum_{n=1}^{\infty} |c_n(f)|^{2-\varepsilon} = \infty \quad \text{for every } \varepsilon > 0,$$

$$c_n(f) = \int_0^1 f(x)\varphi_n(x)\,dx. \tag{26}$$

It is easily seen that the hypothesis of completeness in Theorem 3 cannot be omitted. At the same time, it is shown below that for a broad class of O.N.S. (not necessarily complete) the conclusion of Theorem 3 remains valid and can even be significantly sharpened.

THEOREM 4. *Let the O.N.S.* $\Phi = \{\varphi_n(x)\}_{n=1}^{\infty}$, $x \in (0, 1)$, *satisfy the condition*

$$\|\varphi_n\|_1 \geq \gamma > 0, \qquad n = 1, 2, \ldots \qquad (27)$$

*Then for every sequence* $\{d_n\}$ *with* $\sum_{n=1}^{\infty} d_n^2 = \infty$ *there is a function* $f(x)$, *continuous on* $[0, 1]$, *for which*

$$\sum_{n=1}^{\infty} |d_n c_n(f)| = \infty, \qquad c_n(f) = \int_0^1 f(x)\varphi_n(x)\,dx, \qquad n = 1, 2, \ldots. \qquad (28)$$

REMARK 1. Condition (27) in Theorem 4 is necessary. More precisely, if the O.N.S. $\{\varphi_n(x)\}$ does not satisfy (27), there is a sequence $\{d_n^0\}_{n=1}^{\infty}$ with $\sum_{n=1}^{\infty} (d_n^0)^2 = \infty$ such that $\sum_{n=1}^{\infty} |d_n^0 c_n(f)| < \infty$ for every bounded $f(x)$. To establish this, it is sufficient, in view of the inequality $|c_n(f)| \leq \|f\|_\infty \cdot \|\varphi_n\|_1$, to choose a sequence $\{n_k\}_{k=1}^{\infty}$ such that $\sum_{k=1}^{\infty} \|\varphi_{n_k}\|_1 < \infty$, and then to take $d_n^0 = 1$ if $n \in \{n_k\}$, and $d_n^0 = 0$ if $n \notin \{n_k\}$.

REMARK 2. It is easily verified (by using Hölder's inequality) that if the sequence $\{d_n\}_{n=1}^{\infty} \in l^{2+\varepsilon}$ for every $\varepsilon > 0$, then every function $f(x)$ that satisfies (28) has the property (26).

PROOF OF THEOREM 4. Suppose the contrary: for some sequence $\{d_n\}_{n=1}^{\infty}$ with $\sum_{n=1}^{\infty} d_n^2 = \infty$,

$$\sum_{n=1}^{\infty} |d_n c_n(f)| = C(f) < \infty \quad \text{if } f \in C(0, 1). \qquad (29)$$

Then it is evident that, for every finite set $\{\varepsilon_n\}_{n=1}^{N}$, $\varepsilon_n = \pm 1$,

$$\left| \sum_{n=1}^{\infty} d_n \varepsilon_n c_n(f) \right| = \left| \int_0^1 f(x) \sum_{n=1}^{N} d_n \varepsilon_n \varphi_n(x)\,dx \right| \leq C(f). \qquad (30)$$

Let $\{t_s\}_{s=1}^{\infty}$ be an arbitrary enumeration of the rational points on $[0, 1]$. It follows from (30) that, for every continuous function $f(x)$ on $[0, 1]$,

$$\left| \int_0^1 f(x) K_{N,s}(x)\,dx \right| \leq C(f),$$

where $K_{N,s}(x) = \sum_{n=1}^{\infty} d_n r_n(t_s) \varphi_n(x)$, $N, s = 1, 2, \ldots \qquad (31)$

($\{r_n(t)\}_{n=1}^{\infty}$ is the Rademacher system).

In other words, the following sequence of bounded linear functionals on $C(0, 1)$:

$$L_{N,s}(f) \equiv \int_0^1 f(x) K_{N,s}(x)\,dx, \qquad N, s = 1, 2, \ldots,$$

has the property that $|L_{n,s}(f)| \le C(f)$, $N, s = 1, 2, \ldots$, for every $f \in C(0, 1)$. Hence, since the norm of $L_{N,s}$ is $\int_0^1 |K_{N,s}(x)| \, dx$, we obtain by applying the Banach–Steinhaus theorem

$$\sup_{1 \le N, s < \infty} \int_0^1 \left| \sum_{n=1}^N d_n r_n(t_s) \varphi_n(x) \right| dx = C < \infty.$$

Consequently, for every $t \in [0, 1]$,

$$\int_0^1 \left| \sum_{n=1}^N d_n r_n(t) \varphi_n(x) \right| dx \le C, \qquad N = 1, 2, \ldots. \tag{32}$$

Integrating (32) with respect to $t$ and applying Fubini's theorem, we obtain

$$\int_0^1 \int_0^1 \left| \sum_{n=1}^N d_n r_n(t) \varphi_n(x) \right| dt \, dx \le C, \qquad N = 1, 2, \ldots.$$

It follows from the preceding inequality and Theorem 2.7 that

$$\int_0^1 \left\{ \sum_{n=1}^N d_n^2 \varphi_n^2(x) \right\}^{1/2} dx \le C' < \infty, \qquad N = 1, 2, \ldots,$$

and consequently

$$F(x) \equiv \sum_{n=1}^\infty d_n^2 \varphi_n^2(x) < \infty \quad \text{a.e. on } [0, 1].$$

Since $F(x)$ is finite almost everywhere, we can find a constant $M$ such that

$$m(E) \ge 1 - \frac{\gamma}{16}, \quad \text{where } E = \left\{ x : \sum_{n=1}^\infty d_n^2 \varphi_n^2(x) \le M \right\}$$

(here the constant $\gamma > 0$ is the same as in (27)). Then

$$\sum_{n=1}^\infty d_n^2 \int_E \varphi_n^2(x) \, dx \le M. \tag{33}$$

But then it follows (see Lemma 1 and Theorem 2.7) from (27) and the normality of $\varphi_n(x)$, $n = 1, 2, \ldots$, in $L^2(0, 1)$ that $m\{x \in (0, 1) : |\varphi_n(x)| > \gamma/4\} \ge \gamma/8$, $n = 1, 2, \ldots$, and therefore

$$\int_E \varphi_n^2(x) \, dx \ge \frac{\gamma^2}{16} \cdot \frac{\gamma}{8}, \qquad n = 1, 2, 3, \ldots.$$

Consequently (see (33)) $\sum_{n=1}^\infty d_n^2 < \infty$, which contradicts the assumed divergence of $\sum_{n=1}^\infty d_n^2$. This contradiction completes the proof of Theorem 4.

If we consider a somewhat narrower class of orthonormal systems than in Theorem 4, we can obtain an even more precise result which is definitive.

THEOREM 5. *Let the O.N.S.* $\Phi = \{\varphi_n(x)\}_{n=1}^{\infty}$, $x \in (0,1)$, *have the property that for some* $p > 2$

$$\|\varphi_n\|_p \leq M < \infty, \qquad n = 1, 2, \ldots. \tag{34}$$

*Then for every sequence* $\{a_n\}_{n=1}^{\infty}$ *with* $a_n \geq 0$, $n = 1, 2, \ldots$, *and* $\sum_{n=1}^{\infty} a_n^2 < \infty$ *there is a function* $F(x)$, *continuous on* $[0,1]$ *and with*

$$\|F\|_C \leq C_{p,M} \left( \sum_{n=1}^{\infty} a_n^2 \right)^{1/2}$$

*for which*

$$|c_n(F)| \geq a_n, \qquad n = 1, 2, \ldots, \quad c_n(F) = \int_0^1 F(x)\varphi_n(x)\, dx.$$

(In the statement of Theorem 5, $C_{p,M}$ is a constant which depends only on $p$ and $M$.)

LEMMA 1. *Let there be given an O.N.S.* $\{\varphi_n\}_{n=1}^{\infty}$ *that satisfies* (34) (*with some* $p > 2$), *and a sequence* $\{a_n\}_{n=1}^{\infty}$ *with* $\sum_{n=1}^{\infty} a_n^2 = 1$. *Then there is a sequence* $\varepsilon_n = \pm 1$, $n = 1, 2, \ldots$, *such that the function*

$$f(x) \overset{L^2}{=} \sum_{n=1}^{\infty} \varepsilon_n a_n \varphi_n(x) \tag{35}$$

*can be represented, for each* $\eta > 0$, *in the form*

$$f(x) = g(x) + h(x)$$

$$g \in C(0,1), \quad \|g\|_C \leq \eta, \quad \|h\|_2 \leq \frac{K}{\eta^{\alpha}} \equiv \omega(\eta), \tag{36}$$

*where* $\alpha = p/2 - 1 > 0$ *and* $K$ *is a constant which depends only on* $p$ *and* $M$.

PROOF. Using Khinchin's inequality for the Rademacher system (see Theorem 2.6 and Lemma 1 for Theorem 2.12) and the inequality 2.(75), we obtain

$$\int_0^1 \int_0^1 \left| \sum_{n=1}^{\infty} r_n(t) a_n \varphi_n(x) \right|^p dx\, dt \leq C_p^p \int_0^1 \left[ \sum_{n=1}^{\infty} a_n^2 \varphi_n^2(x) \right]^{p/2} dx$$

$$\leq C_p^p \left\{ \sum_{n=1}^{\infty} a_n^2 \|\varphi_n\|_p^2 \right\}^{p/2} \leq C_p^p M^p.$$

Hence for some binary-irrational point $t_0 \in (0,1)$

$$\left\{ \int_0^1 \left| \sum_{n=1}^{\infty} r_n(t_0) a_n \varphi_n(x) \right|^p dx \right\}^{1/p} \leq C_p M.$$

This means that the function $f(t)$ defined by (35) with $\varepsilon_n = r_n(t_0)$, $n = 1, 2, \ldots$, satisfies the inequality

$$\|f\|_p \leq C_p M. \tag{37}$$

For a given $\eta > 0$ we set

$$g'(x) = \begin{cases} f(x) & \text{if } |f(x)| \leq \eta, \\ 0 & \text{if } |f(x)| > \eta, \end{cases}$$
$$h'(x) = f(x) - g'(x), \qquad x \in (0, 1).$$

Since $h'(x) = 0$ if $x \notin E \equiv \{x \in (0, 1) : |f(x)| > \eta\}$, we deduce from (37) that

$$(C_p M)^p \geq \int_0^1 |f(x)|^p \, dx \geq \int_E |h'(x)|^p \, dx$$
$$\geq \int_E |h'(x)|^2 |h'(x)|^{p-2} \, dx \geq \eta^{p-2} \int_E |h'(x)|^2 \, dx = \eta^{p-2} \|h'\|_2^2,$$

i.e., $\|h'\|_2 \leq \eta^{-\alpha}(C_p M)^{p/2}$. If we approximate $g'(x)$ sufficiently closely in the $L^2(0, 1)$ metric by a continuous function $g(x)$ ($\|g\|_C \leq \eta$), we obtain the required decomposition (36) with $K = 2(C_p M)^p$.

REMARK. If we are given a sequence $\{c_n\}_{n=1}^{\infty}$ with $\{\sum_{n=1}^{\infty} c_n^2\}^{1/2} = \delta$, then by applying Lemma 1 to the sequence $a_n = \delta^{-1} c_n$, $n = 1, 2, \ldots$, we can find numbers $\varepsilon_n = \pm 1$, $n = 1, 2, \ldots$, such that the function $\sum_{n=1}^{\infty} \varepsilon_n c_n \varphi_n(x)$ can, for some $\eta > 0$, be represented in the form (see (36))

$$f(x) \equiv \sum_{n=1}^{\infty} \varepsilon_n c_n \varphi_n(x) = g(x) + h(x),$$
$$\|g\|_C \leq \eta, \|h\|_2 \leq \delta \omega \left(\frac{\eta}{\delta}\right). \tag{38}$$

PROOF OF THEOREM 5. Take a sequence $\lambda_j = ((1+j)N)^R$, $j = 0, 1, 2, \ldots$, where $R = 2/\alpha$ and $N$ is so large that

$$\sum_{j=0}^{\infty} \frac{\lambda_{j+1}}{\lambda_j} \omega(\lambda_j) = \sum_{j=0}^{\infty} \left(\frac{j+2}{j+1}\right)^R \frac{K}{N^{R\alpha}(1+j)^{R\alpha}} \leq \frac{1}{4} \tag{38'}$$

(we retain the notation of Lemma 1 (see (36))). Set

$$\delta_0 = (2\lambda_0)^{-1}, \qquad \delta_{j+1} = \frac{\omega(\lambda_j)}{\lambda_j}, \qquad \eta_0 = 0, \ \eta_{j+1} = \delta_j \lambda_j, \ j = 0, 1, \ldots. \tag{39}$$

It follows from (38') and (39) that

$$\sum_{j=0}^{\infty} \eta_j = \sum_{j=0}^{\infty} \delta_j \lambda_j = \frac{1}{2} + \sum_{j=1}^{\infty} \frac{\lambda_j}{\lambda_{j-1}} \omega(\lambda_{j-1}) \leq \frac{3}{4}. \tag{40}$$

Therefore, in order to prove Theorem 5 it is enough to verify that for every sequence $a_n \geq 0$, $n = 1, 2, \ldots$, with $\{\sum_{n=1}^{\infty} a_n^2\}^{1/2} \leq \delta_0$, there is a function $G(x) \in C(0, 1)$ for which

$$|c_n(G)| \geq \left(1 - \sum_{j=0}^{\infty} \eta_j\right) a_n \geq \frac{1}{4} a_n, \qquad \|G\|_C \leq 1 \qquad (41)$$

(note that $\delta_0$ depends only on $p$ and $M$).

We construct inductively three sequences of functions on $(0, 1)$: $\{F_j\}_{j=0}^{\infty}$, $\{G_j\}_{j=0}^{\infty}$, and $\{H_j\}_{j=0}^{\infty}$, with $F_j = H_j + G_j$, $j = 0, 1, \ldots$, that satisfy the following conditions:

a) $F_j \in L^2(0, 1)$ and $|c_n(F_j)| \geq (1 - \sum_{s=0}^{j} \eta_s) a_n$, $n = 1, 2, \ldots$; $j = 0, 1, \ldots$;

b) $G_0(x) \equiv 0$, $G_j \in C(0, 1)$, and $\|G_j - G_{j-1}\|_C \leq \eta_j$, $j = 1, 2, \ldots$;

c) $\|H_j\|_2 \leq \delta_j$, $j = 0, 1, \ldots$;

d) $H_j(x) = g_j(x) + h_j(x)$, $g_j \in C(0, 1)$, $\|g_j\|_C \leq \eta_{j+1}$, and $\|h_j\|_2 \leq \delta_j \omega(\eta_{j+1}/\delta_j)$, $j = 0, 1, \ldots$.

It is easily seen that under these conditions the sequence $G_j(x)$, $j = 0, 1, \ldots$, will converge uniformly to a function $G(x) \in C(0, 1)$ for which (41) is valid (see a), b), c), and (40)).

We set $G_0(x) \equiv 0$ and then, using the remark on Lemma 1 (see (38)) we find a sequence $\varepsilon_n = \pm 1$, $n = 1, 2, \ldots$, such that $F_0(x) = H_0(x) = \sum_{n=1}^{\infty} \varepsilon_n a_n \varphi_n(x)$ is representable in the form

$$H_0(x) = g_0(x) + h_0(x),$$

$$g_0 \in C(0, 1), \quad \|g_0\|_C \leq \eta_1, \quad \|h_0\|_2 \leq \delta_0 \omega\left(\frac{\eta_1}{\delta_0}\right).$$

It is clear that conditions a)–d) are satisfied for $j = 0$. Now suppose that for some $j \geq 0$ the functions $\{F_s, G_s, H_s\}_{s=0}^{j}$ already constructed satisfy conditions a)–d). Set

$$G_{j+1}(x) = G_j(x) + g_j(x), \qquad x \in [0, 1]. \qquad (42)$$

In order to determine $H_{j+1}$, we consider the set of integers

$$Q = \left\{n \colon |c_n(G_{j+1})| < a_n \left(1 - \sum_{s=0}^{j+1} \eta_s\right)\right\} \qquad (43)$$

and set

$$H_{j+1}(x) = \sum_{n \in Q} \varepsilon_n 2 a_n \left(1 - \sum_{s=0}^{j+1} \eta_s\right) \varphi_n(x), \qquad (44)$$

where the numbers $\varepsilon_n = \pm 1$, $n \in Q$, are chosen so that we have the decomposition

$$H_{j+1}(x) = g_{j+1}(x) + h_{j+1}(x), \qquad g_{j+1} \in C(0,1), \ \|g_{j+1}\|_C \leq \eta_{j+2},$$

$$\|h_{j+1}\|_2 \leq \|H_{j+1}\|_2 \omega \left( \eta_{j+2} \cdot \frac{1}{\|H_{j+1}\|_2} \right) \tag{45}$$

(this choice is possible by the remark on Lemma 1 (see (38))).

Let us verify that conditions a)–d) are satisfied for $G_{j+1}(x)$, $H_{j+1}(x)$, and $F_{j+1}(x) \equiv G_{j+1}(x) + H_{j+1}(x)$:

a') by (43) and (44), for $n \notin Q$

$$|c_n(F_{j+1})| = |c_n(G_{j+1}) + c_n(H_{j+1})| = |c_n(G_{j+1})|$$

$$\geq a_n \left( 1 - \sum_{s=0}^{j+1} \eta_s \right)$$

and for $n \in Q$

$$|c_n(F_{j+1})| \geq |c_n(H_{j+1})| - |c_n(G_{j+1})|$$

$$\geq 2a_n \left( 1 - \sum_{s=0}^{j+1} \eta_s \right) - a_n \left( 1 - \sum_{s=0}^{j+1} \eta_s \right)$$

$$= a_n \left( 1 - \sum_{s=0}^{j+1} \eta_s \right) ;$$

b') since $H_j(x)$ satisfies condition d), it follows from (42) that

$$\|G_{j+1}(x) - G_j(x)\|_C = \|g_j(x)\|_C \leq \eta_{j+1};$$

c') using Parseval's equation and recalling that $\sum_{s=0}^{j+1} \eta_s \geq \eta_1 = \frac{1}{2}$ (see (39)), we obtain

$$\|H_{j+1}\|_2 = 2 \left( 1 - \sum_{s=0}^{j+1} \eta_s \right) \left\{ \sum_{n \in Q} a_n^2 \right\}^{1/2} \leq \left\{ \sum_{n \in Q} a_n^2 \right\}^{1/2} . \tag{46}$$

Since $F_j(x)$ satisfies condition a), then for $n \in Q$ (see (43))

$$|c_n(F_j - G_{j+1})| = |c_n(F_j) - c_n(G_{j+1})| \geq a_n \eta_{j+1};$$

moreover (see (42) and property d) of $H_j(x)$)

$$F_j(x) - G_{j+1}(x) = |G_j(x) + H_j(x)| - G_{j+1}(x)$$

$$= |G_j(x) + g_j(x) + h_j(x)| - |G_j(x) + g_j(x)| = h_j(x).$$

Therefore

$$\left(\sum_{n\in Q} a_n^2\right)^{1/2} \leq \frac{1}{\eta_{j+1}} \left\{\sum_{n\in Q} c_n^2(F_j - G_{j+1})\right\}^{1/2}$$

$$\leq \frac{1}{\eta_{j+1}} \|F_j - G_{j+1}\|_2 \leq \frac{1}{\eta_{j+1}} \|h_j\|_2.$$

From this and (46), since $\|h_j\|_2 \leq \delta_j \omega(\eta_{n+1}/\delta_j)$, we obtain (see also (39))

$$\|H_{j+1}\|_2 \leq \frac{1}{\eta_{j+1}} \delta_j \omega\left(\frac{\eta_{j+1}}{\delta_j}\right) = \frac{1}{\lambda_j} \omega(\lambda_j) = \delta_{j+1}, \qquad (47)$$

which establishes c).

d') Relation d) follows immediately from (45), (47), and the monotonicity of $\omega(\eta)$:

$$\|h_{j+1}\|_2 \leq \|H_{j+1}\|_2 \omega\left(\eta_{j+2} \frac{1}{\|H_{j+1}\|_2}\right) \leq \delta_{j+1} \omega\left(\frac{\eta_{j+2}}{\delta_{j+1}}\right).$$

Thus we have verified that, for $j = 0, 1, \ldots$, the functions $G_{j+1}$, $H_{j+1}$, and $F_{j+1} = H_{j+1} + G_{j+1}$ satisfy conditions a)–d). As we noticed above, this implies the conclusion of Theorem 5.

In concluding this section we prove a result on coefficients of the expansions of functions of $\mathrm{Lip}\,\alpha$ classes in terms of bases of $L^p(0, 1)$.

THEOREM 6. *Let* $\Psi = \{\psi_n(x)\}_{n=1}^{\infty}$ *be a basis of* $L^p(0, 1)$, $1 < p < \infty$, *and let* $\|\psi_n\|_p = 1$, $n = 1, 2, \ldots$. *There is a function* $F \in \mathrm{Lip}\,\alpha$, $\alpha = \min(1 - 1/p, 2)$, *with* $F(0) = F(1) = 0$, *for which the series of absolute values of the coefficients in its expansion in terms of* $\Psi$ *diverges, i.e.,*

$$F(x) \overset{L^p}{=} \sum_{n=1}^{\infty} a_n \psi_n(x), \qquad \sum_{n=1}^{\infty} |a_n| = \infty. \qquad (48)$$

COROLLARY 2. *For every C.O.N.S.* $\{\varphi_n(x)\}_{n=1}^{\infty}$, $x \in (0, 1)$, *there is a function* $F \in \mathrm{Lip}\,\frac{1}{2}$, $F(0) = F(1) = 0$, *for which*

$$\sum_{n=1}^{\infty} |c_n(F)| = \infty, \qquad c_n(F) = \int_0^1 F(x)\varphi_n(x)\,dx, \quad n = 1, 2, \ldots.$$

REMARK 1. The order of smoothness $\alpha$ of $F(x)$ in Theorem 6 is exact, i.e., cannot be increased for any $p$, $1 < p < \infty$. This is shown by the normalized Haar system in $L^p$ (for $1 < p \leq 2$) and by the trigonometric system (for $2 \leq p < \infty$) (see Theorem 3.1 and Corollary 4.5).

REMARK 2. As we shall see from the proof of Theorem 6, it can be generalized to a wide class of systems of functions that are not bases.

**PROOF OF THEOREM 6.** For each $k$, $k = 1, 2, \ldots$, we consider a block $\{\varphi_k^i(x)\}_{i=1}^{2^k}$ of Faber–Schauder functions, and let $\chi_{\Delta_k^i}(x)$ be the characteristic function of the interval $\Delta_k^i = ((i-1)/2^k, \, i/2^k)$, $i = 1, \ldots, 2^k$. Let us expand $\varphi_k^i(x)$ in terms of the basis $\Psi$:

$$\varphi_k^i(x) \overset{L^p}{=} \sum_{n=1}^{\infty} a_n^{k,i} \psi_n(x), \qquad 1 \le i \le 2^k. \tag{49}$$

We may suppose that

$$\sum_{n=1}^{\infty} |a_n^{k,i}| < \infty, \tag{50}$$

for all $i$ and $k$, since otherwise we can take the required function to be $\varphi_k^i(x) \in \mathrm{Lip}\, 1$. We estimate the coefficients $a_n^{k,i}$:

$$\frac{1}{2} = \sum_{i=1}^{2^k} \int_0^1 \varphi_k^i(x) \chi_{\Delta_k^i}(x)\, dx = \sum_{i=1}^{2^k} \int_0^1 \sum_{n=1}^{\infty} a_n^{k,i} \psi_n(x) \chi_{\Delta_k^i}(x)\, dx$$

$$= \sum_{n=1}^{\infty} \int_0^1 \psi_n(x) \left[ \sum_{i=1}^{2^k} a_n^{k,i} \chi_{\Delta_k^i}(x) \right] dx. \tag{51}$$

If we estimate the integrals in (51) by using Hölder's inequality, and use the equation

$$\left\| \sum_{i=1}^{2^k} a_i \chi_{\Delta_k^i} \right\|_q = \left\{ 2^{-k} \sum_{i=1}^{2^k} |a_i|^q \right\}^{1/q},$$

we deduce from (51) that

$$\frac{1}{2} \le \sum_{n=1}^{\infty} \|\psi_n\|_p \left\{ 2^{-k} \sum_{i=1}^{2^k} |a_n^{k,i}|^q \right\}^{1/q} = \sum_{n=1}^{\infty} \left\{ 2^{-k} \sum_{i=1}^{2^k} |a_n^{k,i}|^q \right\}^{1/q}. \tag{52}$$

Consider the functions

$$F_{k,t}(x) = \sum_{i=1}^{2^k} r_i(t) \varphi_k^i(t), \qquad k = 1, 2, \ldots, \; x, t \in (0, 1), \tag{53}$$

where $r_i(t)$, $i = 1, 2, \ldots$, are the Rademacher functions. By (49),

$$F_{k,t}(x) \overset{L^p}{=} \sum_{n=1}^{\infty} \left[ \sum_{i=1}^{2^k} r_i(t) a_n^{k,i} \right] \psi_n(x) \equiv \sum_{n=1}^{\infty} A_{n,k}(t) \psi_n(x), \tag{54}$$

moreover, $\|F_{k,t}\|_\infty \leq 1$ and (see Theorem 2.7)

$$\int_0^1 \sum_{n=1}^\infty |A_{n,k}(t)|\, dt = \sum_{n=1}^\infty \int_0^1 \left| \sum_{i=1}^{2^k} r_i(t) a_n^{k,i} \right|\, dt \geq c \sum_{n=1}^\infty \left\{ \sum_{i=1}^{2^k} (a_n^{k,i})^2 \right\}^{1/2},$$
$$(55)$$

where $c > 0$ is an absolute constant.

If we use, for $n = 1, 2, \ldots$, the inequality

$$\left\{ \sum_{i=1}^{2^k} (a_n^{k,i})^2 \right\}^{1/2} \geq 2^{k(\alpha - 1/q)} \left\{ \sum_{i=1}^{2^k} |a_n^{k,i}|^q \right\}^{1/q}, \qquad \frac{1}{p} + \frac{1}{q} = 1,$$

we find from (55) and (52) that

$$\int_0^1 \sum_{n=1}^\infty |A_{n,k}(t)|\, dt \geq c \sum_{n=1}^\infty 2^{k(\alpha - 1/q)} \left\{ \sum_{i=1}^{2^k} |a_n^{k,i}|^q \right\}^{1/q} \geq \frac{c}{2} 2^{k\alpha},$$
$$k = 1, 2, \ldots.$$

Consequently, for $k = 1, 2, \ldots$ and for some binary irrational $t_k \in (0, 1)$,

$$\sum_{n=1}^\infty |A_{n,k}(t_k)| \geq \frac{c}{2} 2^{k\alpha}. \qquad (56)$$

Set

$$F(x) = \sum_{s=1}^\infty P_{k_s}(x), \quad \text{where } P_k(x) = 2^{-k\alpha} F_{k,t_k}(x), \ x \in (0, 1),$$
$$k = 1, 2, \ldots; \quad (57)$$

here the sequence $k_s$, $s = 1, 2, \ldots$, can be taken to increase so fast that the coefficients $a_n(P_{k_s})$ of the expansion of $P_{k_s}(x)$ in terms of $\Psi$ are "concentrated in different places," or more precisely

$$\sum_{n=1}^\infty |a_n(F)| \geq \frac{1}{2} \sum_{s=1}^\infty \sum_{n=1}^\infty |a_n(P_{k_s})|. \qquad (58)$$

This is possible if we construct the numbers $k_s$, $s = 1, 2, \ldots$, successively, using (50) and the inequality $|a_n(P_k)| \leq C\|P_k\|_p$ (see Theorem 1.6), by which $a_n(P_k) \to 0$ as $k \to \infty$, for each $n$.

It follows from (54), (56), and (58) that

$$\sum_{n=1}^\infty |a_n(F)| \geq \frac{1}{2} \sum_{s=1}^\infty 2^{-k_s \alpha} \cdot \frac{c}{2} 2^{k_s \alpha} = \infty.$$

Moreover, $F(0) = F(1) = 0$, and since the coefficients $c_{k,i}(F)$ of the expansion of $F$ in terms of the Faber–Schauder system satisfy the inequality $|c_{k,i}(F)| \leq 2^{-k\alpha}$, $k = 1, 2, \ldots$ (see (53) and (57)), we have $F \in \operatorname{Lip}\alpha$ by Theorem 5.2. This completes the proof of Theorem 6.

## §3. Some properties of uniformly bounded orthonormal systems

Before stating the propositions on the properties of the O.N.S., we establish some inequalities which play a fundamental role in the proofs in this section.

On the real line $R^1$, let there be given a measurable function $f(x)$ and, as usual,

$$\lambda_f(t) = m\{x \in R^1 \colon |f(x)| > t\},$$
$$\Delta_h(f, x) = f(x + h) - f(x).$$

We recall (see Appendix 1) that

$$\|f\|_{L^p(R^1)}^p = p \int_0^\infty t^{p-1} \lambda_f(t)\, dt, \qquad 1 \leq p < \infty, \tag{59}$$

and consequently, for each $b > 0$,

$$\int_{\{x \in R^1 \colon |f(x)| > b\}} |f(x)|\, dx = b\lambda(b) + \int_b^\infty \lambda_f(t)\, dt \geq \int_b^\infty \lambda_f(t)\, dt. \tag{60}$$

Set($^1$)

$$N(f, W_2^{1/2}) = \left\{ \int_0^\infty h^{-2} \|\Delta_h(f, x)\|_2^2\, dh \right\}^{1/2}.$$

PROPOSITION 1. *Every function $f(x)$ satisfies the inequality*

$$N^2(f, W_2^{1/2}) \leq 8\|f\|_{L^1(R^1)} \cdot \|f'\|_{L^\infty(R^1)}, \tag{61}$$

*where*

$$\|f'\|_{L^\infty(R^1)} = \sup_{h>0, x\in R^1} h^{-1} |\Delta_h(f, x)|.$$

PROOF. Since both sides of (61) are homogeneous, we may suppose without loss of generality that $\|f'\|_{L^\infty(R')} = 1$. For each $h > 0$, we define the set

$$E(h) = \{x \in R^1 \colon |f(x)| \leq h, |f(x + h)| \leq h\},$$
$$F(h) = R^1 \setminus E(h).$$

---

($^1$)The notation is suggested by the fact that by using the functional $N(f, W_2^{1/2})$ one can introduce a norm on the Sobolev space $W_2^{1/2}$.

It is evident that $m\{F(h)\} \leq 2\lambda_f(h)$ and that the following inequalities hold for all $x \in R^1$ and $h > 0$:

a) $|\Delta_h(f, x)| \leq |f(x)| + |f(x + h)|$;

b) $|\Delta_h(f, x)| \leq h$.

Since

$$\|\Delta_h(f, x)\|^2_{L^2(R^1)} = \|\Delta_h(f, x)\|^2_{L^2(E(h))} + \|\Delta_h(f, x)\|^2_{L^2(F(h))},$$

we obtain, by estimating the first term by a) and the second by b), and using (59) (for $p = 2$),

$$\|\Delta_h(f, x)\|_{L^2(R^1)} \leq 2 \int_{E(h)} [f^2(x) + f^2(x + h)] \, dx + h^2 m\{F(h)\} \leq 4.$$

$$\int_{\{x \in R^1 : |f(x)| \leq h\}} f^2(x) \, dx + 2h^2 \lambda_f(h) = 8 \int_0^h t[\lambda_f(t) - \lambda_f(h)] \, dt + 2h^2 \lambda_f(h)$$

$$\leq 8 \int_0^h t\lambda_f(t) \, dt. \tag{62}$$

Consequently (see (59))

$$N^2(f, W_2^{1/2}) \leq 8 \int_0^\infty h^{-2} \int_0^h t\lambda_f(t) \, dt \, dh = 8 \int_0^\infty \int_t^\infty h^{-2} t\lambda_f(t) \, dh \, dt$$

$$= 8 \int_0^\infty \lambda_f(t) \, dt = 8\|f\|_{L^1(R^1)},$$

as was to be proved.

We shall also need (for the proof of Theorem 11) the following modification of Proposition 1.

*For every function $f(x)$ defined on the finite interval $(0, a)$ with $\|f'\|_{L^\infty(0,a)} \leq 1$, and every number $\beta$, $0 < \beta < \min(1, a)$,*

$$\sigma^2(f, \beta) \equiv \int_\beta^a h^{-2} \int_0^{a-h} [f(x + h) - f(x)]^2 \, dx \, dh$$

$$\leq 8 \left[ \beta \int_0^a |f(x)| \, dx + \int_{\{x : |f(x)| > \beta^2\}} |f(x)| \, dx \right]. \tag{63}$$

The difference between (61) and (63) is that (61) provides a lower bound for the integral $\int |f(x)| \, dx$ on the whole domain of $f$, whereas (63) does this only on the set where $|f(x)| > \beta^2$.

To obtain (63) we notice that the proof of (62) applies unchanged to functions defined on a finite interval, i.e., for every $f(x)$, $x \in (0, a)$, with

$\|f'\|_{L^\infty(0,a)} < 1$, and $h \in (0, a)$,

$$
\int_0^{a-h} |\Delta_h(f, x)|^2 \, dx \le 8 \int_0^h t\lambda_f(t) \, dt,
$$

$$
\lambda_f(t) = m\{x \in (0, a): |f(x)| > t\}. \tag{64}
$$

From (64), we deduce by integrating and using (59) and (60), that

$$
\begin{aligned}
\sigma^2(f, \beta) &\le 8 \int_\beta^a h^{-2} \int_0^h t\lambda_f(t) \, dt \, dh \\
&= 8 \int_0^a t\lambda_f(t) \int_{\max(t,\beta)}^a h^{-2} \, dh \, dt \\
&\le 8 \left[ \int_0^\beta \beta^{-1} t\lambda_f(t) \, dt + \int_\beta^a \lambda_f(t) \, dt \right] \\
&\le 8 \left[ \beta \int_0^{\beta^2} \lambda_f(t) \, dt + \int_{\beta^2}^\beta \lambda_f(t) \, dt + \int_\beta^a \lambda_f(t) \, dt \right] \\
&\le 8 \left[ \beta \int_0^a |f(x)| \, dx + \int_{\{x \in (0,a): \, |f(x)| > \beta^2\}} |f(x)| \, dx \right].
\end{aligned}
$$

This establishes (63).

We shall also need a discrete version of Proposition 1:

*For every set $\{a_n\}_{n=1}^N$, $N = 1, 2, \ldots$, of numbers,*

$$
\left( \max_{1 \le n \le N} |a_n| \right) \sum_{m=1}^N \left| \sum_{n=1}^m a_n \right| \ge c \sum_{\nu=0}^{N-1} \sum_{\mu=\nu+1}^N \frac{\left( \sum_{n=\nu+1}^\mu a_n \right)^2}{(\nu - \mu)^2}, \tag{65}
$$

*where $c > 0$ is an absolute constant.*

To obtain (65), we define a function $f(x)$ on $\{a_n\}_{n=1}^N$ extended continuously to $R^1$ by setting

$$
f(x) = \begin{cases} a_n & \text{if } |x - m| \le 1/3, \, m = 1, \ldots, N; \\ 0 & \text{if } -\infty < x \le 1/3; \\ \text{linear on } [m - 2/3, m - 1/3], \, m = 1, \ldots, N; \end{cases}
$$

$$
f(N + x) = f(N - x).
$$

It is easily seen that

$$
\|f'\|_{L^\infty(R^1)} \le 3 \max_{1 \le n \le N} |a_n|, \qquad \|f\|_{L^1(R^1)} \le C \sum_{m=1}^N \left| \sum_{n=1}^m a_n \right|.
$$

Application of (61) to $f(x)$ yields

$$\left(\max_{1 \le n \le N} |a_n|\right) \sum_{m=1}^{N} \left|\sum_{n=1}^{m} a_n\right|$$

$$\ge c_1 \int_{-\infty}^{\infty} \int_{x}^{\infty} \left[\frac{f(y) - f(x)}{y - x}\right]^2 dy\,dx$$

$$\ge c_1 \sum_{\nu=0}^{N-1} \sum_{\mu=\nu+1}^{N} \int_{\nu-1/3}^{\nu+1/3} \int_{\mu-1/3}^{\mu+1/3} \left[\frac{f(y) - f(x)}{y - x}\right]^2 dy\,dx$$

$$\ge \frac{4}{9}c_1 \sum_{\nu=0}^{N-1} \sum_{\mu=\nu+1}^{N} \frac{\left(\sum_{n=\nu+1}^{\mu} a_n\right)^2}{(\nu - \mu)^2}.$$

This completes the proof of (65).

Now we turn to the consideration of properties of O.N.S. that consist of uniformly bounded functions. It turns out that we can obtain a number of general propositions on the divergence of Fourier series for these systems. We begin by establishing the following theorem.

THEOREM 7. *There are absolute constants* $\gamma_1$, $\gamma_2 > 0$ *such that the following relations hold for every O.N.S.* $\{\varphi_n(x)\}_{n=1}^{N}$, $x \in (0, 1)$, $N = 1, 2, \ldots$, *whose elements satisfy the condition*

$$|\varphi_n(x)| \le M \text{ for almost all } x \in (0, 1), \quad n = 1, 2, \ldots. \tag{66}$$

1) 
$$\frac{1}{N} \sum_{m=1}^{M} \int_{0}^{1} \left|\sum_{n=1}^{m} \varphi_n(x)\right| dx \ge \frac{\gamma_1}{M} \ln N;$$

2) 
$$\frac{1}{N} \sum_{m=1}^{N} L_m(x) \ge \frac{\gamma_2}{M^2} \ln N \quad \text{for } x \in E_N \subset (0, 1),$$

$$m(E_N) \ge \frac{1}{4M^2},$$

*where* $L_m(x)$, $m = 1, 2, \ldots$, *are the Lebesgue functions of* $\{\varphi_n\}_{n=1}^{\infty}$ (*see Definition* 1.11).

PROOF. 1) If we apply (65) to the set of numbers $\{\varphi_n(x)\}_{n=1}^{N}$, $x \in (0, 1)$, and then integrate with respect to $x$ and apply Parseval's equation, we obtain

$$\sum_{m=1}^{N} \int_0^1 \left| \sum_{n=1}^{m} \varphi_n(x) \right| dx$$

$$= \int_0^1 \sum_{m=1}^{N} \left| \sum_{n=1}^{m} \varphi_n(x) \right| dx$$

$$\geq c \left( \max_{1 \leq n \leq N} \|\varphi_n\|_\infty \right)^{-1} \sum_{\nu=0}^{N-1} \sum_{\mu=\nu+1}^{N} \frac{1}{(\nu - \mu)^2} \int_0^1 \left( \sum_{n=\nu+1}^{\mu} \varphi_n(x) \right)^2 dx$$

$$\geq \frac{c}{M} \sum_{\nu=0}^{N-1} \sum_{\mu=\nu+1}^{N} \frac{1}{\mu - \nu} \geq \frac{\gamma_1}{M} N \ln N.$$

2) Let

$$E_N = \{ x \in (0,1) : F_N(x) \geq \tfrac{1}{4} N \},$$

$$F_N(x) = \sum_{\frac{N}{4} < n \leq \frac{3}{4}N + 1} \varphi_n^2(x).$$

Let us estimate the measure of $E_N$. Since $\|F_N\|_1 \geq N/2$ and $\|F_N\|_\infty \leq M^2 N$, we have

$$\frac{N}{2} \leq \int_0^1 F_N(x)\, dx = \int_{E_N} F_N(x)\, dx + \int_{(0,1)\setminus E_N} F_N(x)\, dx$$

$$\leq M^2 N m(E_N) + \frac{1}{4} N,$$

i.e., $m(E_N) \geq 1/(4M^2)$.

Furthermore, if we apply (65), in the same way as in the proof of 1), to the set of numbers $\{\varphi_n(x)\varphi_n(t)\}_{n=1}^{N}$, $x, t \in (0,1)$, we obtain

$$\sum_{m=1}^{N} L_m(x) = \int_0^1 \sum_{m=1}^{N} \left| \sum_{n=1}^{m} \varphi_n(x)\varphi_n(t) \right| dt$$

$$\geq cM^{-2} \sum_{\nu=0}^{N-1} \sum_{\mu=\nu+1}^{N} (\nu - \mu)^{-2} \int_0^1 \left[ \sum_{n=\nu+1}^{\mu} \varphi_n(x)\varphi_n(t) \right]^2 dt$$

$$= cM^{-2} \sum_{\nu=0}^{N-1} \sum_{\mu=\nu+1}^{N} (\nu - \mu)^{-2} \sum_{n=\nu+1}^{\mu} \varphi_n^2(x)$$

$$= cM^{-2} \sum_{n=1}^{N} \varphi_n^2(x) \sum_{\nu,\mu:\, 0 \leq \nu < n \leq \mu \leq N} (\mu - \nu)^2$$

$$\geq c'M^{-2} F_N(x) \ln N,$$

i.e., for each $x \in E_N$ we have

$$\sum_{m=1}^{N} L_m(x) \geq \frac{c'}{4} M^{-2} N \ln N = \gamma_2 M^{-2} N \ln N.$$

This completes the proof of Theorem 7.

REMARK. Under the hypotheses of Theorem 7, inequalities 1) and 2) are best possible. This is shown by the examples of the trigonometric system and the Walsh system. For these systems, $L_m(x) \equiv L_m \leq C \ln m$, $m = 1, 2, \ldots$ (see 4.(11) and 4.(115)).

The following corollary is an immediate consequence of Theorem 7.

COROLLARY 3. *Let the functions in the O.N.S.* $\Phi = \{\varphi_n(x)\}_{n=1}^{\infty}$, $x \in (0, 1)$, *be uniformly bounded (i.e., let (66) hold). Then the Lebesgue functions of the system satisfy*

1) $\max_{1 \leq m \leq N} L_m(x) \geq c_M \ln N > 0$ *for* $x \in E_N \subset (0, 1)$, $m(E_N) > c'_M > 0$;

2) $\overline{\lim}_{m \to \infty} (L_m(x) / \ln m) > c_M > 0$ *for* $x \in E \subset (0, 1)$, $m(E) > c'(M) > 0$.

Here the set $E_N$ is the same as in Theorem 7, and $E = \overline{\lim}_{N \to \infty} E_N$.

It is interesting to notice that for general O.N.S. it is possible, conversely, to deduce propositions of the type of Theorem 7 from formally weaker estimates of the type of Corollary 3. We demonstrate the method by obtaining the following corollary, using only the inequality

$$\max_{1 \leq m \leq N} \int_0^1 \left| \sum_{i=1}^{m} \varphi_m(x) \right| dx \geq \frac{\gamma_1}{M} \ln N$$

$$(\{\varphi_m\}_{m=1}^{N} \text{ is an O.N.S. with } \|\varphi_m\|_\infty \leq M, \ m = 1, \ldots, N) \qquad (67)$$

(which in fact follows from Theorem 7).

COROLLARY 4. *For every bounded (not necessarily normalized) set of functions* $\{\psi_n(x)\}_{n=1}^{N} \in L^2(0, 1)$, *we have the inequality*

$$\int_0^1 \left| \sum_{n=1}^{N} \psi_n(x) \right| dx + \frac{1}{N} \sum_{m=1}^{N} \int_0^1 \left| \sum_{n=1}^{m} \psi_n(x) \right| dx$$

$$\geq \gamma_3 \left( N \max_{1 \leq n \leq N} \|\psi_n\|_\infty \right)^{-1} \sum_{n=1}^{N} \|\psi_n\|_2^2 \ln N \qquad (68)$$

($\gamma_3 > 0$ *is an absolute constant*).

PROOF. For $j = 1, \ldots, N$, set $\psi_{j+N}(x) = \psi_j(x)$, $x \in (0, 1)$, and let $a_n = \|\psi_n\|_2$, $n = 1, \ldots, 2N$. We may suppose without loss of generality that

$\sum_{n=1}^{N} a_n^2 = \sum_{n=1}^{N} \|\psi_n\|_2^2 = N$. We define a system of functions $\{\varphi_n(x)\}_{n=1}^{N}$ by setting

$$\varphi_n(x) = \psi_{n+j} \left( N \left( x - \frac{j-1}{N} \right) \right) \quad \text{if } x \in \left( \frac{j-1}{N}, \frac{j}{N} \right), \tag{69}$$
$$n, j = 1, \ldots, N.$$

It is easily seen that $\{\varphi_n(x)\}_{n=1}^{N}$ is an O.N.S., and that

$$\|\varphi_n\|_\infty \leq M \equiv \max_{1 \leq n \leq N} \|\psi_n\|_\infty, \quad n = 1, \ldots, N.$$

If we apply (67) to this system, we obtain, for some $m$, $1 \leq m \leq N$,

$$\frac{\gamma_1}{M} \ln N \leq \int_0^1 \left| \sum_{n=1}^{m} \varphi_n(x) \right| dx$$

$$= \sum_{j=1}^{N} \int_{(j-1)/N}^{j/N} \left| \sum_{n=1}^{m} \psi_{n+j} \left( N \left( x - \frac{j-1}{N} \right) \right) \right| dx$$

$$= \frac{1}{N} \sum_{j=1}^{N} \int_0^1 \left| \sum_{n=j+1}^{m+j} \psi_n(x) \right| dx$$

$$\leq \frac{1}{N} \sum_{j=1}^{N} \int_0^1 \left| \sum_{n=1}^{m+j} \psi_n(x) \right| dx + \frac{1}{N} \sum_{j=1}^{N} \int_0^1 \left| \sum_{n=1}^{j} \psi_n(x) \right| dx. \tag{70}$$

But since $\psi_{j+N}(x) = \psi_j(x)$, $j = 1, \ldots, N$ and $1 \leq m \leq N$, we have

$$\sum_{j=1}^{N} \int_0^1 \left| \sum_{m=1}^{m+j} \psi_n(x) \right| dx \leq \sum_{j=1}^{2N} \int_0^1 \left| \sum_{n=1}^{j} \psi_n(x) \right| dx$$

$$\leq N \int_0^1 \left| \sum_{n=1}^{N} \psi_n(x) \right| dx + 2 \sum_{j=1}^{N} \int_0^1 \left| \sum_{n=1}^{j} \psi_n(x) \right| dx.$$

The conclusion of Corollary 4 (see (68)) follows from the preceding inequality and (70).

If we use the connection, established in Chapter 1, between the property that an O.N.S. $\Phi$ is a basis in the spaces $C(0,1)$ and $L^1(0,1)$ and the boundedness of the Lebesgue functions of this system (see Theorem 1.8 and the remark on it), we obtain the following corollary by using Corollary 3.

COROLLARY 5. *There are no orthonormal uniformly bounded bases in* $C(0,1)$ *and* $L^1(0,1)$.

In fact, we shall establish a somewhat more precise result.

THEOREM 8. *Let the O.N.S.* $\Phi = \{\varphi_n(x)\}_{n=1}^\infty$, $x \in (0,1)$, *satisfy* (66). *Then*

1) *there is a continuous function with domain* $(0,1)$ *whose Fourier series in the system* $\Phi$ *diverges at some point;*

2) *there is a function* $f \in L^1(0,1)$ *whose Fourier series in the system* $\Phi$ *diverges in* $L^1(0,1)$.

PROOF. 1) Let the point $x_0 \in (0,1)$ be such that, for every $f \in C(0,1)$, the sequence of partial sums (see 1.(14))

$$S_N(f,x_0) = \int_0^1 f(t)K_N(t,x_0)\,dt, \qquad K_N(t,x) = \sum_{n=1}^N \varphi_n(x)\varphi_n(t),$$

$$N = 1,2,\dots,$$

converges. This means that the sequence $S_N(f,x_0)$, $N = 1,2,\dots$, of bounded linear functionals on $C(0,1)$ satisfies the condition

$$\sup_{1\le N<\infty} |S_N(f,x_0)| < \infty$$

for every $f \in C(0,1)$. But then, by the Banach–Steinhaus theorem, these functionals have uniformly bounded norm, i.e., (see 1.(17))

$$\sup_{1\le N<\infty} L_N(x_0) < \infty, \qquad L_N(x_0) \equiv \int_0^1 |K_N(t,x_0)|\,dt.$$

Consequently $x_0 \notin E$, where $E \subset (0,1)$, and $m(E) > 0$ is defined for $\Phi$ by relation 2) of Corollary 3.

Conclusion 2) of Theorem 8 is proved very similarly, using Equation 1.(18).

By using Corollaries 3 and 4, we can obtain an estimate of the rate of growth of $\max_{1\le n\le N} \|\varphi_n\|_\infty$ as $N \to \infty$, for any orthonormal basis $\{\varphi_n(x)\}_{n=1}^\infty$ in the space of continuous functions.

THEOREM 9. *Let the O.N.S.* $\Phi = \{\varphi_n(x)\}_{n=1}^\infty$, $x \in (0,1)$, *with* $\varphi_n \subset L^\infty(0,1)$, $n = 1,2,\dots$, *have the property that the Fourier series in the system* $\Phi$ *of every continuous function with domain* $[0,1]$ *converges uniformly on* $[0,1]$. *Then, for some number* $\alpha = \alpha(\Phi) > 0$,

$$M(N) \equiv \max_{1\le n<N} \|\varphi_n\|_\infty \ge \alpha N^{1/2}, \qquad N = 1,2,\dots \qquad (71)$$

Recall that for the Haar system $\{\chi_n(x)\}_{n=1}^\infty$ and the Franklin system $\{f_n(x)\}_{n=1}^\infty$ we have

$$\|\chi_n\|_\infty \asymp \|f_n\|_\infty \asymp n^{1/2} \qquad (n \to \infty),$$

so that (71) is exact in order.

PROOF OF THEOREM 9. By hypothesis, for every $f \in C(0, 1)$,

$$\sup_{1 \le N < \infty} \|S_N(f)\|_\infty < \infty, \qquad S_N(f, x) = \sum_{n=1}^{N} \int_0^1 f(t) \varphi_n(t) \, dt \varphi_n(x),$$

$$N = 1, 2, \dots$$

Hence it follows by the Banach–Steinhaus theorem that the norms of the operators $S_N(f)$, $N = 1, 2, \dots$, from $C(0, 1)$ to $L^\infty(0, 1)$ are bounded. Consequently (see 1.(17)) the Lebesgue functions of $\Phi$ are bounded:

$$\sup_{1 \le N < \infty} \|L_N(t)\|_\infty = K < \infty, \qquad L_N(t) = \int_0^1 \left| \sum_{n=1}^{N} \varphi_n(x) \varphi_n(t) \right| dx,$$

$$N = 1, 2, \dots \quad (72)$$

From (72) we find at once that, for $n = 1, 2, \dots$ and $t \in (0, 1)$,

$$2K \ge \left\| \int_0^1 |\varphi_n(x) \varphi_n(t)| \, dx \right\|_\infty = \|\varphi_n\|_\infty \cdot \|\varphi_n\|_1. \quad (73)$$

Let us show that

$$\max_{1 \le m \le N} \|L_m\|_\infty \ge c(K) \ln \frac{N}{M^2(N)}, \quad (74)$$

for some constant $c = c(K) > 0$, $N = 1, 2, \dots$; hence we obtain the conclusion of the theorem (see (72) and (71)).

Let $N$ be given. We first show that there is a set $E \subset (0, 1)$ with $m(E) > 0$, for every point of which

$$\begin{cases} I(t) \equiv \sum_{n=1}^{N} \varphi_n^2(t) \ge \dfrac{N}{2}, \\ 4K \cdot I(t) \ge \sum_{n=1}^{N} \|\varphi_n\|_\infty |\varphi_n(t)|. \end{cases} \qquad t \in E. \quad (75)$$

In fact, let $E' = \{t \in (0, 1) : I(t) \ge N/2\}$; then

$$\int_{E'} I(t) \, dt = N - \int_{(0,1) \setminus E'} I(t) \, dt \ge \frac{N}{2}. \quad (76)$$

On the other hand, by (73),

$$\int_{E'} \sum_{n=1}^{N} \|\varphi_n\|_\infty |\varphi_n(t)| \, dt \le 2KN. \quad (77)$$

It follows from (76) and (77) that

$$m(E) > 0, \qquad E = \{t \in E' : 4KI(t) \ge \sum_{n=1}^{N} \|\varphi_n\|_\infty |\varphi_n(t)|\},$$

and furthermore that (75) holds at each point $t \in E$.

To obtain (74) for the given $N$, it is of course enough to verify that

$$\max_{1 \leq m \leq N} L_m(t) \geq c(K) \ln \frac{N}{M^2(N)}, \quad \text{if } t \in E. \tag{78}$$

For this purpose we take an arbitrary point $t \in E$ and define a set of numbers $\{k_i\}_{i=0}^s$, $0 = k_0 < k_1 < \cdots < k_s = N$, in the following way.

Let $k_1$ denote the smallest of the numbers $k = 1, \ldots, N$, for which

$$\left\| \sum_{n=1}^{k} \varphi_n(t)\varphi_n(x) \right\|_\infty \geq \frac{1}{2} M^2(N);$$

if there is no such number, we take $k_1 = N$.

Suppose that the numbers $k_1 < \cdots < k_{i-1} < N$ have already been constructed. We take as $k_i$ the smallest of the numbers $k = k_{i-1}+1, \ldots, N$, for which

$$\left\| \sum_{n=k_{i-1}+1}^{k} \varphi_n(t)\varphi_n(x) \right\|_\infty \geq \frac{1}{2} M^2(N);$$

if there is no such number, we take $k_i = N$.

This process terminates at some stage with an index $s = s(t)$, when $k_s = N$.

Since $|\varphi_n(t)\varphi_n(x)| \leq M^2(N)$, $x \in (0,1)$, $n = 1, \ldots, N$, it follows from the construction that

a) when $i = 1, \ldots, s$

$$\left\| \sum_{n=k_{i-1}+1}^{k_i} \varphi_n(t)\varphi_n(x) \right\|_\infty \leq 2M^2(N); \tag{79}$$

b) for $s > 1$ and $i = 1, \ldots, s-1$,

$$\frac{1}{2} M^2(N) \leq \left\| \sum_{n=k_{i-1}+1}^{k_i} \varphi_n(t)\varphi_n(x) \right\|_\infty. \tag{80}$$

We shall suppose below that $s \geq 3$, since when $s < 3$ we have by (79)

$$\left\| \sum_{n=1}^{N} \varphi_n(t)\varphi_n(x) \right\|_\infty \leq 4M^2(N),$$

from which, using (75), we deduce that

$$L_N(t) \geq \frac{1}{4M^2(N)} \int_0^1 \left| \sum_{n=1}^{N} \varphi_n(t)\varphi_n(x) \right|^2 dx$$

$$= \frac{I(t)}{4M^2(N)} \geq \frac{N}{8M^2(N)} \geq \frac{1}{8} \ln \frac{N}{M^2(N)},$$

i.e., when $s = s(t) < 3$ inequality (78) is automatically satisfied. Hence we let $s = s(t) \geq 3$. Set

$$\psi_i(x) = \sum_{n=k_{i-1}+1}^{k_i} \varphi_n(t)\varphi_n(x), \qquad i = 1, \ldots, s.$$

It is clear that the functions $\psi_i(x)$ are pairwise orthogonal, and by (75), (79), and (80)

1) $$\sum_{i=1}^{s} \|\psi_i\|_2^2 = I(t) \geq \frac{N}{2};$$

2) $$\frac{M^2(N)}{2} \leq \begin{array}{l} \|\psi_i\|_\infty \leq 2M^2(N), \\ \|\psi_s\|_\infty \leq 2M^2(N). \end{array} \quad 1 \leq i \leq s-1,$$

(81)

In addition (see (75))

$$\sum_{i=1}^{s} \|\psi_i\|_\infty \leq \sum_{i=1}^{N} |\varphi_n(t)|\|\varphi_n\|_\infty \leq 4K \cdot I(t).$$ (82)

It follows from (81), part 2), and (82), that

$$\frac{s}{2} < s - 1 \leq \frac{8KI(t)}{M^2(N)}.$$ (83)

We now use Corollary 4: applying (68) to the functions $\psi_i(x)$, $i = 1, \ldots, s$, and using (81), part 1, we obtain

$$\int_0^1 \left| \sum_{i=1}^{s} \psi_i(x) \right| dx + \frac{1}{s} \sum_{m=1}^{s} \int_0^1 \left| \sum_{i=1}^{m} \psi_i(x) \right| dx \geq \frac{\gamma_3 I(t) \ln s}{2M^2(N)_s}.$$ (84)

Since the function $\ln s/s$ decreases monotonically for $s \geq 3$, we deduce from (84) (also see (83) and (75)) that

$$\max_{1 \leq m \leq s} \int_0^1 \left| \sum_{i=1}^{m} \psi_i(x) \right| dx \geq \frac{\gamma_3}{2} \cdot \frac{I(t)}{2M^2(N)} \left[ \frac{16K \cdot I(t)}{M^2(N)} \right]^{-1} \ln \left[ \frac{16K \cdot I(t)}{M^2(N)} \right]$$

$$\geq \frac{\gamma_3}{64K} \ln \frac{8KN}{M^2(N)},$$

and consequently that

$$\max_{1 \leq m \leq s} L_{k_m}(t) \geq c(K) \ln \frac{N}{M^2(N)} > 0.$$

This establishes (78) for an arbitrary point $t \in E$ with $m(E) > 0$. As we noted above, this also establishes Theorem 9.

We saw in §4.3 that there is a summable function whose Fourier series in the trigonometric system diverges a.e. on a period. Here we obtain a generalization.

THEOREM 10. *For every O.N.S.* $\Phi = \{\varphi_n(x)\}_{n=1}^{\infty}$, $x \in (0, 1)$, *that satisfies* (66), *there is a function* $f \in L^1(0, 1)$ *whose Fourier series in the system* $\Phi$ *diverges unboundedly at almost every point of a set* $E \subset (0, 1)$ *of positive measure.*

REMARK. It can be shown that under the hypotheses of Theorem 10 and with the supplementary condition that $\Phi$ is complete, it is in general not possible to assert the existence of a function $f \in L^1(0, 1)$ whose Fourier series in the system $\Phi$ diverges a.e. on $(0, 1)$. At the same time, for certain O.N.S. we can immediately deduce from Theorem 10 that there exist functions whose Fourier series diverge a.e. For example, if $\Phi$ is the trigonometric system,([2]) by choosing (by the use of Theorem 10) a function $f(x)$, $\|f\|_{L^1(-\pi,\pi)} = 1$, whose Fourier series diverges unboundedly on a set $E \subset (-\pi, \pi)$ with $m(E) = \alpha > 0$, we can then define a sequence $\{N_k\}_{k=1}^{\infty}$ with $N_{k+1} > 2N_k$, $k = 1, 2, \ldots$, such that the polynomials $P_k(x) = V_{N_{k+1}}(f, x) - V_{N_k}(f, x)$, $k = 1, 2, \ldots$, satisfy

$$m(E_k) > \frac{\alpha}{2}, \qquad E_k \equiv \{x \in (-\pi, \pi): S^*(P_k, x) > 2^k\}.$$

(Here $V_N(f, x)$ are the Vallée–Poussin means of $f(x)$ (see Definition 4.2), and by the properties of these means we have $\|P_k\|_{L^1(-\pi,\pi)} \leq 6, k = 1, 2, \ldots$ (see 4.(21) and 4.(22)); $S^*(P_k)$ is the majorant of the partial sums of the Fourier series of $P_k$.) It is then easy to see that, for some numbers $\theta_k \in (-\pi, \pi)$, $k = 1, 2, \ldots$, the Fourier series of the function

$$g(x) = \sum_{k=1}^{\infty} 2^{-k} P_{2k}(x - \theta_k) \in L^1(-\pi, \pi)$$

diverges unboundedly for almost for $x \in (-\pi, \pi)$ (the numbers $\theta_k$ must be chosen to satisfy

$$m\left\{ \overline{\lim_{k \to \infty}} (E_k)_{\theta_k} \right\} = 2\pi,$$

where $(E_k)_\theta = \{x + \theta(\bmod 2\pi), x \in E_k\}$).

LEMMA 1. *Let the O.N.S.* $\Phi = \{\varphi_n(x)\}_{n=1}^{\infty}$, $|\varphi_n(x)| \leq M$, *for* $n = 1, 2, \ldots$, *and* $x \in (0, 1)$, *have the property that for each* $R = 1, 2, \ldots$, *there is a function* $g_R(x)$ *with* $\|g_R\|_1 \leq 2^{-R}$ *that satisfies*

$$m\{x \in (0, 1): S^*(g_R, x) > 2^R\} > \alpha > 0;$$
$$S^*(g_R, x) = \sup_{1 \leq N < \infty} |S_N(g_R, x)| \tag{85}$$

---

([2])We consider the trigonometric system on $(-\pi, \pi)$ rather than on $(0, 1)$, but this is evidently not significant.

(in (85), the constant $\alpha$ is independent of R; $S_N(g, x)$ are the partial sums of the Fourier series of g in the system $\Phi$).

Then there is a function $f \in L^1(0, 1)$ for which the Fourier series

$$\sum_{n=1}^{\infty} c_n(f)\varphi_n(x), \qquad c_n(f) = \int_0^1 f(t)\varphi_n(t)\, dt, \qquad n = 1, 2, \ldots,$$

diverges unboundedly at each point of a set of positive measure.

PROOF. We may suppose that the partial sums of $g_R$, $R = 1, 2, \ldots$, satisfy

$$\varlimsup_{N \to \infty} |S_N(g_R, x)| < \infty \qquad \text{for almost all } x \in (0, 1); \qquad (86)$$

if not, we may take one of the functions $g_R$ as the required $f(x)$.

Let us show that, for some sequence $R_1 < R_2 < \cdots$, the requirements of the lemma are satisfied by the function

$$f(x) = \sum_{j=1}^{\infty} g_{R_j}(x). \qquad (87)$$

We shall construct the numbers $R_j$ successively; here we use the notation $G_j(x) = \sum_{s=1}^{j} g_{R_s}(x)$, $x \in (0, 1)$, $j = 1, 2, \ldots$.

Suppose that the numbers $\{R_s\}_{s=1}^{j}$ have already been constructed, with

$$m\left\{x \in (0, 1): \sup_{1 \le N \le B_j} |S_N(G_j, x)| > 2^j \right\} > \frac{\alpha}{2}, \qquad (88)$$

where $B_j$ is a function of $j$ (when $j = 1$ relation (88) is satisfied (see (85)) if $G_1 = g_1$ and $B_1$ is sufficiently large). Notice that

a) since $|c_n(g_R)| \le M\|g_R\|_1 \to 0$ as $R \to \infty$ for $n = 1, 2, \ldots$, then for sufficiently large $R$

$$\sum_{n=1}^{B_j} |c_n(g_R)\varphi_n(x)| \le 2^{-j} \qquad \text{for almost all } x \in (0, 1); \qquad (89)$$

b) since $\varlimsup_{N \to \infty} |S_N(G_j, x)| < \infty$ for almost all $x \in (0, 1)$ (see (86)), then by (85) we have, for sufficiently large $R$ and $B$,

$$m\left\{x \in (0, 1): \sup_{1 \le N \le B} |S_N(G_j + g_R, x)| > 2^{j+1} \right\} > \frac{\alpha}{2}. \qquad (90)$$

We choose $R_{j+1} > R_j$ and $B_{j+1} > B_j$ so that (89) and (90) are satisfied for $R = R_{j+1}$ and $B = B_{j+1}$. Continuing the construction, we obtain a sequence of functions $\{g_{R_j}(x)\}_{j=1}^{\infty}$ and therefore (see (87)) a function $f \in L^1(0,1)$. Moreover, by (89),

$$\sum_{n=1}^{B_j} |c_n(f - G_j)\varphi_n(x)| \leq \sum_{s=j+1}^{\infty} \sum_{n=1}^{B_j} |c_n(g_{R_s})\varphi_n(x)|$$

$$\leq \sum_{s=j+1}^{\infty} 2^{-s} < 1$$

for almost all $x \in (0,1)$, $j = 1, 2, \ldots$

Consequently we have, for $j = 1, 2, \ldots$ (see (88))

$$m\left\{x \in (0,1): \sup_{1 \leq N \leq B_j} |S_N(f,x)| > 2^j - 1\right\}$$

$$\geq m\left\{x \in (0,1): \sup_{1 \leq N \leq B_j} |S_N(G_j,x)| > 2^j\right\} > \frac{\alpha}{2}.$$

It immediately follows from the preceding inequality that the Fourier series of $f(x)$ diverges (unboundedly) on a set of positive measure. This completes the proof of Lemma 1.

LEMMA 2. *Let there be given on the unit cube* $I^N$ *in* $R^N$ $[I^N = \{x = (x_1, \cdots, x_N): 0 \leq x_i \leq 1, i = 1, \ldots, N\}]$ *a sequence of measurable functions* $\{F_p(x)\}_{p=1}^{\infty}$ *such that for almost all* $x \in I^N$

$$0 \leq G(x) \equiv \varlimsup_{p \to \infty} F_p(x) < \infty.$$

*Then there are measurable integral-valued functions* $p_s(x)$, $s = 1, 2, \ldots$, *with* $0 < p_1(x) < p_2(x) < \cdots$, *such that for almost all* $x \in I^N$

$$G(x) = \lim_{s \to \infty} F_{p_s(x)}(x). \tag{91}$$

PROOF. Choose a sequence $\{\varepsilon_s\}$ with $1 > \varepsilon_1 > \varepsilon_2 > \cdots$ and $\lim_{s \to \infty} \varepsilon_s = 0$, and for $x \in \{x \in I^N: G(x) < \infty\}$ set

$$p_1(x) = \inf\{p: F_p(x) > G(x) - \varepsilon_1\}$$

and for $s > 1$

$$p_s(x) = \inf\{p: p > p_{s-1}(x), F_p(x) > G(x) - \varepsilon_s\}.$$

Then it is clear that (91) is satisfied for almost all $x \in I^N$. It remains to show that the functions $p_s(x)$ are measurable. For $S = 1, 2, \ldots$,

$$\{x: p_1(x) = j\} = \left(\bigcap_{1 \leq p \leq j-1} \{x: F_p(x) \leq G(x) - \varepsilon_1\}\right)$$

$$\cap \{x: F_j(x) > G(x) - \varepsilon_1\}. \tag{92}$$

Similarly, for $s > 1$ and $j = 2, 3, \ldots$,

$$\{x: p_s(x) = j\} = \bigcup_{\nu=1}^{j-1} [\{x: p_{s-1}(x) = \nu\}$$

$$\cap \{x: F_p(x) \le G(x) - \varepsilon_s, \nu < p < j\}$$

$$\cap \{x: F_j(x) > G(x) - \varepsilon_s\}]. \quad (93)$$

The measurability of $p_1(x)$, $p_2(x)$, etc., follows immediately from (92) and (93). This completes the proof of Lemma 2.

We also note that if the functions $p(x) > 0$, $f_1(x), f_2(x), \ldots, x \in I^N$, are measurable, and $p(x)$ takes only integral values, then $f_{p(x)}(x)$ is also measurable.

LEMMA 3. *Let the O.N.S.* $\Phi = \{\varphi_n(x)\}_{n=1}^\infty$, $x \in (0, 1)$, *satisfy* (66). *Then for each* $N = 1, 2, \ldots$, *there is a set* $\Omega = \Omega(N)$ *of points* $(t, \theta_1, \ldots, \theta_N)$ *of* $I^{N+1}$ *with*

$$m_{N+1}(\Omega) \ge \alpha > 0 \quad (94)$$

*such that for all points* $(t, \theta_1, \ldots, \theta_N) \in \Omega$ *and some sequence (depending on $t$ but not on $\theta_i$)* $\{m_q(t)\}_{q=1}^\infty$, $Nq \le m_q(t) < N(q+1)$, *of integers, $q = 1, 2, \ldots$, we have the relation*

$$\varlimsup_{q \to \infty} \sum_{i=1}^N \sum_{n=Nq}^{m_q(t)} \varphi_n(t)\varphi_n(\theta_i) \ge BN \ln N > 0$$

*(the constants $B$ and $\alpha$ depend only on the number $M$ in* (66)).

PROOF. Let a number $N = 1, 2, \ldots$ be given, and

$$H_q = \left\{ t \in (0, 1): V_q(t) \ge \frac{1}{2}N \right\}, \qquad V_q(t) = \sum_{n=Nq}^{N(1+q)-1} \varphi_n^2(t),$$

$$q = 1, 2, \ldots.$$

By the normalization of the functions $\varphi_n(t)$ in $L^2(0, 1)$, and (66),

$$N = \int_{H_q} V_q(t)\, dt + \int_{(0,1)\setminus H_q} V_q(t)\, dt \le M^2 N m(H_q) + \frac{1}{2}N.$$

Consequently, $m(H_q) \ge (2M^2)^{-1}$, $q = 1, 2, \ldots$, and therefore

$$m(T) \ge (2M^2)^{-1}, \qquad T = \varlimsup_{q \to \infty} H_q. \quad (95)$$

By the definition of the set $T$, there is, for each $t \in T$, an increasing sequence of indices $\{q_p(t)\}_{p=1}^\infty$ such that

$$\sum_{n=Nq_p(t)}^{N(q_p(t)+1)-1} \varphi_n^2(t) \ge \frac{1}{2}N, \qquad p = 1, 2, \ldots. \quad (96)$$

Hence if we apply Corollary 4 to the system of functions $\psi_n(x) = \varphi_n(t)\varphi_n(x)$, $Nq_p(t) \le n < N(q_p(t)+1)$, $p = 1, 2, \ldots$, we can deduce that for every $t \in T$ there is a sequence $\{m_p'(t)\}_{p=1}^{\infty}$, $Nq_p(t) \le m_p'(t) < N(q_p(t)+1)$, such that

$$\int_0^1 \left| \sum_{n=N_{q_p}(t)}^{m_p(t)} \varphi_n(t)\varphi_n(x) \right| dx \ge B_1 \ln N, \qquad p = 1, 2, \ldots$$

$$(B_1 = B_1(M) > 0). \tag{97}$$

For $t \in T$ and $\theta \in (0, 1)$ we set

$$F_p(t, \theta) = \sum_{n=N_{q_p}(t)}^{m_p'(t)} \varphi_n(t)\varphi_n(\theta), \qquad p = 1, 2, \ldots, \tag{98}$$

and show that, for every $t \in T$ and every sequence $\{p_s\}_{s=1}^{\infty}$, $1 < p_1 < p_2 < \cdots$,

a)      $\varliminf_{s \to \infty} F_{p_s}(t, \theta) > 0$   for almost all $\theta \in (0, 1)$; $\tag{99}$

b)      $\varliminf_{s \to \infty} F_{p_s}(t, \theta) \, d\theta \ge B_2 \ln N$   $(B_2 = B_2(M) > 0)$. $\tag{100}$

In fact, for a fixed $t \in T$ the sequence (98) of functions of $\theta$ forms an orthogonal system, bounded by the constant $NM^2$; consequently (see Theorem 1.5) we have, for every measurable subset $E$ of $(0, 1)$,

$$\lim_{s \to \infty} \int_E F_{p_s}(t, \theta) \, d\theta = 0, \tag{101}$$

and consequently

$$m\{\theta \in (0, 1) : \varliminf_{s \to \infty} F_{p_s}(t, \theta) < 0\} = 0.$$

In proving (100), we denote by $F^+(t, \theta)$, $p = 1, 2, \ldots$, the positive part of $F_p(t, \theta)$:

$$F_p^+(t, \theta) = \max\{F_p(t, \theta), 0\}.$$

Then, if we successively apply (99), Beppo Levi's theorem,($^3$) and (101), we obtain

$$
\begin{aligned}
\int_0^1 \varliminf_{s\to\infty} F_{p_s}(t,\theta)\, d\theta &= \int_0^1 \varliminf_{s\to\infty} F_{p_s}^+(t,\theta)\, d\theta \\
&= \int_0^1 \lim_{s\to\infty}\left\{\sup_{j\ge\varepsilon} F_{p_j}^+(t,\theta)\right\} d\theta \\
&= \lim_{s\to\infty}\int_0^1 \left\{\sup_{j\ge s} F_{p_j}^+(t,0)\right\} d\theta \ge \varliminf_{s\to\infty}\int_0^1 F_{p_s}^+(t,\theta)\, d\theta \\
&= \frac{1}{2}\varliminf_{s\to\infty}\int_0^1 |F_{p_s}(t,\theta)|\, d\theta,
\end{aligned}
$$

from which (100) follows by (97) (also see (98)).

Let us introduce the functions

$$
\Phi_p(t,\theta_1,\ldots,\theta_N) = \sum_{i=1}^{N} F_p(t,\theta_i), \qquad p = 1,2,\ldots, \tag{102}
$$

defined for each given $t \in T$ in $I^N = \{(\theta_1,\ldots,\theta_N): 0 \le \theta_i \le 1,\ i = 1,\ldots,N\}$.

In order to estimate the functions (102), we introduce a family of subsequences of the integers:

$$
\{p_s^{(i)}(t,\theta_1,\ldots,\theta_i)\}_{s=1}^{\infty}, \qquad t \in T,\ 0 \le \theta_i \le 1,\ i = 1,\ldots,N.
$$

We choose the sequences $\{p_s^{(1)}(t,\theta_1)\}_{s=1}^{\infty}$, $\theta_1 \in [0,1]$ (for a given $t \in T$), so that

1) $\varlimsup_{p\to\infty} F_p(t,\theta_1) = \lim_{s\to\infty} F_{p_s^{(1)}(t,\theta_1)}(t,\theta_1)$ for almost all $\theta \in (0,1)$;

2) the functions $p_s^{(1)}(t,\theta_1)$ of $\theta_1$ are measurable.

It follows from Lemma 2 that this choice is possible.

Now, for each $\theta_2 \in [0,1]$, again using Lemma 2, and also the remark after that lemma, we select from $\{p_s^{(1)}(t_1,\theta_2)\}_{s=1}^{\infty}$ a subsequence $\{p_s^{(2)}(t,\theta_1,\theta_2)\}_{s=1}^{\infty}($^4)$ so that

1) $\varlimsup_{s\to\infty} F_{p_s^{(1)}(t,\theta_1)}(t,\theta_2) = \lim_{s\to\infty} F_{p_s^{(2)}(t,\theta_1,\theta_2)}(t,\theta_2)$ for almost all $(\theta_1,\theta_2) \in I^2$;

2) the functions $p_s^{(2)}(t,\theta_1,\theta_2)$ of $\theta_1$ and $\theta_2$, $s = 1,2,\ldots$, are measurable.

---

($^3$)The possibility of applying Levi's theorem follows from the uniform boundedness of the functions $F_p(t,\theta)$.

($^4$)That is, $p_s^{(2)}(t,\theta_1,\theta_2) = p_{\nu_s(\theta_2)}^{(1)}(t,\theta_1)$, $s = 1,2,\ldots$, where $\{\nu_s(\theta_2)\}$ is an increasing sequence of positive integers depending on $\theta_2$.

Continuing this process up to index $N$, we obtain a family of sequences $\{p_s^{(i)}(t, \theta_1, \ldots, \theta_i)\}_{s=1}^{\infty}$, $i = 1, \ldots, N$, for which in each case

$$\{p_s^{(i+1)}(t, \theta_1, \ldots, \theta_{i+1})\}_{s=1}^{\infty} \subset \{p_s^{(i)}(t, \theta_1, \ldots, \theta_i)\}_{s=1}^{\infty}, \qquad i = 1, \ldots, N-1,$$

and the following relation is satisfied:

$$\varliminf_{s \to \infty} F_{p_s^{(i)}(t,\theta_1,\ldots,\theta_i)}(t, \theta_{i+1}) = \lim_{s \to \infty} F_{p_s^{(i+1)}(t,\theta_1,\ldots,\theta_{i+1})}(t, \theta_{i+1})$$

$$\text{for almost all } (\theta_1, \ldots, \theta_{i+1}) \in I^{i+1}. \tag{103}$$

Here the functions $p_s(t, \theta_1, \ldots, \theta_i)$, $s = 1, 2, \ldots$, are measurable functions of $\theta_1, \ldots, \theta_i$.

By (103) (also see (102)), we have for almost all $(\theta_1, \ldots, \theta_N) \in I^N$

$$\varliminf_{p \to \infty} \Phi_p(t, \theta_1, \ldots, \theta_N) \geq \lim_{s \to \infty} \sum_{i=1}^{N} F_{p_s^{(N)}(t,\theta_1,\ldots,\theta_N)}(t, \theta_i)$$

$$= \sum_{i=1}^{N} \lim_{s \to \infty} F_{p_s^{(N)}(t,\theta_1,\ldots,\theta_N)}(t, \theta_i)$$

$$= \sum_{i=1}^{N} \varliminf_{s \to \infty} F_{p_s^{(i-1)}(t,\theta_1,\ldots,\theta_{i-1})}(t, \theta_i) \tag{104}$$

(here we denote by $\{p_s^{(0)}(t)\}$ the sequence $p_s^{(0)}(t) = s, s = 1, 2, \ldots$).

Moreover, we obtain from (99) and (100) [using the measurability of $F_{p_s^{(i-1)}(t,\theta_1,\ldots,\theta_{i-1})}(t, \theta_i)$,] the existence of measurable functions $G_i(t, \theta_1, \ldots, \theta_i)$, $i = 1, \ldots, N$, such that

a) $$0 \leq G_i(t, \theta_1, \ldots, \theta_i) \leq \varliminf_{s \to \infty} F_{p_s^{(i-1)}(t,\theta_1,\ldots,\theta_{i-1})}(t, \theta_i)$$
$$\text{for almost } (\theta_1, \ldots, \theta_i) \in I^i; \tag{105}$$

b) $$\int_0^1 G_i(t, \theta_1, \ldots, \theta_i)\, d\theta_i = B_2 \ln N \geq 0.$$

By (105), part b),

$$\int_0^1 \cdots \int_0^1 \left[ \sum_{i=1}^{N} G_i(t, \theta_1, \ldots, \theta_i) \right] d\theta_1 \ldots d\theta_N = B_2 N \ln N. \tag{106}$$

On the other hand, using (105), part a), (106), and also the inequality $|F_p(t, \theta)| \leq NM^2$, $t, \theta \in [0, 1]$, we obtain

$$
\int_0^1 \cdots \int_0^1 \left[ \sum_{i=1}^N G_i(t, \theta_1, \ldots, \theta_i) \right]^2 d\theta_1 \ldots d\theta_N
$$

$$
= \int_0^1 \cdots \int_0^1 \left[ \sum_{i=1}^N G_i^2(t, \theta_1, \ldots, \theta_i) \right] d\theta_1 \ldots d\theta_N
$$

$$
+ 2 \sum_{i=1}^{N-1} \sum_{j=i+1}^N \int_0^1 \cdots \int_0^1 G_i(t, \theta_1, \ldots, \theta_i)
$$

$$
\times G_j(t, \theta_1, \ldots, \theta_j) \, d\theta_1 \ldots d\theta_N
$$

$$
\leq B_2 N^2 M^2 \ln N + 2 B_2^2 N^2 \ln N \leq B_3 N^2 \ln^2 N,
$$

$$
B_3 = B_3(M) > 0. \quad (107)
$$

It follows from (106) and (107), and Lemma 1 for Theorem 2.7, that for each $t \in T$

$$
m_N \left\{ (\theta_1, \ldots, \theta_N) \colon \sum_{i=1}^N G_i(t, \theta_1, \ldots, \theta_i) \geq \frac{B_2}{4} N \ln N \right\} \geq \frac{B_2^2}{8 B_3},
$$

and therefore (see (105), (104), and (95))

$$
m_{N+1}(\Omega) \equiv m_{N+1} \left\{ (t, \theta_1, \ldots, \theta_N) \colon t \in T \text{ and} \right.
$$

$$
\left. \varlimsup_{p \to \infty} \sum_{i=1}^N F_p(t, \theta_i) \geq \frac{B_2}{4} N \ln N \right\}
$$

$$
\geq m(T) \frac{B_2^2}{8 B_3} \geq \frac{B_2^2}{16 M B_3} \geq 0.
$$

We obtain the conclusion of Lemma 3 from the preceding inequality and the definition of the functions $F_p(t, \theta)$ (see (98)).

PROOF OF THEOREM 10. For $N = 1, 2, \ldots$, we denote by $E^N$ the subset of $I^N$ consisting of the points $\theta = (\theta_1, \ldots, \theta_N)$, $0 < \theta_i < 1$, $i = 1, \ldots, N$, that satisfy

$$
\lim_{h \to 0} h^{-1} \int_{\theta_i}^{\theta_i + h} \varphi_n(x) \, dx = \varphi_n(\theta_i), \qquad i = 1, 2, \ldots, N, \ n = 1, \ldots. \quad (108)
$$

Then $m_N(E^N) = m_N(I^N) = 1$ (by the theorem on the differentiability of an indefinite integral; also see the remark on Theorem 2 in Appendix 1), and by Lemma 3 we may select, for $N = 1, 2, \ldots$, a set $\theta = (\theta_1^{(N)}, \ldots, \theta_N^{(N)}) \in E^N$ such that

$$\lim_{q \to \infty} \frac{1}{N} \sum_{i=1}^{N} \sum_{n=Nq}^{m_q(t)} \varphi_n(t)\varphi_n(\theta_i^{(N)}) \geq B \ln N, \qquad (109)$$

$$Nq \leq m_q(t) < N(q+1), \quad q = 1, 2, \ldots,$$

for every $t \in T(N) \subset (0, 1)$ with $m(T(N)) > \alpha/2 > 0$.

It follows from (109) that, for sufficiently large $q_0 = q_0(N, \Phi)$,

$$m\left\{ t \in (0, 1): \sup_{1 \leq q < q_0} \sup_{Nq \leq m < N(q+1)} \frac{1}{N} \sum_{i=1}^{N} \varphi_n(t)\varphi_n(\theta_i^{(N)}) > \frac{B_2}{2} \ln N \right\} > \frac{\alpha}{3},$$

and therefore

$$m\left\{ t \in (0, 1): \sup_{1 \leq m < Nq_0} \sum_{n=1}^{m} \varphi_n(t) \left[ \frac{1}{N} \sum_{i=1}^{N} \varphi_n(\theta_i^{(N)}) \right] > \frac{B_2}{4} \ln N \right\} > \frac{\alpha}{3}. \qquad (110)$$

For sufficiently small $h > 0$, consider on $(0, 1)$ the function

$$f_{h,N}(x) = \frac{1}{N} \sum_{i=1}^{N} h^{-1} \chi_{(\theta_i^{(N)}, \theta_i^{(N)}+h)}(x), \qquad \|f_{h,N}\|_1 = 1$$

(here $\chi_{(a,b)}(x)$ is the characteristic function of the interval $(a, b)$).

By (108), the Fourier coefficients of $f_{h,N}(x)$ satisfy

$$\lim_{h \to +0} c_n(f_{h,N}) = \frac{1}{N} \sum_{i=1}^{N} \varphi_n(\theta_i^{(N)}), \qquad n = 1, 2, \ldots,$$

and consequently we have, for the partial sums of the Fourier series of the difference $f_{h,N}(x) - S_m(f_{h,N}, x)$, $m = 1, 2, \ldots,$

$$\lim_{h \to +0} \left\| S_m(f_{h,N}, x) - \sum_{n=1}^{m} \varphi_n(x) \left[ \frac{1}{N} \sum_{i=1}^{N} \varphi_n(\theta_i^{(N)}) \right] \right\|_2 = 0. \qquad (111)$$

It follows from (110) and (111) that

$$m\left\{ x \in (0, 1): \sup_{1 \leq m \leq N_{f_0}} S_m(f_{h,N}, x) > \frac{B_2}{5} \ln N \right\} > \frac{\alpha}{4} \qquad (112)$$

for $N = 1$ and $h < h_0 = h_0(N, \Phi)$. It remains only to apply Lemma 1: by (112), for each $R = 1, 2, \ldots$, we can find $N = N(R)$ and $h = h(R) > 0$ such that $g_R(x) = (\ln^{-1/2} N)f_{h,N}(x)$ has norm $\|g_R\|_1 \leq 2^{-R}$ and satisfies (85). This completes the proof of Theorem 10.

The last result of this chapter is concerned with questions of absolute convergence of orthogonal series.

We showed in Chapter 4 (see Corollaries 4.5 and 4.6) that either of the following conditions is sufficient for the absolute convergence of the trigonometric Fourier series of a function $f \in C(-\pi, \pi)$:

a) $f \in \text{Lip}(\frac{1}{2} + \varepsilon)$ for some $\varepsilon > 0$;

b) $f(x)$ has bounded variation and its modulus of continuity $\omega(\delta, f) = O(\log^{-(2+\varepsilon)}(1/\delta))$ as $\delta \to 0$ $(\varepsilon > 0)$.

In §2 of this chapter it was observed (see Corollary 2) that condition a) cannot be significantly weakened, in the sense that for every C.O.N.S. $\{\varphi_n(x)\}_{n=1}^{\infty}$ there is a function $F \in \text{Lip} \frac{1}{2}$ such that $F \overset{L^2}{=} \sum_{n=1}^{\infty} a_n \varphi_n$, $\sum_{n=1}^{\infty} |a_n| = \infty$. As for condition b), for some O.N.S. it is unnecessarily restrictive. For example, for the Fourier–Haar coefficients of a function $f(x)$ of bounded variation we have

$$\sum_{n=2^k+1}^{2^{k+1}} |c_n(f, \chi)| = O(2^{-k/2}) \quad \text{as } k \to \infty \tag{113}$$

(this follows from 3.(17) and the inequality $\omega_1(\delta, f) \le 4\delta V(f)$, where $V(f)$ is the total variation of $f$ (see Appendix 1.1)), and therefore the Fourier–Haar series of a function of bounded variation converges absolutely. It turns out, however, that this property of the Haar system is not possessed by any C.O.N.S. that consists of uniformly bounded functions. Moreover, we have the following theorem.

THEOREM 11. *For every C.O.N.S.* $\Phi = \{\varphi_n(x)\}_{n=1}^{\infty}$, $x \in (0, 1)$, *that satisfies* (66), *there is a continuous function* $F(x)$ *of bounded variation with* $F(0) = F(1) = 0$ *and modulus of continuity* $\omega(\delta, F) = O(\ln^{-2} 1/\delta)$ *as* $\delta \to 0$, *for which the series of absolute values of the Fourier coefficients diverges*:

$$\sum_{n=1}^{\infty} |c_n(F)| = \infty, \qquad c_n(F) = \int_0^1 F(x)\varphi_n(x)\,dx, \qquad n = 1, 2, \dots.$$

PROOF. For $f \in C(0, 1)$, we denote by $\|f\|_{H^\omega}$ the number (which satisfies the axioms for a norm)

$$\|f\|_{H^\omega} = \|f\|_C + \sup_{0 \le x < x' \le 1} |f(x) - f(x')| \ln^2(x' - x),$$

and by $V(f)$ the total variation of $f(x)$ on $(0, 1)$. It is clear that $\|f\|_{H^\omega} < \infty$ if and only if $\omega(\delta, f) = O(\ln^{-2} \delta)$ as $\delta \to 0$.

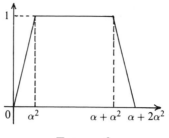

FIGURE 2

For each $\alpha \in (0, 1/4)$, we define a function $\chi_\alpha(x)$, $x \in R^1$, by setting (Figure 2)

$$\chi_\alpha(x) = \begin{cases} 1 & \text{if } \alpha^2 \leq x \leq \alpha + \alpha^2, \\ 0 & \text{if } x \leq 0 \text{ or } x \geq \alpha + 2\alpha^2, \\ \text{linear and continuous on } [0, \alpha^2] \text{ and} \\ \quad [\alpha + \alpha^2, \alpha + 2\alpha^2]. \end{cases}$$

It is easily verified that $V(\chi_\alpha) = 2$ and that, for every $t \in (0, 1/2)$,

$$\|\chi_\alpha(x - t)\|_{H^\omega} = \|\chi\alpha(x)\|_{H^\omega} \leq K \ln^2 \alpha, \tag{114}$$

where $K$ is an absolute constant.

We may suppose that, for every $\alpha \in (0, \frac{1}{4})$ and $t \in (0, \frac{1}{2})$,

$$\sum_{n=1}^{\infty} |c_n(\chi_\alpha(x - t))| < \infty, \tag{115}$$

since otherwise we may take $F(x)$ to be one of the functions $\chi_\alpha(x - t)$. Moreover, if

$$\sup_{0 < \alpha < \frac{1}{4}, 0 < t < \frac{1}{2}} \frac{\sum_{n=1}^{\infty} |c_n(\chi_\alpha(x - t))|}{\ln^2 \alpha} = \infty,$$

i.e.,

$$\sup_{0 < \alpha < \frac{1}{4}, 0 < t < \frac{1}{2}} \frac{\sum_{n=1}^{\infty} |c_n(\chi_\alpha(x - t))|}{\|\chi_\alpha(x - t)\|_{H^\omega}} = \infty, \tag{116}$$

the required function $F(x)$ can be taken in the form

$$F(x) = \sum_{k=1}^{\infty} F_k(x), \qquad F_k(x) = 2^{-k} \ln^{-2} \alpha_k \cdot \chi_{\alpha_k}(x - t_k), \tag{117}$$

where $\{\alpha_k\}$ is a sequence that tends to zero, and $t_k \in (0, \frac{1}{2})$, $k = 1, 2, \ldots$. In fact, since

$$\sum_{n=1}^{N} |c_n(\chi_\alpha(x - t))| \leq NM\|\chi_\alpha\|_1 \leq 2NM\alpha \to 0 \quad \text{as } \alpha \to 0,$$

for $N = 1, 2, \ldots$, we may, by (116) (also see (115)), successively select the numbers $\alpha_k$ and $t_k$ for $k = 1, 2, \ldots$ so that

1) $$\sum_{n=1}^{\infty} |c_n(\chi_{\alpha_k}(x - t_k))| \geq 3^k \ln^2 \alpha_k, \qquad k = 1, 2, \ldots;$$

2) $$\sum_{n=1}^{\infty} \left| c_n \left( \sum_{s=1}^{k} F_s \right) \right| \geq \frac{1}{2} \sum_{n=1}^{\infty} |c_n(F_k)|, \qquad k = 1, 2, \ldots,$$

and then, for the function (117), we have

$$F(0) = F(1) = 0, \qquad \|F\|_{H^\omega} < \infty, \qquad V(f) < \infty, \qquad \sum_{n=1}^{\infty} |c_n(F)| = \infty.$$

Consequently it is enough to prove the theorem in the case when

$$\sup_{0 < \alpha < \frac{1}{4}, 0 < t < \frac{1}{2}} \left\{ \ln^{-2} \alpha \sum_{n=1}^{\infty} |c_n(\chi_\alpha(x - t))| = C_0 < \infty \right. . \tag{118}$$

LEMMA 1 (fundamental lemma). *If* (118) *is satisfied for a system* $\Phi = \{\varphi_n(x)\}_{n=1}^{\infty}$ *that satisfies the hypotheses of Theorem* 11, *then the inequality*

$$\int_0^{1/2} \left\{ \sum_{n:\, |c_n(\chi_\alpha(x-t))| > \alpha^4} |c_n(\chi_\alpha(x - t))| \right\} dt \geq c \ln \frac{1}{\alpha} \tag{119}$$

*holds for* $\alpha \leq \alpha_0 < \frac{1}{4}$ (*the constants* $\alpha_0 > 0$ *and* $c > 0$ *depend on* $\Phi$).

Notice that if $\Phi$ is the trigonometric system then (119) is exact in order, since in this case

$$\sum_{n=1}^{\infty} |c_n(\chi_\alpha(x - t))| \asymp \ln \frac{1}{\alpha}.$$

for every $t$.

Proof of Lemma 1. Supposing that $\alpha \in (0, \frac{1}{4})$ is given, we set, for $n = 1, 2, \ldots$ and $t \in (0, \frac{1}{2})$,

$$f_n(t) = c_n(\chi_\alpha(x - t)) = \int_0^1 \chi_\alpha(x - t)\varphi_n(x) \, dx. \tag{120}$$

Since the functions $\varphi_n(x)$ are bounded (see (66)), for $t, t' \in (0, \frac{1}{2})$ we have

$$|f_n(t) - f_n(t')| \leq \int_0^1 |\chi_\alpha(x - t) - \chi_\alpha(x - t')| \cdot |\varphi_n(x)| \, dx$$

$$\leq M \int_0^1 |\chi_\alpha(x - t) - \chi_\alpha(x - t')| \, dx \leq 4M|t - t'|. \quad (121)$$

It also follows from the definition of $f_n(t)$, by Parseval's equation, that, for $0 \leq t < t' < \frac{1}{2}$,

$$\sum_{n=1}^\infty [f_n(t) - f_n(t')]^2 = \int_0^1 [\chi_\alpha(x - t) - \chi_\alpha(x - t')]^2 \, dx. \quad (122)$$

But it is easily seen that if $2\alpha^2 < t' - t < \alpha$, $0 \leq t < t' < \frac{1}{2}$, then

$$\int_0^1 [\chi_\alpha(x - t) - \chi_\alpha(x - t')]^2 \, dx \geq \frac{1}{3}(t' - t).$$

Hence it follows from (122) that

$$\sum_{n=1}^\infty \int_{2\alpha^2}^\alpha h^{-2} \int_0^{1/2-h} [f_n(t + h) - f_n(t)]^2 \, dt \, dh$$

$$= \int_{2\alpha^2}^\alpha h^{-2} \int_0^{1/2-h} \sum_{n=1}^\infty [f_n(t + h) - f_n(t)]^2 \, dt \, dh$$

$$\geq \frac{1}{3} \int_{2\alpha^2}^\alpha h^{-2} \int_0^{1/2-h} h \, dt \, dh \geq \frac{1}{12} \int_{2\alpha^2}^\alpha h^{-1} \, dh \geq \frac{1}{24} \ln \frac{1}{\alpha}. \quad (123)$$

If we apply (63) to the functions $\frac{1}{4M} f_n(x)$, $n = 1, 2, \ldots$, with $\alpha = \frac{1}{2}$, $\beta = \alpha^2$, we obtain (see (121) and (124))

$$\sum_{n=1}^\infty \left\{ \frac{1}{4M} \int_{\{t \in (0, \frac{1}{2}): \, |f_n(t)| > 4M\alpha^4\}} |f_n(t)| \, dt + \frac{\alpha^2}{4M} \int_0^{1/2} |f_n(t)| \, dt \right\}$$

$$\geq \frac{1}{5000 M^2} \ln \frac{1}{\alpha},$$

and therefore $(M \geq 1)$

$$\sum_{n=1}^\infty \int_0^{1/2} |f_n(t)| \chi \left\{ t \in (0, \tfrac{1}{2}) : |f_n(t)| > \alpha^4 \right\} \, dt$$

$$\geq c' \ln \frac{1}{\alpha} - \int_0^{1/2} \alpha^2 \sum_{n=1}^\infty |f_n(t)| \, dt, \quad (124)$$

where the constant $c' > 0$ depends only on $M$ (in (124), $\chi(E)$ is the characteristic function of $E$).

If we change the order of integration and summation on the left-hand side of (124), and use (118), which is satisfied by the hypotheses of Lemma 1, we obtain

$$\int_0^{1/2} \left\{ \sum_{n:\, |c_n(\chi_\alpha(x-t))|>\alpha^4} |c_n(\chi_\alpha(x-t))| \right\} dt \geq c' \ln \frac{1}{\alpha} - C_0\alpha^2 \ln^2 \alpha$$

$$\geq \frac{c'}{2} \ln \frac{1}{\alpha},$$

if $\alpha \leq \alpha_0$ is sufficiently small. This completes the proof of Lemma 1. We continue the proof of Theorem 11. If we apply the equation

$$\int_0^{1/2} g(t)\, dt = \int_0^{1/2m} \sum_{p=0}^{m-1} g\left(t + \frac{p}{2m}\right) dt, \qquad m = 1, 2, \ldots,$$

which is evidently valid for every $g \in L^1(0,1)$, we find from Lemma 1 that for each $m = 1, 2, \ldots$ and $\alpha \leq \alpha_0$ there is a point $t_0 = t_0(m, \alpha) \in (0, 1/2m)$ such that

$$\frac{1}{m} \sum_{p=0}^{m-1} \left\{ \sum_{n:\, |c_n(\chi_\alpha(x-\frac{p}{2m}-t_0))|>\alpha^4} \left| c_n\left(\chi_\alpha\left(x - \frac{p}{2m} - t_0\right)\right) \right| \right\} \geq c \ln \frac{1}{\alpha} > 0. \tag{125}$$

Notice that if, for arbitrary $\alpha \in (0, \frac{1}{4})$, $m = 1, 2, \ldots$, and $t \in (0, 1/2m)$, we define the set of integers

$$E(t, m, \alpha) = \left\{ n: \max_{0 \leq p < m} \left| c_n\left(\chi_\alpha\left(x - \frac{p}{2m} - t\right)\right) \right| > \alpha^4 \right\}, \tag{126}$$

then by using Parseval's equation

$$\sum_{n=1}^{\infty} c_n^2\left(\chi_\alpha\left(x - \frac{p}{2m} - t\right)\right) = \left\| \chi_\alpha\left(x - \frac{p}{2m} - t\right) \right\|_2^2 \leq C\alpha,$$

we find that $|\operatorname{card} E(t, m, \alpha)|(\alpha^4)^2 \leq C\alpha m$, i.e.,

$$\operatorname{card} E(t, m, \alpha) \leq Cm\alpha^{-7}, \tag{127}$$

where $C$ is an absolute constant.

We define the numbers

$$\alpha_k = (2^{10^k})^{-1}\alpha_0, \qquad m_k = k^{-1/2}10^{2k} \tag{128}$$

and then, for each $k = 1, 2, \ldots$ and $y \in (0, 1)$, we consider the function

$$u_{k,y}(x) = \frac{1}{m_k} \sum_{p=0}^{m_k-1} r_{p+1}(y) \chi_{\alpha_k} \left( x - \frac{p}{2m_k} - t_k \right), \qquad x \in (0, 1), \qquad (129)$$

where $t_k \equiv t_0(m_k, \alpha_k) \in (0, 1/2m_k)$ is such that (125) is valid, and $r_p(y)$ is a Rademacher function. Then for all $y \in (0, 1)$ and $k = 1, 2, \ldots$,

$$u_{k,y}(0) = u_{k,y}(1) = 0 \quad \text{and} \quad V(u_{k,y}) \leq 2. \qquad (130)$$

Moreover, the terms in (129) have disjoint supports (this follows from the inequalities $2\alpha_k < m_k^{-1}$, $k = 1, 2, \ldots$ (see (128))). Therefore it follows at once from the definition of the functions $\chi_\alpha(x)$ that, for every $y \in (0, 1)$ and $k = 1, 2, \ldots$,

a) $\|u_{k,y}\|_c \leq m_k^{-1}$;

b) $\|u'_{k,y}\|_\infty \equiv \sup_{0 \leq x' < x \leq 1} \left| \frac{u_{k,y}(x) - u_{k,y}(x')}{x - x'} \right| \leq \alpha_k^{-2} m_k^{-1}$;

$$(131)$$

c) $\|u_{k,y}\|_1 \leq 2\alpha_k$.

If $E_k$ denotes the set of integers $E_k = E_k(t_k, m_k, \alpha_k)$, $k = 1, 2, \ldots$ (see (126)), then if we use Theorem 2.7 and inequality (125) (see also (126)), we find that, for $k = 1, 2, \ldots$,

$$\int_0^1 \left\{ \sum_{n \in E_k} |c_n(u_{k,y})| \right\} dy$$

$$= \sum_{n \in E_k} \int_0^1 \left| \frac{1}{m_k} \sum_{p=0}^{m_k-1} r_{p+1}(y) c_n \left( \chi_{\alpha_k} \left( x - \frac{p}{2m_k} - t_k \right) \right) \right| dy$$

$$\geq \frac{c}{m_k} \sum_{n \in E_k} \left\{ \sum_{p=0}^{m_k-1} c_n^2 \left( \chi_{\alpha_k} \left( x - \frac{p}{2m_k} - t_k \right) \right) \right\}^{1/2}$$

$$\geq \frac{c}{m_k^{3/2}} \sum_{n \in E_k} \sum_{p=0}^{m_k-1} \left| c_n \left( \chi_{\alpha_k} \left( x - \frac{p}{2m_k} - t_k \right) \right) \right|$$

$$\geq \frac{c}{m_k^{1/2}} \ln \frac{1}{\alpha_k} \geq c \left( \frac{k^{1/2}}{10^{2k}} \right)^{1/2} 10^k = ck^{1/4},$$

where $c > 0$ is an absolute constant.

Consequently, for $k = 1, 2, \ldots$ and some $y_0 = y_0(k) \in (0, 1)$, the function $g_k(x) \equiv u_{k,y_0}(x)$ satisfies (also see (127))

$$\sum_{n \in E_k} |c_n(g_k)| > ck^{1/4} > 0, \qquad \operatorname{card} E_k \le Cm_k \alpha_k^{-7} < C' \alpha_k^{-8}. \tag{132}$$

Let us show finally that, for some sequence $\varepsilon_k = \pm 1$, $k = 1, 2, \ldots$, function

$$F(x) = \sum_{k=1}^{\infty} \frac{\varepsilon_k}{k^{5/4}} g_k(x) \tag{133}$$

is what was required.

In the first place, by (131), part a), and (130), we have, independently of the choice of the numbers $\varepsilon_k = \pm 1$ in (133),

$$F \in C(0, 1), \qquad \|F\|_c \le 2, \qquad F(0) = F(1) = 0,$$

$$V(F) \le 2 \sum_{k=1}^{\infty} k^{-5/4} < \infty. \tag{134}$$

Let us prove that, for arbitrary $\varepsilon_k = \pm 1$,

$$\|F\|_{H^\omega} \le C = C(\Phi) < \infty. \tag{135}$$

Let $h \in (0, \alpha_1)$ and $s$ satisfy $\alpha_{s+1}^2 \le h < \alpha_s^2$. Then since $|\ln \alpha_k| \le C'(\Phi) \cdot 10^k$, $k = 1, 2, \ldots$ (see (128)), we obtain, by (131), part b),

$$\left| \sum_{k=1}^{s} \frac{\varepsilon_k}{k^{5/4}} g_k(x+h) - \sum_{k=1}^{s} \frac{\varepsilon_k}{k^{5/4}} g_k(x) \right| \le \sum_{k=1}^{s} k^{-5/4} m_k^{-1} \alpha_k^{-2} h$$

$$\le \alpha_s^2 \sum_{k=1}^{s} k^{-5/4} m_k^{-1} \alpha_k^{-2}$$

$$\le C \alpha_s^2 (s^{-5/4} m_s^{-1} \alpha_s^{-2})$$

$$< C 10^{-2s} < C_1 \ln^{-2} \alpha_{s+1}$$

$$\le C_1 \ln^{-2} h, \qquad x, x+h \in [0, 1]. \tag{136}$$

Furthermore (see (133), part a), and (128)),

$$\left| \sum_{k=s+1}^{\infty} \frac{\varepsilon_k}{k^{5/4}} g_k(x+h) - \sum_{k=s+1}^{\infty} \frac{\varepsilon_k}{k^{5/4}} g_k(x) \right| \le 2 \sum_{k=s+1}^{\infty} \frac{1}{k^{5/4}} \|g_k\|_c$$

$$\le 2 \sum_{k=s+1}^{\infty} k^{-3/4} 10^{-2k}$$

$$\le 10^{-2s} < C_2 \ln^{-2} h,$$

$$x, x+h \in [0, 1]. \tag{137}$$

Combining (136) and (137) (also see the definition of the norm $\|F\|_{H^\omega}$ and (134)), we obtain (135).

It remains only to choose the numbers $\varepsilon_k = \pm 1$ so that

$$\sum_{n=1}^{\infty} |c_n(F)| = \infty.$$

For this purpose we observe that, for all $n$ and $k$,

$$|c_n(g_k)| \leq M\|g_k\|_1 \leq 2M\alpha_k = B_1(2^{10^k})^{-1} \tag{138}$$

(see (66) and (131), part c)). Hence by the inequality

$$\operatorname{card} E_{k^-} \leq C'\alpha_k^{-8} = C \cdot 2^{8 \cdot 10^k}$$

(see (132)) we deduce that, when $k' > k$,

$$\sum_{n \in E_k} |c_n(g_{k'})| \leq [\operatorname{card} E_k] \cdot B_1(2^{10^k})^{-1} \leq B_2(2^{10^{k'}-1})^{-1},$$

and consequently

$$\sum_{k=1}^{\infty} \sum_{k'=k+1}^{\infty} \sum_{n \in E_k} |c_n(g_{k'})| \leq B_2 \sum_{k=1}^{\infty} \sum_{k'=k+1}^{\infty} (2^{10^{k'}-1})^{-1} \leq B_3 < \infty. \tag{139}$$

Moreover, if we set $\Omega_k = E_k \setminus \bigcup_{s=1}^{k-1} E_s$ for $k = 2, 3, \ldots$, we have, for $k > k_0$, by (132) and (138),

$$\sum_{n \in \Omega_k} |c_n(g_k)| \geq ck^{1/4} - \sum_{\substack{n \in \bigcup_{s=1}^{k-1} E_s}} |c_n(g_k)|$$

$$\geq ck^{1/4} - \left[\operatorname{card}\left(\bigcup_{s=1}^{k-1} E_s\right)\right] B_1 \cdot (2^{10^k})^{-1}$$

$$\geq ck^{1/4} - B_4 \cdot 2^{8 \cdot 10^{k-1}} \cdot (2^{10^k})^{-1} \geq B_5 k^{1/4} > 0. \tag{140}$$

We may choose $\varepsilon_k = \pm 1$ successively so that at each step (i.e., for $k = 2, 3, \ldots$) we have the inequality

$$\sum_{n \in \Omega_k} \left| c_n \left( \sum_{s=1}^{k-1} \frac{\varepsilon_s}{k^{5/4}} g_s(x) + \frac{\varepsilon_k}{k^{5/4}} g_k(x) \right) \right| \geq \frac{1}{k^{5/4}} \sum_{n \in \Omega_k} |c_n(g_k)| \tag{141}$$

(this is always possible, since $\max(\|a + a'\|, \|a - a'\|) \geq \|a'\|$ in any normed space). Then by using (139)–(141) we obtain

$$\sum_{n=1}^{\infty} |c_n(F)| \geq \sum_{k=2}^{\infty} \sum_{n \in \Omega_k} |c_n(F)|$$

$$\geq \sum_{k=k}^{\infty} k^{-5/4} \sum_{n \in \Omega_k} c_n(g_k) - B_3$$

$$\geq B_5 \sum_{k=k_0}^{\infty} \frac{1}{k} - B_3 = \infty.$$

This completes the proof of Theorem 11.

## CHAPTER X

# Some Theorems on the Representation of Functions by Orthogonal Series

In this chapter we consider an additional aspect of the theory of orthogonal series, namely theorems on the representation of measurable functions by series.

The problems which will be discussed below can be described schematically as follows. Let there be given a system $\Phi = \{\varphi_n(x)\}_{n=1}^\infty$ of functions and a class $F = \{f(x)\}$ of measurable functions, and also some specified kind of convergence, which we denote by $\rho$ (for example, convergence a.e., convergence in some metric space, etc.) We are to investigate the possibility of finding, for each $f \in F$, a series

$$\sum_{n=1}^\infty a_n \varphi_n(x),$$

that converges to $f(x)$ in the sense of $\rho$-convergence.

Representation theorems can be considered as analogs of one of the very first results of the theory of orthogonal series, Theorem 1.4, according to which for every C.O.N.S. $\Phi$ and every $f \in L^2$ there is a series in the system $\Phi$ which converges to $f(x)$ in $L^2$. Here we must emphasize the difference between representation theorems and results (which are natural when $\rho$ is a metric) on the completeness of $\Phi$ in the class $F$, i.e., on the possibility of arbitrarily close approximation by polynomials in $\Phi$ to each $f$ in $F$ in the $\rho$-metric. The necessity of obtaining series whose partial sums converge to $f(x)$ changes the nature of the problem in an essential way and determines the character of theorems on the representation of the functions by series.

All the results in Chapter 10, except for Theorem 5, are related to the fundamental (but of course not the only possible) case of this theme when $\Phi$ is a C.O.N.S.

We should remark that, although the proofs presented below are rather lengthy, we think that Chapter 10 is no more complex than the preceding chapters.

## §1. Representation of functions
### by series that converge in measure

THEOREM 1. *Let* $\{\varphi_n(x)\}_{n=1}^\infty$, $x \in (0,1)$, *be a C.O.N.S. Then for every measurable function* $f(x)$, $-\infty \le f(x) \le +\infty$, $x \in (0,1)$, *there is a series*

$$\sum_{n=1}^\infty a_n \varphi_n(x),$$

*that converges to* $f(x)$ *in measure.*

REMARK. In Theorem 1 we allow $f(x)$ to take the values $+\infty$ and $-\infty$ on measurable sets $B$ and $C$. Here a sequence of measurable functions $f_n(x)$, $n = 1, 2, \ldots$, that are finite a.e. is said to converge in measure to $f(x)$ if it converges in measure in the usual sense on $(0,1)\backslash(B \cup C)$, and for $M = 1, 2, \ldots,$

$$\lim_{n\to\infty} m\{x \in B: f_n(x) < M\} = \lim_{n\to\infty} m\{x \in C: f_n(x) > -M\} = 0.$$

The proof of Theorem 1 is based on two lemmas.

LEMMA 1. *Let the functions* $\psi_0(x)$, $\psi_1(x), \ldots, \psi_m(x)$ $(m = 1, 2, \ldots)$ *be defined on an interval* $\omega \subset (0,1)$ *and belong to* $L^2(\omega)$. *Then for arbitrary numbers* $\varepsilon_0(0,1)$ *and* $\varepsilon > 0$ *we can define a function* $\psi(x) \in L^\infty(\omega)$ *and a set* $e$, *with the following properties*:
1) $\psi(x) = \psi_0(x)$ *if* $x \notin e$, $e \subset \omega$, $m(e) \le \varepsilon_0|\omega|$.
2) $\|\psi\|_{L^2(\omega)} \le 2\varepsilon_0^{-1/2}\|\psi_0\|_{L^2(\omega)}$.
3) $|(\psi, \psi_i)_\omega| \le \varepsilon$, $i = 1, 2, \ldots, m$, *where* $(f, g)_\omega \equiv \int_\omega f(x)g(x)\,dx$.

PROOF. Let $r$ be a positive integer. We partition the interval $\omega$ into $2^r$ equal intervals $\omega_1, \ldots, \omega_{2^r}$, and set, for $g \in L^2(\omega)$,

$$g^{(r)}(x) = |\omega_k|^{-1} \int_{\omega_k} g(t)\,dt$$

if $x \in \omega_k$, $k = 1, \ldots, 2^r$. Then

$$\lim_{r\to\infty} \|g^{(r)} - g\|_{L^2(\omega)} = 0, \qquad \|g^{(r)}\|_{L^2(\omega)} \le \|g\|_{L^2(\omega)},$$

$$r = 1, 2, \ldots. \quad (1)$$

In fact, if $\omega = (0,1)$, then $g^{(r)}(x)$ coincides with a partial sum $S_{2^r}(g, x)$ of the Fourier–Haar series of $g$ (see 3.(8) and 3.(1)), whence (1) follows when

$\omega = (0, 1)$. The case of an arbitrary $\omega$ reduces to that already considered by means of a similarity transformation.

It follows from (1) that

$$\lim_{r \to \infty} (\psi_i^{(r)}, \psi_0^{(r)})_\omega = (\psi_i, \psi_0)_\omega, \qquad i = 1, 2, \ldots, m. \tag{2}$$

Inside each interval $\omega_k$, $k = 1, \ldots, 2^r$, we select an interval $\delta_k$ of length

$$|\delta_k| = \varepsilon_0 |\omega_k|, \qquad k = 1, \ldots, 2^r, \tag{3}$$

and set $e_r = \bigcup_{i=1}^{2^r} \delta_k$ and

$$f_r(x) = \begin{cases} \varepsilon_0^{-1} \psi_0^{(r)}(x) & \text{if } x \in \delta_k, \ k = 1, \ldots, 2^r, \\ 0 & \text{if } x \notin e_r. \end{cases}$$

We shall show that $e_r$ and the function

$$\psi(x) = \psi_0(x) - f_r(x) \tag{4}$$

satisfy conditions 1)–3) if $r$ is sufficiently large. It is easily seen that for $i = 1, \ldots, m$

$$(\psi_i^{(r)}, \psi_0^{(r)})_\omega = (\psi_i^{(r)}, f_r)_\omega, \qquad r = 1, 2, \ldots. \tag{5}$$

Moreover, for $r = 1, 2, \ldots$,

$$\|f_r\|_{L^2(\omega)}^2 = \sum_{k=1}^{2^r} |\delta_k| \left\{ \varepsilon_0^{-1} |\omega_k|^{-1} \int_{\omega_k} \psi_0(x) \, dx \right\}^2$$

$$\leq \sum_{k=1}^{2^r} \varepsilon_0^{-1} \int_{\omega_k} \psi_0^2(x) \, dx = \varepsilon_0^{-1} \|\psi_0\|_{L^2(\omega)}^2. \tag{6}$$

Consequently (also see (1))

$$|(\psi_i^{(r)}, f_r)_\omega - (\psi_i, f_r)_\omega| \leq \|\psi_i^{(r)} - \psi_i\|_{L^2(\omega)} \varepsilon_0^{-1/2} \|\psi_0\|_{L^2(\omega)} \to 0$$
$$\text{as } r \to \infty. \quad (7)$$

From (7) and (5) we obtain, for $i = 1, \ldots, m$,

$$|(\psi_i, f_r)_\omega - (\psi_i^{(r)}, \psi_0^{(r)})_\omega| \to 0 \quad \text{as } r \to \infty,$$

and therefore (see (2) and (4)) $\psi(x)$ satisfies condition 3) for sufficiently large $r$. That conditions 1) and 2) are satisfied follows from the definition of $\psi(x)$ (see (3), (4)) and the inequality (6). This completes the proof of Lemma 1.

LEMMA 2. *Let $\{\varphi_n(x)\}_{n=1}^{\infty}$, $x \in (0,1)$, be a C.O.N.S., and let $f(x)$, $x \in (0,1)$, be a measurable finite function. Then for arbitrary numbers $0 < \delta < 1$ and $k = 1, 2, \ldots$, there are a set $E$ and numbers $a_{k+1}, a_{k+2}, \ldots, a_m$ such that*

1) $E \subset (0,1)$, $m(E) \le \delta$;

2) $\|\sum_{n=k+1}^{m} a_n \varphi_n(x) - f(x)\|_{L^2(G)} \le \delta$, $G = (0,1)\backslash E$;

3) $\|\sum_{n=k+1}^{s} a_n \varphi_n(x)\|_{L^2(D)} \le \delta + \|f\|_{L^2(D)}, s = k+1, \ldots, m$,

*where $D \subset G$ is an arbitrary measurable set.*

PROOF. Since $|f(x)| < \infty$ for $x \in (0,1)$, we may select a function $F \in L^{\infty}(0,1)$ for which

$$m(P) < \frac{\delta}{2}, \qquad P \equiv \{x \in (0,1): f(x) \ne F(x)\}. \tag{8}$$

Choosing a sufficiently large $N$, we partition the interval $(0,1)$ into $N$ disjoint intervals $\omega_1, \ldots, \omega_N$ so that

$$\|F\|_{L^2(\omega_i)} \le \frac{1}{2}\left(\frac{\delta}{4}\right)^{3/2}, \qquad i = 1, \ldots, N. \tag{9}$$

We set $n_1 = k$ and apply Lemma 1 to the functions $F(x), \varphi_1(x), \ldots, \varphi_{n_1}(x)$, considered on $\omega_1$. By that lemma, if we take $\varepsilon_0 = \delta/4$ and $0 < \varepsilon < \delta$, we can find a function $F_1 \in L^{\infty}(\omega_1)$ and a set $e_1$ with the following properties:

(1) $F_1(x) = F(x)$ if $x \in \omega_1\backslash e_1$, $e_1 < \omega_1$, $m(e_1) \le (\frac{\delta}{4})|\omega_1|$;

(2) $\|F_1\|_{L^2(\omega_1)} \le 2(\frac{\delta}{4})^{-1/2}\|F\|_{L^2(\omega_1)}$;

(3) $|\int_{\omega_1} F_1(x)\varphi_n(x)\,dx| \le \varepsilon$, $n = 1, \ldots, n_1$.

From these and (9), we find that if $\varepsilon$ $(0 < \varepsilon < \delta)$ is sufficiently small, the function

$$\Phi_1(x) = \begin{cases} F_1(x) & \text{if } x \in \omega_1, \\ 0 & \text{if } x \in (0,1)\backslash\omega_1, \end{cases}$$

satisfies the inequalities

a) $\|\Phi_1\|_{L^2(0,1)} \le \frac{\delta}{4}$;

b) $|c_n(\Phi_1)| \le \frac{\delta}{4Nk}$, $n = 1, \ldots, n_1$;

c) $\|S_n(\Phi_1)\|_{L^2(0,1)} \le \frac{\delta}{4N}$, $n = 1, \ldots, n_1$.

Here we use the usual notation: if $f \in L^2(0,1)$,

$$c_n(f) = \int_0^1 f(x)\varphi_n(x)\,dx, \qquad S_n(f) = \sum_{j=1}^{n} c_j(f)\varphi_j(x), \qquad n = 1, 2, \ldots.$$

Since $\{\varphi_n(x)\}$ is a C.O.N.S., then by a)

c′) $\|S_n(\Phi_1)\|_{L^2(0,1)} \le \frac{\delta}{4}, n = 1, 2, \ldots,$

and if $n_2$ is a sufficiently large number, then

d) $\|S_n(\Phi_1) - \Phi_1\|_{L^2(0,1)} \leq \frac{\delta}{4N}$ for $n \geq n_2$.

In a similar way, if we apply Lemma 1 successively for $i = 2, \ldots, N$ to the systems $\{F(x), \varphi_n(x), 1 \leq n \leq n_i\}$, $x \in \omega_i$ (where $n_i$ was defined in the preceding step), we obtain functions $\Phi_1(x)$, $\Phi_2(x), \ldots, \Phi_N(x)$, $x \in (0, 1)$, a sequence $k = n_1 < \cdots < n_{N+1}$ and sets $e_1, \ldots, e_N$, with the following properties.

A) $\Phi_i(x) = F(x)$ if $x \in \omega_i \backslash e_i$, $e_i \subset \omega_i$, $m(e_i) \leq \frac{\delta}{4}|\omega_i|$, and $\Phi_i(x) = 0$ if $x \in (0, 1) \backslash \omega_i$, $i = 1, \ldots, N$;

B) $|c_n(\Phi_i)| \leq \frac{\delta}{4Nn_i} \leq \frac{\delta}{4Nk}$, $n = 1, \ldots, n_i$, $i = 1, \ldots, N$;

C) $\|S_n(\Phi_i)\|_{L^2(0,1)} \leq \frac{\delta}{4N}$, $n = 1, \ldots, n_i$, and $\|S_n(\Phi_i)\|_{L^2(0,1)} \leq \frac{\delta}{4}$, $n = 1, 2, \ldots$, $i = 1, E \ldots, N$;

D) $\|S_n(\Phi_i) - \Phi_i\|_{L^2(0,1)} \leq \frac{\delta}{4}$ for $n \geq n_{i+1}$, $i = 1, \ldots, N$.

We set $m = n_{N+1}$, $\Phi(x) = \sum_{i=1}^{N} \Phi_i(x)$ and (see (8) and A))

$$E = P \cup \left( \bigcup_{i=1}^{N} e_i \right) \tag{10}$$

and show that the set $E$ and the numbers $a_n = c_n(\Phi)$, $n = k + 1$, $k + 2, \ldots, m$, satisfy the requirements 1)–3) of Lemma 2. The relation 1) follows immediately from (8), A), and the definition of $E$ (see (10)). In addition, by B),

$$|c_n(\Phi)| \leq \sum_{i=1}^{N} |c_n(\Phi_i)| \leq \frac{\delta}{4k}, \qquad n = 1, \ldots, k, \tag{11}$$

and by D)

$$\left\| \Phi(x) - \sum_{n=1}^{m} c_n(\Phi)\varphi_n(x) \right\|_{L^2(0,1)} \leq \sum_{i=1}^{N} \frac{\delta}{4N} = \frac{\delta}{4}. \tag{12}$$

Using inequalities (11) and (12), and the fact that (see (8), A)) $\Phi(x) = F(x) = f(x)$ for $x \in (0, 1) \backslash E = G$, we obtain

$$\left\| \sum_{n=k+1}^{m} a_n \varphi_n(x) - f(x) \right\|_{L^2(G)}$$

$$\leq \left\| \sum_{n=1}^{k} a_n \varphi_n(x) \right\|_{L^2(G)} + \left\| \sum_{n=1}^{m} a_n \varphi_n(x) - \Phi(x) \right\|_{L^2(G)}$$

$$\leq \sum_{n=1}^{k} |a_n| + \frac{\delta}{4} < \delta.$$

It remains only to establish the validity of inequality 3) for an arbitrary measurable set $D \subset G$. Having chosen a number $s$, $n < s \leq m$, we find an index $i_0$, $1 \leq i_0 \leq N$, such that

$$n_{i_0} < s \leq n_{i_0+1}. \tag{13}$$

Then

$$\|S_s(\Phi)\|_{L^2(D)} \leq \left\| \sum_{i=1}^{i_0-1} S_s(\Phi_i) \right\|_{L^2(D)}$$

$$+ \|S_s(\Phi_{i_0})\|_{L^2(D)} + \left\| \sum_{i=i_0+1}^{N} S_s(\Phi_i) \right\|_{L^2(D)} \tag{14}$$

(when $i_0 = 1$ the first term, and for $i_0 = N$, the third, is missing from the right-hand side of (14)). It follows from D), A), and (13) that

$$\left\| \sum_{i=1}^{i_0-1} S_s(\Phi_i) \right\|_{L^2(D)} \leq \sum_{i=1}^{i_0-1} \|S_s(\Phi_i) - \Phi_i\|_{L^2(D)} + \left\| \sum_{i=1}^{i_0-1} \Phi_i \right\|_{L^2(D)}$$

$$\leq \frac{\delta}{4} + \|f\|_{L^2(D)}. \tag{15}$$

We estimate the third term in (14) by using C) (also see (13)):

$$\left\| \sum_{i=i_0+1}^{N} S_s(\Phi_i) \right\|_{L^2(D)} \leq \sum_{i=i_0+1}^{N} \|S_s(\Phi_i)\|_{L^2(0,1)} \leq \frac{\delta}{4}. \tag{16}$$

Finally, by the second inequality in C),

$$\|S_s(\Phi_{i_0})\|_{L^2(D)} \leq \frac{\delta}{4}. \tag{17}$$

As a result (see (14)–(17)) we obtain

$$\|S_s(\Phi)\|_{L^2(D)} \leq \frac{3}{4}\delta + \|f\|_{L^2(D)}.$$

Moreover (see (11))

$$\|S_k(\Phi)\|_{L^2(D)} \leq \|S_k(\Phi)\|_{L^2(0,1)} \leq \sum_{n=1}^{k} |a_n| \leq \frac{1}{4}\delta.$$

Consequently

$$\left\| \sum_{n=k+1}^{s} a_n \varphi_n \right\|_{L^2(D)} = \|S_s(\Phi) - S_k(\Phi)\|_{L^2(D)} \leq \delta + \|f\|_{L^2(D)}.$$

This completes the proof of Lemma 2.

REMARK. It follows from Lemma 2 that for every C.O.N.S. $\{\varphi_n(x)\}_{n=1}^\infty$ and $k = 1, 2, \ldots$, the system of functions $\{\varphi_n(x)\}_{n=k+1}^\infty$ is complete in $L^0(0, 1)$ with the metric of convergence in measure (see 1.(1)), i.e., for every $f \in L^0(0, 1)$ and $\varepsilon > 0$ there is a polynomial $P(x) = \sum_{n=k+1}^m a_n \varphi_n(x)$ with $\rho(P, f) \leq \varepsilon$.

PROOF OF THEOREM 1. It is enough to consider the case of nonnegative functions $f(x)$; for functions of arbitrary sign the required series can then be obtained as the difference of series that converge in measure to $f^+(x)$ and $f^-(x)$, where $f^+(x) = \max\{f(x), 0\}$ and $f^-(x) = f^+(x) - f(x)$.

We consider a measurable function $f(x)$, $0 \leq f(x) \leq +\infty$, $x \in (0, 1)$, and write

$$A = \{x \in (0, 1): f(x) < \infty\},$$
$$B = \{x \in (0, 1): f(x) = +\infty\}.$$

We choose a sequence $\{\varepsilon_i\}_{i=1}^\infty$ with $\varepsilon_1 > \varepsilon_2 > \ldots$, $\sum_{i=1}^\infty \varepsilon_i < \infty$, and use it to define the coefficients $a_n$ of the required series, $n = 1, 2, \ldots$. We carry out the construction of $a_n$ for successive blocks $m_{i-1} < n \leq m_i$ of indices $n$, $i = 1, 2, \ldots$. Let $m_0 = 0$, $m_1 = 1$, $a_1 = 0$. Now suppose that $a_1, \ldots, a_{m_{i-1}}$ have been defined for some $i > 1$. Applying Lemma 2 for $\delta = \varepsilon_i$, $k = m_{i-1}$, and the function

$$f_i(x) = \begin{cases} f(x) - \sum_{n=1}^{m_{i-1}} a_n \varphi_n(x) & \text{if } x \in A, \\ 1 & \text{if } x \in B, \end{cases} \tag{18}$$

we find a measurable set $E_i$ and numbers $a_{m_{i-1}+1}, \ldots, a_{m_i}$ such that

$\alpha)$ $E_i \subset (0, 1)$,    $m(E_i) \leq \varepsilon_i$;

$\beta)$ $\|\sum_{n=m_{i-1}+1}^{m_i} a_n \varphi_n(x) - f_i(x)\|_{L^2((0,1)\backslash E_i)} \leq \varepsilon_i$;

$\gamma)$ $\|\sum_{n=m_{i-1}+1}^{s} a_n \varphi_n(x)\|_{L^2(D)} \leq \varepsilon_i + \|f_i\|_{L^2(D)}$ $(m_{i-1} < s < m_i)$,

where $D \subset (0, 1)\backslash E_k$ is a measurable set.

Continuing this process, we successively define $\{a_n\}_{n=1}^\infty$. Let us show that the series

$$\sum_{n=1}^\infty a_n \varphi_n(x) \tag{19}$$

converges in measure to $f(x)$

Let $\delta$ be an arbitrarily small positive number. Choose a number $i_0 > 1$ so that $\sum_{i_0}^\infty \varepsilon_i \leq \delta/4$, and set $A_\delta = A\backslash(\bigcup_{i=i_0}^\infty E_i)$. Then $m(A_\delta) > m(A) - \delta/4$. For arbitrary $m$, $m_i \leq m_{i+1}$, $i > i_0$, we have

$$\left\|\sum_{n=1}^m a_n \varphi_n - f\right\|_{L^2(A_\delta)} \leq \left\|\sum_{n=1}^{m_i} a_n \varphi_n - f\right\|_{L^2(A_\delta)} + \left\|\sum_{n=m_i+1}^m a_n \varphi_n\right\|_{L^2(A_\delta)}. \tag{20}$$

But by (18) and $\beta$)

$$\left| \sum_{n=1}^{m_i} a_n\varphi_n(x) - f(x) \right| = \left| \sum_{n=m_{i-1}+1}^{m_i} a_n\varphi_n(x) - f_i(x) \right|,$$

if $x \in A_\delta$, and

$$\|f_{i+1}\|_{L^2(A_\delta)} = \left\| f_i - \sum_{n=m_{i-1}+1}^{m_i} a_n\varphi_n \right\|_{L^2(A_\delta)} \leq \varepsilon_i.$$

From this and (20), also using the inequality $\gamma$), we obtain

$$\left\| \sum_{n=1}^{m} a_n\varphi_n - f \right\|_{L^2(A_\delta)} \leq \left\| \sum_{n=m_{i-1}+1}^{m_i} a_n\varphi_n - f_i \right\|_{L^2(A_\delta)} + \varepsilon_i + \|f_{i+1}\|_{L^2(A_\delta)}$$

$$\leq 2\varepsilon_i + \varepsilon_{i+1} < 3\varepsilon_i.$$

Consequently, we have, by Chebyshev's inequality, for every $\varepsilon > 0$,

$$m\left\{ x \in A : \left| \sum_{n=1}^{m} a_n\varphi_n(x) - f(x) \right| > \varepsilon \right\}$$

$$\leq m(A\backslash A_\delta) + \frac{1}{\varepsilon^2} \left\| \sum_{n=1}^{m} a_n\varphi_n - f \right\|_{L^2(A_\delta)}^2 \leq \frac{\delta}{4} + \left( \frac{3\varepsilon_i}{\varepsilon} \right)^2 < \delta,$$

if $m$ is sufficiently large. Hence we have proved that (19) converges in measure to $f(x)$ on the set $A$.

Let us now show that (19) converges in measure to $+\infty$ on the set $B$. For this purpose we select a positive integer $M$, and $\delta$, $0 < \delta < 1$, and find an index $N$ such that

$$m\left\{ x \in B : \sum_{n=1}^{m} a_n\varphi_n(x) < M \right\} < \delta \qquad \text{if } m \geq N. \tag{21}$$

Let $i_1$ and $K$ be positive integers such that

$$\sum_{i=i_1+1}^{\infty} \varepsilon_i < \frac{\delta}{4},$$

$$m\left\{ x \in B : \left| \sum_{n=1}^{m_{i_1}} a_n\varphi_n(x) \right| > K \right\} < \frac{\delta}{4}, \qquad \frac{4}{K^2} < \frac{\delta}{4}. \tag{22}$$

Then (see (18) and $\beta$)) for $i > i_1$

$$\left\| \sum_{n=m_{i_1}+1}^{m_i} a_n\varphi_n(x) - (i - i_1) \right\|_{L^2(B)} \leq \sum_{i=i_1+1}^{\infty} \varepsilon_i < \frac{\delta}{4},$$

where $B_\delta = B \backslash \bigcup_{i=i_1+1}^{\infty} E_i$, $m(B_\delta) \geq m(B) - \frac{\delta}{4}$ (see $\alpha$)). Consequently

$$m\left\{x \in B_\delta: \left|\sum_{n=m_{i_1}+1}^{m_i} a_n \varphi_n(x) - (i - i_1)\right| > 1\right\} < \left(\frac{\delta}{4}\right)^2 < \frac{\delta}{4}, \qquad i > i_1.$$

From this and (22) we find that when $i > i_1$

$$m\left\{x \in B_\delta: \sum_{n=1}^{m_i} a_n \varphi_n(x) < i - i_1 - K - 1\right\} < \frac{\delta}{2}. \qquad (23)$$

Moreover, if $m_i < m \leq m_{i+1}$, $i > i_1$, then by $\gamma$)

$$\left\|\sum_{n=m_i}^{m} a_n \varphi_n\right\|_{L^2(B_\delta)} \leq \varepsilon_{i+1} + \|f_{i+1}\|_{L^2(B_\delta)} \leq 2,$$

and therefore (also see (22))

$$m(Q) \equiv m(Q_m) \equiv m\left\{x \in B_\delta: \left|\sum_{n=m_i+1}^{m} a_n \varphi_n(x)\right| > K\right\}$$

$$\leq \frac{4}{K^2} < \frac{\delta}{4}; \qquad m_i < m \leq m_{i+1}, \quad i > i_1. \qquad (24)$$

Set $i_2 = i_1 + 2K + 1 + M$. Then, using (22)–(24), we find that when $m > N = m_2$ ($m_i < m \leq m_{i+1}$, $i \geq i_2$)

$$m\left\{x \in B: \sum_{n=1}^{m} a_n \varphi_n(x) < M\right\}$$

$$\leq m(B \backslash B_\delta) + m(Q) + m\left\{x \in B_\delta \backslash Q: \sum_{n=1}^{m_i} a_n \varphi_n(x) < i_2 - i_1 - 2K - 1\right\}$$

$$\leq \frac{\delta}{4} + \frac{\delta}{4} + m\left\{x \in B_\delta \backslash Q: \sum_{n=1}^{m_i} a_n \varphi_n(x) < i - i_1 - K - 1\right\} < \delta$$

i.e., (21) holds. This shows that (19) converges in measure to $f(x)$, and therefore Theorem 1 is fully established.

We introduce the following definition in order to state a corollary of Theorem 1.

DEFINITION 1. An orthogonal series of the form (19) is called a *null series in the sense of convergence in measure* (or *almost everywhere*) if it converges in measure (or almost everywhere) to the function $f(x) \equiv 0$ and $a_{n_0} \neq 0$ for some index $n_0$.

COROLLARY 1. *For every C.O.N.S.* $\{\varphi_n(x)\}_{n=1}^{\infty}$ *there is a null series in the sense of convergence in measure.*

In fact, the series (19) constructed in the proof of Theorem 1 that converges in measure to a given function $f(x)$ has its first coefficient $a_1 = 0$. Hence if we take $f(x) = \varphi_1(x)$ we obtain a series

$$\sum_{n=2}^{\infty} a'_n \varphi_n(x),$$

that converges in measure to $\varphi_1(x)$. Then

$$\varphi_1(x) - \sum_{n=2}^{\infty} a'_n \varphi_n(x)$$

is a null series in the sense of convergence in measure. This establishes Corollary 1.

## §2. Representation of functions by series that converge almost everywhere

THEOREM 2. *There is a C.O.N.S.* $\Psi = \{\psi_n(x)\}_{n=1}^{\infty}$, $x \in (0,1)$, *such that every series of the form*

$$\sum_{n=1}^{\infty} a_n \varphi_n(x), \qquad \sum_{n=1}^{\infty} a_n^2 = \infty, \qquad (25)$$

*diverges on a set of positive measure.*

An immediate consequence of Theorem 2 is the following corollary.

COROLLARY 2. *Let the C.O.N.S.* $\Psi$ *be the same as in Theorem 2, and let* $\mathcal{F}$ *be the set of functions represented by a.e. convergent series in this system. Then* $\mathcal{F} \subset L^2(0,1)$.

Corollary 2 shows that there is no result as general as Theorem 1 for convergence a.e. Looking ahead, we note that one can give (see Theorem 4) extensive sufficient conditions on a C.O.N.S. $\Phi$ that guarantee the possibility of representing every almost everywhere finite $f(x)$ by a convergent series in the system $\Phi$.

We also notice that for the system $\Psi$ of Theorem 2 there is no full series in the sense of convergence a.e., since null series cannot be Fourier series of functions in $L^2(0,1)$.

PROOF OF THEOREM 2. We shall use properties of the orthonormal sets

$$\Psi(N) = \{\psi_k^N(x)\}_{k=1}^{N} \subset \mathcal{D}_{4N}, \qquad N = 1, 2, \dots, \qquad (26)$$

constructed in Lemma 1 for Theorem 8.2 (see 8.(12)).

It is easily deduced from the definition of $\psi_k^N(x)$ (by arguing as in the deduction of 8.(13)) that for $p = 3, 4, \ldots$ and arbitrary numbers $c_k$, $k = 1, \ldots, 2^p$, we have the inequality

$$\int_0^1 \left| \sum_{k=1}^{j_p(x)} c_k \psi_k^{2^p}(x) \right| dx \geq \frac{cp}{2^{p/2}} \left| \sum_{k=1}^{a(p)} c_k \right|, \tag{27}$$

where $a(p) = 2^p - [2^{p/2}]$ and $j_p(x) = \min\{i, a(p)\}$, if $x \in ((i-1)/2^{p+1}, i/2^{p+1})$, $i = 1, \ldots, 2^{p+1}$ (here $c > 0$ is an absolute constant).

In addition, for $N = 3, 4, \ldots$ we define an orthonormal matrix $A_N = \{a_{\nu,r}^N\}_{\nu,r=1}^N$ by setting

$$a_{\nu,r}^N = \begin{cases} 1 - (N-1)^{-1} & \text{if } \nu = r < N, \\ 0 & \text{if } \nu = r = N, \\ -(N-1)^{-1} & \text{if } \nu \neq r, \ \nu \text{ and } r < N, \\ (N-1)^{-1/2} & \text{if } \nu = N \text{ or } r = N, \ \nu \neq r. \end{cases} \tag{28}$$

Now we can turn to the construction of the required system $\Psi$. It is convenient to define $\psi_n(x)$ on the interval $(0, 2)$ (then, by the transformation $f(x) \to \sqrt{2} f(2x)$ the functions $\psi_n(x)$ can, of course, be carried over to $(0, 1)$). We first construct an auxiliary O.N.S. $\{\varphi_n(x)\}$, $x \in (0, 2)$: let $\varphi_n(x) = 2^{-1/2} r_n(x)$, $x \in (0, 2)$, $n = 1, \ldots, 8$, and

$$\varphi_n(x) = \begin{cases} \frac{1}{\sqrt{2}} r_{s_p}(x) \psi_{n-2^p}^{2^p}(x) & \text{if } x \in (0, 1), \\ \frac{1}{\sqrt{2}} r_n(x) & \text{if } x \in [1, 2), \\ 2^p < n < 2^{p+1} & p = 3, 4, \ldots, \end{cases} \tag{29}$$

where $r_n(x)$ are Rademacher functions, extended to the whole real line from $(0, 1)$ with period 1, and the sequence $s_p$, $p = 3, 4, \ldots$, in (29) is taken to increase so fast that the functions $\{\varphi_n(x)\}$, $n \neq 2^p$, $p = 4, 5, \ldots$, form an O.N.S. (this is possible, since $\psi_k^{2^k} \in \mathscr{D}_{2^{p+2}}$—see (26)). For $n = 2^p$, $p = 4, 5, \ldots$, we do not define $\varphi_n(x)$.

Now choose a set $\{u_p(x)\}_{p=3}^\infty$, $x \in (0, 2)$, such that the system $\{\varphi_n(x)\} \cup \{u_p(x)\}$ is an O.N.S. which is complete in $L^2(0, 2)$, and set

$$\psi_n(x) = \varphi_n(x) = -\frac{1}{\sqrt{2}} r_n(x), \qquad n = 1, \ldots, 8, \ x \in (0, 2),$$

and, for $n = 2^p + \nu$, $\nu = 1, \ldots, 2^p$, $p = 3, 4, , \ldots$,

$$\psi_n(x) = \sum_{r=1}^{2^p - 1} a_{\nu,r}^{2^p} \varphi_{2^p+r}(x) + a_{\nu,2^p}^{2^p} u_p(x). \tag{30}$$

The construction of the C.O.N.S. $\Psi = \{\psi_n(x)\}_{n=1}^{\infty}$, $x \in (0, 2)$, is possible since the matrices $A_{2^p}$ are orthonormal and the functions $\varphi_n(x)$ and $u_p(x)$ can be expressed in terms of the functions $\psi_n(x)$. Let us show that $\Psi$ satisfies the requirements of Theorem 2: we consider any series of the form (25) and show that it diverges on a set of positive measure.

For $p = 3, 4, \ldots$, let

$$\beta_p = \left\{ \sum_{n=2^p+1}^{2^{p+1}} a_n^2 \right\}^{1/2}, \qquad d_p = 2^{-p/2} \left| \sum_{n=2^p+1}^{2^{p+1}} a_n \right|.$$

By Cauchy's inequality, $\beta_p \geq d_p$, $p = 3, 4, \ldots$. We choose an absolute constant

$$c_0 = \frac{1}{20} c_1^{1/2} \leq \frac{1}{20},$$

where $c_1$ is the constant in Theorem 2.7 (i.e., $c = c_M$, $M = 1$), and partition the integers $p \geq 3$ into two blocks. To the first block $S_1$ we assign the numbers $p$ for which

$$c_0 \beta_p \leq d_p. \tag{31}$$

The remaining numbers form the block $S_2$.

There are two possible cases.

1) $\sum_{p \in S_1} \beta_p^2 = \infty$. Then we shall show that the series (25) diverges at each point of a set $E \subset (0, 1)$ with $m(E) > 0$.

2) $\sum_{p \in S_1} \beta_p^2 = \infty$. Then (see (25)) $\sum_{p \in S_2} \beta_p^2 = \infty$, and we shall show that (25) does not converge in measure on the interval [1,2].

Consider case 1). Let $p \in S_1$. By the definition of the matrices $A_{2^p}$ (see (28)) and equation (30), it follows at once that, for $2^p < n < 2^{p+1}$, $p = 3, 4, \ldots$,

$$\psi_n(x) = \varphi_n(x) + \Delta_n(x),$$

where

$$\int_0^2 |\Delta_n(x)| \, dx \leq 2 \left\{ \int_0^2 \Delta_n^2(x) \, dx \right\}^{1/2} \leq \frac{8}{2^{p/2}}.$$

Therefore, for every integral-valued measurable function $N(x)$ such that $2^p + 1 \leq N(x) \leq 2^{p+1}$, we have the inequality

$$\int_0^1 \left| \sum_{n=2^p+1}^{N(x)} a_n \psi_n(x) \right| dx$$

$$= \int_0^1 \left| \sum_{n=2^p+1}^{N(x)} a_n \varphi_n(x) + \sum_{n=2^p+1}^{N(x)} a_n \Delta_n(x) \right| dx$$

$$\geq \int_0^1 \left| \sum_{n=2^p+1}^{N(x)} a_n \varphi_n(x) \right| dx - \sum_{n=2^p+1}^{N(x)} |a_n| \int_0^1 |\Delta_n(x)| \, dx$$

$$\geq \int_0^1 \left| \sum_{n=2^p+1}^{N(x)} a_n \varphi_n(x) \right| dx - \frac{8}{2^{p/2}} \sum_{n=2^p+1}^{2^{p+1}} |a_n|$$

$$\geq \int_0^1 \left| \sum_{n=2^p+1}^{N(x)} a_n \varphi_n(x) \right| dx - 8\beta_p. \tag{32}$$

Let $c_k = a_{2^p+k}$, $k = 1, \ldots, 2^p$. Then it is clear that

$$\sum_{k=1}^{2^p} c_k^2 = \beta_p^2, \qquad 2^{-p/2} \left| \sum_{k=1}^{2^p} c_k \right| = d_p, \quad p = 3, 4, \ldots. \tag{33}$$

Moreover, by Cauchy's inequality,

$$\left| \sum_{k=a(p)+1}^{2^p} c_k \right| \leq 2^{p/4} \left\{ \sum_{k=1}^{2^p} c_k^2 \right\}^{1/2} = \beta_p \cdot 2^{p/4},$$

and therefore, for sufficiently large $p \in S_1$ (see (31))

$$2^{-p/2} \left| \sum_{k=1}^{a(p)} c_k \right| \geq 2^{-p/2} \left| \sum_{k=1}^{2^p} c_k \right| - \beta_p \cdot 2^{-p/4} = d_p - \beta_p \cdot 2^{-p/4} \geq \frac{c_0}{2} \beta_p.$$

Consequently (see (32), (29), and (27)), for sufficiently large $p$,

$$\int_0^1 \left| \sum_{n=2^p+1}^{2^p+j_p(x)} a_n \psi_n(x) \right| dx \geq \int_0^1 \left| \sum_{k=1}^{j_p(x)} c_k \psi_k^{2^p}(x) \right| dx - 8\beta_p$$

$$\geq cp 2^{-p/2} \left| \sum_{k=1}^{a(p)} c_k \right| - 8\beta_p$$

$$\geq c\frac{c_0}{2} p\beta_p - 8\beta_p \geq c'p\beta_p > 0. \tag{34}$$

On the other hand, by Lemma 1 and Theorem 8.1 (the Men'shov–Rademacher lemma)

$$\int_0^1 \left| \sum_{n=2^p+1}^{2^p+j_p(x)} a_n \psi_n(x) \right|^2 dx \le K(\ln 2^p)\beta_p < Kp\beta_p. \tag{35}$$

From (34) and (35) we can deduce from Theorem 2.7 by using Lemma 1 that

$$m\left\{ x \in (0,1): \left| \sum_{n=2^p+1}^{2^p+j_p(x)} a_n \psi_n(x) \right| > \frac{c'}{4}p\beta_p \right\} \ge c'' > 0 \tag{36}$$

for every sufficiently large $p \in S_1$. Since we assumed that $\sum_{p \in S_1} \beta_p^2 = \infty$, we have $\overline{\lim}_{p \in S_1} p\beta_p = \infty$, and it follows from (36) that (25) diverges on a set $E \subset (0,1)$ of positive measure ($m(E) \ge c''$) in case 1).

We now consider case 2). Let $p \in S_2$; then by (30) and the orthonormality of the matrix $A_{2^p}$ we have

$$\sum_{n=2^p+1}^{2^{p+1}} a_n \psi_n(x) = \sum_{n=2^p+1}^{2^{p+1}-1} \gamma_n \varphi_n(x) + \omega_p u_p(x),$$

where

$$\beta_p^2 = \sum_{n=2^p+1}^{2^{p+1}} a_n^2 = \omega_p^2 + \sum_{n=2^p+1}^{2^{p+1}-1} \gamma_n^2.$$

Moreover, it follows from the definition of $A_{2^p}$ that

$$|\omega_p| \le (2^p-1)^{-1/2} \left| \sum_{n=2^p+1}^{2^{p+1}} a_n \right| + (2^p-1)^{-1/2} \max_{2^p < n \le 2^{p+1}} |a_n|$$

$$\le 2d_p + 2^{1-p/2}\beta_p,$$

$$\sum_{n=2^p+1}^{2^{p+1}-1} \gamma_n^2 = \beta_p^2 - \omega_p^2 \ge \beta_p^2 - (2d_p + 2^{1-p/2}\beta_p)^2.$$

From these inequalities (since by the construction of $S_2$ we have $d_p < c_0\beta_p \le \frac{1}{20}\beta_p$ for $p \in S_2$) we can deduce that, for sufficiently large $p \in S_2$,

$$\sum_{n=2^p+1}^{2^{p+1}-1} \gamma_n^2 \ge \frac{1}{2}\beta_p^2, \quad \left( \sum_{n=2^p+1}^{2^{p+1}-1} \gamma_n^2 \right) \frac{c_1}{20} \ge \omega_p^2. \tag{37}$$

Now choose a sequence $m_1 < m_2 < \dots$ of positive integers such that (37) holds for all $p \ge m_1$, $p \in S_2$ and

$$\sum_{p=m_s}^{m_{s+1}-1} \beta_p^2 = \sum_{n=2^{m_s}+1}^{2^{m_{s+1}}} a_n^2 \ge s+1, \quad s = 1,2,\dots. \tag{38}$$

Let us consider the sums

$$\sum_{n=2^{m_s}+1}^{2^{m_{s+1}}} a_n \psi_n(x) = I_s'(x) + I_s''(x), \qquad (39)$$

where

$$I_s'(x) = \sum_{p \in S_1 \cap [m_s, m_{s+1})} \sum_{n=2^p+1}^{2^{p+1}} a_n \psi_n(x),$$

$$I_s''(x) = \sum_{p \in S_2 \cap [m_s, m_{s+1})} \sum_{n=2^p+1}^{2^{p+1}} a_n \psi_n(x).$$

Since, by hypothesis, $\sum_{p \in S_1} \beta_p^2 < \infty$, we have

$$\sum_{s=1}^{\infty} \int_0^2 [I_s'(x)]^2 \, dx \le \sum_{p \in S_1} \beta_p^2 < \infty. \qquad (40)$$

Moreover,

$$I_s''(x) = \sum_{p \in S_2 \cap [m_s, m_{s+1})} \sum_{n=2^p+1}^{2^{p+1}} a_n \psi_n(x)$$

$$= \sum_{p \in S_2 \cap [m_s, m_{s+1})} \left[ \omega_p u_p(x) + \sum_{n=2^p+1}^{2^{p+1}-1} \gamma_n \varphi_n(x) \right].$$

Hence by (37) and (38)

$$J_s \equiv \sum_{p \in S_2 \cap [m_s, m_{s+1})} \sum_{n=2^p+1}^{2^{p+1}-1} \gamma_n^2 \ge \frac{20}{c_1} \sum_{p \in S_2 \cap [m_s, m_{s+1})} \omega_p^2$$

and for sufficiently large $s$ (also see (40))

$$J_s \ge \frac{1}{2} \sum_{p \in S_2 \cap [m_s, m_{s+1})} \beta_p^2 \ge \frac{s}{2}.$$

Since $\varphi_n(x) = (1/\sqrt{2}) r_n(x)$ for $x \in (1, 2)$ and $n \ne 2^p$, $p = 4, 5, \ldots$, we have the equation

$$I_s''(x) = P_s(x) + Q_s(x), \qquad Q_s(x) = \sum_{p \in S_2 \cap [m_s, m_{s+1})} \omega_p u_p(x),$$

where $P_s(x)$ is a polynomial in the Rademacher system and $(s > s_0)$

$$\int_1^2 P_s^2(x) \, dx = \frac{1}{2} J_s \ge \frac{s}{4}, \qquad \sum_{p \in S_2 \cap [m_s, m_{s+1})} \omega_p^2 \le J_s \frac{c_1}{20}.$$

Using Chebyshev's inequality and the pairwise orthogonality of the functions $u_p(x)$, we obtain

$$m\left\{x \in (1,2): |Q_s(x)| > \frac{J_s^{1/2}}{3}\right\} \leq \frac{9}{J_s}\|Q_s\|_s^2 \leq \frac{9}{20}c_1 < \frac{c_1}{2}. \qquad (41)$$

On the other hand, by Theorem 2.7,

$$m\left\{x \in (1,2): |P_s(x)| > \frac{1}{2\sqrt{2}}J_s^{1/2}\right\} \geq c_1. \qquad (42)$$

It follows from inequalities (41) and (42) that, for $s > s_0$,

$$m\left\{x \in (1,2): |I_s''(x)| \geq \frac{1}{100}J_s^{1/2} \geq \frac{1}{200}s^{1/2}\right\} \geq \frac{c_1}{2}.$$

Hence if we use (40) (also see (39)) we obtain at once that in case 2) the series (25) converges in measure on $(1,2)$. This completes the proof of Theorem 2.

DEFINITION 2. An O.N.S. $\Phi = \{\varphi_n(x)\}_{n=1}^\infty$, $x \in (0,1)$, is said to be a *system with strong convergence* if the condition

$$\sum_{n=1}^\infty a_n^2 < \infty$$

is necessary and sufficient for the convergence a.e. of the series

$$\sum_{n=1}^\infty a_n\varphi_n(x)$$

(the Rademacher system is an example of a system with strong convergence).

Theorem 2 can be strengthened:

THEOREM 2'. *There is an O.N.S., complete in $L^2(0,1)$, that has strong convergence.*

The proof of this result is much like the proof of Theorem 2, but technically more complicated, and therefore we do not present it here. We confine ourselves to the proof of the following proposition, which plays an auxiliary role in the proof of Theorem 2', and is also of independent interest.

THEOREM 3. *There is an orthonormal convergence system $\Phi_0 = \{\varphi_n(x)\}_{n=1}^\infty$, $x \in (0,1)$ (see Definition 1.2), such that corresponding to every* *measurable set* $E \subset (0, 1)$ $m(E) > 0$, *there is a sequence* $\{a_n\}_{n=1}^{\infty} \in l^2$ *for which*

$$S_{\Phi_0}^*(\{a_n\}) \equiv \sup_{1 \leq N < \infty} \left| \sum_{n=1}^{N} a_n \varphi_n(x) \right| \notin L^2(E).$$

The functions in the system $\Phi_0$ replace in the construction of a complete system with strong convergence, the functions $\psi_k^N(x)$, $k = 1, 2, \ldots$ (see (26) and (27)) which are used in the proof of Theorem 2. Recall also that the statement of Theorem 3 was given at the end of Chapter 7 to justify the definitive nature of Theorem 7.9, which we proved.

PROOF OF THEOREM 3. In constructing the system $\Phi_0$ we again use the orthonormal set $\Psi(N) = \{\psi_k^N(x)\}_{k=1}^N$ which was defined in Chapter 8 (see Lemma 1 and Theorem 8.2). We recall some properties of these functions (see 8.(11) and 8.(13)):

a) $\quad \psi_k^N \in \mathscr{D}_{4N}, \quad \int_0^1 \psi_k^N(x)\,dx = 0, \qquad k = 1, \ldots, N;$

b) $\quad m\left\{ x \in (0, 1): \sup_{1 \leq j < N} \left| \sum_{k=1}^{j} \frac{1}{\sqrt{N}} \psi_k^N(x) \right| > c_0 \ln N \right\} \geq \frac{1}{2} - \frac{1}{\sqrt{N}}$  (43)

($c_0 > 0$ is an absolute constant).

For each $N$ of the form $N = 2^{2q}$, $q \geq 4$, we define a set $G(N) = \{g_k^N(x)\}_{k=1}^N$ on $(-\infty, \infty)$ by first taking

$$g_k^N(x) = \begin{cases} (\log_2 N)^{-1} 2^{s/2} \psi_k^N(2^s(x - 2^{-s})), \\ \quad \text{if } 2^{-s} < x \leq 2^{-(s-1)}, \ s = 1, 2, \ldots, \log_2^2 N, \\ 0 \quad \text{if } 0 < x \leq 2^{-\log_2^2 N}, \end{cases}$$  (44)

for $x \in (0, 1]$ and $k = 1, \ldots, N$, and then setting $g_k^N(x + r) = g_k^N(x)$ if $r$ is an integer and $x \in (0, 1]$.

Using the orthonormality of $\Psi(N)$ and the relation (43), part a), we obtain

a) $g_k^N \in \mathscr{D}_{4N 2^{\log_2^2 N}}, \quad \int_0^1 g_k^N(x)\,dx = 0, \qquad k = 1, \ldots, N;$

b) $\int_0^1 g_k^N(x) g_{k'}^N(x)\,dx = \begin{cases} 1, & \text{if } k = k', \\ 0, & \text{if } k \neq k', \end{cases} \quad 1 \leq k, k' \leq N;$

c) $\int_{2^{-s}}^{2^{-(s-1)}} \left[ \sum_{k=1}^N a_k g_k^N(x) \right]^2 dx = \left( \sum_{k=1}^N a_k^2 \right) \log_2^{-2} N,$

$\hspace{8cm} s = 1, \ldots, \log_2^2 N.$  (45)

In addition, if

$$g_0^N \equiv \sup_{1 \leq j < N} \left| \sum_{k=1}^{j} N^{-1/2} g_k^N(x) \right|,$$

it follows from (43), part b), that for every set $A \subset (2^{-s}, 2^{-(s-1)})$ with $m(A) \geq \frac{3}{5} 2^{-s}$ the integral

$$\int_A [g_0^N(x)]^2 \, dx \geq c > 0 \quad \text{if } 1 \leq s \leq \log_2^2 N. \tag{46}$$

We establish a property of $G(N)$.

LEMMA 1. *For all* $y > 0$, $N = 2^q$, $q = 4, 5, \ldots$, *and every sequence* $a = \{a_k\}_{k=1}^N$, *we have the inequality*

$$m\left\{ x \in (0,1): g(a,x) \equiv \sup_{1 \leq j \leq N} \left| \sum_{k=1}^j a_k g_k^N(x) \right| > y \right\} \leq \frac{C}{y^2} \sum_{k=1}^N a_k^2. \tag{47}$$

PROOF. Supposing that the sequence $a = \{a_k\}$ is given, we set

$$F(x) = (\log_2^{-1} N) \sup_{1 \leq j \leq N} \left| \sum_{k=1}^j a_k \psi_k^N(x) \right|,$$

$$\lambda(t) = m\left\{ x \in (0,1): F^2(x) > t \right\}, \quad t > 0.$$

It follows from the definition of $g_k^N(x)$ (see (44)) that when $s = 1, \ldots,$ $\log_2^2 N$,

$$m\left\{ x \in (2^{-s}, 2^{-(s-1)}): g(a,x) > y \right\} = 2^{-s} m\{ x \in (0,1): F(x) > 2^{-s/2} y \}$$
$$= 2^{-s} \lambda(2^{-s} y^2).$$

Summing on $s$ and using the monotonicity of $\lambda(t)$, we obtain

$$m\{ x \in (2^{-\log_2^2 N}, 1): g(a,x) > y \} = \sum_{s=1}^{\log_2^2 N} 2^{-s} \lambda(2^{-s} y^2)$$

$$\leq \frac{2}{y^2} \sum_{s=1}^{\infty} 2^{-s-1} y^2 \lambda(2^{-s} y^2)$$

$$\leq \frac{2}{y^2} \sum_{s=1}^{\infty} \int_{2^{-s-1} y^2}^{2^{-s} y^2} \lambda(t) \, dt$$

$$\leq \frac{2}{y^2} \int_0^{\infty} \lambda(t) \, dt = \frac{2}{y^2} \int_0^1 F^2(x) \, dx. \tag{48}$$

We obtain the conclusion of the lemma by estimating the right-hand side of (48) by Lemma 1 from Theorem 8.1, according to which

$$\int_0^1 F^2(x) \, dx \leq K \sum_{k=1}^N a_k^2,$$

and recalling that $g_k^N(x) = 0$ for $x \in (0, 2^{-\log_2^2 N})$.

We can now construct the required system $\Phi_0$. We choose a sequence $\{s_j\}_{j=1}^{\infty}$ of positive integers such that $s_1 = 0$, the numbers $N_j = s_{j+1} - s_j$ $(j = 1, 2, \ldots)$ are representable in the form $N_j = 2^q$ $(q \geq 4)$, and for each $j = 1, 2, \ldots$,

$$4N_j^2 2^{\log_2^2 N_j} < N_{j+1}. \tag{49}$$

Then for $n \in (s_j, s_{j+1}]$, $j = 1, 2, \ldots$, we set

$$\varphi_n(x) = g_{n-s_j}^{N_j}(N_j x), \qquad x \in (0, 1). \tag{50}$$

By (45), part a), and (49), it follows that for $s_j < n \leq s_{j+1}$, $j = 1, 2, \ldots$, the functions $\varphi_n(x)$ are constant on each interval of the form

$$\left( \frac{i-1}{2^q}, \frac{i}{2^q} \right), \qquad 1 \leq i \leq 2^q, \ 2^q = 4N_j^2 2^{\log_2^2 N_j}.$$

At the same time, the period of $\varphi_n(x)$ with $s_{j+1} < n \leq s_{j+2}$ is $N_{j+1}^{-1} = 2^{-q'}$, $q' > q$ (see (50)). Therefore if we use (45), parts a) and b), we find that $\Phi_0 = \{\varphi_n(x)\}_{n=1}^{\infty}$ is an O.N.S.

Let us prove that $\Phi_0$ is a convergence system. Let there be given the series

$$\sum_{n=1}^{\infty} a_n \varphi_n(x), \qquad \sum_{n=1}^{\infty} a_n^2 < \infty. \tag{51}$$

By what was said above, the functions

$$f_j(x) = \sum_{n=s_j+1}^{s_{j+1}} a_n \varphi_n(x), \qquad j = 1, 2, \ldots,$$

are "nonintersecting" polynomials in the Haar system (i.e., $f_j(x) = \sum_{k=m_j+1}^{m_{j+1}} a_k \chi_k(x)$, $m_1 < m_2 < \cdots$), the series of which converges in $L^2(0, 1)$. Hence we obtain, by using Theorem 3.4, that the sequence

$$s_{s_j}(x) = \sum_{n=1}^{s_j} a_n \varphi_n(x) = \sum_{\nu=1}^{j-1} f_\nu(x), \qquad j = 2, 3, \ldots,$$

converges for almost all $x \in (0, 1)$. Therefore it is enough to show that the sum

$$\sum_{j=1}^{\infty} m \left\{ x \in (0, 1) : \sup_{s_j < N \leq s_{j+1}} \left| \sum_{n=s_j+1}^{N} a_n \varphi_n(x) \right| > y \right\} < \infty \tag{52}$$

is finite for every $y > 0$. But by (50) and Lemma 1,

$$m \left\{ x \in (0, 1) : \sup_{s_j < N \leq s_{j+1}} \left| \sum_{n=s_j+1}^{N} a_n \varphi_n(x) \right| > y \right\} \leq \frac{C}{y} \sum_{n=s_j+1}^{s_{j+1}} a_n^2,$$

from which (see (51)) relation (52) follows.

To complete the proof of Theorem 3 we need to determine, for each set $E$ (which we may suppose closed) with $E \subset (0,1)$, $m(E) > 0$, a series of the form (51) such that

$$\int_E \sup_{1 \le N < \infty} \left| \sum_{n=1}^N a_n \varphi_n(x) \right|^2 dx = \infty. \tag{53}$$

It is enough to verify that

$$\varlimsup_{j \to \infty} \int_E \sup_{s_j < N \le s_{j+1}} \left| \sum_{n=1}^N N_j^{-1/2} \varphi_n(x) \right|^2 dx = \infty \tag{54}$$

(the required series can then be found in the form

$$\sum_{\nu=1}^\infty \frac{1}{\nu^2} \sum_{n=s_{j_\nu}+1}^{s_{j_\nu+1}} N_{j_\nu}^{-1/2} \varphi_n(x),$$

where $\{j_\nu\}$, $1 < j_1 < j_2 < \ldots$, is a sequence of integers).

For this purpose we choose a number $\varepsilon$, $0 < \varepsilon < \frac{1}{100}$, and find a finite set of disjoint intervals $\{I_l\}$ with binary-rational endpoints, such that

$$E \subset \bigcup_l I_l \equiv I \text{ and } m(E) > (1 - \varepsilon)m(I). \tag{55}$$

For $j = 1, 2, \ldots$ and $\nu = 1, 2, \ldots$, we define the set

$$G_{j,\nu} = \bigcup_{i=1}^{N_j} \omega_i(j, \nu);$$

$$\omega_i(j, \nu) \left( \frac{i-1}{N_j} + \frac{1}{2^\nu N_j}, \frac{i-1}{N_j} + \frac{1}{2^{\nu-1} N_j} \right].$$

It is clear that $m\{\omega_i(j, \nu)\} = 2^{-\nu} N_j^{-1}$, $m(G_{j,\nu}) = 2^{-\nu}$, $j, \nu = 1, 2, \ldots$. It is also easily seen that for $\nu = 1, 2, \ldots$

$$\lim_{j \to \infty} N_j^{-1} \cdot \operatorname{card}\{i: \omega_i(j, \nu) \subset I\} = m(G_{j,\nu}) \cdot m(I) = 2^{-\nu} m(I). \tag{56}$$

It follows from (56) that for sufficiently large $j$ and $\nu = 1, 2, \ldots$, $[\frac{1}{2} \log_2(1/10\varepsilon)]$

$$N_j^{-1} \operatorname{card}\{i: m[\omega_i(j, \nu) \cap E] > \tfrac{3}{5} m[\omega_i(j, \nu)]\} \ge \tfrac{1}{2} 2^{-\nu} m(I) \tag{57}$$

(otherwise we would obtain, for $j > j_0$,

$$m(I \backslash E) \ge \tfrac{2}{5} m[\omega_1(j, \nu)] \cdot N_j \cdot \tfrac{1}{4} 2^{-\nu} m(I) = \tfrac{1}{10} 2^{-2\nu} m(I) \ge \varepsilon m(i),$$

which would contradict (55)).

Now if we use the definition of $\varphi_n(x)$ (see (50)) and (46), we obtain, by using (57), that for $\nu = 1, \ldots, [\frac{1}{2} \log_2(1/10\varepsilon)]$,

$$\varlimsup_{j \to \infty} \int_{G_{j,\nu} \cap E} \sup_{s_j < N \leq s_{j+1}} \left| \sum_{n=s_j+1}^{N} N_j^{-1/2} \varphi_n(x) \right|^2 dx \geq c(E) > 0.$$

Combining these inequalities for $\nu = 1, 2, \ldots, [\frac{1}{2} \log_2(1/10\varepsilon)]$, we obtain

$$\varlimsup_{j \to \infty} \int_{E} \sup_{s_j < N \leq s_{j+1}} \left[ \sum_{n=s_j+1}^{N} N_j^{-1/2} \varphi_n(x) \right]^2 dx \geq c(E) \ln \frac{1}{\varepsilon}. \qquad (58)$$

Since $\varepsilon$ can be arbitrarily small, (54) follows from (58). This completes the proof of Theorem 3.

The question of the boundedness of the majorant operator on the partial sums, $S_\Phi^* : l^2 \to L^2(0, 1)$, is very important for problems on the representation of functions by series in the system $\Phi$. As indicated above (see Theorem 2′), the fact that a C.O.N.S. $\Phi$ is a convergence system does not guarantee the possibility of representing a function $f \notin L^2(0, 1)$ by an almost everywhere convergent series in this system. However, if we know just a little more about $\Phi$, we can obain a positive result.

THEOREM 4. *Let the C.O.N.S.* $\Phi = \{\varphi_n(x)\}_{n=1}^{\infty}$, $x \in (0, 1)$, *have the property that the majorant operator* $S_\Phi^*$ *on the partial sums (see* 1.(5) *and Proposition* 1.1) *is a bounded operator from* $l^2$ *to* $L^2(0, 1)$. *Then for every* $g(x)$ *which is measurable and a.e. finite on* $(0, 1)$ *there is a series*

$$\sum_{n=1}^{\infty} a_n \varphi_n(x),$$

*which converges to* $f(x)$ *a.e. on* $(0, 1)$.

REMARK 1. It is evident that in the statement of Theorem 4 the condition that the operator $S_\Phi^*$ is bounded can be presented as follows: For every function $f \in L_2(0, 1)$

$$\|S_\Phi^*(f, x)\|_2 \leq K \|f\|_2, \qquad (59)$$

where the constant $K$ is independent of $f$ and

$$S_\Phi^*(f, x) = \sup_{1 \leq N < \infty} \left| \sum_{n=1}^{N} c_n(f) \varphi_n(x) \right|,$$
$$c_n(f) = \int_0^1 f(x) \varphi_n(x) \, dx, \quad n = 1, 2, \ldots. \qquad (60)$$

REMARK 2. In contrast to Theorem 1, Theorem 4 has nothing to say about the representation of functions that take the value $\infty$ on a set of positive measure. The conclusion of Theorem 4 fails for such functions. In fact, it was shown in Chapter 3 (see Theorem 3.14) that a series in the Haar system (for which (59) holds by Theorem 3.4) cannot converge to $+\infty$ on a set of positive measure.

REMARK 3. Besides the Haar system, the trigonometric, Walsh, and Franklin systems satisfy the condition of Theorem 4. One can also say that complete O.N.S. that do not satisfy the condition of Theorem 4 are exceptions to the general rule (this statement can be given a rigorous formulation).

The proof of Theorem 4 is rather long, and will be divided into several subsections.

1°. Here we show that it is sufficient to represent by an almost everywhere convergent series in the system $\Phi$ only (a.e. finite) functions $g$ of the form

$$g(x) = \lim_{n \to \infty} g_n(x) \quad \text{for almost all } x \in (0,1), \quad g_n \in C(0,1),$$

$$0 < g_n(x) \le g_{n+1}(x) - 2^{-n}, \quad n-1, 2, \ldots \quad (61$$

For this purpose we verify that for every function $f \in L^0(0,1)$ there are functions $g'(x)$ and $g''(x)$ of the form (61) such that $f(x) = g'(x) - g''(x$ for almost all $x \in (0,1)$. By Luzin's theorem we can choose a sequence of continuous functions $\{f_n(x)\}_{n=1}^\infty$ such that $f_1(x) \equiv 0$ and

$$m\{x \in (0,1): f_n(x) \ne f(x)\} \le \frac{1}{n^2}, \quad n = 2, 3, \ldots.$$

Then

$$f(x) = \sum_{n=1}^\infty F_n(x) \quad \text{for almost all } x \in (0,1),$$

$$F_n(x) = f_{n+1}(x) - f_n(x), \quad n = 1, 2, \ldots$$

and if, for $F \in C(0,1)$, we write $F^+(x) \equiv \max\{F(x), 0\}$, $F^-(x) \equiv -\min\{F(x), 0\}$, we have, for almost every $x \in (0,1)$,

$$f(x) = \sum_{n=1}^\infty F_n^+(x) - \sum_{n=1}^\infty F_n^-(x)$$

$$= \sum_{n=1}^\infty (F_n(x) + 2^{-n}) - \sum_{n=1}^\infty (F_n^-(x) + 2^{-n})$$

$$= g'(x) - g''(x).$$

**2°. LEMMA 1.** *Let the functions* $f_p \in L^1(0, 1)$, $p = 1, 2, \ldots$, *have uniformly bounded norms*

$$\|f_p\|_1 \leq M, \qquad p = 1, 2, \ldots. \tag{62}$$

*Then there is a sequence* $\{p_s\}_{s=1}^{\infty}$ *of positive integers,* $p_1 < p_2 < \cdots$ *such that*

$$\sum_{s=1}^{\infty} m\{x \in (0, 1): |f_{p_s}(x)| \geq s\} < \infty. \tag{63}$$

PROOF. We construct inductively a family of sequences $\{p_s(k)\}_{s=1}^{\infty}$ of integers, $k = 1, 2, \ldots$, subject to the condition

$$\{p_s(k)\}_{s=1}^{\infty} \subset \{p_s(k-1)\}_{s=1}^{\infty}, \qquad k = 2, 3, \ldots, \tag{64}$$

and numbers $\mu_k \in [0, 1]$, $k = 1, 2, \ldots$, such that

$$\mu_k \leq m\{x \in (0, 1): |f_{p_s(k)}(x)| \geq k\} \leq \mu_k + \frac{1}{k^2}, \qquad s = 1, 2, \ldots. \tag{65}$$

Set $p_s(1) = s$, $s = 1, 2, \ldots$; $\mu_1 = 0$; and if the sequence $\{p_s(k-1)\}_{s=1}^{\infty}$ and the number $\mu_{k-1}$ have already been constructed, then, using the evident relation

$$m\{x \in (0, 1): |f_{p_s(k-1)}(x)| \geq k\} \in [0, 1], \qquad s = 1, 2, \ldots,$$

we can select a subsequence $\{p_s(k)\}$ of the sequence $\{p_s(k-1)\}$ and a number $\mu_k$ such that (65) is satisfied.

Let us show that the sequence $p_s = p_s(s)$, $s = 1, 2, \ldots$, is the required sequence. In fact, by (64) and (65) we have, for $s = 1, 2, \ldots$,

$$\mu_k \leq m\{x \in (0, 1): |f_{p_s}(x)| \geq k\} \leq \mu_k + \frac{1}{k^2}, \qquad k = 1, \ldots, s. \tag{66}$$

On the other hand, by (62), for $s = 1, 2, \ldots$,

$$\sum_{k=1}^{\infty} m\{x \in (0, 1): |f_{p_s}(x)| \geq k\}$$

$$= \sum_{k=1}^{\infty} \sum_{i=k}^{\infty} m\{x \in (0, 1): i \leq |f_{p_s}(x)| < i+1\}$$

$$= \sum_{i=1}^{\infty} im\{x \in (0, 1): i \leq |f_{p_s}(x)| < i+1\}$$

$$\leq \sum_{i=1}^{\infty} \int_{\{x \in (0,1): i \leq |f_{p_s}(x)| < i+1\}} |f_{p_s}(x)| \, dx \leq \int_0^1 |f_{p_s}(x)| \, dx \leq M,$$

and, using (66), we find that for $s = 1, 2, \ldots$,

$$\sum_{k=1}^{s} m \{x \in (0, 1) : |f_{p_s}(x)| \geq k\}$$

$$\leq \sum_{k=1}^{s} \left( \mu_k + \frac{1}{k^2} \right)$$

$$\leq \sum_{k=1}^{s} m \{x \in (0, 1) : |f_{p_s}(x)| \geq k\} + \sum_{k=1}^{\infty} \frac{1}{k^2}$$

$$\leq M + \sum_{k=1}^{\infty} \frac{1}{k^2}.$$

This completes the proof of Lemma 1.

3°. Here we construct an auxiliary system $\Psi = \{\psi_k(x)\}_{k=1}^{\infty}$, $x \in (0, 1)$ (very similar to the Haar system), and a sequence $\{N_k\}_{k=1}^{\infty}$, $N_1 < N_2 < \cdots$, of integers, and note for later use a number of their properties (see a)–e) below). As we successively construct, for $k = 1, 2, \ldots$, the functions $\psi_k(x)$, and then the numbers $N_{k+1}$, we will see to it that the following relations are satisfied (see (60)):

a) $$\sum_{n=1}^{N_k} |c_n(\psi_k)| \leq 2^{-k}, \quad k = 2, 3, \ldots;$$

b) $$\sum_{l=1}^{k} \left[ \sum_{n=N_{k+1}+1}^{\infty} c_n^2(\psi_l) \right]^{1/2} \leq 2^{-k}, \quad k = 1, 2, \ldots;$$

c) $\quad \psi_k(x) \in \mathscr{D}_{2^\nu}$ for $k = 1, 2, \ldots$, and some $\nu = \nu(k)$.[1]

We set $\psi_1(x) \equiv 1$, $N_1 = 0$, and select, using Bessel's inequality, a number $N_2 > N_1$ for which b) is satisfied for $k = 1$. We construct the functions $\psi_k(x)$, $k > 1$, in groups (or blocks) $\{\psi_k(x)\}_{k=k_p+1}^{k_{p+1}}$ (the numbers $k_p$, $1 = k_1 < k_2 < \cdots$, will be defined later). The first block (i.e., $k_2 = 2$) consists of the single function

$$\psi_2(x) = r_{j_2}(x), \quad x \in (0, 1),$$

where the index $j_2$ of the Rademacher function $r_{j_2}(x)$ is chosen so large that the inequality

$$\sum_{n=1}^{N_2} |c_n(\psi_2)| \equiv \sum_{n=1}^{N_2} \left| \int_0^1 r_{j_2}(x) \varphi_n(x) \, dx \right| \leq 2^{-2}$$

---

[1] Here and later, when we consider equations between functions or sets, we disregard the binary-rational points of $[0, 1]$.

is satisfied. Then, using Bessel's inequality, we select $N_3 > N_2$ so that b) holds for $k = 2$.

Now suppose that the functions $\{\psi_k(x)\}_{k=1}^{k_p}$ $(p = 2, 3, \dots)$ and the numbers $N_k$, $k = 1, \dots, k_p + 1$, have already been constructed so that relations a)–c) are satisfied. Let us construct the $p$th block $\{\psi_k(x)\}_{k=k_p+1}^{k_{p+1}}$. For this purpose we consider a partition of the interval $(0, 1)$ into binary intervals.

$$\delta_1, \dots, \delta_s, \qquad \delta_i \cap \delta_j = \varnothing \text{ for } i \neq j, \qquad \bigcup_{i=1}^{s} \delta_i = (0, 1), \qquad (67)$$

such that the functions $\psi_k(x)$, $k = 1, \dots, k_p$, are constant on each interval $\delta_i$, $i = 1, \dots, s$ (see c)); the $p$th block will consist of $s$ functions (i.e., $k_{p+1} = k_p + s$), and $\operatorname{supp} \psi_{k_p+i}(x) = \delta_i$, $i = 1, \dots, s$.

If the functions $\psi_1(x), \dots, \psi_{k-1}(x)$, $k = k_p + i$, $i = 1, \dots, s$, and the numbers $N_1, \dots, N_k$ have already been constructed (and conditions a)–c) are satisfied), then $\psi_k(x)$ and $N_{k+1}$ are defined as follows: we set

$$\psi_k(x) = \chi_{\delta_i}(x) r_{j_k}(x), \qquad (68)$$

where $\chi_\delta(x)$ is the characteristic function of $\delta$, and the index $j_k$ of the Rademacher function $r_{j_k}(x)$ is chosen so large that

$$j_k > j_{k-1}, \qquad \int_{\delta_i} r_{j_k}(x)\, dx = 0,$$

$$\sum_{n=1}^{N_k} |c_n(\psi_k)| = \sum_{n=1}^{N_k} \left| \int_{\delta_i} r_{j_k}(x)\varphi_n(x)\, dx \right| \leq 2^{-k}$$

(the possibility of such choices follows from the fact that

$$\lim_{j \to \infty} \int_\delta r_j(x)\varphi_n(x)\, dx = 0 \qquad (n = 1, 2, \dots)$$

for every interval $\delta \subset (0, 1)$). Then we determine (again using Bessel's inequality) the number $N_{k+1}$ so that condition b) is satisfied. It follows immediately from (68) that c) holds for $\psi_k(x)$.

We have now constructed a system $\Psi = \{\psi_k(x)\}_{k=1}^{\infty}$ with properties a)–c). It follows at once from the construction (see (67) and (68)) that if $\Delta(k)$ denotes the binary interval supporting $\psi_k(x)$, $k = 1, 2, \dots$, then we have, for $p = 1, 2, \dots$,

d) $$\bigcup_{k=k_p+1}^{k_{p+1}} \Delta(k) = (0, 1),$$

$$\Delta(k) \cap \Delta(k') = \varnothing, \quad \text{if } k_p < k < k' \leq k_{p+1}.$$

We now observe that by construction (see (68)) each $\psi_k(x)$, $k = 2, 3, \ldots$, is a polynomial in the Haar system of the form

$$\psi_k(x) = \sum_{i:\Delta^i_{j_k-1} \subset \Delta(k)} 2^{-(j_k-1)/2} \chi^{(i)}_{j_k-1}(x), \qquad x \in (0, 1) \tag{69}$$

(the Haar functions in (69) are from the block with index $j_k - 1$), and moreover

$$|\psi_k(x)| = \chi_{\Delta(k)}(x), \qquad x \in (0, 1). \tag{70}$$

Finally we note the following properties of $\Psi$ which follow from d) and (69) (using the properties that the numbers $j_k$, $k = 2, 3, \ldots$, in (69) increase, and that the supports of the Haar functions that appear in a single block are disjoint):

e) to every series

$$\sum_{k=1}^{\infty} d_k \psi_k(x) \tag{71}$$

in the system $\Psi$ there corresponds a Haar series

$$\sum_{n=1}^{\infty} A(d_n)\chi_n(x) \equiv d_1 + \sum_{k=2}^{\infty} \sum_{i:\Delta^i_{j_k-1} \subset \Delta(k)} d_k 2^{-(j_k-1)/2} \chi^{(i)}_{j_k-1}(x) \tag{72}$$

such that the partial sums of (71) and (72) and of the corresponding series

$$\sum_{k=1}^{\infty} d_k^2 \psi_k^2(x) \quad \text{and} \quad \sum_{n=1}^{\infty} [A(d_n)]^2 \chi_n^2(x)$$

have common upper and lower limits at each binary-irrational point $x \in (0, 1)$.

4°. For $p = 1, 2, \ldots$ (see (60)) let

$$S_p^*(x) \equiv \left\{ \sum_{k=k_p+1}^{k_{p+1}} [S_\Phi^*(\psi_k, x)]^2 \right\}^{1/2}. \tag{73}$$

By (59), d) and (70),

$$\int_0^1 [S_p^*(x)]^2 \, dx \leq K^2 \sum_{k=k_p+1}^{k_{p+1}} \int_0^1 \psi_k^2(x) \, dx = K^2.$$

Applying Lemma 1 with $f_p(x) = [S_p^*(x)]^2$, $p = 1, 2, \ldots$, $M = K^2$, we can find a subsequence $\{p_s\}_{s=1}^{\infty}$ of the positive integers such that

$$\sum_{s=1}^{\infty} m\left\{ x \in (0, 1) : [S_{p_s}^*(x)]^2 \geq s \right\} < \infty. \tag{74}$$

It follows from (74) that

$$\lim_{s\to\infty} (s\ln s)^{-1/2} S_{p_s}^*(x) = 0 \quad \text{for a.e. } x \in (0,1). \tag{75}$$

Introducing the notation

$$\Omega = \bigcup_{s=2}^{\infty} \{k : k_{p_s} < k \le k_{p_s+1}\}, \tag{76}$$

we consider the following series in the system $\Psi$:

$$\sum_{k\in\Omega} b_k \psi_k(x) \equiv \sum_{s=2}^{\infty} \sum_{k=k_{p_s}+1}^{k_{p_s+1}} (s\ln s)^{-1/2} \psi_k(x) \tag{77}$$

(for convenience, we also take $b_k = 0$ if $f \notin \Omega$). Since (see d) and (70))

$$\sum_{k=k_p+1}^{k_{p+1}} \psi_k^2(x) = 1, \quad x \in (0,1), \ p = 1, 2, \ldots,$$

then from the definition of the coefficients $b_k$ (see (77)) we obtain

$$\sum_{k=1}^{\infty} b_k^2 \psi_k^2(x) \equiv \sum_{s=2}^{\infty} (s\ln s)^{-1} = \infty, \quad x \in (0,1),$$

and according to e) (see (72))

$$\sum_{n=1}^{\infty} [A(b_n)]^2 \chi_n^2(x) = +\infty, \quad x \in (0,1).$$

Hence it follows by Theorems 3.13 and 3.14 that

$$\overline{\lim_{N\to\infty}} \sum_{n=1}^{N} A(b_n)\chi_n(x) = +\infty \quad \text{for almost all } x \in (0,1)$$

and (see e))

$$\overline{\lim_{N\to\infty}} \sum_{k=1}^{N} b_k \psi_k(x) = +\infty \quad \text{for almost all } x \in (0,1). \tag{78}$$

We have the following lemma.

LEMMA 2. *For every a.e. finite function $g(x)$ of the form* (61) *there is a subseries of* (77),

$$\sum_{k\in\Omega_1} b_k \psi_k(x), \quad \Omega_1 \subset \Omega, \tag{79}$$

*which converges to $g(x)$ a.e. on $(0, 1)$ :*

$$\sum_{k \in \Omega_1} b_k \psi_k(x) = g(x) \quad \text{for almost all } x \in (0, 1). \tag{80}$$

PROOF. Using (61), for the given function $g(x)$ we can easily construct functions $\{G_n(x)\}_{n=1}^{\infty}$ with

$$G_0(x) \equiv 0, \qquad G_n(x) \in D_{2^{k_n}}, \qquad k_n < k_{n+1}, \; n = 1, 2, \ldots, \tag{81}$$

such that

$$g_n(x) \le G_n(x) \le g_{n+1}(x), \qquad G_{n+1}(x) \ge G_n(x) + \frac{1}{2^{n+1}},$$
$$x \in (0, 1), \; n = 1, 2, \ldots. \tag{82}$$

Then for almost every $x \in (0, 1)$

$$0 < G_n(x) < g(x), \qquad n = 1, 2, \ldots, \quad \text{and} \quad \lim_{n \to \infty} G_n(x) = g(x). \tag{83}$$

We now define inductively sequences of integers $\{N_l\}_{l=1}^{\infty}$, $N_1 < N_2 < \cdots$ ; $\{\varepsilon_k\}_{k=1}^{\infty}$, $\varepsilon_k = 0$ or 1; and sets $E_1 \subset (0, 1)$, $l = 1, 2, \ldots$, such that

1)
$$P_l(x) \equiv \sum_{k=1}^{N_l} \varepsilon_k b_k \psi_k(x) \ge G_{l-1}(x),$$

$$\text{if } x \in E_l, \; m(E_l) \ge 1 - 1/l, \; l = 1, 2, \ldots;$$

2)
$$\sum_{k=1}^{N} \varepsilon_k b_k \psi_k(x) < G_l(x),$$

$$x \in (0, 1), \; N = 1, \ldots, N_l, \; l = 1, 2, \ldots.$$

Set $N_1 = 1$, $\varepsilon_1 = 0$, $E_1 = (0, 1)$. Clearly, conditions 1) and 2) are satisfied for $l = 1$.

Suppose that we have already defined the numbers $N_i$, $i = 1, \ldots, l$; $\varepsilon_k$, $k = 1, \ldots, N_l$; and sets $E_i$, $i = 1, \ldots, l$, satisfying conditions 1) and 2). We now define $N_{l+1}$, $E_{l+1}$, and $\varepsilon_k$, $N_1 < k \le N_{l+1}$. In doing this, we shall need an auxiliary sequence $\{\delta_k\}_{k=N_l+1}^{\infty}$ of numbers $\delta_k = 0$ or 1, which is constructed as follows.

Supposing that the support of $\psi_k(x)$ is the binary interval $\Delta(k)$ whose length tends to zero as $k \to \infty$ (see (70)), we find an index $q > N_l$ such that (see (81))

$$G_l(x) = \text{const} \quad \text{for } x \in \Delta(k), \; k = q, q+1, \ldots, \tag{84}$$

and also (see (77))

$$0 \le b_k \le 2^{-l-1}, \qquad k = q, q+1, \ldots. \tag{85}$$

Set

$$\delta_k = 0, \qquad k = N_l + 1, \ldots, q. \tag{86}$$

If the numbers $\delta_k$, $k = N_1 + 1, \ldots, k_0$ ($k_0 \geq q$) have already been constructed, and also

$$I_{k_0}(x) \equiv P_l(x) + \sum_{k=N_l+1}^{k_0} \delta_k b_k \psi_k(x) < G_{l+1}(x), \qquad x \in (0,1) \tag{87}$$

(for $k_0 = q$ this follows from 2)), we then set

$$\delta_{k_0+1} = \begin{cases} 1 & \text{if } I_{k_0}(x) < G_l(x), \ x \in \Delta(k_0 + 1), \\ 0 & \text{if } I_{k_0}(x) \geq G_l(x), \ x \in \Delta(k_0 + 1). \end{cases} \tag{88}$$

(Recall that the functions $I_{k_0}(x)$ and $G_l(x)$ are constant on $\Delta(k_0 + 1)$ (see (67) and (84).) Since $G_{l+1}(x) \geq G_l(x) + 2^{-l-1}$, $x \in (0,1)$ (see (82)), it follows from (70), (85), and (88) that $I_{k_0+1}(x) < G_{l+1}(x)$, $x \in (0,1)$. Continuing the construction, we define the sequence $\{\delta_k\}_{k=N_j+1}^{\infty}$. Moreover, by construction (also see (86)), the inequality (87) holds for every $k_0 \geq N_l + 1$.

For $s = q, q + 1, \ldots$, we consider the set (see (87))

$$Q_s = \{x \in (0,1) : I_s(x) < G_l(x)\}. \tag{89}$$

By the construction of $\delta_k$ (see (88)), $Q_s \supset Q_{s+1}$ for $s = q, q + 1, \ldots$. Let

$$Q = \bigcap_{s=q}^{\infty} Q_s.$$

If $x$ is any point of $Q$, then by (88) and (89) it follows from the inclusion $x \in \Delta(k)$, $k > q$, that $\delta_k = 1$ and therefore (also see (86) and (87))

$$\sum_{k=q}^{s} b_k \psi_k(x) = \sum_{k=N_l+1}^{s} \delta_k b_k \psi_k(x) = I_{k_0}(x) - P_l(x)$$

$$< G_{l+1}(x) - P_l(x), \qquad x \in Q, \ s > q.$$

From this and (78) we find that $m(Q) = 0$, and if $s_0$ is sufficiently large, then

$$m(Q_{s_0}) < \frac{1}{l+1}. \tag{90}$$

Set $N_{l+1} = s_0$, $\varepsilon_k = \delta_k$, $k = N_l + 1, \ldots, N_{l+1}$, $E_{l+1} = (0,1) \backslash Q_{s_0}$; then conditions 1) and 2) will be satisfied (see (87) and (89)).

The sequences $\{N_l\}$, $\{\varepsilon_k\}$, and $\{E_l\}$ are determined by properties 1) and 2). Let us show that the series

$$\sum_{k=1}^{\infty} \varepsilon_k b_k \psi_k(x) \equiv \sum_{k \in \Omega_1} b_k \psi_k(x), \tag{91}$$

$$\Omega_1 = \{k : b_k \varepsilon_k \neq 0\},$$

converges a.e. on $(0, 1)$ to $g(x)$, thus completing the proof of Lemma 2. It follows from 2) and (83) that the partial sums of (91) are bounded above by $g(x)$. But then (see e))

$$\varlimsup_{N \to \infty} \left| \sum_{n=1}^{N} A(\varepsilon_n b_n) \chi_n(x) \right| \leq g(x), \qquad x \in (0, 1),$$

and by Theorem 3.14 the series $\sum_{n=1}^{\infty} A(\varepsilon_n b_n) \chi_n(x)$, and consequently also (91), converge a.e. on $(0, 1)$. Now the convergence of (91) to $g(x)$ follows (see 1), 2), (82), and (83)) from the relation

$$\varlimsup_{N \to \infty} \sum_{k=1}^{N} \varepsilon_k b_k \psi_k(x) = g(x) \quad \text{for almost all } x \in (0, 1).$$

This completes the proof of Lemma 2.

5°. We now complete the proof of Theorem 4. We choose a function $g(x)$ of the form (61) and consider a series of the form (79) which converges to $g(x)$ a.e. on $(0, 1)$. By (77),

$$b_k > 0, \qquad k \in \Omega, \qquad b \equiv \sup_{k \in \Omega} b_k = (2 \ln 2)^{-1/2} < 1.$$

Consequently (see property a) of the system $\{\psi_k\}$) the sum

$$\sum_{k \in \Omega_1} b_k c_n(\psi_k) \equiv a_n \tag{92}$$

is finite for $n = 1, 2, \ldots$, and moreover

$$|a_1| \leq b \sum_{k=2}^{\infty} |c_1(\psi_k)| \leq b \sum_{k=2}^{\infty} 2^{-k} < 1 \tag{93}$$

(inequality (93) will be needed only in proving a corollary of Theorem 4). Let us show that

$$\sum_{n=1}^{\infty} a_n \varphi_n(x) = g(x) \qquad \text{for almost all } x \in (0, 1). \tag{94}$$

By (92) we have

$$S_{N_k}(x) \equiv \sum_{n=1}^{N_k} a_n \varphi_n(x) = \sum_{n=1}^{N_k} \sum_{l \in \Omega_1} b_l c_n(\psi_l) \varphi_n(x)$$

$$= \sum_{l \in \Omega_1} b_l \sum_{n=1}^{N_k} c_n(\psi_l) \varphi_n(x)$$

for $k = 2, 3, \ldots$. Using this equation, the completeness of $\Phi$, and properties a) and b) of $\Psi$, we obtain ($b_l = 0$ for $l \notin \Omega_1$)

$$
\left\| \sum_{l=1}^{k-1} b_l \psi_l(x) = S_{N_k}(x) \right\|_2 \leq \left\| \sum_{l=1}^{k-1} b_l \left[ \psi_l(x) - \sum_{n=1}^{N_k} c_n(\psi_l)\varphi_n(x) \right] \right\|_2
$$

$$
+ \left\| \sum_{l=k}^{\infty} b_l \sum_{n=1}^{N_k} c_n(\psi_l)\varphi_n(x) \right\|_2
$$

$$
\leq b \sum_{l=1}^{k-1} \left\{ \sum_{n=N_k+1}^{\infty} c_n^2(\psi_l) \right\}^{1/2} + b \sum_{l=k}^{\infty} \sum_{n=1}^{N_k} |c_n(\psi_l)|
$$

$$
\leq b \cdot 2^{-k+1} + b \sum_{l=k}^{\infty} 2^{-l}
$$

$$
= 4b2^{-k} < 2^{-k+2}, \qquad k = 2, 3, \ldots.
$$

From our estimates and Beppo Levi's theorem, it follows at once that for almost all $x \in (0, 1)$

$$
\lim_{k \to \infty} \left| \sum_{l=1}^{k} b_l \psi_l(x) - S_{N_k}(x) \right| = 0
$$

and (see (80))

$$
\lim_{k \to \infty} S_{N_k}(x) = g(x).
$$

Consequently we will have established the convergence a.e. of $\sum_1^{\infty} a_n \varphi_n(x)$ to $g(x)$ as soon as we verify that

$$
\lim_{k \to \infty} S_{\Phi}^*\left( \sum_{n=N_k+1}^{N_k+1} a_n \varphi_n, x \right) = 0 \quad \text{for almost all } x \in (0, 1). \tag{95}
$$

According to (92) for $k = 2, 3, \ldots$,

$$
\sigma_k(x) \equiv \sum_{n=N_k+1}^{N_k+1} a_n \varphi_n(x) = \sum_{l=1}^{\infty} \left\{ b_l \sum_{n=N_k+1}^{N_k+1} c_n(\psi_l)\varphi_n(x) \right\}
$$

$$
= \sum_{l<k} + \sum_{l=k} + \sum_{l>k} \equiv \sigma_k^{(1)}(x) + \sigma_k^{(2)}(x) + \sigma_k^{(3)}(x). \tag{96}
$$

By relation b) and Parseval's equation,

$$
\left\| \sigma_k^{(1)}(x) \right\|_2 \leq b \sum_{l=1}^{k-1} \left\{ \sum_{n=N_k+1}^{\infty} c_n^2(\psi_l) \right\}^{1/2} \leq 2^{-k+1}, \qquad k = 2, 3, \ldots.
$$

Consequently (see (59))

$$\sum_{k=2}^{\infty} \left\| S_{\Phi}^*(\sigma_k^{(1)}, x) \right\|_2 \le K \sum_{k=2}^{\infty} 2^{-k+1} < \infty$$

and, by Beppo Levi's theorem,

$$\lim_{k \to \infty} S_{\Phi}^*(\sigma_k^{(1)}, x) = 0 \quad \text{for a.e. } x \in (0, 1). \tag{97}$$

Similarly, by the estimate (see a))

$$\left\| \sigma_k^{(3)}(x) \right\|_2 \le b \sum_{l=k+1}^{\infty} \sum_{n=N_k+1}^{N_{k+1}} |c_n(\psi_l)|$$

$$\le b \sum_{l=k+1}^{\infty} \sum_{n=1}^{N_l} |c_n(\psi_l)| \le \sum_{l=k+1}^{\infty} 2^{-l} = 2^{-k}, \quad k = 2, 3, \ldots,$$

we obtain

$$\lim_{k \to \infty} S_{\Phi}^*(\sigma_k^{(3)}, x) = 0 \quad \text{for almost all } x \in (0, 1). \tag{98}$$

Finally we estimate the majorant $S_{\Phi}^*(\sigma_k^{(2)}, x)$. Here we may suppose that $k \in \Omega_1$, since $b_k = 0$ if $k \notin \Omega_1$. In this case (see (76), (77), and (79)) $k_{p_s} < k < k_{p_s+1}$, $b_k = (s \ln s)^{-1/2}$ for some $s = s(k)$, and

$$S_{\Phi}^*(\sigma_k^{(2)}, x) \le 2(s \ln s)^{-1/2} S_{\Phi}^*(\psi_k, x) \le 2(s \ln s)^{-1/2} S_{p_s}^*(x),$$

where $S_p^*(x)$, $p = 1, 2, \ldots$, were defined in (73). From the last inequality and (75) we deduce, recalling that $s = s(k) \to \infty$ as $k \to \infty$,

$$\lim_{k \to \infty} S_{\Phi}^*(\sigma_k^{(2)}, x) = 0 \quad \text{for almost all } x \in (0, 1). \tag{99}$$

Therefore (see (96)–(99)) we have established (95). This completes the proof of (94) and hence of Theorem 4.

COROLLARY 3. *For every C.O.N.S.* $\Phi = \{\varphi_n(x)\}$ *which satisfies the hypotheses of Theorem 4 there is a null series in the sense of convergence a.e.*

In fact, in the proof of Theorem 4 we constructed, for every function $g \in L^0(0, 1)$, a series

$$\sum_{n=1}^{\infty} a_n\{g\}\varphi_n(x), \quad |a_1\{g\}| < 1,$$

that converges to $g(x)$ a.e. on $(0, 1)$. Let $g(x) \equiv \varphi_1(x)$; then the series

$$(a_1\{\varphi_1\} - 1)\varphi_1(x) + \sum_{n=2}^{\infty} a_n\{\varphi_1\}\varphi_n(x)$$

converges to zero for almost all $x \in (0, 1)$, although $a_1\{\varphi_1\} = 1 \neq 0$. This establishes Corollary 3.

## §3. Two theorems on universal series

DEFINITION 3. Let $X$ be a Banach space. A series

$$\sum_{n=1}^{\infty} x_n, \qquad x_n \in X, \ n = 1, 2, \ldots, \tag{100}$$

is said to be *universal (in $X$) with respect to rearrangements* if for each $y \in X$ there is a permutation $\{\sigma(n)\}_{n=1}^{\infty}$ of the positive integers such that

$$\sum_{n=1}^{\infty} x_{\sigma(n)} \overset{X}{=} y.$$

We may also consider series which are universal with respect to rearrangements for other kinds of convergence (in particular, for convergence almost everywhere, in measure, etc.). The term "universal" is evidently explained by the idea that by means of transformations (in this case, rearrangements) a series (100) can represent every element of $X$.

We could also consider other kinds of universal series (for example, universality with respect to changes of the signs of the terms, etc.). Moreover, the very existence of universal series in infinite-dimensional spaces is not, as a rule, obvious. The next theorem (also see the example following it) provides a sufficient condition for a series

$$\sum_{n=1}^{\infty} f_n(x), \qquad f_n \in L^2(0, 1), \ n = 1, 2, \ldots, \tag{101}$$

to be universal in $L^2(0, 1)$ with respect to rearrangements.

THEOREM 5. *Let a sequence* $\{f_n(x)\}_{n=1}^{\infty} \subset L^2(0, 1)$ *of functions satisfy the following conditions:*

a) *the series* (101) *converges in the norm of* $L^2(0, 1)$;
b) $\sum_1^{\infty} \|f_n\|_2^2 < \infty$;
c) *for every function* $g \in L^2(0, 1)$ *with* $\|g\|_2 > 0$

$$\sum_{n=1}^{\infty} |(g, f_n)| \equiv \sum_{n=1}^{\infty} \left| \int_0^1 g(x) f_n(x) \, dx \right| = \infty.$$

*Then the series* (101) *is universal in* $L^2(0, 1)$ *with respect to rearrangements.*

We need three lemmas for the proof of Theorem 5.

LEMMA 1. *Let there be given functions* $f_n \in L^2(0,1)$, $n = 1, \ldots, N$, *and numbers* $\lambda_n$, $0 \le \lambda_n \le 1$, $n = 1, \ldots, N$. *Then there is a set* $\{\varepsilon_n\}_{n=1}^N$, $\varepsilon_n = 0$ *or* 1, *such that*

$$\left\| \sum_{n=1}^N \lambda_n f_n - \sum_{n=1}^N \varepsilon_n f_n \right\|_2 \le \frac{1}{2} \left\{ \sum_{n=1}^N \|f_n\|_2^2 \right\}^{1/2}. \tag{102}$$

PROOF. If we apply Lemma 1 from Theorem 2.14 to the numbers $\lambda_n' = 2\lambda_n - 1$ ($|\lambda_n'| \le 1$), $n = 1, \ldots, N$, we can find numbers $\delta_n = \pm 1$, $n = 1, \ldots, N$, for which

$$\left\| \sum_{n=1}^N \lambda_n' f_n - \sum_{n=1}^N \delta_n f_n \right\|_2 \le \left\{ \sum_{n=1}^N \|f_n\|_2^2 \right\}^{1/2} \equiv I.$$

Then

$$\left\| \sum_{n=1}^N \frac{\lambda_n' + 1}{2} f_n - \sum_{n=1}^N \frac{\delta_n + 1}{2} f_n \right\|_2 \le \frac{1}{2} I.$$

If we set $\varepsilon_n = \frac{1}{2}(\delta_n + 1)$ ($\varepsilon_n = 0$ or 1), $n = 1, \ldots, N$, and use the equation $\frac{1}{2}(\lambda_n' + 1) = \lambda_n$, we obtain (102). Thus Lemma 1 is established.

LEMMA 2. *Let there be given functions* $g_n \in L^2(0,1)$, $n = 1, \ldots, N$, *and* $\sum_{n=1}^N g_n(x) = g(x)$. *There is a permutation* $\sigma = \{\sigma(n)\}_{n=1}^N$ *of the numbers* $(1, \ldots, N)$ *such that*

$$q(\sigma) \equiv \max_{1 \le m \le N} \left\| \sum_{n=1}^m g_{\sigma(n)} \right\|_2^2 \le C \left( \sum_{n=1}^N \|g_n\|_2^2 + \|g\|_2^2 \right)$$

(*C is an absolute constant*).

PROOF. If we use Corollary 2.8 (for $p = 2$), we find (here, as usual, $S(N)$ is the group of permutations of the numbers $(1, \ldots, N)$)

$$\min_{\sigma \in S(N)} q(\sigma) \le \frac{1}{N!} \sum_{\sigma \in S(N)} q(\sigma)$$

$$\le \frac{1}{N!} \sum_{\sigma \in S(N)} \int_0^1 \left( \max_{1 \le m \le N} \left| \sum_{n=1}^m g_{\sigma(n)}(x) \right| \right)^2 dx$$

$$= \int_0^1 \frac{1}{N!} \sum_{\sigma \in S(N)} \left( \max_{1 \le m \le N} \left| \sum_{n=1}^m g_{\sigma(n)}(x) \right| \right)^2 dx$$

$$\le C \left\{ \sum_{n=1}^N \int_0^1 g_n^2(x) \, dx + \int_0^1 g^2(x) \, dx \right\}.$$

LEMMA 3. *Let $B$ be a convex subset of $L^2(0,1)$ such that, for every function $g \in L^2(0,1)$ and every number $T$, there is a function $p \in B$ for which $(p, g) > T$. Then $B$ is dense in $L^2(0,1)$.*

PROOF. Suppose the contrary: the closure $\overline{B}$ of $B$ does not coincide with $L^2(0,1)$. Then there are $f_0 \in L^2(0,1)$ and $r > 0$ such that

$$K = \{f \in L^2(0,1): \|f - f_0\|_2 \le r\} \subset L^2(0,1)\backslash\overline{B}.$$

By a familiar corollary of the Hahn–Banach theorem, the convex sets $K$ and $\overline{B}$ are separated by linear functionals: for some $g_0 \in L^2(0,1)$ and a constant $T$, $(p, g) \le T \le (g_0, f)$ if $p \in \overline{B}$ and $f \in K$ (in the present case, the reference to the Hahn–Banach theorem could be replaced by simple geometric considerations). In particular, for all $p \in B$ we have the inequality $(g_0, p) \le T < \infty$, which contradicts the hypothesis of Lemma 3.

PROOF OF THEOREM 5. Let there be given functions $p \in L^2(0,1)$, an integer $N \ge 0$, and $\varepsilon > 0$. Let us show that there is a polynomial

$$Q(X) = Q(p, N, \varepsilon, x) = \sum_{n=N+1}^{N+M} \varepsilon_n f_n(x), \quad \varepsilon_n = 0, 1 \ (M = 1, 2, \dots), \quad (103)$$

for which

$$\|Q(p, N, \varepsilon, x) - p(x)\|_2 \le \varepsilon + A_N, \quad (104)$$

where

$$A_N = \left\{ \sum_{n=N+1}^{\infty} \|f_n\|_2^2 \right\}^{1/2} < \infty. \quad (105)$$

Let $\Omega_N$ be the set of polynomials of the form (103) and

$$\Lambda_N = \left\{ P(x): P(x) = \sum_{n=N+1}^{N+M} \lambda_n f_n(x), \ 0 \le \lambda_n \le 1, \ M = 1, 2, \dots \right\}. \quad (106)$$

It is easily seen that $\Lambda_N$ is convex and $\Lambda_N \supset \Omega_N$. It follows from conditions a) and c) that, for every $g \in L^2(0,1)$,

$$\sum_{n=1}^{\infty} (g, f_n)^+ = \infty, \quad (g, f_n)^+ = \max\{(g, f_n), 0\},$$

i.e.,

$$\sup_{Q \in \Omega_N} (g, Q) = \infty \text{ and hence } \sup_{p \in \Lambda_N} (g, p) = \infty. \quad (107)$$

It follows from (107) and Lemma 4 that $\Lambda_N$ is dense in $L^2(0,1)$, and we can find a polynomial $P \in \Lambda_N$ such that

$$\|P - p\|_2 \le \varepsilon. \quad (108)$$

Then, using Lemma 1, we can find a polynomial $Q \in \Omega_N$ for which

$$\|Q - P\|_2 \le \left\{ \sum_{n=N+1}^{M} \|f_n\|_2^2 \right\}^{1/2} \le A_N. \tag{109}$$

Combining (108) and (109), we obtain (104).

Now we take an arbitrary function $g \in L^2(0,1)$ and a sequence $\rho_k > 0$, $k = 1, 2, \ldots$, $\lim_{k \to \infty} \rho_k = 0$. Let $Q_1(x) = Q(g, 0, \rho_1, x)$. Then (see (103) and (104))

$$\|Q_1 - g\|_2 \le \rho_1 + A_0.$$

If $Q_1$ has the form $Q_1(x) = \sum_{n=1}^{M_1} \varepsilon_n^{(1)} f_n(x)$, $\varepsilon_n^{(1)} = 0$ or 1, we set

$$Q_1'(x) = \sum_{\{n \in [1, M_1] : \varepsilon_n^{(1)} = 1\}} f_n(x) + \delta_1 f_1(x),$$

where $\delta_1 = 1 - \varepsilon_1^{(1)}$ (i.e., the function $f_1(x)$ always "enters into" $Q_1'(x)$). It is clear that

$$\|g - Q_1'\|_2 \le \rho_1 + A_0 + \|f_1\|_2.$$

At the next step we consider the polynomial

$$Q_2(x) = Q(g - Q_1', M_1, \rho_2, x) = \sum_{n=M_1+1}^{M_2} \varepsilon_n^{(2)} f_n(x), \qquad \varepsilon_n^{(2)} = 0, 1,$$

and set

$$Q_2'(x) = \sum_{\{n \in [M_1+1, M_2] : \varepsilon_n^{(2)} = 1\}} f_n(x) + \delta_2 f_2(x),$$

where $\delta_2 = 0$ if $\varepsilon_2^{(1)} + \varepsilon_2^{(2)} > 0$ (i.e., if $f_2(x)$ entered into either $Q_1(x)$ or $Q_2(x)$), and $\delta_2 = 1$ otherwise. By construction (see (104))

$$\|g - Q_1' - Q_2''\|_2 \le \rho_2 + A_{M_1} + \|f_2\|_2.$$

Continuing this construction, we obtain a sequence of sums $Q_k'(x)$ of the form

$$Q_k'(x) = \sum_{i=1}^{i_k} f_{n_i}(x), \qquad k = 1, 2, \ldots. \tag{110}$$

Here each function $f_n(x)$, $n = 1, 2, \ldots$, enters into the sum (110) for exactly one value of $k$ and (see (104))

$$\|g - (Q_1' + Q_2' + \cdots + Q_k')\|_2 \le \rho_k + A_{M_k} + \|f_k\|_2, \\ k = 1, 2, \ldots \ (M_1 < M_2 < \cdots). \tag{111}$$

By b) the right-hand side of (111) tends to zero as $k \to \infty$; therefore it follows from (111) that there is a rearrangement $\sum_{n=1}^{\infty} f_{\sigma'(n)}(x)$ of (101) for which

$$\lim_{k \to \infty} \left\| g - \sum_{n=1}^{N_k} f_{\sigma'(n)} \right\|_2 = 0, \qquad N_1 < N_2 < \cdots . \tag{112}$$

Finally we use Lemma 2: we rearrange the functions $f_{\sigma'(n)}(x)$ by using the permutation $\{\sigma(n)\}$ inside the block $\{f_{\sigma'(n)}: N_k < n \leq N_{k+1}\}$, $k = 1, 2, \ldots,$ so that we have the estimate

$$\max_{N_k < m \leq N_{k+1}} \left\| \sum_{n=N_k}^{m} f_{\sigma(n)} \right\|_2^2 \leq C \left\{ \left\| \sum_{n=N_k+1}^{N_{k+1}} f_{\sigma'(n)} \right\|_2^2 + \sum_{n=N_k+1}^{N_{k+1}} \| f_{\sigma'(n)} \|_2^2 \right\},$$

$$k = 1, 2, \ldots \tag{113}$$

$[\{\sigma(n)\}_{n=N_k+1}^{N_{k+1}} = \{\sigma'(n)\}_{n=N_k+1}^{N_{k+1}}, k = 1, 2, \ldots]$. Then (see (112) and (113), b)) the series $\sum_{n=1}^{\infty} f_{\sigma(n)}(x)$ will converge in $L^2$ to $g(x)$. This establishes the universality of (101).

EXAMPLE. Let $\{\varphi_n(x)\}_{n=1}^{\infty}$ be a C.O.N.S. For $k = 0, 1, \ldots,$ we set $N_k = \frac{k(k+1)}{2}$ (i.e., $N_k - N_{k-1} = k$, $k = 1, 2, \ldots$) and consider the series

$$\sum_{n=1}^{\infty} f_n(x), \quad f_n(x) = \frac{(-1)^k}{k \ln(k+1)} \varphi_{n-N_{k-1}}(x),$$

$$\text{if } N_{k-1} < n \leq N_k, \quad k = 1, 2, \ldots . \tag{114}$$

Then (114) is a universal series for $L^2(0, 1)$ for rearrangements. To establish this, we verify the conditions of Theorem 5 for the sequence $\{f_n(x)\}_{n=1}^{\infty}$.

Since, for $k = 1, 3, 5, \ldots,$

$$\sum_{n=N_{k-1}+1}^{N_k} f_n(x) + \sum_{n=N_k+1}^{N_{k+1}} f_n(x) = \frac{(-1)^{k+1}}{(k+1)\ln(k+2)} \varphi_{k+1}(x) + \Delta_k(x),$$

where $\sum_k \| \Delta_k \|_2 < \infty$, and by Parseval's equation

$$\lim_{k \to \infty} \left\{ \max_{N_{k-1} < m \leq N_k} \left\| \sum_{n=N_{k-1}+1}^{m} f_n \right\|_2 \right\} = 0,$$

the series (114) converges in $L^2(0, 1)$. Moreover,

$$\sum_{n=1}^{\infty} \| f_n \|_2^2 = \sum_{k=1}^{\infty} \frac{k}{k^2 \ln^2(k+1)} < \infty.$$

Finally, since $\|g\|_2 > 0$ for every $g \in L^2(0,1)$, then since the system $\{\varphi_n\}$ is complete, some Fourier coefficient $(g, \varphi_{k_0})$, $1 \le k_0 < \infty$, is different from zero, and we obtain

$$\sum_{n=1}^{\infty} |(g, f_n)| \ge |(g, \varphi_{k_0})| \sum_{k=k_0}^{\infty} \frac{1}{k \ln k} = \infty.$$

This establishes that (114) is universal.

THEOREM 6. *For every C.O.N.S.* $\{\varphi_n(x)\}_{n=1}^{\infty}$, $x \in (0,1)$, *whose elements satisfy, for some $p > 2$, the condition*

$$\|\varphi_n\|_p \le M < \infty, \qquad n = 1, 2, \ldots, \tag{115}$$

*there is a series*

$$\sum_{n=1}^{\infty} a_n \varphi_n(x) \tag{116}$$

*which is universal with respect to rearrangements, for convergence a.e. in the space of measurable functions on $(0,1)$.*

REMARK. In Theorem 6, as in Theorem 1, we admit the possibility that the measurable functions take the values $+\infty$ and $-\infty$ on sets of positive measure.

The proof of the following auxiliary proposition is based on Lemma 1 for Theorem 1 and on the properties of random permutations of sets of numbers that were set forth in §2.4.

LEMMA 1. *Let the C.O.N.S.* $\{\varphi_n(x)\}_{n=1}^{\infty}$ *satisfy condition (115) with $p > 2$, and let $f(x)$ be measurable and finite a.e. For an arbitrary number $\delta \in (0,1)$ and $N = 1, 2, \ldots$, there are a polynomial*

$$P(x) = \sum_{N+1 \le n \le m} a_n \varphi_n(x),$$

*a permutation $\sigma = \{\sigma(n)\}_{n=N+1}^{m}$, and a set $E \subset (0,1)$, such that*
1) $m(E) \le \delta$;
2) $|P(x) - f(x)| \le \delta$ for $x \in (0,1)\backslash E$; and
3) $P^*(\sigma, x) \equiv \sup_{N < k \le m} |\sum_{n=N+1}^{k} a_{\sigma(n)} \varphi_{\sigma(n)}(x)| \le f(x) + \delta$ for $x \in (0,1)\backslash E$.

PROOF. Choose a function $F \in L^{\infty}(0,1)$ for which

$$m(E_0) \le \delta/4, \qquad E_0 \equiv \{x \in (0,1): f(x) \ne F(x)\},$$

and let $R = R(\delta, p, M, \|F\|_\infty)$ be a sufficiently large positive integer ($R$ will be specified later). We construct inductively a sequence of polynomials

$$P_r(x) = \sum_{n=N_{r-1}+1}^{N_r} a_n \varphi_n(x), \quad r = 1, \ldots, R, \ N = N_1 < N_1 < \cdots < N_R,$$

and sets $E_r$, $r = 1, \ldots, R$, that satisfy the following conditions:

I) $|P_r(x) - R^{-1}F(x)| \leq \delta/(2R)$, $x \in (0,1)\backslash E_r$, $m(E_r) \leq \delta/(2R)$;

II) $\|P_r\|_2 = (\sum_{n=N_{r-1}+1}^{N_r} a_n^2)^{1/2} \leq 4\|F\|_\infty (R\delta)^{-1/2}$.

The polynomial $P_r$ is constructed in the following way: applying Lemma 1 for Theorem 1 to the system

$$\left\{\frac{1}{R}F(x), \varphi_1(x), \ldots, \varphi_{N_{r-1}}(x)\right\},$$

$$\omega = (0,1), \quad \varepsilon_0 = \frac{\delta}{4R}, \quad \varepsilon = \frac{\delta^{3/2}}{4R^{3/2}(N_{r-1}+1)},$$

we obtain a function $F_r' \subset L^\infty(0,1)$ and a set $E_r' \subset (0,1)$ for which

a) $F_r'(x) = \frac{1}{R}F(x)$ for $x \in (0,1)\backslash E_r'$, $m(E_r') \leq \frac{\delta}{4R}$;

b) $\|F_r'\|_2 \leq 2(\frac{4R}{\delta})^{1/2}\frac{1}{R}\|F\|_2 \leq 4\|F\|_\infty/(\delta R)^{1/2}$;

c) $|c_n(F_r')| \leq \varepsilon$, $n = 1, \ldots, N_{r-1}$, where $c_n(f) \equiv \int_0^1 f\varphi_n\, dx$. We now choose, using the completeness of $\{\varphi_n\}$, a number $N_r > N_{r-1}$ so large that

$$\left\|\sum_{n=1}^{N_r} c_n(F_r')\varphi_n - F_r'\right\|_2 \leq \varepsilon, \tag{117}$$

and set

$$P_r(x) = \sum_{n=N_{r-1}+1}^{N_r} a_n \varphi_n(x), \quad \text{where } a_n = c_n(F_r'), \ n = N_{r-1}+1, \ldots, N_r.$$

$$\tag{118}$$

Then (see c) and (117))

$$\|P_r - F_r'\|_2 \leq (N_{r-1}+1)\varepsilon = \frac{\delta^{3/2}}{4}R^{-3/2},$$

and consequently, by Chebyshev's inequality,

$$m(E_r'') \equiv m\left\{x \in (0,1)\colon |P_r - F_r'| > \frac{\delta}{2R}\right\} \leq \frac{4R^2}{\delta^2}\frac{\delta^3}{16R^3} = \frac{\delta}{4R}.$$

If we set $E_r = E_r' \cup E_r''$, the preceding inequality and a) ensure that condition I) is satisfied. Condition II) is satisfied because of b) (also see (118)).

Let $m = N_r$ and

$$P(x) = \sum_{n=N+1}^{m} a_n \varphi_n(x) = \sum_{r=1}^{R} P_r(x). \tag{119}$$

We can now construct the required permutation $\sigma = \{\sigma(n)\}_{n=N+1}^{m}$. For this purpose we consider, for $r = 1, \ldots, R$, the set $S_r$ of all permutations of $\{N_{r-1} + 1, \ldots, N_r\}$, and for $\sigma = \{\sigma(n)\} \in S_r$ we set

$$P_r^*(\sigma, x) = \sup_{N_{r-1} \leq k \leq N_r} \left| \sum_{n=N_{r-1}+1}^{k} a_{\sigma(n)} \varphi_{\sigma(n)}(x) \right|.$$

By Corollary 2.8,

$$\sum \equiv \frac{1}{(N_r - N_{r-1})!} \sum_{\sigma \in S_r} \int_{(0,1)\backslash E_r} |P_r^*(\sigma, x)|^p \, dx$$

$$\leq C_p \int_{(0,1)\backslash E_r} (|P_r(x)| + Q_r(x))^p \, dx$$

$$\leq 2^p C_p \left[ \int_{(0,1)\backslash E_r} |P_r(x)|^p + \int_{(0,1)\backslash E_r} Q_r^p(x) \, dx \right], \tag{120}$$

where

$$Q_r(x) = \left( \sum_{n=N_{r-1}+1}^{N_r} a_n^2 \varphi_n^2(x) \right)^{1/2}, \qquad r = 1, \ldots, R,$$

and $C_p$ is the constant in Corollary 2.8. In addition, when $x \in (0,1)\backslash E_r$ (see property 1) of $P_r$)

$$|P_r(x)| \leq R^{-1}(\|F\|_\infty + \delta)$$

and (see the inequality 2.(75) and property II))

$$\int_{(0,1)\backslash E_r} Q_r^p \, dx \leq \left( \sum_{n=N_{r-1}+1}^{N_r} \|a_n \varphi_n\|_p^2 \right)^{p/2}$$

$$\leq M^p \left( \sum_{n=N_{r-1}+1}^{N_r} a_n^2 \right)^{p/2} \leq M^p [4\|F\|_\infty (R\delta)^{-1/2}]^p.$$

As a result,

$$\sum \leq 2^p C_p \left[ \left( \frac{\|F\|_\infty + \delta}{R} \right)^p + \left( \frac{M \cdot 4\|F\|_\infty}{\sqrt{R\delta}} \right)^p \right] \leq \frac{\delta^{p+1}}{2^{p+2}R}, \tag{121}$$

if the number $R = R(\delta, p, M, \|F\|_\infty)$ is sufficiently large. The inequality (121) ensures the existence of a permutation

$$\sigma_r = \{\sigma_r(n)\}_{n=N_{r-1}+1}^{N_r} \in S_r$$

for which

$$\int_{(0,1)\backslash E_r} |P_r^*(\sigma_r, x)|^p \, dx \le \frac{\delta^{p+1}}{2^{p+2}R},$$

and consequently, by Chebyshev's inequality,

$$m(G_r) \equiv m\left\{x \in (0,1): P_r^*(\sigma_r, x) > \frac{\delta}{2}\right\} \le \left(\frac{2}{\delta}\right)^p \frac{\delta^{p+1}}{2^{p+2}R} = \frac{\delta}{4R}. \quad (122)$$

Set $E = E_0 \cup [\bigcup_{r=1}^R (E_r \cup G_r)]$; then the permutation $\sigma = \{\sigma(n)\}_{n=N+1}^m$ can be defined by $\sigma(n) = \sigma_r(n)$ if $N_{r-1} < n \le N_r$, $r = 1, \ldots, R$. Then by (122) and property 1) we have $m(E) \le \delta$ and (also see (119)) $|P(x) - f(x)| \le \delta/2$ if $x \in (0,1)\backslash E$. Hence we have verified relations 1) and 2) from the statement of Lemma 1. In addition, by the construction of $\sigma$, we have, taking account of property 1 and (122), when $x \in (0,1)\backslash E$,

$$P^*(\sigma, x) \le \sum_{r=1}^R |P_r(x)| + \max_{1 \le r \le R} P_r^*(\sigma_r, x) \le |f(x)| + \delta,$$

i.e., 3) is also satisfied. This completes the proof of Lemma 1.

PROOF OF THEOREM 6. Let the sequence of functions $\{V_k(x)\}_{k=1}^\infty$ have the properties that $V_k \subset L^\infty(0,1)$ and $\|V_k\|_\infty = M_k$, $k = 1, 2, \ldots$, and that for every measurable $F(x)$ some subsequence $\{V_{k_p}\}$ converges to $F$ almost everywhere:

$$\lim_{p\to\infty} V_{k_p}(x) = F(x) \quad \text{for almost all } x \in (0,1) \quad (123)$$

(for example, we may take $\{V_k\}$ to be the system of Haar polynomials with rational coefficients; we notice, incidentally, that it follows easily from Theorem 1 that the necessary property is possessed also by the set of polynomials with rational coefficients in any C.O.N.S. that is made up of bounded functions).

We select even numbers $A_k$, $k = 1, 2, \ldots$, such that

$$A_1 < A_2 < \cdots, \quad \lim_{k\to\infty} M_k \cdot A_k^{-1} = 0, \quad \sum_{k=1}^\infty A_k^{-1} < 1, \quad (124)$$

and set $\Omega_k = \{B_{k-1} + 1, \ldots, B_k\}$, where $B_k = \sum_{j=1}^k A_j$, $k = 1, 2, \ldots$, $B_0 = 0$, and

$$v_i(x) = \frac{(-1)^i V_k(x)}{A_k} \quad \text{for } i \in \Omega_k, \ k = 1, 2, \ldots.$$

Then

$$\sum_{i\in\Omega_k} v_i(x) = 0, \quad \max_{B_{k-1} < N \le B_k} \left\| \sum_{i=B_{k-1}+1}^N V_i(x) \right\|_\infty \le \frac{M_k}{A_k}. \quad (125)$$

We first show that the series

$$\sum_{i=1}^{\infty} v_i(x) \qquad (126)$$

is universal, and then, using this result and Lemma 1, we establish the existence of a universal series for the system $\{\varphi_n\}$.

Let $F(x)$, $-\infty \le F(x) \le \infty$, be an arbitrary measurable function, and let (123) hold for the sequence $\{k_p\}_{p=1}^{\infty}$. Since the series

$$\sum_{i \notin \Omega} v_i(x), \qquad \Omega \equiv \bigcup_{p=1}^{\infty} \Omega_{k_p},$$

converges to zero everywhere (see (125)), it is enough to show that the series $\sum_{i \in \Omega} v_i(x)$, after a rearrangement $\sigma_0 = \{\sigma_0(i)\}_{i \in \Omega}$, converges to $F(x)$ almost everywhere (and then we set $\sigma_0(i) = i$ if $i \notin \Omega$).

Let us introduce the notation

$$b = \min\{i : i \in \Omega\} = B_{k_1 - 1} + 1; \qquad \alpha_i = \frac{(-1)^i}{A_k} \quad \text{if } i \in \Omega_k, \ k = 1, 2, \dots. \qquad (127)$$

We construct the rearrangement $\sigma_0$ of the set $\Omega$ of positive integers in the following way: $\sigma_0(b) = b + 1$ (notice that $\alpha_{b+1} > 0$), and if when $i \in \Omega$ the set $I = \{\sigma_0(j), j \in \Omega, \ j < i\}$ has already been defined, we set

$$\sigma_0(i) = \begin{cases} \min\{j : j \in \Omega \backslash I, \alpha_j > 0\} & \text{if } \sum_{j \in I} \alpha_j \le 1; \\ \min\{j : j \in \Omega \backslash I, \alpha_j < 0\} & \text{in the opposite case.} \end{cases}$$

(Notice also that, by construction, in the case when $i < j$ and $\operatorname{sgn} \alpha_{\sigma_0(i)} = \operatorname{sgn} \alpha_{\sigma_0(j)}$ we have $\sigma_0(i) < \sigma_0(j)$.) It is easily seen that $\{\sigma_0(i)\}_{i \in \Omega}$ is a permutation of the elements of $\Omega$ with the following properties:

1) the series $\sum_{i \in \Omega} \alpha_{\sigma_0(i)}$ converges and its sum $s = 1$;

2) for $\nu \ge b$,

$$s(\nu) \equiv \sum_{i \in \Omega, i \le \nu} \alpha_{\sigma_0(i)} = \sum_{p=1}^{p(\nu)} \frac{d(p, \nu)}{A_{k_p}}, \qquad d(p, \nu) \ge 0,$$

where $d(p, \nu)$ is the difference between the number of positive and the number of negative terms of modulus equal to $(A_{k_p})^{-1}$ that appear in the sum $s(\nu)$; moreover, $s(\nu) \to 1$ when $\nu \to \infty$.

Hence, since $\sum_{i \in \Omega_k} \alpha_i = 0$, $k = 1, 2, \dots$, we have

$$p'(\nu) \equiv \min\{p : d(p, \nu) \ne 0\} \to \infty \quad \text{as } \nu \to \infty. \qquad (128)$$

Using properties 1) and 2) of $\sigma_0$, we obtain

$$S(\nu, x) \equiv \sum_{i \in \Omega, i \leq \nu} v_{\sigma_0(i)}(x) = \sum_{p=p'(\nu)}^{p(\nu)} \frac{d(p, \nu)}{A_{k_p}} V_{k_p}(x), \qquad (129)$$

and consequently, for such $x$, we have $|F(x)| < \infty$,

$$|S(\nu, x) - s(\nu)F(x)| \leq 2 \sup_{p \geq p'(\nu)} |V_{k_p}(x) - F(x)|. \qquad (129')$$

It follows from (123), (129'), and (128) that, for almost all $x$ in the set where $|F(x)| < \infty$, the series

$$\sum_{i \in \Omega} v_{\sigma_0(i)}(x) \qquad (130)$$

convergence a.e. to $F(x)$. The convergence a.e. of (130) to $+\infty$ or $-\infty$ on the sets where $F(x)$ takes one of the values $+\infty$ or $-\infty$ follows easily from (123), (129), and (128). Hence we have shown that the series (126) is universal.

If we now apply Lemma 1, we can find polynomials in the system $\{\varphi_n(x)\}$,

$$P_i(x) = \sum_{n=\nu_{i-1}+1}^{\nu_i} a_n \varphi_n(x), \qquad 0 = \nu_0 < \nu_1 < \cdots, \quad i = 1, 2, \ldots,$$

and permutations $\sigma_i$ of the sets of numbers $\{\nu_{i=1} + 1, \ldots, \nu_i\}$, $i = 1, 2, \ldots$, such that

$$m\{x: |v_i(x) - P_i(x)| \geq 2^{-i}\} \leq 2^{-i}, \qquad (131)$$

$$m\{x: |P_i^*(\sigma_i, x) \geq |v_i(x)| + 2^{-i}\} \leq 2^{-i}, \qquad (132)$$

and we can prove that therefore for the given coefficients $a_n$, $n = 1, 2, \ldots$, the series (116) is universal. In fact, by (131)

$$\sum_{i=1}^{\infty} |v_i(x) - P_i(x)| < \infty \quad \text{for almost all } x \in (0, 1),$$

and therefore the convergence a.e. of the series (130) to $F(x)$ implies the convergence a.e. of the series $\sum_{i=1}^{\infty} P_{\sigma_0(i)}(x)$ to the function $F(x) + G_0(x)$, where

$$G_0(x) = \sum_{i=1}^{\infty} (P_i(x) = v_i(x)).$$

Moreover, by using (132) and the fact that $\|v_i\|_\infty \to 0$ as $i \to \infty$ (see (124)), we can assert that every rearrangement of the series (116) converges a.e. to $F + G_0$. It follows that the series (116) is universal. This completes the proof of Theorem 6.

# Some Topics from the Theory of Functions of a Real Variable and Functional Analysis

## §1. Equations for integrals. Moduli of continuity

**1°. PROPOSITION 1.** *Let $f \in L^1(0,1)$, and for $t \in R^1$ let*

$$\lambda_f(t) = m\{x \in (0,1): |f(x)| > t\}; \tilde{\lambda}_f(t) = m\{x \in (0,1): f(x) > t\}.$$

*Then*

$$\int_0^1 f(x)\,dx = -\int_{-\infty}^{\infty} t\,d\tilde{\lambda}_f(t), \tag{1}$$

*and if $f \in L^p(0,1)$, $0 < p < \infty$, then*

$$\int_0^1 |f(x)|^p\,dx = -\int_0^{\infty} t^p\,d\lambda_f(t) = p\int_0^{\infty} t^{p-1}\lambda_f(t)\,dt. \tag{2}$$

PROOF. Equation (1) follows from the definitions of the Lebesgue and Lebesgue–Stieltjes integrals:

If $\{t_i^{(n)}\}_{i=-\infty}^{\infty}$, $n = 1, 2, \ldots$, is a sequence of partitions of the real line:

$$\cdots < t_{-k}^{(n)} < \cdots < t_{-1}^{(n)} < t_0^{(n)} < t_1^{(n)} < \cdots < t_k^{(n)} < \cdots,$$

$$\lim_{n\to\infty}\left[\sup_{-\infty < i < \infty}(t_{i+1}^{(n)} - t_i^{(n)})\right] = 0,$$

then the integrals $\int_0^1 f_n(x)\,dx$, where $f_n(x) = t_i^{(n)}$ if $x \in E_i^{(n)} \equiv \{x \in (0,1): t_i^{(n)} < f(x) \le t_{i+1}^{(n)}\}$, $i = 0, \pm 1, \pm 2, \ldots$, tend (as $n \to \infty$) to $\int_0^1 f(x)\,dx$. On the other hand,

$$\int_0^1 f_n(x)\,dx = \sum_{i=-\infty}^{\infty} t_i^{(n)} m(E_i^{(n)})$$

$$= -\sum_{i=-\infty}^{\infty} t_i^{(n)}[\tilde{\lambda}_f(t_{i+1}^{(n)}) - \tilde{\lambda}_f(t_i^{(n)})] \to \int_{-\infty}^{\infty} t\,d\tilde{\lambda}_f(t) \quad \text{as } n \to \infty,$$

which proves (1).

Now let $f \in L^p(0,1)$, $0 < p < \infty$. By (1), if we use the equation $\tilde{\lambda}_{|f|}(t) = 1$ for $t < 0$, we obtain

$$\int_0^1 |f(x)|^p \, dx = -\int_{-\infty}^{\infty} t \, d\tilde{\lambda}_{|f|^p}(t) = -\int_0^{\infty} t \, d\tilde{\lambda}_{|f|^p}(t). \tag{2'}$$

Since

$$\tilde{\lambda}_{|f|^p}(t) = m\{x \in (0,1) : |f(x)|^p > t\}$$
$$= m\{x \in (0,1) : |f(x)| > t^{1/p}\} = \lambda_f(t^{1/p}),$$

for $t > 0$, then from (2') (making the change of variable $t = (t')^p$) we at once obtain the first equation (2). In addition, for every $A > 0$,

$$-\int_0^A t^p \, d\lambda_f(t) = \left. -t^p \lambda_f(f) \right|_0^A + \int_0^A \lambda_f(t) \, dt^p$$

$$= -A^p \lambda_f(A) + p \int_0^A \lambda_f(t) \cdot t^{p-1} \, dt.$$

To prove the second equation in (2), it is enough to let $A \to \infty$ in the preceding relation and use the inequality

$$A^p \lambda_f(A) \leq \int_{\{x : |f(x)| > A\}} |f(x)|^p \, dx = o(1) \quad \text{as } A \to \infty.$$

REMARK 1. It is easy to see that the finiteness of the integral

$$\int_0^{\infty} t^{p-1} \lambda_f(t) \, dt$$

is equivalent to having $f(x)$ belong to $L^p(0,1)$.

REMARK 2. If $f(x)$ is defined on $R^1$ then by equation (2) for $f_k(x) \equiv f(x+k)$, $x \in (0,1)$, $k = 0, \pm 1, \ldots$, if we use the equation

$$\lambda_f(t) = m\{x \in R^1 : |f(x)| > t\} = \sum_{k=-\infty}^{\infty} \lambda_{f_k}(t)$$

we obtain

$$\int_{R^1} |f(x)|^p \, dx = \sum_{k=-\infty}^{\infty} \int_0^1 |f_k(x)|^p \, dx$$

$$= \sum_{k=-\infty}^{\infty} p \int_0^{\infty} t^{p-1} \lambda_{f_k}(t) \, dt$$

$$= p \int_0^{\infty} t^{p-1} \lambda_f(t) \, dt, \qquad 0 < p < \infty. \tag{3}$$

**2°. Moduli of continuity.**   Let there be given a function $f \in C(0,1)$ (or $f(x) \in L^p(0,1)$, $1 \le p < \infty$). The *modulus of continuity* and the *integral modulus of continuity* are defined by the equations

$$\omega(\delta, f) = \sup_{\substack{0 \le x < x+h \le 1 \\ 0 < h \le \delta}} |f(x+h) - f(x)|, \qquad 0 < \delta < 1,$$

$$\left( \omega_p(\delta, f) = \left\{ \sup_{0 < h \le \delta} \int_0^{1-h} |f(x+h) - f(x)|^p \, dx \right\}^{1/p}, \qquad 0 < \delta < 1 \right).$$
$$(4)$$

The modulus of continuity of second order of a function $f \in C(0,1)$, and the integral modulus of continuity of second order of a function $f \in L^p(0,1)$, $1 \le p < \infty$, are defined by

$$\omega^{(2)}(\delta, f) = \sup_{\substack{0 \le x-h < x+h \le 1 \\ 0 < h \le \delta}} |f(x+h) + f(x-h) - 2f(x)|, \qquad (5)$$

$0 < \delta \le \frac{1}{2}$, and

$$\omega_p^{(2)}(\delta, f) = \left\{ \sup_{0 < h < \delta} \int_h^{1-h} |f(x+h) + f(x-h) - 2f(x)|^p \, dx \right\}^{1/p}. \qquad (5')$$

For periodic functions the moduli of continuity are defined somewhat differently:

If $f(x)$ is defined on $R^1$ and has period $T > 0$, then

a)

$$\omega(\delta, f) = \sup_{\substack{0 < h \le \delta \\ x \in [0,T]}} |f(x+h) - f(x)|,$$

$$\omega^{(2)}(\delta, f) = \sup_{\substack{0 < h \le \delta \\ x \in [0,T]}} |f(x+h) + f(x-h) - 2f(x)|,$$

$$\delta > 0, \quad f \in C(0,T); \quad (6)$$

b)

$$\omega_p(\delta, f) = \left\{ \sup_{0 < h \le \delta} \int_0^T |f(x+_h) - f(x)|^p \, dx \right\}^{1/p},$$

$$\omega_p^{(2)}(\delta, f) = \left\{ \sup_{0 < h \le \delta} \int_0^T |f(x+h) + f(x-h) - 2f(x)|^p \, dx \right\}^{1/p},$$

$$\delta > 0, \quad f \in L^p(0,T). \quad (6')$$

We record the fundamental properties of moduli of continuity:

(I) $\quad \lim_{\delta \to 0} \omega(\delta, f) = \lim_{\delta \to 0} \omega^{(2)}(\delta, f)$

$$= \lim_{\delta \to 0} \omega_p(\delta, f) = \lim_{\delta \to 0} \omega_p^{(2)}(\delta, f) = 0,$$

(II) $\qquad \omega(2\delta, f) \leq 2\omega(\delta, f),$

$$\omega^{(2)}(2\delta, f) \leq 4\omega^{(2)}(\delta, f),$$

$$\omega_p(2\delta, f) \leq 2\omega_p(\delta, f),$$

$$\omega_p^{(2)}(2\delta, f) \leq 4\omega_p^{(2)}(\delta, f), \qquad (7)$$

(III) $\qquad \omega^{(2)}(\delta, f) \leq 2\omega(\delta, f),$

$$\omega_p^{(2)}(\delta, f) \leq 2\omega_p(\delta, f).$$

Relations (II) and (III) in (7) follow immediately from the definitions. When $f \in C(0,1)$, equations (I) follow at once from the uniform continuity of $f$ on $[0,1]$. If $f \in L^p(0,1)$, $1 \leq p < \infty$, then if we find, for a given $\varepsilon > 0$, a function $g \in C(0,1)$ such that $\|g - f\|_p \leq \varepsilon/3$, we will have, for sufficiently small $h > 0$,

$$\|f(x - h) - f(x)\|_{L^p(0,1-h)} \leq \|f(x + h) - g(x + h)\|_{L^p(0,1-h)}$$
$$+ \|g(x + h) - g(x)\|_{L^p(0,1-h)}$$
$$+ \|g(x) - f(x)\|_{L^p(0,1-h)} \leq \varepsilon,$$

which proves that $\lim_{\delta \to 0} \omega_p(\delta, f) = 0$.

PROPOSITION 2. *Let* $f(x)$, $x \in R^1$, *have period* $T$ *and bounded variation on* $[0, T]$. *Then*

$$\omega_1(\delta, f) \leq 4\delta V(f), \qquad 0 < \delta < T,$$

*where* $V(f) \equiv V_0^T(f)$ *is the total variation of* $f(x)$ *on* $[0, T]$.

PROOF. Let $f_1(x) = V_0^x(f)$, $f_2(x) = f_1(x) - f(x)$, $x \in [0, T]$. Since $f_1(x)$ and $f_2(x)$ are nondecreasing functions, we obtain for $h \in (0, T)$, $i = 1, 2$,

$$\int_0^{T-h} |f_i(x + h) - f_i(x)| \, dx = \int_0^{T-h} f_i(x + h) \, dx - \int_0^{T-h} f_i(x) \, dx$$
$$= \int_h^T f_i(x) \, dx - \int_0^{T-h} f_i(x) \, dx$$
$$= \int_{T-h}^T f_i(x) \, dx - \int_0^h f_i(x) \, dx$$
$$\leq [f_i(T) - f_i(0)]h.$$

Consequently for $h \in (0, T)$

$$\int_0^T |f(x+h) - f(x)| \, dx = \int_0^{T-H} + \int_{T-h}^T$$
$$\leq [f_1(T) - f_1(0) + f_2(T) - f_2(0)] \cdot h$$
$$+ h \sup_{x,x' \in [0,T]} |f(x) - f(x')|$$
$$\leq [2\{f_1(T) - f_1(0)\} + f(0) - f(T)]h + h \cdot V(f)$$
$$\leq 4hV(f),$$

as was to be proved.

If $\omega(\delta)$ is convex upward and $\omega(0) = 0$, we set

$$H_\omega = \{ f \in C(0,1) : \omega(\delta, f) = O(\omega(\delta)) \text{ as } \delta \to 0 \}. \tag{8}$$

The class $H_{\delta^\alpha}$, $0 < \alpha \leq 1$ (Lipschitz class of order $\alpha$), is denoted by Lip $\alpha$ in the text.

$3°$. In studying properties of the partial sums of Fourier series in orthonormal systems, there arise the operators

$$S(f) = S(f, x) = \int_0^1 f(t) K(x, t) \, dt.$$

It is clear that if the kernel $K(x, t) \in C([0, 1] \times [0, 1])$, then $S(f)$ acts from $C(0, 1)$ to $C(0, 1)$. Moreover, it is easy to see that

$$\|S\|_{C \to C} \equiv \sup_{\|f\|_{C(0,1)} \leq 1} \|S(f)\|_{C(0,1)} \|_{C(0,1)} = \max_{x \in [0,1]} \int_0^1 |K(x,t)| \, dt. \tag{9}$$

It is also clear that when $K(x, t) \in L^\infty([0, 1] \times [0, 1])$, the operator $S(f)$ acts from $L^1(0, 1)$ to $L^\infty(0, 1)$. Let us verify that in this case

$$\|S\|_{L^1 \to L^1} \equiv \sup_{\|f\|_1 \leq 1} \|S(f)\|_1 = \left\| \int_0^1 |K(x,t)| \, dx \right\|_\infty. \tag{10}$$

Set $I(t) = \int_0^t |K(x,t)| \, dx$, where $t \in [0, 1]$. If $\|f\|_1 \leq 1$, then

$$\|S(f)\|_1 \leq \int_0^1 |f(t)| \int_0^1 |K(x,t)| \, dx \, dt \leq \|I(t)\|_\infty,$$

i.e.,

$$\|S\|_{L^1 \to L^1} \leq \left\| \int_0^1 |K(x,t)| \, dx \right\|_\infty. \tag{11}$$

To prove an inequality in the opposite direction, we consider the set $E$ of the points $y \in (0, 1)$ for which

$$\lim_{n \to \infty} 2n \int_{y-1/n}^{y+1/n} K(x,t)\,dt = K(x,y) \quad \text{for almost all } x \in (0,1). \quad (12)$$

By Lebesgue's theorem on the differentiation of absolutely continuous functions and Fubini's theorem,

$$m(E) = 1. \quad (13)$$

For a given $y \in E$ we set

$$f_n(t) = \begin{cases} 2n & \text{if } t \in (y - 1/n, y + 1/n), \\ 0 & \text{if } t \in (0, 1) \backslash (y - 1/n, y + 1/n), \end{cases} \quad n = 1, 2, \dots .$$

Then $\|f_n(t)\|_1 = 1$, $n = 1, 2, \dots$, and by applying Fatou's theorem we find that for every $y \in E$

$$I(y) = \int_0^1 |K(x,y)|\,dy \leq \varlimsup_{n \to \infty} \int_0^1 \left| 2n \int_{y-1/n}^{y+1/n} K(x,t)\,dt \right| dx$$

$$= \varlimsup_{n \to \infty} \int_0^1 \left| \int_0^1 f_n(t) K(x,t)\,dt \right| dx = \varlimsup_{n \to \infty} \|S(f_n)\|_1 \leq \|S\|_{L^1 \to L^1}.$$

Therefore (see (13)) $\|I(y)\|_\infty < \|S\|_{L^1 \to L^1}$ and (also see (11)) we obtain inequality (10), as required.

## §2. The maximal function and interpolation theorems

### 1°. Marcinkiewicz's interpolation theorem.

DEFINITION 1. An operator $T$ from $L^p(R^1)$ ($1 \leq p < \infty$) to $L^0(R^1)$ is an *operator of weak type* $(p,p)$ if

$$m\{x \in R^1 : |T(f,x)| > y\} \leq \frac{M}{y^p} \|f\|_{L^p(R^1)}^p, \qquad f \in L^p(R^1), \quad (14)$$

for every $y > 0$, and an *operator of type* $(p,p)$ if

$$\|T(f)\|_{L^p(R^1)} \leq M\|f\|_{L^p(R^1)}, \qquad f \in L^p(R^1) \quad (15)$$

(the constants $M$ in (14) and (15) are independent of $f(x)$). If $p = \infty$, we say that $T$ is of *weak type* $(p,p)$ if it is of type $(p,p)$.

It follows immediately from Chebyshev's inequality that an operator of type $(p,p)$ is also of weak type $(p,p)$. We have the following theorem.

THEOREM 1. *If a linear operator $T$ has weak type $(p_1, p_1)$ and also weak type $(p_2, p_2)$ $(1 \leq p_1 \leq p_2 \leq \infty)$, then $t$ has type $(p, p)$ for every $p$ on the interval $(p_1, p_2)$.*

REMARK. We assume that $T$ can be extended to a linear operator on the space

$$L^{p_1}(R^1) \oplus L^{p_2}(R^1) = \{f(x): f(x) = f_1(x) + f_2(x), \\ f_1(x) \in L^{p_1}(R^1), f_2(x) \in L^{p_2}(R^1)\}, \tag{16}$$

i.e., (see (16)) $T(f, x) = T(f_1, x) + T(f_2, x)$ for every function $f \in L^{p_1}(R^1) \oplus L^{p_2}(R^1)$. It is easy to see that this definition of $T(f, x)$ is independent of the choice of $f_1(x)$ and $f_2(x)$ in (16). Since $L^p(R_1) \subset L^{p_1}(R^1) \oplus L^{p_2}(R^1)$, $p_1 < p < p_2$, it follows that $T$ is also defined on $L^p(R^1)$.

PROOF. We carry out the proof of Theorem 1 on the assumption that $p_2 < \infty$ (it is only this case that we actually use in the text). If $p_2 = \infty$ the proof requires a few changes (in fact, the necessary discussion is given below in the proof of Theorem 2).

We consider a function $f \in L^p(0, 1)$ and a number $t \geq 0$, and estimate the function

$$\lambda(t) = m\{x \in R^1 : |T(f, x)| > t\}.$$

For $x \in R^1$, set

$$h(x) = \begin{cases} f(x) & \text{if } |f(x)| > t, \\ 0 & \text{if } |f(x)| \leq t, \end{cases} \qquad g(x) = f(x) - h(x),$$

Then $h \in L^{p_1}(R^1)$, $g \in L^{p_2}(R^1)$, and $T(f, x) = T(h, x) + T(g, x)$. Consequently

$$\lambda(t) \leq m\{x \in R^1 : |T(h, x)| > t/2\} + m\{x \in R^1 : |T(g, x)| > t/2\}.$$

Hence, using the inequalities for weak types $(p_1, p_1)$ and $(p_2, p_2)$ (see (14)), we find that when $t > 0$

$$\lambda(t) \leq \frac{2^{p_1} M_1}{t^{p_1}} \int_{R^1} |h(x)|^{p_1} dx + \frac{2^{p_2} M_2}{t^{p_2}} \int_{R^1} |g(x)|^{p_2} dx$$

$$= \frac{2^{p_1} M_1}{t^{p_1}} \int_{\{x \in R^1 : |f(x)| > t\}} |f(x)|^{p_1} dx$$

$$+ \frac{2^{p_2} M_2}{t^{p_2}} \int_{\{x \in R^1 : |f(x)| \leq t\}} |f(x)|^{p_2} dx.$$

From the preceding inequality and (3) we obtain

$$
\begin{aligned}
\|T(f)\|^p_{L^p(R^1)} &= p \int_0^\infty t^{p-1}\lambda(t)\,dt \\
&\le p2^{p_1}M_1 \int_0^\infty \frac{t^{p-1}}{t^{p_1}} \int_{\{x\in R^1:\,|f(x)|>t\}} |f(x)|^{p_1}\,dx\,dt \\
&\quad + p2^{p_2}M_2 \int_0^\infty \frac{t^{p-1}}{t^{p_2}} \int_{\{x\in R^1:\,|f(x)|>t\}} |f(x)|^{p_2}\,dx\,dt \\
&= p2^{p_1}M_1 \int_{R^1} |f(x)|^{p_1} \int_0^{|f(x)|} t^{p-p_1-1}\,dt\,dx \\
&\quad + p2^{p_2}M_2 \int_{R^1} |f(x)|^{p_2} \int_{|f(x)|}^\infty t^{p-p_2-1}\,dt\,dx \\
&= \frac{p\cdot 2^{p_1}M_1}{p-p_1} \int_{R^1} |f(x)|^p\,dx + \frac{p\cdot 2^{p_2}M_2}{|p_2|-|p|} \int_{R^1} f(x)|^p\,dx \\
&= M\cdot \|f\|^p_{L^p(R^1)},
\end{aligned}
$$

i.e., $T$ has type $(p,p)$. This completes the proof of Theorem 1.

**2°. The maximal function.**

DEFINITION 2. Let $f(x)$, $x \in R^1$, be summable on every interval $(-A,A)$, $A > 0$. The *maximal function* of $f(x)$ is the function

$$
M(f,x) = \sup_{I\ni x} |I|^{-1} \int_I |f(t)|\,dt, \qquad x \in R^1,
$$

where the supremum is taken over all intervals $I$ that contain $x$.

THEOREM 2. a) *If* $f(x) \in L^1(R^1)$, *then for every* $t > 0$

$$
m\{x \in R^1 : M(f,x) > t\} \le \frac{5}{t} \int_{R^1} |f(x)|\,dx.
$$

b) *If* $f(x) \in L^p(R^1)$, $1 < p \le \infty$, *then*

$$
\|M(f)\|_{L^p(R^1)} \le C_p \|f\|_{L^p(R^1)},
$$

*where* $C_p$ *is a constant that depends only on* $p$.

LEMMA 1. *Let* $E \subset R^1$ *be a measurable set and* $\{I_\alpha\}$ *a family of intervals of bounded length which cover* $E$: $E \subset \bigcup_\alpha I_\alpha$. *Then we can extract from this family a sequence* $I_1, I_2, \ldots$ *of pairwise disjoint intervals (finite or infinite) such that*

$$
\sum_k |I_k| \ge \frac{1}{5} m(E). \tag{17}
$$

PROOF OF LEMMA 1. We will construct the intervals $I_1, I_2,\ldots$, inductively. At the first step we select from $\{I_\alpha\}$ an interval $I_1$ for which

$|I_1| \geq \frac{1}{2} \sup_\alpha |I_\alpha|$. Now suppose that $I_1, \ldots, I_k$ have already been constructed; we find an interval $I_{k+1}$, disjoint from $I_1, \ldots, I_k$, with

$$|I_{k+1}| \geq \tfrac{1}{2} \sup\{|I_\alpha| \colon I_\alpha \cap I_j = \varnothing, \ j = 1, \ldots, k\}.$$

If there is no such interval, the construction stops.

We have now defined the sequence $I_1, I_2, \ldots$. Let us prove (17). We need to consider only the case when $\sum_k |I_k| < \infty$. Let $I_k^*$ be an interval with the same midpoint as $I_k$ but five times as long: $|I_k^*| = 5|I_k|$. Let us show that

$$E \subset \bigcup_k I_k^*. \tag{18}$$

It is enough to verify that every $I$ in $\{I_\alpha\}$ is contained in $\bigcup_k I_k^*$.

Since $\sum_k |I_k| < \infty$, we have $\lim_{k \to \infty} |I_k| = 0$ and, for sufficiently large $k$,

$$|I_{k+1}| > |I|/2. \tag{19}$$

Let $k_0$ $(1 \leq k_0 < \infty)$ be the smallest $k$ satisfying (19). By construction, $I$ intersects one of the intervals $I_1, \ldots, I_{k_0}$, i.e., for some $j$ we have $1 \leq j \leq k_0$, $I \cap I_j \neq \varnothing$, and $\frac{1}{2}|I| \leq |I_j|$. But then $I \subset I_j^*$ and we have obtained (18). Finally, from (18) we have

$$m(E) \leq \sum_k |I_k^*| = 5 \sum_k |I_k|.$$

PROOF OF THEOREM 2. Let $f(x) \in L^1(R^1)$. Choose a number $t > 0$ and set

$$E_t = \{x \in R^1 \colon M(f, x) > t\}.$$

By the definition of the maximal function, for every $x \in E_t$ there is an interval $I_x$ such that

$$I_x \ni x \quad \text{and} \quad \int_{I_x} |f(u)| \, du > t|I_x|. \tag{20}$$

The family $\{I_x\}_{x \in E_t}$ of bounded intervals (see (20)) covers $E_t$. By Lemma 1 there is a sequence $\{I_k\}$ of disjoint intervals with

$$\sum_k |I_k| \geq \frac{1}{5} m(E_t). \tag{21}$$

Since (see (20))

$$\sum_k |I_k| < \frac{1}{t} \sum_k \int_{I_k} |f(u)| \, du \leq \frac{1}{t} \int_{R^1} |f(u)| \, du,$$

it follows from (21) that $m(E_t)$ is finite and

$$m(E_t) < \frac{5}{t} \int_{R^1} |f(u)| \, du.$$

Now we can prove Proposition b). For $p = \infty$ the inequality b) (with $C_\infty = 1$) is evident. If $1 < p < \infty$, then with a given $t > 0$ we obtain

$$h(x) = \begin{cases} f(x) & \text{if } |f(x)| > \dfrac{t}{2}, \\ 0 & \text{if } |f(x)| \le \dfrac{t}{2}, \end{cases} \quad x \in R^1.$$

Then $|f(x)| \le |h(x)| + t/2$ and $M(f, x) \le M(h, x) + t/2$, $x \in R^1$. Consequently

$$E_t \equiv \{x \in R^1 : M(f, x) > t\} \subset \{x \in R^1 : M(h, x) > t/2\}$$

and (see a))

$$m(E_t) \le \frac{10}{t} \|h\|_{L^1(R^1)} = \frac{10}{t} \int_{\{x \in R^1 : |f(x)| > t/2\}} |f(x)| \, dx.$$

Finally, we use (3):

$$\int_{R^1} M^p(f, x) \, dx = p \int_0^\infty t^{p-1} m(E_t) \, dt$$

$$\le 10p \int_0^\infty t^{p-1} \cdot \frac{1}{t} \int_{\{x \in R^1 : |f(x)| > t/2\}} |f(x)| \, dx$$

$$= 10p \int_{R^1} |f(x)| \int_0^{2|f(x)|} t^{p-2} \, dt \, dx = C_p \int_{R^1} |f(x)|^p \, dx.$$

REMARK. Lebesgue's theorem on the differentiability a.e. of an indefinite integral is easily deduced from part a) of Theorem 2.

## §3. Some topics from functional analysis

**1°. DEFINITION 3.** Let $\Phi = \{\varphi_n(x)\}_{n-1}^N \subset L^2(E)$ be a system of functions defined on a set $E$ ($E \subset R^2, 1 \le N \le \infty$). The *Gram matrix* of $\Phi$ is the matrix

$$G = G_\Phi = \{u_{n,j}\}_{n,j=1}^N, \quad \text{where } u_{n,j} = \int_E \varphi_n(x)\varphi_j(x) \, dx.$$

It is not difficult to verify (this is done for all Euclidean spaces in courses on linear algebra) the following proposition.

**PROPOSITION 3.** *A system $\Phi = \{\varphi_n\}_{n=1}^N \subset L^2(E)$, $1 \le N < \infty$, is linearly independent if and only if the determinant $\det G_\Phi \ne 0$.*

**PROPOSITION 4.** *Let there be given in $L^2(E)$ two systems $\Phi = \{\varphi_n(x)\}_{n=1}^N$ and $\Psi = \{\psi_n(x)\}_{n=1}^\infty$ of functions, $1 \le N < \infty$, and let $L_\Phi$, $L_\Psi$ be the* *subspaces spanned by the functions* $\{\varphi_n\}_{n=1}^N$ *and* $\{\psi_n\}_{n=1}^N$, *respectively. Then if* $\Phi$ *and* $\Psi$ *have equal Gram matrices* $(G_\Phi = G_\Psi)$, *there is an isometry* $T: L_\Phi \to L_\Psi$ *(i.e.,* $\|T(f)\|_{L^2(E)} = \|f\|_{L^2(E)}$ *if* $f \in L_\Phi$*) such that*

$$T(\varphi_n) = \psi_n, \qquad n = 1, \ldots, N. \tag{22}$$

### 2°. Minimax theorem.

THEOREM (VON NEUMANN). *Let* $A = \{a_{i,j}\}$ $(a_{i,j} \geq 0; 1 \leq i \leq m, 1 \leq j \leq n)$ *be a matrix with nonnegative elements and*

$$\sigma_m = \{x \in R^m : x = (x_1, \ldots, x_m), x_i \geq 0, \sum_{i=1}^m x_i = 1\}.$$

*Then the quadratic form*

$$F(x, y) = \sum_{i=1}^m \sum_{j=1}^n a_{i,j} x_i y_j \qquad (x = \{x_i\}, y = \{y_j\})$$

*satisfies*

$$\min_{y \in \sigma_n} \max_{x \in \sigma_m} F(x, y) = \max_{x \in \sigma_m} \min_{y \in \sigma_n} F(x, y). \tag{23}$$

PROOF. Let $\gamma$ and $\rho$ denote the left-hand and right-hand sides of (23). The inequality $\gamma \geq \rho$ can be verified very simply. In fact, if $x_0 \in \sigma_m$ is a point such that $\min_{y \in \sigma_n} F(x_0, y) = \rho$, then by using the inequality $\max_{x \in \sigma_m} F(x, y) \geq F(x_0, y)$, $y \in \sigma_n$, we obtain

$$\gamma = \min_{\nu \in \sigma_n} \max_{x \in \sigma_m} F(x, y) \geq \min_{y \in \sigma_n} F(x_0, y) = \rho.$$

To prove the converse inequality $\gamma \leq \rho$ we consider the closed sets

$$A(x, \varepsilon) = \{y \in \sigma_n : F(x, y) \leq \rho + \varepsilon\}, \qquad x \in \sigma_m, \quad \varepsilon > 0.$$

It is enough to show that for each $\varepsilon > 0$

$$\bigcap_{\nu=1}^s A(x^\nu, \varepsilon) \neq \varnothing, \qquad s = 1, 2, \ldots, \tag{24}$$

where $x^\nu = \{x_1^\nu, \ldots, x_m^\nu\}$, $\nu = 1, 2, \ldots$, is a sequence which is everywhere dense in $\sigma_m$. In fact, if follows from (24) and the fact that $A(x, \varepsilon)$ is closed that there is a point $y_0 \in \sigma_n$ such that $y_0 \in A(x^\nu, \varepsilon)$ for $\nu = 1, 2, \ldots$, i.e.,

$$\max_{1 \leq \nu < \infty} F(x^\nu, y_0) \leq \rho + \varepsilon. \tag{25}$$

Since $F(x, y)$ is continuous, it follows from (25) that $\max_{x \in \sigma_m} F(x, y_0) \leq \rho + \varepsilon$, whence, since $\varepsilon > 0$ is arbitrary, we obtain the inequality $\gamma \leq \rho$.

Let the numbers $s = 1, 2, \ldots$ and $\varepsilon > 0$ be given, and let

$$P = \{(\xi_1, \ldots, \xi_s) \in R^s : \xi_\nu \leq \rho + \varepsilon, \nu = 1, \ldots, s\}.$$

In addition, let $Q \subset R^s$ be the set of vectors of the form

$$(F(x^1, y), F(x^2, y), \ldots, F(x^s, y)), \qquad y \in \sigma_n.$$

It is clear that $Q$ is convex, closed and bounded. Let us show that

$$P \cap Q \neq \varnothing. \tag{26}$$

Suppose, on the contrary, that $P \cap Q = \varnothing$. Then by a corollary of the Hahn–Banach theorem the convex sets $P$ and $Q$ are separated by a linear functional: There are a vector $(\alpha_1, \ldots, \alpha_s) \in R^s$ and a number $\alpha \in R^1$ such that

$$1) \sum_{\nu=1}^{s} |\alpha_\nu| = 1;$$

$$2) \sum_{\nu=1}^{s} \alpha_\nu \xi_\nu \leq \alpha \quad \text{if } (\xi_1, \ldots, \xi_s) \in P; \tag{27}$$

$$3) \sum_{\nu=1}^{s} \alpha_\nu \xi_\nu \geq \alpha \quad \text{if } (\xi_1, \ldots, \xi_s) \in Q.$$

Since the vectors $(0, \ldots, 0, \xi_\nu, 0, \ldots, 0) \in P$ if $\xi_\nu \leq \rho$ and $1 \leq \nu \leq s$, it follows from (27), 2) that $\alpha_\nu \geq 0$, $\nu = 1, \ldots, s$. Consider the vector

$$x^0 = (x_1^0, \ldots, x_m^0) = \sum_{\nu=1}^{s} \alpha_\nu x^\nu$$

(i.e., $x_i^0 = \sum_{\nu=1}^{s} \alpha_\nu x_i^\nu$). Since $\alpha_\nu \geq 0$ and $x^\nu \in \sigma_m$, $1 \leq \nu \leq s$, then $x_0^i \geq 0$ and (see 1))

$$\sum_{i=1}^{m} x_i^0 = \sum_{\nu=1}^{s} \alpha_\nu \sum_{i=1}^{m} x_i^\nu = \sum_{\nu=1}^{s} \alpha_\nu = 1,$$

i.e., $x^0 \in \sigma_m$.

On the other hand, since the $s$-component vector $(\rho + \varepsilon, \ldots, \rho + \varepsilon) \in P$, we can deduce from 2) and 3) that for every $y$

$$F(x^0, y) = \sum_{\nu=1}^{s} \alpha_\nu F(x^\nu, y) \geq \alpha \geq \sum_{\nu=1}^{s} \alpha_\nu (\rho + \varepsilon) > \rho,$$

which contradicts the definition of $\rho$. Consequently (26) is established.

We now take an arbitrary vector $(\xi_1^0, \ldots, \xi_s^0) \in P \cap Q$. Then $\xi_\nu^0 \leq \rho \uparrow \varepsilon$ and $\xi_\nu^0 = F(x^\nu, y^0)$, $\nu = 1, \ldots, s$, $y^0 \in \sigma_n$ (see the definition of $P$ and $Q$), i.e., $y^0 \in A(x^\nu, \varepsilon)$, $\nu = 1, \ldots, s$. This establishes (24) and hence the theorem is proved.

3°. *Proof of Theorem* 1.6 (*see* §1.4).

*Necessity.* Let $\{x_n\}_{n=1}^{\infty}$ be a basis in the Banach space $X$, i.e., there exists, for every $x \in X$, a unique series

$$\sum_{n=1}^{\infty} a_n(x)x_n = x,$$

that converges to $x$ in the norm of $X$.

It is clear that $\{x_n\}$ is dense in $X$. Consider the set $Y$ of sequences of numbers $A = \{a_n\}_{n=1}^{\infty}$ for which the series $\sum_{n=1}^{\infty} a_n x_n$ converges in the norm of $X$. For $A = \{a_n\} \in Y$, set

$$\|A\|_Y = \sup_{1 \le N < \infty} \left\| \sum_{n=1}^{N} a_n x_n \right\|_X. \tag{28}$$

It is easily seen that (28) defines a norm on the linear space $Y$. Let us show that $Y$ is a Banach space under the norm (28).

Let $A^{(k)} = \{a_n^{(k)}\}_{n=1}^{\infty} \in Y$, $k = 1, 2, \ldots$, and

$$\|A^{(k)} - A^{(k+p)}\|_Y \to 0 \quad \text{if } k \to \infty, \ p = 1, 2, \ldots.$$

Then by (28), for every $\varepsilon > 0$ there is a number $k_0$ such that, for $N, R$, $p = 1, 2, \ldots$,

$$\left\| \sum_{n=N}^{N+R} a_n^{(k)} x_n - \sum_{n=N}^{N+R} a_n^{(k+p)} x_n \right\|_X \le \varepsilon, \quad \text{if } k \ge k_0. \tag{29}$$

It follows, in particular, from (29) that the limit $\lim_{k\to\infty} a_n^{(k)} = a_n$ exists for $n = 1, 2, \ldots$. If we fix $k$, $N$, and $R$ in (29) and let $p$ tend to infinity, we find

$$\left\| \sum_{n=N}^{N+R} a_n^{(k)} x_n - \sum_{n=N}^{N+R} a_n x_n \right\|_X \le \varepsilon, \quad k \ge k_0, \ N, R = 1, 2, \ldots. \tag{30}$$

Since $\{a_n^{k_0}\} \in Y$, we have, for sufficiently large $N_0 = N_0(\varepsilon)$,

$$\left\| \sum_{n=N}^{N+R} a_n^{(k_0)} x_n \right\|_X \le \varepsilon, \quad \text{if } N \ge N_0, \ R = 1, 2, \ldots.$$

It follows from this and from (30) that when $N > N_0$

$$\left\| \sum_{n=N}^{N+R} a_n x_n \right\|_X \le 2\varepsilon, \quad R = 1, 2, \ldots,$$

i.e., the series $\sum_{n=1}^{\infty} a_n x_n$ converges in $X$, and that $A = \{a_n\} \in Y$. Inequality (30) shows that $\lim \|A^{(k)} - A\|_Y = 0$. Therefore $Y$ is a Banach space.

Now we consider the linear operator $T\colon Y \to X$ which assigns to each element $A = \{a_n\} \in Y$ the sum (in $X$) $\sum_{n=1}^{\infty} a_n x_n$; that is,

$$T(A) \equiv T(\{a_n\}) = \sum_{n=1}^{\infty} a_n x_n.$$

From the fact that $\{x_n\}_{n=1}^{\infty}$ is a basis in $X$ and from the definition of the norm in $Y$ (see (28)) it follows immediately that $T$ is a bounded linear operator which provides a one-to-one mapping of $Y$ onto $X$. By Banach's theorem on inverse operators, $T^{-1}$ is also bounded, i.e., there is a constant $M > 0$ such that

$$\sup_{1 \leq N < \infty} \left\| \sum_{n=1}^{N} a_n(x) \cdot x_n \right\|_X \leq M \|x\|_X, \qquad x \in X. \tag{31}$$

Hence we have shown that the basis $\{x_n\}_{n=1}^{\infty}$ satisfies condition c) (see the statement of Theorem 1.6).

In addition, it follows from (31) that the linear functionals $a_n(x)$, $x \in X$, $n = 1, 2, \ldots$, are bounded, and therefore (see Theorem 1.2) that $\{x_n\}_{n=1}^{\infty}$ is minimal.

*Sufficiency.* Let $\{x_n\}_{n=1}^{\infty}$ be complete and minimal in $X$, and let $\{y_n\}_{n=1}^{\infty} \subset X^*$ be the system dual to $\{x_n\}$ (the existence and uniqueness of $\{y_n\}$ was established in Chapter 1). In addition, let there be a number $M$ such that for every $x \in X$

$$\left\| \sum_{n=1}^{N} \langle y_n, x \rangle x_n \right\|_X \leq M \|x\|_X, \qquad n = 1, 2, \ldots \tag{32}$$

Let us show that $\{x_n\}$ is a basis in $X$. Choose an element $x$ in $X$ and an $\varepsilon > 0$. Since $\{x_n\}$ is complete, there is a polynomial $P = \sum_{n=1}^{R} a_n x_n$ in this system for which

$$\|P - x\|_X \leq \varepsilon. \tag{33}$$

Since $\sum_{n=1}^{N} \langle y_n, P \rangle x_n = P$ if $N \geq R$, we find from (32) and (33) that

$$\left\| \left\{ \sum_{n=1}^{N} \langle y_n, x \rangle x_n \right\} - x \right\|_X \leq \left\| \left\{ \sum_{n=1}^{N} \langle y_n, x \rangle x \right\} - P \right\|_X + \|P - x\|_X$$

$$\leq \left\| \sum_{n=1}^{N} \langle y_n, x - P \rangle x_n \right\|_X + \varepsilon$$

$$\leq M \|x - P\|_X + \varepsilon \leq (M + 1)\varepsilon, \qquad \text{if } N \geq R,$$

i.e., the series $\sum_{n=1}^{\infty} \langle y_n, x \rangle x_n$ converges to $x$ in $X$. The uniqueness of this series follows from the biorthonormality of the system $\{x_n, y_n\}$: if $\sum_{n=1}^{\infty} a_n x_n$ converges to $x$, then for $n = 1, 2, \ldots$,

$$\langle y_n, x \rangle = \left\langle y_n, \sum_{k=1}^{\infty} a_k x_k \right\rangle = \sum_{k=1}^{\infty} a_k \langle y_n, x_k \rangle = a_n.$$

This completes the proof of the theorem.

# Some Topics from the Theory of Functions
# of a Complex Variable

Here we prove some results from the theory of functions of a complex variable which are used in the text (in §§5.2, 5.3, and 6.5), but which fall outside the limits of the usual university curriculum.

## §1. The Poisson integral

If $f(x)$ and $g(x)$, $x \in R^1$, are summable on $[-\pi, \pi]$, periodic with period $2\pi$, and complex-valued, we denote by $f * g(x)$ the convolution

$$f * g(x) = \frac{2}{2\pi} \int_{-\pi}^{\pi} f(x + t)g(t)\,dt.$$

It follows easily from Fubini's theorem that a convolution of summable functions is also summable on $[-\pi, \pi]$ and

$$c_n(f * g) = c_n(f) \cdot c_{-n}(g), \qquad n = 0, \pm 1, \pm 2, \ldots, \tag{1}$$

where $\{c_n(f)\}$ are the Fourier coefficients of $f(x)$:

$$c_n(f) = \frac{1}{2\pi} \int_{-\pi}^{\pi} f(t)e^{-int}\,dt, \qquad n = 0, \pm 1, \pm 2, \ldots.$$

Let $f \in L^1(-\pi, \pi)$. For $0 \le r < 1$ we consider the function

$$f_r(x) = \sum_{n=-\infty}^{\infty} c_n(f)r^{|n|}e^{inx}, \qquad x \in [-\pi, \pi], \tag{2}$$

where the series on the right of (2) converges uniformly in $x$ for every given $r$, $0 \le r < 1$. The Fourier coefficients of $f_r(x)$ are $c_n(f_r) = c_n(f) \cdot r^{|n|}$, $n = 0, \pm 1, \pm 2, \ldots$; and this means, because of (1), that $f_r(x)$ can be represented as a convolution:

$$f_r(x) = \frac{1}{2\pi} \int_{-\pi}^{\pi} f(x + t)P_r(t)\,dt, \tag{3}$$

where

$$P_r(t) = \sum_{n=-\infty}^{\infty} r^{|n|} e^{int}, \qquad t \in [-\pi, \pi]. \tag{4}$$

The function $P_r(t)$ of two variables, $0 \le r < 1$, $t \in [-\pi, \pi]$, is the *Poisson kernel*, and the integral (3) is the *Poisson integral*. It is easily verified that

$$P_r(t) = \text{Re}\left[1 + 2\sum_{n=1}^{\infty} r^n e^{int}\right] = \text{Re}\,\frac{1 + re^{it}}{1 - re^{it}}.$$

Consequently

$$P_r(t) = \frac{1 - r^2}{1 + r^2 - 2r\cos t}, \qquad 0 \le r < 1,\ t \in [-\pi, \pi]. \tag{5}$$

If $f \in L^1(-\pi, \pi)$ is a real function, then since $c_{-n}(f) = \overline{c_n(f)}$, $n = 0, \pm 1, \pm 2, \ldots$, we find from (2) that

$$\begin{aligned}
f_r(x) &= c_0(f) + \sum_{n=1}^{\infty} c_n(f) r^n e^{inx} + \sum_{n=1}^{\infty} \overline{c_n(f)} r^n e^{-inx} \\
&= \tfrac{1}{2}[F(re^{ix}) + \overline{F(re^{ix})}] = \text{Re}\,F(re^{ix}),
\end{aligned} \tag{6}$$

where

$$F(z) = c_0(f) + 2\sum_{n=1}^{\infty} c_n(f) r^n e^{inx} \qquad (z = re^{ix}) \tag{7}$$

is analytic in the unit disk. Equation (6) shows that for every real $f \in L^1(-\pi, \pi)$ the Poisson integral (3) defines a harmonic function in the unit disk,

$$u(z) = f_r(e^{ix}), \qquad z = re^{ix},\ \ 0 \le r < 1,\ \ x \in [-\pi, \pi].$$

Moreover, the harmonic conjugate $v(z)$ of $u(z)$, with $v(0) = 0$, is given by the formula

$$v(z) = \text{Im}\,F(z) = \sum_{n=-\infty}^{\infty} [-i(\text{sgn}\,n) r^{|n|} c_n(f)] e^{inx}, \tag{8}$$

where, as usual,

$$\text{sgn}\,\alpha = \begin{cases} \dfrac{\alpha}{|\alpha|} & \text{if } \alpha \ne 0, \\[2mm] 0 & \text{if } \alpha = 0, \end{cases} \qquad \alpha \in R^1 \tag{9}$$

**PROPOSITION 1.** *Let $u(z)$ be harmonic (or analytic) in the disk $|z| < 1 + \varepsilon$ ($\varepsilon > 0$), and let $f(x) = u(e^{ix})$, $x \in [-\pi, \pi]$. Then*

$$u(z) = \frac{1}{2\pi} \int_{-\pi}^{\pi} f(x + t) P_r(t)\,dt \qquad (z = re^{ix},\ |z| < 1). \tag{10}$$

Since the Poisson kernel is a real function, it is enough to verify (10) in the case when $u(z)$ is an analytic function:

$$u(z) = \sum_{n=0}^{\infty} c_n z^n, \qquad |z| < 1 + \varepsilon.$$

But then

$$c_n(f) = \frac{1}{2\pi} \int_{-\pi}^{\pi} u(e^{ix}) e^{-inx} \, dx = \begin{cases} c_n & \text{if } n \geq 0, \\ 0 & \text{if } n < 0, \end{cases}$$

and (10) follows at once from (2) and (3).

Before we turn to the investigation of the behavior of $f_r(x)$ as $r \to 1$, we note some properties of the Poisson kernel:

a) $P_r(t) \geq 0, \quad \leq r < 1, \quad t \in [-\pi, \pi]$.

b) $\dfrac{1}{2\pi} \displaystyle\int_{-\pi}^{\pi} P_r(t) \, dt = 1, \qquad 0 \leq r < 1.$  (11)

c) for every $\delta > 0$

$$\mu_\delta(r) = \sup_{\delta \leq |t| \leq \pi} P_r(t) = o(1) \quad \text{as } r \to 1.$$

Relations a) and c) follow at once from (5); and to prove (b) it is enough to set $f(x) \equiv 1$ in (2) and (3).

THEOREM 1. *For every (complex-valued) function* $f \in L^p(-\pi, \pi)$, $1 \leq p < \infty$, *we have the equation*

$$\lim_{r \to 1} \| f - f_r \|_{L^p(-\pi, \pi)} = 0$$

[*see* (3)]; *if also* $f(x)$ *is continuous on* $[-\pi, \pi]$ *and* $f(-\pi) = f(\pi)$, *then*

$$\lim_{r \to 1} \| f - f_r \|_{C(-\pi, \pi)} = 0.$$

PROOF. By (3) and property b) of the Poisson kernel,

$$f_r(x) - f(x) = \frac{1}{2\pi} \int_{-\pi}^{\pi} [f(x+t) - f(x)] P_r(t) \, dt.$$  (12)

For every $g \in L^q(-\pi, \pi)$, $\frac{1}{p} + \frac{1}{q} = 1$, $\|g\|_{L^q(-\pi,\pi)} = 1$, we can use Hölder's inequality and the positivity of the Poisson kernel (see a)) to obtain

$$\int_{-\pi}^{\pi} [f_r(x) - f(x)] g(x) \, dx$$

$$= \frac{1}{2\pi} \int_{-\pi}^{\pi} \left\{ \int_{-\pi}^{\pi} [f(x+t) - f(x)] g(x) \, dx \right\} P_r(t) \, dt$$

$$\leq \frac{1}{2\pi} \int_{-\pi}^{\pi} \| f(x+t) - f(x) \|_{L^p(-\pi,\pi)} \| g \|_{L^q(-\pi,\pi)} P_r(t) \, dt$$

$$\leq \frac{1}{2\pi} \int_{-\pi}^{\pi} \omega_p(|t|, f) P_r(t) \, dt.$$

Consequently

$$\|f_r - f\|_{L^p(-\pi,\pi)} \le \frac{1}{2\pi} \int_{-\pi}^{\pi} \omega_p(|t|, f) P_r(t) \, dt,$$

and if for a given $\varepsilon > 0$ we find a number $\delta = \delta(\varepsilon)$ such that $\omega_p(t, f) \le \varepsilon/2$ for $t \in (0, \delta)$, then for $r$ sufficiently close to 1 we obtain (see a)–c))

$$\|f_r - f\|_{L^p(-\pi,\pi)} \le \frac{1}{2\pi} \int_{|t|<\delta} \omega_p(|t|, f) P_r(t) \, dt$$

$$+ \frac{1}{2\pi} \int_{\delta \le |t| \le \pi} \omega_p(|t|, f) P_r(t) \, dt$$

$$\le \frac{\varepsilon}{2} + 2\|f\|_{L^p(-\pi,\pi)} \cdot \mu_\delta(r) < \varepsilon.$$

Similarly, the second statement in Theorem 1 follows from the inequality (see (12), and a), b))

$$\|f_r - f\|_{C(-\pi,\pi)} \le \frac{1}{2\pi} \int_{-\pi}^{\pi} \omega(|t|, f) p_r(t) \, dt.$$

This completes the proof of Theorem 1.

THEOREM 2. (FATOU). *Let $f(x)$ be a complex-valued element of $L^1(-\pi, \pi)$. Then*

$$\lim_{r \to 1} f_r(x) = f(x) \quad \text{for almost all } x \in [-\pi, \pi].$$

PROOF. Let us show that, for $x \in [-\pi, \pi]$ and $0 < r < 1$,

$$f_r(x) \le C M(f, x), \tag{13}$$

where $C$ is an absolute constant and $M(f, x)$ is the maximal function for $f(x)$[1] (see Definition 2 in Appendix 1). For this purpose we use the inequality

$$P_r(t) \le K \frac{\varepsilon}{\varepsilon^2 + t^2}, \qquad \varepsilon = 1 - r, \ 0 < r < 1, \ t \in [-\pi, \pi]$$

which is easily derived from (5) ($K$ is an absolute constant).

---

[1] We suppose that $f(x)$ has been continued with preservation of periodicity to $[-2\pi, 2\pi]$ (i.e., $f(x) = f(y)$ if $x, y \in [-2\pi, 2\pi]$, $x - y = 2\pi$, and $f(x) = 0$ if $|x| > 2\pi$.

Let $k_0 = k_0(\varepsilon)$ be a number such that $2^{k_0}\varepsilon < \pi \leq 2^{k_0+1}\varepsilon$. Then for $x \in [-\pi, \pi]$

$$|f_r(x)| = \frac{1}{2\pi} \left| \int_{-\pi}^{\pi} f(x+t) P_r(t) \, dt \right|$$

$$\leq \frac{1}{2\pi} \sum_{k=-\infty}^{k_0-1} \int_{2^k\varepsilon < |t| \leq 2^{k+1}\varepsilon} |f(x+t)| P_r(t) \, dt$$

$$+ \frac{1}{2\pi} \int_{2^{k_0}\varepsilon < |t| \leq \pi} |f(x+t)| P_r(t) \, dt$$

$$\leq \frac{K}{2\pi} \sum_{k=-\infty}^{\infty} \int_{2^k\varepsilon < |t| \leq 2^{k+1}\varepsilon} |f(x+t)| \frac{\varepsilon}{\varepsilon^2 + t^2} \, dt$$

$$\leq \frac{K}{2\pi} \sum_{k=-\infty}^{\infty} \frac{\varepsilon}{\varepsilon^2 + 2^{2k}\varepsilon^2} \int_{0 \leq |t| \leq 2^{k+1}\varepsilon} \varepsilon |f(x+t)| \, dt$$

$$\leq M(f,x) \frac{K}{2\pi} \sum_{k=-\infty}^{\infty} \frac{1}{\varepsilon(1+2^{2k})} 2^{k+2}\varepsilon \leq CM(f,x).$$

This completes the proof of (13). Now, using the operator $f(x) \to M(f,x)$ of weak type $(1,1)$ (see Theorem 2 in Appendix 1), we can find a sequence $\{f^{(n)}(x)\}_{n=1}^{\infty}$, $x \in R^1$, such that

$$f^{(n)}(x) \in C(-2\pi, 2\pi),$$
$$f^{(n)}(x) = 0 \quad \text{if } |x| > 2\pi, \; n = 1, 2, \ldots,$$
$$\lim_{n \to \infty} f^{(n)}(x) = f(x) \quad \text{and} \quad \lim_{n \to \infty} M(f^{(n)} - f, x) = 0 \tag{14}$$
$$\text{for almost all } x \in (-2\pi, 2\pi).$$

By (13), for $x \in (-\pi, \pi)$,

$$|f_r(x) - f(x)| \leq |f_r(x) - f_r^{(n)}(x)|$$
$$+ |f_r^{(n)}(x) - f^{(n)}(x)| + |f^{(n)}(x) - f(x)|$$
$$\leq C \cdot M(f^{(n)} - f, x) + |f_r^{(n)}(x) - f^{(n)}(x)|$$
$$+ |f^{(n)}(x) - f(x)|.$$

We obtain the conclusion of Theorem 2 by using the preceding inequality and (14), and taking account of the fact that, by Theorem 1, $\lim_{r \to 1} f_r^{(n)}(x) = f^{(n)}(x)$ for each $x \in [-\pi, \pi]$.

REMARK. If we use, instead of (13), the stronger inequality (59) which we shall prove in §3, we can show that $f_r(t) \to f(x)$ for almost all $x \in [-\pi, \pi]$ when $re^{it}$ approaches $e^{ix}$ on a path that is not tangent to the circle $|z| = 1$ (for details see §3).

## §2. $H^p$ spaces

DEFINITION 1. The space $H^p$ ($1 \leq p \leq \infty$) is the collection of functions $F(z)$ that are analytic in the unit disk and have finite norm

$$\|F\|_{R^p} = \sup_{0 < r < 1} \left\{ \int_{-\pi}^{\pi} |F(re^{it})|^p \, dt \right\}^{1/p}. \tag{15}$$

Let the complex-valued function $\Phi(x) \in L^p(-\pi, \pi)$ satisfy the condition

$$\int_{-\pi}^{\pi} \Phi(t) e^{int} \, dt = 0, \qquad n = 1, 2, \ldots, \tag{16}$$

then the function $F(z)$ defined by

$$F(z) = \Phi_r(x) = \frac{1}{2\pi} \int_{-\pi}^{\pi} \Phi(x + t) P_r(t) \, dt \qquad (z = re^{ix}), \tag{17}$$

belongs to $H^p$, and

$$\|F\|_{H^p} = \|\Phi\|_{L^p(-\pi, \pi)}. \tag{18}$$

In fact, it follows from (16) and (2) that $F(z)$ is analytic. Moreover, by the inequality

$$\|f * g\|_{L^p(-\pi, \pi)} \leq \frac{1}{2\pi} \|f\|_{L^p(-\pi, \pi)} \|g\|_{L^1(-\pi, \pi)}$$

(see, for example, the proof of 4.(16)), we have

$$\|F\|_{H^p} \leq \frac{1}{2\pi} \|\Phi\|_{L^p(-\pi, \pi)} \cdot \int_{-\pi}^{\pi} P_r(t) \, dt = \|\Phi\|_{L^p(-\pi, \pi)}.$$

Finally, by Theorem 1 (and for $p = \infty$, by Theorem 2) $\|\Phi\|_{L^p(-\pi, \pi)} \leq \|F\|_{H^p}$, and we obtain (18).

We shall show below that every function $F \in H^p$ ($1 \leq p \leq \infty$) can be represented in the form (17). For this purpose we shall need the following theorem.

THEOREM 3. *Let the complex-valued function $\varphi(t)$ have bounded variation on $[-\pi, \pi]$, be continuous on the left on $(-\pi, \pi]$, and satisfy*

$$\int_{-\pi}^{\pi} e^{int} d\varphi(t) = 0 \qquad \text{for } n = 1, 2, \ldots. \tag{19}$$

*Then $\varphi(t)$ is absolutely continuous on $[-\pi, \pi]$.*

REMARK. In (19), and later, we use a Lebesgue–Stieltjes integral with respect to a complex-valued function $\varphi(t)$ of bounded variation. We say that $\varphi(t) = u(t) + iv(t)$ has bounded variation (or is absolutely continuous) if both $u(t)$ and $v(t)$ have bounded variation (or are absolutely continuous). Then the integral

$$\int_{-\pi}^{\pi} f(t)\, d\varphi(t) = \int_{-\pi}^{\pi} f(t)\, du(t) + i \int_{-\pi}^{\pi} f(t)\, dv(t)$$

is defined for every $f(t)$ which is continuous on $[-\pi, \pi]$, and also if $f(t) = \chi_F(t)$ is the characteristic function of a closed set $F \subset [-\pi, \pi]$.

PROOF OF THEOREM 3. It is enough to verify that

$$\int_{-\pi}^{\pi} \chi_F(t)\, d\varphi(t) = 0 \tag{20}$$

for every closed set $F \subset (-\pi, \pi)$ with $m(F) = 0$. (In fact, it follows from (20) that $\varphi$ is absolutely continuous on $(-\pi, \pi)$, and consequently $\varphi(t) = \varphi_1(t) + s(t)$, where $\varphi_1$ is absolutely continuous on $[-\pi, \pi]$, $s(t) = 0$ for $t \in (-\pi, \pi]$, and $s(-\pi) = \lambda$. But then

$$0 = \int_{-\pi}^{\pi} e^{int}\, d\varphi(t) = \int_{-\pi}^{\pi} e^{int}\, d\varphi_1(t) - \lambda e^{-in\pi} = o(1) - \lambda e^{-in\pi}, \qquad n \to \infty;$$

that is, $\lambda = 0$ and $\varphi(t)$ is absolutely continuous on $[-\pi, \pi]$.)

We need the following lemma.

LEMMA 1. *Let $F$ be a closed set, $V$ an open set, and let $F \subset V \subset (-\pi, \pi)$ and $m(F) = 0$. Then for every $\varepsilon$, $0 < \varepsilon < 1/3$, there is a function $g(t) \in C^\infty(-\pi, \pi)$ of the form*

$$g(t) = \sum_{n=1}^{\infty} c_n(g) e^{int}, \tag{21}$$

*with the following properties:*

a)  $|g(t) - 1| \le 3\varepsilon$   *if $t \in F$;*
b)  $|g(t)| \le 3\varepsilon$   *if $t \notin V$;*  (22)
c)  $\|g\|_{C(-\pi,\pi)} \le 3.$

We first deduce (20) from Lemma 1, and then prove Lemma 1.

Let $(-\pi, \pi) \setminus F = \bigcup_k (a_k, b_k)$, where $(a_k, b_k)$, $k = 1, 2, \dots$, is the finite or infinite sequence of complementary intervals of $F$, and for $r = 1, 2, \dots$ and $\nu = 3, 4, \dots$, let

$$V_{r,\nu} = (-\pi, \pi) \setminus \bigcup_{k \le r} \left( a_k + \frac{1}{\nu}(b_k - a_k), b_k - \frac{1}{\nu}(b_k - a_k) \right).$$

It is evident that $V_{r,\nu}$ is an open set and $F \subset V_{r,\nu}$. For a given $\varepsilon > 0$ and $r, \nu = 3, 4, \ldots$, consider the function $g(t) = g_{\varepsilon,r,\nu}(t)$, constructed in Lemma 1 for the number $\varepsilon$ and the set $V_{r,\nu}$. Then it is easy to verify that if $r \to \infty$, $\nu \to \infty$, and $\varepsilon \to 0$, then the difference

$$\int_{-\pi}^{\pi} \chi_F(t)\, d\varphi(t) - \int_{-\pi}^{\pi} g_{\varepsilon,r,\nu}(t)\, d\varphi(t) \to 0. \tag{23}$$

But by (19) and the uniform convergence of (21),

$$\int_{-\pi}^{\pi} g_{\varepsilon,r,\nu}(t)\, d\varphi(t) = \sum_{n=1}^{\infty} c_n(g_{\varepsilon,r,\nu}) \int_{-\pi}^{\pi} e^{int}\, d\varphi(t) = 0,$$

and (see (23)) we obtain (20).

PROOF OF LEMMA 1. Let $f(t)$ be a continuous function on $[-\pi, \pi]$ for which $\|f\|_{C(-\pi,\pi)} \leq 1$, $f(t) = 1$ if $t \in F$, and $f(t) = 0$ if $t \notin V$. Since the Fejér means $\sigma_N(f, t)$ (see (4.12)), $N = 1, 2, \ldots$, converge uniformly to $f(t)$ and $\|\sigma_N(f, t)\|_{C(-\pi,\pi)} \leq \|f\|_{C(-\pi,\pi)}$, there is a trigonometric polynomial

$$G(t) = \sum_{|n| \leq m} \alpha_n e^{int} \tag{24}$$

such that

$$\|G\|_{C(-\pi,\pi)} \leq 1,$$
$$|G(t) - 1| < \varepsilon \quad \text{if } t \in F, \tag{25}$$
$$|G(t)| < \varepsilon \quad \text{if } t \notin V.$$

Let $\varepsilon = e^{-A}(A > 1)$. For each $\delta > 0$ consider a function $h_\delta(t) \in C(-\pi, \pi)$ such that

$$h_\delta(t) = -2A + \frac{\varepsilon}{2} \quad \text{if } t \in F,$$

$$-2A + \frac{\varepsilon}{2} \leq h_\delta(t) \leq 0 \quad \text{for all } t \in (-\pi, \pi),$$

$$\|h_\delta(t)\|_{L^1(-\pi,\pi)} \leq \delta.$$

The function $h(t)$ can be constructed in the following way: select a closed set $B \subset [-\pi, \pi]$, $B \cap F = \varnothing$, $\pi$, $-\pi \in B$, with the measure $m(B)$ sufficiently close to $2\pi$, and set

$$h_\delta(t) = \left(-2A + \frac{\varepsilon}{2}\right) \frac{\rho(t, B)}{\rho(t, F) + \rho(t, B)},$$

$$\rho(t, E) = \inf_{y \in E} |t - y|).$$

Since

$$\sum_{|n| \leq m} |c_n(h_\delta)| \leq (2m + 1)\|h_\delta\|_{L^1(-\pi,\pi)} \leq (2m + 1)\delta$$

(here $m$ is the same as in (24)), for sufficiently small $\delta > 0$ the function $h(t) = h_\delta(t) - \sum_{|n| \le m} c_n(h_\delta)e^{int}$ satisfies

$$|h(t) + 2A| < \varepsilon \quad \text{if } t \in F,$$
$$-2A < h(t) < \varepsilon, \qquad t \in [-\pi, \pi]. \tag{26}$$

Moreover, $c_n(h) = 0$ if $|n| \le m$. Then the Fejér means $\sigma_N(h, t)$ of $h(t)$ have the form

$$P(t) \equiv \sigma_N(h, t) = \sum_{n=-N}^{-m-1} \beta_n e^{int} + \sum_{n=m+1}^{N} \beta_n e^{int}$$

and for sufficiently large $N$ (see (26))

$$|P(t) + 2A| < \varepsilon, \qquad t \in F, \ -2A < P(t) < \varepsilon, \ t \in [-\pi, \pi]. \tag{27}$$

Set

$$P^+(t) = \sum_{n=m+1}^{N} \beta_n e^{int}, \ P^-(t) = \sum_{n=-N}^{-m-1} \beta_n e^{int}. \tag{28}$$

Since $h(t)$ is real-valued, $c_n(h) = \overline{c_{-n}(h)}$ and $\beta_n = \overline{\beta}_{-n}$, $n = 0, \pm 1, \pm 2, \ldots$. Therefore

$$P^+(t) = \overline{P^-(t)} \quad \text{and} \quad \operatorname{Re} P^+(t) = \frac{P(t)}{2}, \qquad t \in [-\pi, \pi]. \tag{29}$$

We can now define the required function $g(t)$:

$$g(t) = G(t)[1 - \exp\{P^+(t)\}] = -G(t) \sum_{\nu=1}^{\infty} \frac{[P^+(t)]^\nu}{\nu!}.$$

It is clear that $g(t) \in C^\infty(-\pi, \pi)$, and it follows from (24) and (28) that $c_n(g) = 0$ for $n < 0$, i.e.,

$$g(t) = \sum_{n=1}^{\infty} c_n(g)e^{int}. \tag{30}$$

For $t \in F$, by (25), (27), and (29) we have

$$|g(t) - 1| \le |g(t) - G(t)| + |G(t) - 1|$$
$$\le \exp\{\operatorname{Re} P^+(t)\} + \varepsilon$$
$$\le \exp\left\{\frac{P(t)}{2}\right\} + \varepsilon \le \exp\left\{-A + \frac{\varepsilon}{2}\right\} + \varepsilon < 3\varepsilon,$$

and for $t \in [-\pi, \pi] \backslash V$,

$$|g(t)| \le \varepsilon[1 + \exp\{\operatorname{Re} P^+(t)\}] = \varepsilon\left[1 + \exp\left\{\frac{P(t)}{2}\right\}\right] < 3\varepsilon.$$

Finally, for every $t \in [-\pi, \pi]$

$$|g(t)| \leq 1 + \exp\{\operatorname{Re} P^+(t)\} < 3.$$

Consequently $g(t)$ has the required properties (see (22)). Lemma 1, and with it Theorem 3, are established.

THEOREM 4. *Let* $F \in H^p$ $(1 \leq p \leq \infty)$. *Then the limit*

$$\lim_{r \to 1} F(re^{ix}) = \Phi(x). \tag{31}$$

*exists for almost every* $x \in [-\pi, \pi]$. *Moreover,*

1)      $F(re^{ix}) = \Phi_r(x) = \frac{1}{2\pi} \int_{-\pi}^{\pi} \Phi(x + t) P_r(t) \, dt,$
         $0 \leq r < 1, \quad x \in [-\pi, \pi];$
2)      $\lim_{r \to 1} \|F(re^{ix}) - \Phi(x)\|_{L^p(-\pi,\pi)} = 0 \ (1 \leq p < \infty);$
3)      $\|F\|_{H^p} = \|\Phi\|_{L^p(-\pi,\pi)} \ (1 \leq p \leq \infty).$

PROOF. It is enough to show that for every $F \in H^1$ there is a function $\Phi \in L^1(-\pi, \pi)$ such that 1) holds. In fact, if $f \in H^p$, $1 \leq p \leq \infty$, then also $F \in H^1$ and (31) follows from 1) and Theorem 2 for almost all $x \in [-\pi, \pi]$. Moreover,

$$\|\Phi\|_{L^p(-\pi,\pi)} \leq \sup_{0 < r < 1} \|F(re^{ix})\|_{L^p(-\pi,\pi)} < \infty,$$

and by Theorem 1

$$\lim_{r \to 1} \|F(re^{ix}) - \Phi(x)\|_{L^p(-\pi,\pi)} = 0 \qquad (1 \leq p < \infty).$$

Finally, it follows from 1) (also see (2)) that

$$\int_{-\pi}^{\pi} \Phi(x) e^{inx} \, dx = 0, \qquad n = 1, 2, \ldots,$$

and then (see (18))

$$\|F\|_{H^p} = \|\Phi\|_{L^p(-\pi,\pi)}.$$

Let $F \in H^1$. To construct the required function $\Phi \in L^1(-\pi, \pi)$, we set

$$\varphi_r(t) = \int_{-\pi}^{\pi} F(re^{ix}) \, dx, \qquad t \in [-\pi, \pi], \ 0 < r < 1.$$

The function $\varphi_r(t)$, $0 < r < 1$, has bounded variation on $[-\pi, \pi]$:

$$V_{[-\pi,\pi]}(\varphi_r) \leq \int_{-\pi}^{\pi} |F(re^{it})| \, dt \leq \|F\|_{H^1}.$$

By using the definition of the Stieltjes integral and Helly's theorem, it is easy to deduce from the preceding relation that there are a function $\varphi(t)$ of bounded variation, continuous on the left on $(-\pi, \pi]$, and a sequence

$\{r_k\}$, $0 < r_k < 1$, $k = 1, 2, \ldots, \lim_{k \to \infty} r_k = 1$, such that $\varphi_{r_k}(t) \to \varphi(t)$ at each $t \in [-\pi, \pi]$ and

$$\int_{-\pi}^{\pi} g(t) \, d\varphi(t) = \lim_{k \to \infty} \int_{-\pi}^{\pi} g(t) \, d\varphi_{r_k}(t) \equiv \lim_{k \to \infty} \int_{-\pi}^{\pi} g(t) F(r_k e^{it}) \, dt \quad (32)$$

for every function $g \in C(-\pi, \pi)$. Moreover (see (32)), for $n = 1, 2, \ldots,$

$$\int_{-\pi}^{\pi} e^{int} \, d\varphi(t) = \lim_{k \to \infty} \int_{-\pi}^{\pi} e^{int} F(r_k e^{it}) \, dt = 0$$

(we used the analyticity of $F(z)$ in the unit disk) and consequently (see Theorem 3) $\varphi(t)$ is absolutely continuous: there is a function $\Phi \in L^1(-\pi, \pi)$ for which

$$\varphi(t) = \int_{-\pi}^{\pi} \Phi(u) \, du, \qquad t \in [-\pi, \pi].$$

Then (see (32))

$$\int_{-\pi}^{\pi} g(t) \Phi(t) \, dt = \lim_{k \to \infty} \int_{-\pi}^{\pi} g(t) F(r_k e^{it}) \, dt, \qquad g \in C(-\pi, \pi). \quad (33)$$

Select a number $r$, $0 < r < 1$. The functions $F_k(z) = F(r_k z)$, $k = 1, 2, \ldots,$ are analytic in the disk $|z| < 1/r_k$ ($1/r_k > 1$), and consequently, by Proposition 1,

$$F(r_k r e^{ix}) = \frac{1}{2\pi} \int_{-\pi}^{\pi} F(r_k e^{it}) P_r(t - x) \, dt, \qquad x \in [-\pi, \pi].$$

In the limit as $k \to \infty$, it follows from the preceding equation that

$$F(r e^{ix}) = \frac{1}{2\pi} \int_{-\pi}^{\pi} P_r(t - x) \Phi(t) \, dt$$

$$= \frac{1}{2\pi} \int_{-\pi}^{\pi} \Phi(x + t) P_r(t) \, dt, \qquad 0 \le r < 1, \ x \in [-\pi, \pi].$$

This establishes equation 1), and with it also Theorem 4.

We denote by $H^p$ ($1 \le p \le \infty$) the class of functions $\Phi(x)$, $x \in [-\pi, \pi]$, which are boundary values of elements of $H^p$, i.e., can be represented in the form (see (31))

$$\Phi(x) = \lim_{r \to 1} F(r e^{ix}) \quad \text{for almost all } x \in [-\pi, \pi], \ F \in H^p.$$

By parts 2) and 3) of Theorem 4, $\mathcal{H}^p \subset L^p(-\pi, \pi)$, and every function $\Phi \in \mathcal{H}^p$ satisfies (16). On the other hand, we showed above that, for any $\Phi \in L^p(-\pi, \pi)$, under condition (16) the Poisson integral (17) defines an element of $H^p$. Consequently (also the Theorem 2)

$$\mathcal{H}^p = \left\{ \Phi \in L^p(-\pi, \pi): \int_{-\pi}^{\pi} \Phi(x) e^{inx} \, dx = 0, \ n = 1, 2, \ldots \right\}. \quad (34)$$

It follows from (34) that $\mathcal{H}^p$ is a (closed) subspace of $L^p(-\pi, \pi)$, and $H^p$ is a Banach space with the norm (15) (also see Theorem 4, 3).

Let $1 \le p \le \infty$. Set

$$H_0^p = \{F \in H^p : \operatorname{Im} F(0) = 0\},$$

$$\mathcal{H}_0^p = \left\{ \Phi \in \mathcal{H}^p : \int_{-\pi}^{\pi} \operatorname{Im} \Phi(x)\, dx = 0 \right\}$$

and

$$\operatorname{Re} \mathcal{H}^p = \{f(x) : f(x) = \operatorname{Re} \Phi(x),\ x \in [-\pi, \pi],\ \Phi \in \mathcal{H}^p\}. \tag{35}$$

We have the following theorem.

THEOREM 5. *For* $1 \le p \le \infty$, *the following conditions are equivalent :*

a) $$\Phi(x) \in \mathcal{H}_0^p;$$
b) $$\Phi(x) \in L^p(-\pi, \pi), \qquad \int_{-\pi}^{\pi} \operatorname{Im} \Phi(x)\, dx = 0,$$
$$\int_{-\pi}^{\pi} \Phi(x) e^{inx}\, dx = 0, \qquad n = 1, 2, \ldots ;$$
b) $$F(re^{ix}) = \Phi * P_r(x) \equiv \frac{1}{2\pi} \int_{-\pi}^{\pi} \Phi(x + t) P_r(t)\, dt \in H_0^p;$$
d) $$\Phi(x) = f(x) + i\tilde{f}(x), \quad \text{where } f \in L^p(-\pi, \pi)$$

*is a real function whose conjugate* $\tilde{f}(x)$ *(see Definition 5.2) also belongs to* $L^p(-\pi, \pi)$ :

$$\tilde{f}(x) = \frac{1}{\pi} \int_{-\pi}^{\pi} \frac{f(t)}{2\tan\frac{x-t}{2}}\, dt = \frac{1}{\pi} \int_{-\pi}^{\pi} \frac{f(t) - f(x)}{2\tan\frac{x-t}{2}}\, dt \in L^p(-\pi, \pi). \tag{36}$$

PROOF. The equivalence of a) and b) follows immediately from (34), and the equivalence of a) and c) follows from Theorems 4 and 2.

Let us show that d) implies b). It is enough to show that when the function and its conjugate are summable (i.e., $f, \tilde{f} \in L^1(-\pi, \pi)$) we have the equations

$$c_n(\tilde{f}) = i(\operatorname{sgn} n) c_n(f), \qquad n = 0, \pm 1, \pm 2, \ldots \tag{37}$$

(see (9)). A direct calculation from (36) shows that $\widetilde{\cos nx} = \sin nx$, $n = 0, 1, \ldots$, and that $\widetilde{\sin nx} = -\cos nx$, $n = 1, 2 \ldots$ —in this case

$$\frac{f(t) - f(x)}{2\tan\frac{1}{2}(t - x)} \in L^1.$$

Consequently (37) is satisfied for every trigonometric polynomial.

Let the number $n = 0, \pm1, \pm2, \ldots$ be given. For an arbitrary $f \in L^1(-\pi, \pi)$ and $k = 1, 2, \ldots$, we set

$$I_{f,k}(t) = \frac{1}{2k} \sum_{j=-k+1}^{k} f(x_j + t), \qquad t \in [-\pi, \pi],$$

$$I_{f,k}^*(t) = \frac{1}{2k} \sum_{j=-k+1}^{k} \widetilde{f}(x_j + t)e^{-in(x_j+t)},$$

where $x_j = \frac{\pi j}{k}$, $j = -k+1, -k+2, \ldots, k$.

Let us show that for a given index $n$ equation (37) is a consequence of the following properties of $I_{f,x}(t)$ and $I_{f,k}^*(t)$ (that $I_{f,k}(t)$ and $I_{f,k}^*(t)$ have these properties will be established later):

1) $m\{t \in (-\pi, \pi): |I_{f,k}(t)| > \varepsilon\} \geq \frac{1}{\varepsilon}\|f\|_{L^1(-\pi,\pi)}$, $\varepsilon > 0$, $k = 1, 2, \ldots$.

2) As $k \to \infty$ the function $I_{f,k}(t)$, $t \in (-\pi, \pi)$, converges in measure to $I(t) = \frac{1}{2\pi} \int_{-\pi}^{\pi} f(x)\,dx = \text{const}$.

3) $m\{t \in (-\pi, \pi): |I_{j,k}^*(t)| > \varepsilon\} \leq \frac{C}{\varepsilon}\|f\|_{L^1(-\pi,\pi)}$, $\varepsilon > 0$, $k = 1, 2, \ldots$, where $C > 0$ is an absolute constant.

Thus we assume that 1) and 3) hold.

It is easily seen that $I_{f,k}^*(t) = I_{g,k}(t)$, where $g(x) = \widetilde{f}(x)e^{-inx} \in L^1(-\pi, \pi)$; hence 2) implies the convergence in measure of the sequence of functions $I_{f,k}^*(t)$, $k = 1, 2, \ldots$:

$$I_{f,k}^*(t) \to c_n(\widetilde{f}) \quad \text{in measure } (k \to \infty). \tag{38}$$

For an arbitrary $\varepsilon \in (0, 1)$, we can find a trigonometric polynomial $f_1(x)$ such that

$$f(x) = f_1(x) + f_2(x), \qquad \|f_2\|_{L^1(-\pi,\pi)} \leq \frac{\varepsilon^2}{4(C+1)}. \tag{39}$$

Then by 3)

$$m\left\{t \in (-\pi, \pi): I_{f_2,k}^*(t)| > \frac{\varepsilon}{2}\right\} \leq \frac{2C}{\varepsilon}\|f_2\|_{L^1(-\pi,\pi)} \leq \frac{\varepsilon}{2} \tag{40}$$

and (see (38)) for $k > k_0$

$$m\left\{t \in (-\pi, \pi): |I_{f_1,k}^*(t) - c_n(\widetilde{f_1})| > \frac{\varepsilon}{4}\right\} \leq \frac{\varepsilon}{2}. \tag{41}$$

Since $f_1(x)$ is a polynomial, we have $c_n(\widetilde{f_1}) = -i(\operatorname{sgn} n)c_n(f_1)$ and (see (39))

$$|c_n(\widetilde{f_1}) + i(\operatorname{sgn} n)c_n(f)| \leq |c_n(f_1) - c_n(f)|$$
$$= |c_n(f_2)| \leq \|f_2\|_{L^1(-\pi,\pi)} \leq \frac{\varepsilon}{4}. \tag{42}$$

Using the equation $I_{f,k}^* = I_{f_1,k}^* + I_{f_2,k}^*$, and inequalities (40)–(42), we obtain

$$m\{t \in (-\pi, \pi): |I_{f,k}^*(t) + i(\text{sgn} n)c_n(f)| > \varepsilon\} \le \varepsilon, \qquad k > k_0,$$

which, with (38), establishes (37).

We now show that every function $f \in L^1(-\pi, \pi)$ satisfies 1)–3). Inequality 1) follows at once from Tchebycheff's inequality, since

$$\|I_{f,k}(t)\|_{L^1(-\pi,\pi)} \le \frac{1}{2k} \sum_{j=-k+1}^{k} \|f(x_j + t)\|_{L^1(-\pi,\pi)} = \|f\|_{L^1(-\pi,\pi)}. \tag{43}$$

To prove 2), we take an arbitrary $\varepsilon \in (0, 1)$ and represent $f(x)$ in the form

$$f(x) = g(x) + h(x), \qquad g \in C(-\pi, \pi), \quad \|h\|_{L^1(-\pi,\pi)} \le \frac{\varepsilon^2}{2}.$$

Since $g$ is continuous, it follows easily that

$$\lim_{k \to \infty} I_{g,k}(t) = \frac{1}{2\pi} \int_{-\pi}^{\pi} g(x)\, dx$$

uniformly in $t \in (-\pi, \pi)$. Therefore, for sufficiently large $k$ $(k > k')$ we will have, taking account of (43),

$$\left| I_{g,k}(t) - \frac{1}{2\pi} \int_{-\pi}^{\pi} f(x)\, dx \right| \le \left| I_{g,k}(t) - \frac{1}{2\pi} \int_{-\pi}^{\pi} g(x)\, dx \right|$$
$$+ \frac{1}{2\pi} \left| \int_{-\pi}^{\pi} h(x)\, dx \right| \le \frac{\varepsilon}{2}, \qquad t \in (-\pi, \pi). \tag{44}$$

In addition, by 1) and (43)

$$m\left\{ t \in (-\pi, \pi): |I_{h,k}(t)| > \frac{\varepsilon}{2} \right\} \le \frac{2}{\varepsilon} \|h\|_{L^1(-\pi,\pi)} \le \varepsilon;$$

it follows form this inequality and (44) that when $k > k'$

$$m\left\{ t \in (-\pi, \pi): \left| I_{f,k}(t) - \frac{1}{2\pi} \int_{-\pi}^{\pi} f(x)\, dx \right| > \varepsilon \right\}$$
$$\le m\left\{ t \in (-\pi, \pi): |I_{h,k}(t)| > \frac{\varepsilon}{2} \right\} \le \varepsilon.$$

To prove 3), we observe that

$$|I_{f,k}^*(t)| = \left| \frac{1}{2k} \sum_{j=-k+1}^{k} \tilde{f}(x_j + t)e^{-inx_j} \right| = |\tilde{F}(t)|,$$

where

$$F(t) = \frac{1}{2k} \sum_{j=-k+1}^{k} f(x_j + t)e^{-inx_j}.$$

We obtain 3) by applying inequality a) from Theorem 5.3 to $F(t)$ and using the inequality $\|F\|_{L^1(-\pi,\pi)} \leq \|f\|_{L^1(-\pi,\pi)}$.

Properties 1)–3) are now established. Hence we have established that condition d) of Theorem 5 implies b). To complete the proof of Theorem 5, it is enough to show that c) implies d).

Let $\Phi(x) = f(x) + ig(x)$ $(x \in (-\pi, \pi)$, $f = \operatorname{Re}\Phi$, $g = \operatorname{Im}\Phi)$ and $F(re^{ix}) = \Phi * P_r(x) \in H_0^p$. Then $f, g \in L^p(-\pi, \pi)$ (see Theorem 4), and we need only prove that $g(x) = \tilde{f}(x)$ for almost all $x \in (-\pi, \pi)$.

Since the Poisson kernel is a real function, we may suppose that for $0 \leq r < 1$ and $x \in [-\pi, \pi]$ (see (3))

$$u(re^{ix}) = \operatorname{Re} F(re^{ix}) = f_r(x), \qquad v(re^{ix}) = \operatorname{Im} F(re^{ix}) = g_r(x).$$

On the other hand, it follows from (2), (8), and (37) that for every $r$, $0 < r < 1$,

$$g_r(x) = \tilde{f}_r(x), \qquad x \in (-\pi, \pi). \tag{45}$$

By Theorem 1,

$$\lim_{r \to 1} \|f - f_r\|_{L^1(-\pi,\pi)} = \lim_{r \to 1} \|g - g_r\|_{L^1(-\pi,\pi)} = 0. \tag{46}$$

In addition, by Theorem 5.3, it follows from the limit $f_r \overset{L^1}{\to} f$ $(r \to 1)$ that $\tilde{f}_r(x)$ converges to $\tilde{f}(x)$ in measure. Therefore (see (45)) $g_r(x) \to \tilde{f}(x)$ in measure $(r \to 1)$, and hence (see (46)) $g(x) = \tilde{f}(x)$ for almost all $x \in [-\pi, \pi]$. This completes the proof of Theorem 5.

COROLLARY 1. a) *If $f \in \operatorname{Re} \mathcal{H}^1$, then $\tilde{f} \in \operatorname{Re} \mathcal{H}^1$.*
b) *If $f \in \operatorname{Re} \mathcal{H}^1$ and $\int_{-\pi}^{\pi} f(x)\,dx = 0$ then $f = -(\tilde{\tilde{f}})$.*
c) *If $f \in \operatorname{Re} \mathcal{H}^p$ and $g \in \operatorname{Re} \mathcal{H}^q$, where $\frac{1}{p} + \frac{1}{q} = 1$ and $1 \leq p \leq \infty$, then*

$$\int_{-\pi}^{\pi} f(x)\tilde{g}(x)\,dx = -\int_{-\pi}^{\pi} \tilde{f}(x)g(x)\,dx. \tag{47}$$

PROOF. Relations a) and b) follow at once from the equivalence of conditions a) and d) of Theorem 5.

To obtain c), we set

$$F(re^{ix}) = \{f + i\tilde{f}\} * P_r(x),$$
$$G(re^{ix}) = \{g + i\tilde{g}\} * P_r(x).$$

According to Theorem 5, we have $F \in H_0^p$ and $G \in H_0^q$, and consequently $F \cdot G \in H_0^1$. But then (for almost all $x \in (-\pi, \pi)$)

$$\Phi(x) = (f + i\tilde{f})(g + i\tilde{g}) = \lim_{r \to 1} F(re^{ix})G(re^{ix}) \in \mathcal{H}_0^1,$$

and we obtain (see the definition of $\mathcal{H}_0^p$)

$$\int_{-\pi}^{\pi} \operatorname{Im} \Phi(x)\,dx = 0. \tag{48}$$

Equation (47) follows immediately from (48).

REMARK 1. If $1 < p < \infty$, then by, Theorem 5, d) and Theorem 5.3, the space $\operatorname{Re}\mathcal{H}^p$ coincides with $L^p(-\pi, \pi)$. For $p = 1$ this is not the case. The space $\operatorname{Re}\mathcal{H}^1$ is narrower than $L^1(-\pi, \pi)$ and consists (see Theorem 5, d)) of the functions for which $\tilde{f} \in L^1(-\pi, \pi)$. The space $\operatorname{Re}\mathcal{H}^1$ is a Banach space with the norm

$$\|f\|_{\operatorname{Re}\mathcal{H}^1} = \|f\|_{L^1(-\pi,\pi)} + \|\tilde{f}\|_{L^1(-\pi,\pi)}. \tag{49}$$

(The completeness of $\operatorname{Re}\mathcal{H}^1$ with the norm (49) follows from Theorem 5.3 and the completeness of $L^1(-\pi, \pi)$: if $\|f_n - f_m\|_{\operatorname{Re}\mathcal{H}^1} \to 0$ as $n, m \to \infty$, then $f_n \overset{L^1}{\to} f$, $\tilde{f}_n \overset{L^1}{\to} g$, and $f, g \in L^1(-\pi, \pi)$; and since $f_n \to \tilde{f}$ in measure as $n \to \infty$, we have $g = \tilde{f}$ and $\|f_n - f\|_{\operatorname{Re}\mathcal{H}^1} \to 0$ as $n \to \infty$.)

REMARK 2. According to Remark 1, equation (47) is satisfied, in particular, in the case when $f \in L^p(-\pi, \pi)$ and $g \in L^q(-\pi, \pi)$, $1/p + 1/q = 1$, $1 < p < \infty$.

We also notice that if we take $\tilde{f}(x)$ in (47) instead of $f(x)$ and use b), we obtain

$$\int_{-\pi}^{\pi} f(x)g(x)\,dx = \int_{-\pi}^{\pi} \tilde{f}(x)\tilde{g}(x)\,dx, \quad \text{if } \int_{-\pi}^{\pi} f(x)\,dx = 0. \tag{50}$$

## §3. The Blaschke product
## and the nontangential maximal function

Let a sequence $\{a_n\}_{n=1}^{\infty}$ of nonzero complex numbers (not necessarily all different) satisfy the condition

$$|a_n| < 1, \qquad n = 1, 2, \ldots, \qquad \sum_{n=1}^{\infty}(1 - |a_n|) < \infty. \tag{51}$$

We consider the product (*Blaschke product*)

$$B(z) = \prod_{n=1}^{\infty} \frac{a_n - z}{1 - \bar{a}_n z} \cdot \frac{\bar{a}_n}{|a_n|} \equiv \prod_{n=1}^{\infty} b(z, a_n). \tag{52}$$

For a given $r$, $0 < r < 1$, we have, for $|z| < r$,

$$|1 - b(z, a_n) \cdot |a_n|| = \frac{1 - |a_n|^2}{|1 - \bar{a}_n z|} \le \frac{2(1 - |a_n|)}{1 - r}. \tag{53}$$

Since the series in (51) converges, it is easily seen from (53) that the product (52) converges absolutely and uniformly in the disk $|z| \leq r$, i.e., $B(z)$ is analytic in the unit disk and has zeros at the points $a_n$, $n = 1, 2, \ldots$, and only at these points. Moreover, by using the inequality $|b(z, a_n)| \leq 1$ ($|z| < 1, n = 1, 2, \ldots$) we obtain

$$|B(z)| \leq 1, \qquad |z| \leq 1. \tag{54}$$

Suppose now that $a_1, a_2, \ldots$ ($|a_n| < 1$) are the zeros of a function $F(z) \in H^1$ with $F(0) \neq 0$, and that each $a_n$ is repeated according to its multiplicity. Let us show that then (51) converges. Set

$$B_m(z) = \prod_{n=1}^{m} \frac{z - a_n}{1 - \overline{a}_n z}, \qquad m = 1, 2, \ldots.$$

The function $B_m(z)$ ($m = 1, 2, \ldots$) is analytic in a disk of radius greater than 1, and $|B_m(z)| = 1$ if $|z| = 1$. Consequently $F_m(z) = F(z) \cdot (B_m(z))^{-1} \in H^1$ and (see Theorem 4, 3)) $\|F_m\|_{H^1} = \|F\|_{H^1}$. But then

$$\left| \frac{F(0)}{\prod_{n=1}^{m} a_n} \right| = |F_m(0)| = \frac{1}{2\pi} \left| \int_{-\pi}^{\pi} F_m \left( \frac{1}{2} e^{it} \right) dt \right| \leq \|F_m\|_{H^1} = \|F\|_{H^1}$$

and

$$\prod_{n=1}^{m} |a_n| \geq \frac{|F(0)|}{\|F\|_{H^1}} > 0, \qquad m = 1, 2, \ldots. \tag{55}$$

Since $|a_n| < 1$, $n = 1, 2, \ldots$, it follows from (55) that the product $\pi_1^{\infty} |a_n|$ converges, and therefore that the series (51) converges.

Let $F(z)$ be analytic in the disk $|z| < 1$ and let $a_n$, $n = 1, 2, \ldots$ ($0 < |a_n| < 1$) be its zeros, counted according to multiplicity. Also let $p \geq 0$ be the multiplicity of the zero of $F$ at $z = 0$. The product (see (52))

$$B(z) = z^p \prod_{n} b(z, a_n) \tag{56}$$

is the Blaschke product of $F(z)$.

We have the following theorem.

THEOREM 6. *Every function $F \in H^1$ can be represented in the form*

$$F(z) = B(z) \cdot G(z),$$

*where $G(z)$ has no zeros in the disk $|z| < 1$ and*

$$G \in H^1, \|G\|_{H^1} = \|F\|_{H^1},$$

*and $B(z)$ is the Blaschke product of $F(z)$.*

PROOF. Let $a_n$, $n = 1, 2, \ldots$ $(0 < |a_n| < 1)$, be the zeros of $F$ (or, equivalently, of $F/z^p \in H^1$). Then, as we noticed above, $B(z)$ is analytic in the disk $|z| < 1$, and

$$|B(z)| < 1, \qquad |z| < 1. \tag{57}$$

Moreover, $G(z) = F(z)/B(z)$ is also analytic in the unit disk and has no zeros there, and (see (57)) $\|G\|_{H^1} \geq \|F\|_{H^1}$.

To prove the inequality in the opposite direction, we consider the partial products of (56):

$$B_m(z) = z^p \prod_{n=1}^{m} b(z, a_n), \qquad m = 1, 2, \ldots, |z| \leq 1.$$

Since $|B_m(e^{ix})| = 1$ for every $x \in [-\pi, \pi]$, then by Theorem 4

$$\left\| \frac{F}{B_m} \right\|_{H^1} = \|F\|_{H^1}$$

and

$$\int_{-\pi}^{\pi} \left| \frac{F(re^{ix})}{B_m(re^{ix})} \right| dx \leq \|F\|_{H^1}, \quad \text{if } 0 < r < 1.$$

If we let $m$ tend to infinity in the preceding inequality and use the fact that $B_m(re^{ix}) \to B(re^{ix})$ as $m \to \infty$, uniformly for $x \in [-\pi, \pi]$, we obtain

$$\int_{-\pi}^{\pi} \left| \frac{F(re^{ix})}{B(re^{ix})} \right| dx \leq \|F\|_{H^1}, \qquad 0 < r < 1,$$

i.e., $G \in H^1$ and $\|G\|_{H^1} = \|F\|_{H^1}$. This completes the proof of Theorem 6.

Let $\sigma$, $0 \leq \sigma < 1$, be an arbitrary number. We denote by $\Omega_\sigma(x)$, $x \in [-\pi, \pi]$, the region bounded by two tangents to the circle $|z| = \sigma$ from $e^{ix}$ and the longer arc of the circle included between the points of tangency (when $\sigma = 0$, $\Omega_\sigma(x)$ degenerates to a radius of the unit disk). For $f \in L^1(-\pi, \pi)$ we set

$$f_\sigma^*(x) = \sup_{re^{i\theta} \in \Omega_\sigma(x)} |f_r(\theta)|, \qquad x \in [-\pi, \pi],$$

where $f_r(x)$ is the Poisson integral of $f(x)$ (see (3)).

The function $f^*(x)$ is the nontangential maximal function of $f(x)$. By Theorem 2,

$$|f(x)| \leq f_\sigma^*(x) \quad \text{for almost all } x \in [-\pi, \pi]. \tag{58}$$

We shall show that for any $f \in L^1(-\pi, \pi)$ the number $f^*_\sigma(x)$ does not exceed (in order) the value of the maximal function $M(f)(^2)$ at $x$, i.e.,

$$f^*_\sigma(x) \le C_\sigma M(f, x), \qquad x \in [-\pi, \pi]. \tag{59}$$

Let $re^{i\theta} \in \Omega_\sigma(x)$ and $x - \theta = \xi$. By the definition of the Poisson integral,

$$f_r(\theta) = \frac{1}{2\pi} \int_{-\pi}^{\pi} f(\theta + t) P_r(t)\, dt = \frac{1}{2\pi} \int_{-\pi}^{\pi} f(x + t) P_r(t + \xi)\, dt.$$

Let $F(t) = \int_0^t f(x + u)\, du$. Then we will have

$$2\pi f_r(\theta) = \int_{-\pi}^{\pi} P_r(t + \xi)\, dF(t) = P_r(\pi + \xi) F(\pi)$$

$$- P_r(-\pi + \xi) F(-\pi) - \int_{-\pi}^{\pi} F(t) P'_r(t + \xi)\, dt$$

and, by the inequality $|F(t)| \le |t| \cdot M(f, x)$, $t \in [-\pi, \pi]$, and the periodicity of $P_r(t)$,

$$2\pi |f_r(\theta)| \le M(f, x) \left[ 2\pi P_r(\pi + \xi) + \int_{-\pi}^{\pi} |t P'_r(t + \xi)|\, dt \right]. \tag{60}$$

Since the functions $g(t) = -t$ and $g_1(t) = P'_r(t)$ are positive for $t \in (-\pi, 0)$ and negative for $t \in (0, \pi)$ (see (5)), then, assuming without loss of generality that $\xi > 0$, we obtain

$$\int_{-\pi}^{\pi} |t P'_r(t + \xi)|\, dt = - \int_{-\pi}^{-\xi} + \int_{-\xi}^{0} - \int_{0}^{\pi - \xi} + \int_{\pi - \xi}^{\pi} t P'_r(t + \xi)\, dt. \tag{61}$$

For $-\pi \le a < b \le \pi$ we have the inequalities

$$2\pi \ge \int_a^b P_r(t + \xi)\, dt = P_r(t + \xi) t \big|_a^b - \int_a^b t P'_r(t + \xi)\, dt,$$

$$\left| \int_a^b t P'_r(t + \xi)\, dt \right| \le 2\pi + |b| P_r(b + \xi) + |a| P_r(a + \xi).$$

Consequently (see (60) and (61)), in order to prove (59) it is enough to verify that

$$|t| P_r(t + \xi) \le C_\sigma \quad \text{for } t = -\pi, -\xi, 0, \pi - \xi, \text{ and } \pi \tag{62}$$

if $re^{i\xi} \in \Omega_\sigma(x)$. Let $t = -\xi$; then

$$|t| P_r(t + \xi) = \xi P_r(0) \le 2 \frac{x - \theta}{1 - r} \le C_\sigma.$$

---

$(^2)$Since $M(f, x)$ was defined in Appendix 1 for a function $f$ defined on $R^1$, we also set $f(x) = 0$ if $|x| > 2\pi$, $f(x) = f(x + 2\pi)$ for $-2\pi \le x < \pi$, and $f(x) = f(x - 2\pi)$ for $\pi < x \le 2\pi$.

In the remaining cases, (62) is evident. It follows from (58), (59), and Theorem 2 of Appendix 1 that for every $f \in L^p(-\pi, \pi)$, $1 < p < \infty$,

$$\|f\|_{L^p(-\pi,\pi)} \leq \|f_\sigma^*\|_{L^p(-\pi,\pi)} \leq C_{\sigma,p}\|f\|_{L^p(-\pi,\pi)}, \tag{63}$$

where $C_{\sigma,p}$ is a constant depending only on $\sigma$ and $p$.

THEOREM 7. *Let $F(z) \in H^p$ $(1 \leq p < \infty)$, $0 \leq \sigma < 1$, and*

$$F_\sigma^*(x) = \sup_{z \in \Omega_\sigma(x)} |F(z)|, \qquad x \in [-\pi, \pi].$$

*Then $F_\sigma^*(x) \in L^p(-\pi, \pi)$ and*

$$\|F\|_{H^p} \leq \|F_\sigma^*\|_{L^p(-\pi,\pi)} \leq C_{p,\sigma}\|F\|_{H^p}. \tag{64}$$

PROOF. In the case $1 < p < \infty$, Theorem 7 is a direct corollary of (63) and Theorem 4. Now let $F(z) \in H^1$. By Theorem 6 we have $F(z) = B(z) \cdot G(z)$, where $|B(z)| \leq 1$, $G(z) \neq 0$ if $|z| < 1$, and $\|G\|_{H^1} = \|F\|_{H^1}$. We may take the square root of $G(z)$: there is a function $E(z) \in H^2$ such that $E^2(z) = G(z)$ and consequently (see (64) for $p = 2$)

$$\int_{-\pi}^{\pi} F_\sigma^*(x)\, dx \leq \int_{-\pi}^{\pi} G_\sigma^*(x)\, dx = \int_{-\pi}^{\pi} [E_\sigma^*(x)]^2\, dx$$

$$\leq C_{\sigma,2}^2 \|E\|_{H^2}^2 = C_\sigma \|G\|_{H^1} = C_\sigma \|F\|_{H^1}.$$

The lower bound for $\|F_\sigma^*\|_{L^1(-\pi,\pi)}$ follows from (58). This completes the proof of Theorem 7.

# Notes

First of all we observe that the theory of orthogonal series is treated in a number of books: Kaczmarz and Steinhaus [64] (first edition in 1935, in German; Russian translation, supplemented by a wider survey of the literature, by R. S. Guter and P. L. Ul'yanov, in 1958), Alexits [2] and Olevskiĭ [38]. These books throw light on several topics on which we hardly touch (for example, summability of orthogonal series, and series and Fourier coefficients of functions in $L^p$, $p \neq 2$, with respect to general O.N.S.). In [38] there is also a survey of a number of directions in the theory of orthogonal series that were developed between 1960 and 1975. In this connection, we should say that the monographic literature on the theory of general orthogonal series can hardly be called extensive. It does not fill all the gaps in the material that has been investigated, and this book, in particular, represents only to a small extent the work of the Hungarian school of the theory of orthogonal series.

We now give bibliographic references to the material that we do consider.

## Chapter 1

This chapter is introductory; the results presented here we obtained before 1935. We restrict ourselves to brief indications.

Theorem 1 was proved by Orlicz [143]. Theorem 2 was probably first published in [64] (see [64], p. 264). Theorems 3 and 4 belong to the foundations of the theory of real functions and are always included in university curricula; for their history see [2]. Theorem 5 was obtained as early as 1909 by Lebesgue (see [2], Chapter IV, §1). Theorem 6 was proved by Banach [9], to whom also belongs (in essentials) Theorem 7 (for details see [170] and [35]). Theorem 8 was actually proved by Haar [54] as early as 1910, although the concept of a basis was itself introduced later by Schauder (see [9] and [170]). A special case of Theorem 9 (for the Haar system; see Chapter 3) was discussed by Schauder [164]; the proof in the general case is similar. Theorem 10 is a simple corollary of Theorems 6 and 1. Finally, for Theorem 11 see the notes on Chapter 2.

## Chapter 2

In §1 and §2 we introduce and study sequences of independent functions. In the language of probability theory, the fundamental definition of this chapter, Definition 1, is nothing but the definition of a set of independent random variables $\{f_n\}$ defined on a probability space $(\Omega, F, P)$, where $\Omega = (0, 1)$, $F$ is the system of Borel sets, and $P$ is Lebesgue measure on $(0, 1)$. All the theorems of §1 and §2 are very often formulated in probabilistic language. For example, Theorem 4 states that the expectation of the product of two independent random variables is the product of their expectations. The method that we used for constructing sequences of independent functions were used in the 1930's in the work of the Polish school (Steinhaus, Marcinkiewicz, Zygmund, etc.; for details see [105], p. 235, and [64]); and for the discrete case by Khinchin and Kolmogorov [83]. This method (possibly somewhat out of date) is closely related to the theory of functions and has the advantage that it does not require any additional knowledge of measure theory. Theorems 1–10 of §1 and §2 are related to the general theory of probability; for detailed comments on them we refer the reader to monographs on this theory. We restrict ourselves to a few remarks.

The Rademacher functions $r_n(x)$, $n = 1, 2, \ldots$, were introduced in 1922 by Rademacher [157], but the behavior of the sums $\sum_1^N r_n(x)$ as $N \to \infty$ and for $x \in (0, 1)$ was investigated considerably earlier in connection with binary decompositions of the real line (see, for example, [97], p. 42). Theorems 5 and 6 were obtained for the special case of the Rademacher system by Khinchin [82] in 1923, and later were extended to other systems of independent functions (in the most general form, Khinchin's inequality (Theorem 6) was proved in 1937 by Marcinkiewicz and Zygmund (see [105], p. 257). Theorem 7 was obtained by Paley and Zygmund [147] as an immediate corollary of Khinchin's inequality and the simple, but important, Lemma 1 (see Theorem 7), which was probably first formulated in [147]. Theorem 8 is due to Kolmogorov [83] (inequality (47) was not mentioned in [83], but, like the statement of Theorem 7, is easily deduced form the results of §3 of [83]). Theorem 9 was also proved by Kolmogorov [87] (the case of the Rademacher system had already been discussed in

[157]). In [87] there appears the inequality (Kolmogorov's inequality)

$$m\{x \in (0, 1): S_\psi^*(\{a_n\}) > y\} \leq \frac{1}{y^2} \sum a_n^2; y > 0, \qquad (*)$$

which is hardly less precise than (47). It should be noticed that the method of estimating the majorant, proposed in [87] for proving (*), has turned out to be very important in both probability theory and function theory. We often use it in Chapter 3. Inequality (47) in Theorem 9 was proved by Marcinkiewicz and Zygmund (see [105], p. 238), but follows at once from a version of (*) that was obtained in [88].

The application of properties of systems of independent functions to the study of series of functions was initiated almost simultaneously by Orlicz [140]–[142] and Paley and Zygmund [146], [147] (see also Littlewood [100], [101]). At the present time, random series play a fundamental role in the theory of orthogonal series and in functional analysis (see in particular [65], [17], [169], and [107]), and are frequently used in the present book.

In Theorem 11 the implication 2) $\Rightarrow$ 3) was proved by Orlicz [142], and 3) $\Rightarrow$ 1), by Paley and Zygmund [146]. Corollary 4 was given by Orlicz [142]. In Theorem 12 the necessity of condition (72) for the convergence of the series $\sum_1^\infty f_n(x)$ in $L^p$ for almost all choices of signs was obtained by Orlicz [143], and the sufficiency (in the slightly weaker form of Corollary 5) by Paley and Zygmund [146]. Theorem 1.11, stated in Chapter 1 and proved in §2.3, follows easily from Theorems 12 and 1.10. Theorem 12 is Pełczyński's [151] (our proof is close to that given in [122]). Theorem 14 and Corollary 6 were proved by Kashin [70]. The similar result about convergence in measure (see the remark at the end of §3) was obtained in [70] and independently by Maurey and Pisier [113]. Lemma 1 from Theorem 14 (with an inexact constant in (84)) is contained in Drobot [38] (also see Theorem 10.5). Our proof provides the exact constant in (84).

The study of random rearrangements of series of functions (see §4) was initiated by Garsia [48], [49]. Theorem 16 was proved by Garsia in [48], and Corollary 8 and part b) of Corollary 7, in [49]. Theorem 15 and Corollary 7, a) (see [80], Russian p. 386, English p. 51), represent a sharpening of estimates in [49]. The approach used here (and in [49]) for the study of random permutations of sets of numbers, based on analogy with the proof of Kolmogorov's inequality (*), was suggested by Rosén [159].

## Chapter 3

The Haar system was introduced in Haar's dissertation (see [54]) and is now widely used in the theory of functions, as well as in probability theory and numerical mathematics. From the point of view of probability theory the partial sums $S_N(x)$, $N = 1, 2, \ldots$, of an arbitrary series $\sum_1^\infty a_n\chi_n(x)$ in the Haar system are a special case of a sequence of random variables that form a martingale (on martingales see, for example, [36], [169]). Martingales are extensively used in probability theory, and some of the theorems proved in Chapter 3 were first established in probabilistic terms for martingales. We also notice that in many cases propositions on properties of the Haar system can be generalized to martingales without the introduction of essentially new ideas.

The formulas for the partial sums of Fourier–Haar series were obtained by Haar himself [54]. Inequalities for Fourier–Haar coefficients (see Theorem 1) were noticed for $f \in C(0, 1)$ by Ciesielski [24], and for $f \in L^p(0, 1)$, $1 \le p < \infty$, by Ul'yanov [189], who initiated the systematic study of the Haar system in the USSR. The uniform convergence of the Fourier–Haar series of continuous functions (see Theorem 2) was established by Haar (the construction of an O.N.S. with this property was one of Haar's original objectives; see [54]); inequality (14) in Theorem 2 was noticed by B. Szökefalvi–Nagy in [175], which was devoted to general O.N.S. That the Haar system forms a basis in $L^p(0, 1)$, $1 \le p < \infty$ (see Theorem 3) was established by Schauder [164], and inequality (17) by Ul'yanov [189]. The convergence a.e. of arbitrary Fourier–Haar series (see Theorem 4) was established by Haar [54], and the properties of the majorants of the partial sums of these series were observed by Marcinkiewicz (see [105], p. 310). Theorem 5 was proved by Golubov [50], and Theorem 6 by Bochkarev [11] (also see [17]).

The study of unconditional convergence of Fourier–Haar series in $L^p(0, 1)$ (see §3) was initiated by Marcinkiewicz (see [105], p. 308), who obtained the statement of Theorem 8 and inequality 2) of Theorem 9 as direct consequences of Paley's results on the Walsh system. Theorem 7 and inequality 1) in Theorem 9 were established by Yano [200]; our proof of Theorem 7 was given by Watari [196] (also see [61]). In Theorem 10, statement 2) follows from Theorem 2.13, and 3) was obtained by Ul'yanov [191] and then extended by Olevskii (see [138], p. 75) to general C.O.N.S. Theorem 11 was proved at the same time for matrices by Burkholder and Gundy [18] and Davis [34]. We note that the method of proof of Theorem 11 (see, in particular, Lemma 1) is a typical method of martingale theory, and has features in common with the classical proof of Kolmogorov's inequality (see Theorem 2.9 and the notes to Chapter 2). Theorem 12 can be obtained as a corollary of general interpolation theorems; the direct proof in the text was suggested by Saakyan. Theorem 13 was proved independently by Arutyunyan [4] and (for martingales) by Gundy [53]. Theorem 14 follows from results of Chow [23] on martingales and was also obtained by Arutyunyan [4] (also see [53] and [180]).

The first result on the unconditional convergence a.e. of Fourier–Haar series was obtained by Ul'yanov: in [184] he proved Corollaries 6 and 7 (see §5). Theorem 15 was proved by Nikishin and Ul'yanov [130], and Theorem 16 by Olevskiĭ [132], [133]. Theorem 17 and Corollary 8 were established by Ul'yanov [187], [188], [192]; and Theorem 18, by Bochkarev [12].

Transformations of the Haar system similar to those considered in §6 were already used by Schauder [164] for the construction of bases in $L^p(\Omega)$ (where $\Omega$ is a bounded domain in $R^n$). Corollary 9 was obtained by Olevskiĭ (see [138], p. 61, and also the remarks on Chapter 9, below).

In conclusion, we note that a number of results on the Haar system, not included here, are given in the survey by Golubov [51].

## Chapter 4

We consider the trigonometric system in §§1–4. There is immense literature on trigonometric series, including the important monographs of Bari [10] and Zygmund [201], both of which contain extensive bibliographies.

In Chapter 4, §1 consists of standard material. Theorem 5 (see §2) was proved by Jackson in 1911. Jackson actually considered only continuous functions, but his proof applies without essential changes to functions in $L^p(-\pi, \pi)$, $1 \le p < \infty$ (for details see [201] and [156]). Our proof of Theorem 3 is close to that in [22], and is based on properties of the Vallée-Poussin means. We note that the Vallée-Poussin means, introduced by Vallée–Poussin in 1918, play an extremely important role in approximation theory (see [202], "Vallée–Poussin method of summation," and [39]).

Theorem 4 was proved by M. Riesz (see [10], Chapter VIII, §19). Corollary 3 was proved independently by Konyagin [93] and McGehee, Pigno and Smith [114]. Theorem 5 was also obtained in [114], whose method we follow. Carleson's theorem was proved in [19]. Later, Hunt [60] proved by Carleson's method that the Fourier series of every function in $L^p(-\pi, \pi)$, $p > 1$, converges almost everywhere. The existence of a function $f \in L^1(-\pi, \pi)$ whose Fourier series diverges a.e. was discovered by Kolmogorov as early as 1923 (see [10], Chapter V, §17). Theorem 6 was obtained in 1925 by Kolmogorov and Seliverstov, and also by Plessner (see [10], Chapter V, §2).

Theorem 7 was obtained in the last century by Dini and Lipschitz (see [202], "Dini–Lipschitz test"). The study of the strong summability of Fourier series was initiated by Hardy and Littlewood; the fundamental results on strong summability were obtained by Marcinkiewicz; for details see [10], Chapter VII, §6; Theorem 8 was proved by Alexits and Králik [3] (also see [66]). Theorem 9 was proved by H. Bohr by complex-variable methods; our proof was suggested by Saakyan [160] (also see [161] and [66]). Theorem 10 was obtained by Paley and Zygmund (see [10], Chapter IV, §13); generalizations of Theorem 10 and Proposition 2 were given in [106]. Theorem 11 was proved independently by Shapiro and by Rudin (see [67], p. 133). Corollary 5 was established by S. N. Bernstein, and Corollary 6 by Zygmund; generalizations of the results of Theorem 12 were obtained by Szász (for details see [10], Chapter IX, §2).

The Walsh system was introduced by Walsh [195] in 1923; at present, it is used in many branches of mathematics, and especially in applied problems. The enumeration of the Walsh functions used in §5 (and in most of the work on the Walsh system) was suggested by Paley [145], who proved many of the fundamental properties of the Walsh system (in particular, Theorem 15). Inequality (115) for the Lebesgue functions of the Walsh system was obtained by Vilenkin [194], and Corollary 7 by Fine [44]. Many properties of the Walsh system (see, for example, Theorems 13 and 14) are analogs of properties of the trigonometric system, a fact which is often explained by the observation that both systems are character systems of locally compact Abelian groups. More information on the Walsh system and character systems can be found in the survey by Balashov and Rubinshteĭn [8]; also in [1].

## Chapter 5

The Hilbert transform and matrix (see (27)) were introduced by Hilbert around 1900. Hilbert himself (see [57], [56]) proved that the Hilbert transform is a bounded operator from $L^2(R^1)$ to $L^2(R^1)$. Subsequently the concept of conjugate function, which is closely related to the Hilbert transform, was introduced by Luzin [102] in the theory of trigonometric series. In the problems that we consider, the study of the Hilbert transform and of the operator of conjugation are equivalent.

The existence and finiteness for a.e. $x \in R^1$ of the Hilbert transform $T(f, x)$ of every $f \in L^1(R^1)$ (see Theorem 1) was proved by Privalov [155], and inequality (5), by Kolmogorov [86] (also see [10] and [201]).

Our proof of Theorem 1 resembles that in [49]; we note that the essential point in the proof, Lemma 1, was obtained by Boole in 1857 (see [99], p. 68).

Theorem 2 was proved by M. Riesz [158]. Theorem 3 is a combination of the results of Privalov, Kolmogorov, and Riesz already cited.

The space BMO was introduced by John and Nirenberg in [62] and is much used in analysis at present; Theorem 7 is also proved in [62]. Theorems 4 and 5 were obtained by Fefferman (see [42] and [31]), and are probably better known in connection with the long series of results on the Hardy spaces $H^p$ and their higher-dimensional analogs that were established in the seventies (we touch on this extensive topic mainly in connection with the Franklin system in Chapter 6; for details see [32] and [43]). In the presentation of Theorem 4 we use the reasoning of [198], simplified for the one-dimensional case by Oswald.

Theorem 6 is due to Coifman and Weiss [32].

## Chapter 6

The Faber–Schauder system was introduced in 1910 by Faber [41], who constructed it by integrating the Haar functions (see equation (2)). Theorem 1 was also proved in [41]. In 1927 Schauder [163] rediscovered the Faber–Schauder system: this system is the simplest member of the family of bases of $C(0,1)$ that were constructed in [163].

In Corollary 1, inequality a) follows immediately from Theorem 1, and b) was obtained by Matveev [108].

Theorem 2 was established by Karlin [69]; our proof was suggested by Arutyunyan [5]. Theorem 3 was proved by Ciesielski [25] (see also [162]). Proposition 1 and Corollary 2 were obtained by Saakyan [160]. For the systems of Faber–Schauder type discussed in §2, see [64], p. 50, and [163].

The Franklin system was introduced in 1928 by Franklin as the first example of an orthonormal basis in the space of continuous functions (see Theorem 6). Afterwards the Franklin system was not studied until the work of Ciesielski, who originated the systematic investigation of this system. Ciesielski in [26] and [27] proved Theorems 5 and 7–9, and Corollary 4. The proofs of Theorems 8 and 9 in the text are simpler than those in [27] and were suggested by Ciesielski in 1977; they use the functions $N_j(t)$ (see (30)), which are special cases of the $B$-splines which were introduced in [33]. We note that Theorem 9 plays a fundamental role in many papers on the Franklin system.

Theorems 10 and 11, and Corollary 5, were established by Wojtaszczyk [199], making essential use of Carleson's arguments [20]. Before Carleson and Wojtaszchyk, Maurey [112] had discovered (by a nonconstructive method) that there is an unconditional basis in $H^1$. It should be noted in connection with Wojtaszcyk's theorem that as early as 1969 Ciesielski [28] and Schonefeld [166] had applied the Franklin system to the construction of a basis in the space $C^1(I^2)$ of continuously differentiable functions on a square. Then Bochkarev [14] (by transforming the functions in the Franklin system into analytic functions by the rule given in Corollary 5.6; see 5.(90)) applied this system to the construction of a basis in the space $C_A$ of functions that are analytic in the disk $|z| < 1$ and continuous up to the boundary (the question of the existence of bases in $C^1(I^2)$ and $C_A$ was mentioned in [9]). Hence the Franklin system has proved to be very useful in constructing bases in various function spaces.

Theorem 12 was proved by Bochkarev [14]. The stronger form of the theorem, mentioned after the proof of Theorem 12, was obtained in [30]. Theorem 13 was proved by Chang and Ciesielski [21].

Finally we note that at present there are rather detailed investigations and generalizations of the Franklin system, namely the orthogonal splines; for details, see [29] and [168].

## Chapter 7

The orthogonalization theorems discussed in §1 are usually helpful in the construction of various orthonormal systems. Theorem 1 was proved by I. Schur ([167]; see also [64]). Theorem 2 was obtained by Men'shov [121], who in its proof a method of orthogonalization which was probably first applied by Kolmogorov and Men'shov [90]. Criteria for the extendability of a system of functions to a complete orthonormal system (see Theorem 3) were given by Kozlov [94] and (in the form given in the text) by Olevskiĭ (see [137] and [138], p. 57).

Corollary 1 has probably been known for thirty years; Hobby and Rice [58] obtained a more precise result: they showed that the required function $\varepsilon(x)$ (see Corollary 1) can be chosen to have $\leq m$ changes of sign. Theorem 4 follows immediately from Corollary 1.

In §§2–4 we consider factorization theorems and their applications to the theory of orthogonal series. This subject was opened by the work of Grothendieck [52]; one of his results is mentioned at the beginning of this chapter. In addition, here we must mention Kolmogorov [86]. In [86] it was shown by the explicit example of the Hilbert operator (transformation) $f(x) \to T(f, x)$ (see §5.1) how by using the information that $T(f, x)$ is finite almost everywhere for every $f \in L^1$ one can obtain substantially more: the inequality 5.5 for the weak type. Later Kolmogorov's approach was generalized by Stein [172].

Essential developments of factorization theorems were obtained by Nikishin in [126], [127], and [129], and later by Maurey, Pisier and others (see [110]). Nikishin, in contrast to Grothendieck, considered operators that act not only on Banach spaces, but also on the space of all measurable, a.e. finite functions; this significantly extends the domain of applicability of factorization theorems. At present, factorization theorems have found a variety of applications in the theory of orthogonal series, probability theory, and functional analysis (see, in particular, the bibliographies in [68] and [84]).

Theorems 5 and 6 were proved by Nikishin in [127] and [126]. Theorem 7 was published by Maurey, but can be immediately obtained by successive applications of theorems of Nikishin (see [127], Theorem 4) and Grothendieck [52]. In connection with Theorem 7, we notice the following result (see [113], [144], and also [111]):

For the series $\sum_1^\infty f_n(x)$, $x \in (0, 1)$, to be unconditionally convergent in measure, it is necessary and sufficient that for every $\varepsilon > 0$ there are a set $E_\varepsilon \subset (0, 1)$, $m(E_\varepsilon) > 1 - \varepsilon$, a constant $K_\varepsilon$, and a series

$$\sum_{n=1}^\infty a_n \varphi_n(x), \qquad \sum_{n=1}^\infty a_n^2 < \infty, \qquad \{\varphi_n(x)\}_{n=1}^\infty \quad \text{an O.N.S. (on } (0,1))$$

such that $f_n(x) = K_\varepsilon a_n \varphi_n(x)$ for $x \in E_\varepsilon$ and $n = 1, 2, \ldots$ (we did not include the proof of this proposition in the text, since it depends on methods that are somewhat remote from the subject matter of the book).

Theorem 8 is a corollary of results of Olevskiĭ [135], and Theorem 9 was established by Nikishin [128], [129].

## Chapter 8

Theorem 1 is the famous Men'shov–Rademacher theorem ([117], [157]); Men'shov [117] also proved Theorem 2. Theorem 3 was obtained by Kashin [72]; the simpler proof given in the text was suggested by Tandori [183]. Theorem 4 was established by Kaczmarz [63], but the main idea of the proof appeared in papers of Kolmogorov and Seliverstov, and of Plessner, on the convergence a.e. of trigonometric series (see the notes to Chapter 4 and [2], Chapter III, §1, Theorem 3.1.5). In the proof of Theorem 4 we follow [2]; on the definitiveness of this result, see [182]. On Theorems 1 and 4, see also Schipp [165].

Corollary 3 was established by Men'shov [118]; this was probably the first result on the unconditional convergence of series of functions. Corollary 2 was obtained by Orlicz [139]; Theorem 5 is also proved by the method of [139]. The definitive form of the sufficient conditions for unconditional convergence a.e. of orthogonal series (conditions imposed on the coefficients of the series) was given by Tandori [181]: see Theorem 5.

Theorem 6 and Corollary 4 were proved by Tandori [181]. Strictly speaking, Tandori proved somewhat more, namely that for a given sequence $a_1 \geq a_2 \geq \cdots$ the series (47) converges a.e. for every O.N.S. $\{\varphi_n(x)\}_{n=1}^{\infty}$ if and only if condition (48) is satisfied. Later, Tandori gave a simpler proof; in the text we present Tandori's discussion in a modified form.

Corollary 5 was obtained by Men'shov [119] and Marcinkiewicz [104] (also see [105], p. 164). Theorem 7 was proved by Marcinkiewicz [104] and a little later by Men'shov [120]; actually [120] practically repeats the discussion in [119]. We should say that the methods of proof in [119] and [104], in spite of superficial differences, are very similar.

In connection with Corollary 5 we note a result of Komlós [92], where it was proved that from every O.N.S. one can select a subsystem with unconditional convergence.

Theorem 8 and Corollary 6 were established by Erdös [40] for the special case of subsystems of the trigonometric system, and in the general case (by a similar method) by Stechkin (see [47] and [190]).

The inequality for the density of a Sidon subsystem of the trigonometric system (see (99)) was obtained by Stechkin [171]. The result mentioned in §4 on the possibility of selecting a Sidon subsystem from every collectively bounded O.N.S. was established by Gaposhkin [47]. Theorem 9 was proved by Kashin [79] (for applications of propositions of the type of Theorem 9 to problems on the geometry of normed spaces, see [123]). Theorem 10 was also obtained by Kashin and is a generalization of a theorem of Nisio (see [97], p. 102; Nisio proved that every series of the form (120)2) converges, for almost all $t$, uniformly (in $y$) on $(0,1)$). Finally, proposition 1 was obtained by McLaughlin [115].

## Chapter 9

The subject matter of Chapter 9 follows the work of Ul'yanov [186], [185] and Olevskiĭ [131], in which, using Corollary 3.6, established earlier in [184], Theorem 1 was proved. Among earlier work on divergence of orthogonal series and series of Fourier coefficients in general systems, we note the papers of Orlicz [139], [141], and Kozlov [95].

The systematic study of problems of divergence of orthogonal series and series of Fourier coefficients in general systems was begun by Bochkarev (see [17], [12]), who obtained many definitive results.

Corollary 1 was established by Kozlov [95] and strengthens a result of Orlicz [137], according to which, for every complete O.N.S. $\{\varphi_n(x)\}_{n=1}^{\infty}$, the sum $\sum_1^{\infty} \varphi_n^2(x) = \infty$ for almost all $x$. Theorems 2 and 3 were proved by Olevskii [132], [133]. Theorem 4 was first established by Makhmudov [103], although his proof practically repeats the discussion by Orlicz ([141]; see also [10], Chapter IV, §16), who obtained a similar result for uniformly bounded O.N.S.

The statement of Theorem 5, in the special case when $\{\varphi_n\}$ is the trigonometric system, was proved in [98]; then S. V. Khrushchev noticed that the method of [98] is completely sufficient for obtaining results on general O.N.S., including Theorem 5 (also see [85]). Theorem 6 was established by Kashin [71], [75]; Corollary 2 had been obtained earlier by Mityagin [124] and generalizes a result of S. N. Bernstein (see [10], Chapter IX, §4) for the case of the trigonometric system. The proof of Theorem 6 in the text is different from that in [75] and is close to the work of Bochkarev (see, for example, [17]).

In §3 of Chapter 9 we consider uniformly bounded O.N.S. The first lower bounds for the Lebesgue functions of uniformly bounded systems, with which §3 begins, were obtained by Olevskii [134] (there he established Theorem 8 and Corollary 5). Subsequently in [136], [138] he also obtained Corollary 3. Bochkarev [15], [16] proved a more precise result, Theorem 7, by a different method. Bochkarev's approach was modified in [77], following which we established Proposition 1 and inequality (63), which play a fundamental role in §3. In connection with Corollary 5 we note the weaker result of Szarek [174], who proved (using, in particular, Olevskii's method) that for an arbitrary basis $\{\psi_n(x)\}_{n=1}^{\infty}$ in $L^1(0,1)$ with $\|\psi_n\|_1 = 1$, $n = 1, 2, \ldots$, and for every $p > 1$, there is the relation

$$\varlimsup_{n \to \infty} \|\psi_n\|_p = \infty.$$

Theorem 9 was proved by Krantsberg [96]. Theorem 10 was proved by Bochkarev [16]; later, Kazaryan [81] showed that under the additional requirement of completeness one cannot assert the existence of a Fourier series that diverges almost everywhere.

Concerning inequality (113), see [189]. Theorem 11 was obtained by Bochkarev [12]. In [13] (also see [17]) Bochkarev also established a more precise result: under the hypotheses of Theorem 11, for every modulus of continuity $\omega(\delta)$ with

$$\sum_{n=1}^{\infty} \frac{\sqrt{\omega(1/n)}}{n} = \infty,$$

the required function $F(x)$ of bounded variation whose series of Fourier coefficients does not converge absolutely can be found in the class $H_\omega$ (as the example of the trigonometric system shows (see [10], Chapter IX, §3) a further increase in the smoothness of $F(x)$ is in general not possible).

Our presentation of the proof of Theorem 11 follows [75] and [77]; also see Wik [197], where the most important special case (the trigonometric system) is considered.

## Chapter 10

In Chapter 10 we discuss metric theorems on the representation of functions by general orthogonal series (we do not say much about the old subject of the representation of functions by everywhere convergent trigonometric series).

The development of this direction of the theory of orthogonal series was initiated by Luzin (see [102]) and continued by Men'shov, who in 1916, in [116], constructed a null series for the trigonometric system. Originally considered only for trigonometric series, the problem of the representation of functions by general orthogonal series was developed in the middle thirties by Marcinkiewicz (see [105], p. 312). At present, theorems on the representation of functions occupy an appreciable part of the theory of general orthogonal series. For more details on this topic see the surveys by Talalyan [178] and Ul'yanov [193], and also [78], [153], [179], [6], and [125].

Theorem 1 and Corollary 1 were proved by Talalyan (see [176], [177]). Theorems 2, 2′, and 3 were proved by Kashin (see [73], [74], and [76]). The proof of Theorem 2′ clarified (see [74]) the existence of a difference, for problems of representing functions by series, between general convergence systems and the slightly narrower class $Q$ of such systems (a C.O.N.S. $\Phi \in Q$ if the operator $S_\Phi^*$ of majorants of the partial sums is a bounded operator from $L^2$ to $L^2$). In this connection it is natural to raise the question of whether one can always represent a measurable function, finite almost everywhere, by an almost everywhere convergent series in a system of class $Q$. Theorem 4 (Arutyunyan and Pogosyan) provides a positive answer to this question. Theorem 4 was announced by Arutyunyan [7] and Pogosyan [154] in notes that appeared simultaneously, and proved by them in seminars. A proof of this result is published here for the first time. In our exposition of Theorem 4 we follow, in the main, Arutyunyan's presentation. Lemma 1 in Theorem 4 follows immediately from the results of [91], and was first applied to related problems by Pogosyan [152]. The use of the Haar system in problems on the representation of functions by general orthogonal series was initiated by Arutyunyan (see [6]).

We should remark that the possibility of representing every measurable function, finite a.e., by an almost everywhere convergent trigonometric series, and by a Haar series, had been established much earlier by the proofs of Theorems 1 and 4, respectively, by Men'shov and Bari (for details see [178]).

Theorem 5 was proved by Drobot [38]. A more general approach to similar theorems was suggested by Pecherskiĭ [149], [150]; also see [45]. For Lemma 1 and Theorem 5 see the notes on Theorem 2.14.

Theorem 6 was established by Pogosyan [153], and his paper [154] contains a proposition on universal series for an arbitrary C.O.N.S. (without the supplementary condition (115)). The case of the trigonometric and Haar systems was considered earlier by G. M. Mushegyan (see [154]).

## Appendices

The concept of the maximal function was introduced by Hardy and Littlewood [55] in 1930. In that paper they also gave the first application of this concept to problems on the convergence of series of functions. In §2 of Appendix 1 we follow [172]; also see [201].

For the minimax theorem see, for example, [148].

More details on the contents of Appendix 2 are given in [59] and [201]. Theorem 7 in this appendix, which was established by the brothers F. and M. Riesz in 1916, is presented along the lines of [37].

# Bibliography

1. G. N. Agaev et al., *Multiplicative systems of functions and harmonic analysis on zero-dimensional groups*, "Èlm", Baku, 1981. (Russian)

2. G. Alexits, *Konvergenzprobleme der Orthogonalreihen*, Akad. Kiadó, Budapest, 1960; English transl., Pergamon Press, 1961.

3. G. Alexits and D. Králik, *Über die Approximation mit starken de la Vallée-Poussinschen Mitteln*, Acta Math. Acad. Sci. Hungar. **16** (1965), 43–49.

4. F. G. Arutyunyan, *Series with respect to the Haar system*, Akad. Nauk Armyan. SSR Dokl. **42** (1966), 134–140. (Russian)

5. ____, *Bases of the spaces $L_1[0, 1]$ and $C[0, 1]$*, Mat. Zametki **11** (1972), 241–249; English transl. in Math. Notes **11** (1972).

6. ____, *Representation of measurable functions almost everywhere by convergent series*, Mat. Sb. **90(132)** (1973), 483–520; English transl. in Math. USSR Sb. **19** (1973).

7. ____, *Representation of functions by multiple series*, Akad. Nauk Armyan. SSR Dokl. **64** (1977), 72–76. (Russian)

8. L. A. Balashov and A. I. Rubinshteĭn, *Series in the Walsh system, and their generalizations*, Itogi Nauki i Tekhniki: Mat. Anal. 1970, VINITI, Moscow, 1971, pp. 147–202; English transl. in J. Soviet Math. **1** (1973), no. 6.

9. Stefan Banach, *Théorie des opérations linéaires*, PWN, Warsaw, 1932.

10. N. K. Bari, *Trigonometric series*, Fizmatgiz, Moscow, 1961. (Russian)

11. S. V. Bochkarev, *On the coefficients of Fourier series with respect to the Haar system*, Mat. Sb. **80(122)** (1969), 97–116; English transl. in Math. USSR Sb. **9** (1969).

12. ____, *Absolute convergence of Fourier series with respect to complete orthonormal systems*, Uspekhi Mat. Nauk **27** (1972), no. 2 (164), 53–76; English transl. in Russian Math. Surveys **27** (1972).

13. ____, *On the absolute convergence of Fourier series in bounded complete orthonormal systems of functions*, Mat. Sb. **93(135)** (1974), 203–217; English transl. in Math. USSR Sb. **22** (1974).

14. ____, *Existence of a basis in the space of functions analytic in the disk, and some properties of Franklin's system*, Mat. Sb. **95(137)** (1974), 3–18; English transl. in Math. USSR Sb. **24** (1974).

15. ____, *Logarithmic growth of arithmetic means of Lebesgue functions of bounded orthonormal systems*, Dokl. Akad. Nauk SSSR **223** (1975), 16–19; English transl. in Soviet Math. Dokl. **16** (1975).

16. ____, *A Fourier series in a bounded orthonormal system that diverges on a set of positive measure*, Mat. Sb. **98(140)** (1975), 436–449; English transl. in Math. USSR Sb. **27** (1975).

17. ____, *A method of averaging in the theory of orthogonal series, and some problems in the theory of bases*, Trudy Mat. Inst. Steklov. **146** (1978); English transl., Proc. Steklov Inst. Math. **1980**, no. 3(146).

18. D. L. Burkholder and R. F. Gundy, *Extrapolation and interpolation of quasi-linear operators on martingales*, Acta Math. **124** (1970), 249–304.

19. Lennart Carleson, *On covergence and growth of partial sums of Fourier series*, Acta Math. **116** (1966), 135–157.

20. ____, *An explicit unconditional basis in $H^1$*, Bull. Sci. Math. (2) **104** (1980), 405–416.

21. Sun-Yung, A. Chang and Z. Ciesielski, *Spline characterizations of $H^1$*, Studia Math. **75** (1983), 183–192.

22. E. W. Cheney, *Introduction to approximation theory*, McGraw-Hill, New York, 1966.

23. Y. S. Chow, *Convergence theorems of martingales*, Z. Wahrsch. Verw. Gebiete **1** (1962/63), 340–346.

24. Z. Ciesielski, *On Haar functions and on the Schauder basis of the space $C_{\langle 0,1 \rangle}$*, Bull. Acad. Polon. Sci. Sér. Sci. Math. Astr. Phys. **7** (1959), 227–232.

25. ____, *Some properties of Schauder basis of the space $C_{\langle 0,1 \rangle}$*, Bull. Acad. Polon. Sci. Sér. Sci. Math. Astr. Phys. **8** (1960), 141–144.

26. ____, *Properties of the orthonormal Franklin system*. I, Studia Math. **23** (1963), 141–157.

27. ____, *Properties of the orthonormal Franklin system*. II, Studia Math. **27** (1966), 289–323.

28. ____, *A construction of basis in $C^{(1)}(I^2)$*, Studia Math. **33** (1969), 243–247.

29. ___, *Convergence of spline expansions*, Linear Spaces and Approximation (Proc. Conf., Oberwolfach, 1977; P. L. Butger and B. Sz.-Nagy, editors), Birkhäuser, 1978, pp. 433–448.

30. Z. Ciesielski, P. Simon, and P. Sjölin, *Equivalence of Haar and Franklin bases in Lp spaces*, Studia Math. **60** (1977), 195–210.

31. Ronald R. Coifman, *A real variable characterization of $H^p$*, Studia Math. **51** (1974), 269–274.

32. Ronald R. Coifman and Guido Weiss, *Extensions of Hardy spaces and their use in analysis*, Bull. Amer. Math. Soc. **83** (1977), 569–645.

33. H. B Curry and I. J. Schoenberg, *On Pólya frequency functions. IV*, J. Analyse Math. **17** (1966), 71–107.

34. Burgess Davis, *On the integrability of the martingale square function*, Israel J. Math. **8** (1970), 187–190.

35. Mahlon M. Day, *Normed linear spaces*, Springer-Verlag, Berlin, 1958.

36. J. L. Doob, *Stochastic processes*, Wiley, New York, and Chapman & Hall, London, 1953.

37. Raouf Doss, *Elementary proof of the Rudin-Carleson and the F. and M. Riesz theorems*, Proc. Amer. Math. Soc. **82** (1981), 599–602.

38. Vladimir Drobot, *Rearrangements of series of functions*, Trans. Amer. Math. Soc. **142** (1969), 239–248.

39. V. K. Dzyadik, *Introduction to the theory of uniform approximation of functions by polynomials*, "Nauka", Moscow, 1977. (Russian)

40. P. Erdős, *On the convergence of trigonometric series*, J. Math. and Phys. **22** (1943), 37–39.

41. Georg Faber, *Über die Orthogonalfunktionen des Herrn Haar*, Jber. Deutsch. Math.-Verein. **19** (1910), 104–112.

42. Charles Fefferman, *Characterizations of bounded mean oscillation*, Bull. Amer. Math. Soc. **77** (1971), 587–588.

43. C. Fefferman and E. M. Stein, *$H^p$ spaces of several variables*, Acta Math. **129** (1972), 137–193.

44. N. J. Fine, *On the Walsh functions*, Trans. Amer. Math. Soc. **65** (1949), 372–414.

45. V. P. Fonf, *Conditionally convergent series in a uniformly smooth Banach space*, Mat. Zametki **11** (1972), 209–214; English transl. in Math. Notes **11** (1972).

46. Philip Franklin, *A set of continuous orthogonal functions*, Math. Ann. **100** (1928), 522–529.

47. V. F. Gaposhkin, *Lacunary series and independent functions,* Uspekhi Mat. Nauk **21** (1966), no. 6 (132), 3–82; English transl. in Russian Math. Surveys **21** (1966).

48. Adriano M. Garsia, *Existence of almost everywhere convergent rearrangements for Fourier series of L₂ functions,* Ann. of Math. (2) **79** (1964), 623–629.

49. ____, *Topics in almost everywhere convergence,* Markham, Chicago, Ill., 1970.

50. B. I. Golubov, *On Fourier series of continuous functions in the Haar system,* Izv. Akad. Nauk SSSR Ser. Mat. **28** (1964), 1271–1296. (Russian)

51. ____, *Series in the Haar system,* Itogi Nauki i Tekhniki: Mat. Anal. 1970, VINITI, Moscow, 1971, pp. 109–146; English transl. in J. Soviet Math. **1** (1973), no. 6.

52. A. Grothendieck, *Résumé de la théorie métrique des produits tensoriels topologiques,* Bol. Soc. Mat. São Paulo **8** (1953), 1–79 (1956).

53. Richard F. Gundy, *Martingale theory and pointwise convergence of certain orthogonal series,* Trans. Amer. Math. Soc. **124** (1966), 228–248.

54. Alfred Haar, *Zur Theorie der orthogonalen Funktionensysteme.* I, Math. Ann. **69** (1910), 331–371.

55. G. H. Hardy and J. E. Littlewood, *A maximal theorem with function-theoretic applications,* Acta Math. **54** (1930), 81–116.

56. G. H. Hardy, J. E. Littlewood, and G. Pólya, *Inequalities,* Cambridge Univ. Press, 1934.

57. David Hilbert, *Grundzüge einer allgemeinen Theorie der linearen Integralgleichungen,* Teubner, Leipzig, 1912.

58. Charles R. Hobby and John R. Rice, *A moment problem in L₁ approximation,* Proc. Amer. Math. Soc. **16** (1965), 665–670.

59. Kenneth Hoffman, *Banach spaces of analytic functions,* Prentice-Hall, Englewood Cliffs, N.J., 1962.

60. Richard A. Hunt, *On the convergence of Fourier series,* Orthogonal Expansions and Their Continuous Analogues (Proc. Conf., Edwardsville, Ill., 1967; D. T. Haimo, editor), Southern Illinois Univ. Press, Edwardsville, Ill., and Feffer & Simons, London, 1968, pp. 235–255.

61. Satoru Igari, *An extension of the interpolation theorem of Marcinkiewicz.* II, Tôhoku Math. J. (2) **15** (1963), 343–358.

62. F. John and L. Nirenberg, *On functions of bounded mean oscillation,* Comm. Pure Appl. Math. **14** (1961), 415–426.

63. S. Kaczmarz, *Sur la convergence et la sommabilité des développements orthogonaux,* Studia Math. **1** (1929), 87–121.

64. Stefan Kaczmarz and Hugo Steinhaus, *Theorie der Orthogonalreihen*, PWN, Warsaw, 1935.

65. Jean-Pierre Kahane, *Some random series of functions*, Heath, Lexington, Mass., 1968.

66. J.-P. Kahane and Y. Katznelson, *Séries de Fourier des fonctions bornées*, Studies in Pure Math.: To the Memory of Paul Turán (P. Erdős et al., editors), akad. Kiadó, Budapest, and Birkhäuser, Basel, 1983, pp. 395–410.

67. Jean-Pierre Kahane and Raphael Salem, *Ensembles parfaits et séries trigonométriques*, Actualités Sci. Indust., no. 1301, Hermann, Paris, 1963.

68. N. J. Kalton, N. T. Peck, and J. W. Roberts, $L_0$-*valued vector measures are bounded*, Proc. Amer. Math. Soc. **85** (1982), 575–582.

69. S. Karlin, *Bases in Banach spaces*, Duke Math. J. **15** (1948), 971–985.

70. B. S. Kashin, *The stability of unconditional almost everywhere convergence*, Mat. Zametki **14** (1973), 645–654; English transl. in Math. Notes **14** (1973).

71. ____, *On some properties of series of functions and orthogonal series*, Candidate's Dissertation, Steklov Math. Inst. Acad. Sci. USSR, Moscow, 1976. (Russian)

72. ____, *On Weyl's multipliers for almost everywhere convergence of orthogonal series*, Anal. Math. **2** (1976), 249–266.

73. ____, *On a complete orthonormal system*, Mat. Sb. **99(141)** (1976), 356–365; English transl. in Math. USSR Sb. **28** (1976).

74. ____, *On orthogonal systems of convergence*, Dokl. Akad. Nauk SSSR **228** (1976), 285–286; English transl. in Soviet Math. Dokl. **17** (1976).

75. ____, *On the coefficients in the expansion of a certain class of functions in terms of complete systems*, Sibirsk. Mat. Zh. **18** (1977), 122–131; English transl. in Siberian Math. J. **18** (1977).

76. ____, *On some properties of orthogonal systems of convergence*, Trudy Mat. Inst. Steklov. **143** (1977), 68–87; English transl. in Proc. Steklov Inst. Math. **1980**, no. 1 (143).

77. ____, *Remarks on estimating the Lebesgue functions of an orthonormal system*, Mat. Sb. **106(148)** (1978), 380–385; English transl. in Math. USSR Sb. **35** (1979), no. 1.

78. ____, *General orthonormal systems and certain problems of approximation theory* (author's summary of doctoral dissertation), Mat. Zametki **26** (1979), 299–315; English transl. in Math. Notes **26** (1979).

79. ____, *On some properties of the space of trigonometric polynomials with uniform norm*, Trudy Mat. Inst. Steklov. **145** (1980), 111–116; English transl. in Proc. Steklov Inst. Math. **1981**, no. 1 (145).

80. ____, *Some properties of matrices of bounded operators $l_2^n$ to $l_2^m$*, Izv. Akad. Nauk Armyan. SSR Ser. Mat. **15** (1980), 379–394; English transl. in Soviet J. Contemp. Math. Anal. **15** (1980), no. 5.

81. K. S. Kazaryan, *On some questions in the theory of orthogonal series*, Mat. Sb. **119(161)** (1982), 278–294; English transl. in Math. USSR Sb. **47** (1984).

82. A. Khintchine [Khinchin], *Über dyadische Brüche*, Math. Z. **18** (1923), 109–116.

83. A. Khintchine [Khinchin] and A. Kolmogoroff [Kolmogorov], *Über Konwergenz von Reihen, deren Glieder durch den Zufall bestimmt werden*, Mat. Sb. **23** (1924/25), 668–677.

84. S. V. Kislyakov, *p-absolutely summing operators*, Geometry of Linear Spaces and Operator Theory (B. S. Mityagin, editor), Yaroslav. Gos. Univ., Yaroslavl, 1977, pp. 114–174. (Russian)

85. ____, *The Fourier coefficients of the boundary values of functions analytic in the disk and in the bidisk*, Trudy Mat. Inst. Steklov. **155** (1981), 77–94; English transl. in Proc. Steklov Inst. Math. **1983**, no. 1 (155).

86. A. Kolmogoroff [A. N. Kolmogorov], *Sur les fonctions harmoniques conjuguées et les séries de Fourier*, Fund. Math. **7** (1925), 24–29.

87. ____, *Über die Summen durch den Zufall bestimmter unabhängiger Grossen*, Math. Ann. **99** (1928), 309–319.

88. ____, *Über das Gesetz des iterierten Logarithmus*, Math. Ann. **101** (1929), 126–135.

89. A. N. Kolmogorov and S. V. Fomin, *Elements of the theory of functions and of functional analysis*, 3rd ed., "Nauka", Moscow, 1972; English transl. of 1st ed., Vols. 1, 2, Graylock Press, Albany, N.Y., 1957, 1961.

90. A. Kolmogoroff [Kolmogorov] and D. Menchoff [Men'shov], *Sur la convergence des séries de fonctions orthogonales*, Math. Z. **26** (1927), 432–441.

91. János Komlós, *A generalization of a problem of Steinhaus*, Acta Math. Acad. Sci. Hungar. **18** (1967), 217–229.

92. ____, *Every sequence converging to O weakly in $L_2$ contains an unconditional convergence sequence*, Ark. Mat. **12** (1974), 41–49.

93. S. V. Konyagin, *On a problem of Littlewood*, Izv. Akad. Nauk SSSR **45** (1981), 243–265; English transl. in Math. USSR Izv. **18** (1982).

94. V. Ya. Kozlov, *On a local property of a complete orthonormal system of functions*, Mat. Sb. **23(65)** (1948), 441–474. (Russian)

95. ____, *On the distribution of positive and negative values of orthonormal functions that form a complete system*, Mat. Sb. **23(65)** (1948), 475–480. (Russian)

96. A. S. Krantsberg, *On convergence systems in C and bases in L*, Mat. Zametki **26** (1979), 183–200; English transl. in Math. Notes **26** (1979).

97. John Lamperti, *Probability*, Benjamin, New York, 1966.

98. Karel de Leeuw, Yitzhak Katznelson, and Jean-Pierre Kahane, *Sur les coefficients de Fourier des fonctions continues*, C. R. Acad. Sci. Paris Sér. A-B**285** (1977), A1001–A1003.

99. Norman Levinson, *Gap and density theorems*, Amer. Math. Soc., Providence, R.I., 1940.

100. J. E. Littlewood, *On the mean values of power series*, Proc. London Math. Soc. (2) **25** (1926), 328–337.

101. ____, *On bounded bilinear forms in an infinite number of variables*, Quart. J. Math. Oxford Ser. **1** (1930), 164–174.

102. N. N. Luzin, *Integral and trigonometric series*, Lissner i Sovko, Moscow, 1915; 2nd ed., GITTL, Moscow, 1951. (Russian)

103. A. S. Makhmudov, *On Fourier coefficients of continuous functions*. III, Some problems of Functional Analysis and Its Applications, Izdat. Akad. Nauk Azerbaĭdzhan. SSR, Baku, 1965, pp. 103–117. (Russian)

104. Józef Marcinkiewicz, *Sur la convergence des séries orthogonales*, Studia Math. **6** (1936), 39–45.

105. ____, *Collected papers*, PWN, Warsaw, 1964.

106. Michael B. Marcus and Gilles Pisier, *Random Fourier series with applications to harmonic analysis*, Princeton Univ. Press, Princeton, N.J., and Univ. of Tokyo Press, Tokyo, 1981.

107. A. I. Markushevich, *The theory of analytic functions, a short course*, 4th ed., "Nauka", Moscow, 1978; rev. English transl., "Mir", Moscow, 1983.

108. V. A. Matveev, *Series in a Schauder system*, Mat. Zametki **2** (1967), 267–278; English transl. in Math. Notes **2** (1967).

109. Bernard Maurey, *Sur une application de la théorie des opérateurs p-sommants*, C. R. Acad. Sci. Paris Sér. A-B**274** (1982), A1304–A1307.

110. ____, *Théorèmes de factorisation pour les opérateurs linéaires à valeurs dans les espaces $L^p$*, Astérisque, no. 11, Soc. Math. France, Paris, 1974.

111. \_\_\_\_, *Quelques problèmes de factorisation d'opérateurs linéaires*, Proc. Internat. Congr. Math. (Vancouver, 1974), Vol. 2, Canad. Math. Congr., Montréal, 1975, pp. 75–79.

112. \_\_\_\_, *Isomorphismes entre espaces $H_1$*, Acta Math. **145** (1980), 79–120.

113. Bernard Maurey and Gilles Pisier, *Un théorème d'extrapolation et ses conséquences*, C.R. Acad. Sci. Paris Sér. A-B **277** (1973), A39–A42.

114. O. Carruth McGehee, Louis Pigno, and Brent Smith, *Hardy's inequality and the $L^1$ norm of exponential sums*, Ann. of Math. (2) **113** (1981), 613–618.

115. James R. McLaughlin, *Integrated orthonormal series*, Pacific J. Math. **42** (1972), 469–475.

116. D. Menchoff [D. E. Men'shov], *Sur l'unicité du développement trigonométrique*, C. R. Acad. Sci. Paris **163** (1916), 433–436.

117. \_\_\_\_, *Sur les séries de fonctions orthogonales*, Fund. Math. **4** (1923), 82–105.

118. \_\_\_\_, *Sur la convergence des séries de fonctions orthogonales*, C. R. Acad. Sci. Paris **178** (1924), 301–303.

119. \_\_\_\_, *Sur la convergence et la sommation des séries de fonctions orthogonales*, Bull. Soc. Math. France **64** (1936), 147–170.

120. \_\_\_\_, *Sur la sommation des séries de fonctions orthogonales par les méthodes linéaires*, Izv. Akad. Nauk SSSR Ser. Mat. **1** (1937), 203–229. (Russian; French summary)

121. \_\_\_\_, *Sur les séries de fonctions orthogonales bornées dans leur ensemble*, Mat. Sb. **3(45)** (1938), 103–120.

122. V. D. Mil'man, Geometric theory of Banach spaces. II, Uspekhi Mat. Nauk **26** (1971), no. 6 (162), 73–149; English transl. in Russian Math. Surveys **26** (1971).

123. V. D. Mil'man and H. Wolfson, *Minkowski spaces with extremal distance from the Euclidean space*, Israel J. Math. **29** (1978), 113–131.

124. B. S. Mityagin, *Absolute convergence of the series of Fourier coefficients*, Dokl. Akad. Nauk SSSR **157** (1964), 1047–1050; English transl. in Soviet Math. Dokl. **5** (1964).

125. G. M. Mushegyan, *On the coefficients of everywhere convergent series in some rearranged orthonormal systems*, Izv. Akad. Nauk SSSR Ser. Mat. **42** (1978), 807–832; English transl. in Math. USSR Izv. **13** (1979).

126. E. M. Nikishin, *Systems of absolute convergence*, Mat. Sb. **74(116)** (1967), 544–553; English transl. in Math. USSR Sb. **3** (1967).

127. ____, *Resonance theorems and superlinear operators*, Uspekhi Mat. Nauk **25** (1970), no. 6 (156), 129–191; English transl. in Russian Math. Surveys **25** (1970).

128. ____, *Resonance theorems and series of functions*, Doctoral Dissertation, Moscow State Univ., Moscow, 1971; Author's summary published in Mat. Zametki **10** (1971), 583–595; English transl. in Math. Notes **10** (1971).

129. ____, *A resonance theorem and series in eigenfunctions of the Laplacian*, Izv. Akad. Nauk SSSR Ser. Math. **36** (1972), 795–813; English transl. in Math. USSR Izv. **6** (1972).

130. E. M. Nikishin and P. L. Ul'yanov, *On absolute and unconditional convergence*, Uspekhi Mat. Nauk **22** (1967), no. 3 (135), 240–242. (Russian)

131. A. M. Olevskiĭ, *Divergent series in $L^2$ with respect to complete systems*, Dokl. Akad. Nauk SSSR **138** (1961), 545–548; English transl. in Soviet Math. Dokl. **2** (1961).

132. ____, *Divergent Fourier series for continuous functions*, Dokl. Akad. Nauk SSSR **141** (1961), 28–31; English transl. in Soviet Math. Dokl. **2** (1961).

133. ____, *Divergent Fourier series*, Izv. Akad. Nauk SSSR Ser. Mat. **27** (1963), 343–366. (Russian)

134. ____, *Fourier series of continuous functions relative to bounded orthonormal systems*, Izv. Akad. Nauk SSSR Ser. Mat. **30** (1966), 387–432; English transl. in Amer. Math. Soc. Transl. (2) **67** (1968).

135. ____, *On an orthonormal system and its applications*, Mat. Sb. **71(113)** (1966), 297–336; English transl. in Amer. Math. Soc. Transl. (2) **76** (1968).

136. ____, *Fourier series and Lebesgue functions*, Uspekhi Mat. Nauk **22** (1967), no. 3 (135), 237–239. (Russian)

137. ____, *Extension of a sequence of functions to a complete orthonormal system*, Mat. Zametki **6** (1969), 737–747; English transl. in Math. Notes **6** (1969).

138. ____, *Fourier series with respect to general orthogonal systems*, Springer-Verlag, Berlin, 1975.

139. Władysław Orlicz, *Zur Theorie der Orthogonalreihen*, Bull. Internat. Acad. Polon. Sci. Lett. Cl. Sci. Math. Nat. Sér. A Sci. Math. **1927**, 81–115.

140. ____, *Beiträge zur Theorie der Orthogonalentwicklungen. II*, Studia Math. **1** (1929), 241–255.

141. ____, *Beiträge zur Theorie der Orthogonalentwicklungen.* III, Bull. Internat. Acad. Polon. Sci. Lett. Cl. Sci. Math. Nat. Sér. A Sci. Math. **1932**, 229–238.

142. ____, *Über die Divergenz von allgemeinen Orthogonalreihen*, Studia Math. **4** (1933), 27–32.

143. ____, *Über unbedingte Konvergenz in Funktionenräumen.* I, Studia Math. **4** (1933), 33–37.

144. Peter Ørno, *A note on unconditionally converging series in $L_p$*, Proc. Amer. Math. Soc. **59** (1976), 252–254.

145. R. E. A. C. Paley, *A remarkable series of orthogonal functions.* I, II, Proc. London Math. Soc. (2) **34** (1932), 241–264, 265–279.

146. R. E. A. C. Paley and A. Zygmund, *On some series of functions.* I, II, Proc. Cambridge Philos. Soc. **26** (1930), 337–357, 458–474.

147. ____, *On some series of functions.* III, Proc. Cambridge Philos. Soc. **28** (1932), 190–205.

148. T. Parthasarathy and T. E. S. Raghavan, *Some topics in two-person games*, Amer. Elsevier, New York, 1971.

149. D. V. Pecherskiĭ, *On rearrangements of terms in series of functions*, Dokl. Akad. Nauk SSSR **209** (1973), 1285–1287; English transl. in Soviet Math. Dokl. **14** (1973).

150. ____, *A theorem on projections of rearranged series with terms in $L_p$*, Izv. Akad. Nauk SSSR Ser. Mat. **11** (1977), 203–214; English transl. in Math. USSR Izv. **11** (1977).

151. A. Pełczyński, *Projections in certain Banach spaces*, Studia Math. **19** (1960), 209–228.

152. N. B. Pogosyan, *On a property of complete orthonormal systems*, Mat. Zametki **17** (1975), 681–690; English transl. in Math. Notes **17** (1975).

153. ____, *Representation of measurable functions by orthogonal series*, Mat. Sb. **98(140)** (1975), 102–112; English transl. in Math. USSR Sb. **27** (1975).

154. ____, *Representation of measurable functions by bases in $L_p[0, 1]$* ($p \geq 2$), Akad. Nauk Armyan. SSR Dokl. **63** (1976), 205–209. (Russian)

155. I. I. Privalov, *The Cauchy integral*, Izv. Saratov. Univ. Fiz.-Mat. Fak. **2** (1918), no. 1; reprinted, under separate covers, as his dissertation, Saratov, 1919. (Russian)

156. E. S. Quade, *Trigonometric approximation in the mean*, Duke Math. J. **3** (1937), 529–543.

157. Hans Rademacher, *Einige Sätze über Reihen von allgemeinen Orthogonalfunktionen*, Math. Ann. **87** (1922), 111–138.

158. Marcel Riesz, *Sur les fonctions conjuguées*, Math. Z. **27** (1927/28), 218–244.

159. Bengt Rosén, *Limit theorems for sampling from finite populations*, Ark. Mat. **5** (1963/65), 383–424.

160. A. A. Saakyan, *On properties of Fourier coefficients of a composition of functions*, Dokl. Akad. Nauk SSSR **248** (1979), 302–306; English transl. in Soviet Math. Dokl. **20** (1979).

161. ____, *Integral moduli of smoothness and the Fourier coefficients of the composition of functions*, Mat. Sb. **110(152)** (1979), 597–608; English transl. in Math. USSR Sb. **38** (1981).

162. T. N. Saburova, *Some properties of Fourier coefficients with respect to a Faber-Schauder system*, Soobshch. Akad. Nauk Gruzin. SSR **82** (1976), 297–300. (Russian)

163. Julius Schauder, *Zur Theorie stetiger Abbildungen in Funktionalräumen*, Math. Z. **26** (1927), 47–65.

164. ____, *Eine Eigenschaft des Haarschen Orthogonalsystems*, Math. Z. **28** (1928), 317–320.

165. F. Schipp, *Maximal inequalities*, Approximation and Function Spaces (Proc. Internat. Conf., Gdánsk, 1979; Z. Ciesielski, editor), PWN, Warsaw, and North-Holland, Amsterdam, 1981, pp. 629–644.

166. Steven Schonefeld, *Schauder bases in spaces of differentiable functions*, Bull. Amer. Math. Soc. **75** (1969), 586–590.

167. I. Schur, *Über endliche Gruppen und Hermitesche Formen*, Math. Z. **1** (1918), 184–207.

168. Zbigniew Semadeni, *Schauder bases in Banach spaces of continuous functions*, Lecture Notes in Math., vol. 918, Springer-Verlag, Berlin, 1982.

169. A. N. Shiryaev, *Probability*, "Nauka", Moscow, 1980; English transl., Springer-Verlag, Berlin, 1984.

170. Ivan Singer, *Bases in Banach spaces*. I, Springer-Verlag, Berlin, 1970.

171. S. B. Stechkin, *On absolute convergence of Fourier series*, Izv. Akad. Nauk SSSR Ser. Mat. **20** (1956), 385–412. (Russian)

172. Elias M. Stein, *On limits of sequences of operators*, Ann. of Math. (2) **74** (1961), 140–170.

173. ____, *Singular integrals and differentiability properties of functions*, Princeton Univ. Press Princeton, N.J., 1970.

174. S. J. Szarek, *Bases and biorthogonal systems in the spaces C and $L^1$*, Ark. Mat. **17** (1979), 255–271.

175. Béla Sz.-Nagy, *Approximation properties of orthogonal expansions*, Acta Sci. Math. (Szeged) **15** (1953/54), 31–37.

176. A. A. Talalyan, *On convergence of orthogonal series*, Dokl. Akad. Nauk SSSR **110** (1956), 515–516. (Russian)

177. \_\_\_\_, *Representation of measurable functions by series*, Uspekhi Mat. Nauk **15** (1960), no. 5 (95), 77–141; English transl. in Russian Math. Surveys **15** (1960).

178. \_\_\_\_, *Questions of representation and uniqueness in the theory of orthogonal series*, Itogi Nauki i Tekhniki: Mat. Anal. 1970, VINITI, Moscow, 1971, pp. 5–64; English transl. in J. Soviet Math. **1** (1973), no. 6.

179. \_\_\_\_, *On approximation properties of certain incomplete systems*, Mat. Sb. **115(157)** (1981), 499–531; English transl. in Math. USSR Sb. **43** (1982).

180. A. A. Talalyan and F. G. Arutyunyan, *On the convergence of Haar series to +∞*, Mat. Sb. **66(108)** (1965), 240–247; English transl. in Amer. Math. Soc. Transl. (2) **72** (1968).

181. Károly Tandori, *Über die orthogonalen Funktionen. X: Unbedingte Konvergenz*, Acta Sci. Math. (Szeged) **23** (1962), 185-221.

182. \_\_\_\_, *Über die Lebesgueschen Funktionen*, II. Acta Math. Acad. Sci. Hungar. **29** (1977), 177–184.

183. \_\_\_\_, *Einfacher Beweis eines Satzes von B. S. Kašin*, Acta Sci. Math. (Szeged) **39** (1977), 175–178.

184. P. L. Ul'yanov, *Divergent Fourier series of class $L^p$ ($p \geq 2$)*, Dokl. Akad. Nauk SSSR **137** (1961), 786–789; English transl. in Soviet Math. Dokl. **2** (1961).

185. \_\_\_\_, *Divergence of series in the Haar system and in bases*, Dokl. Akad. Nauk SSSR **138** (1961), 556–559; English transl. in Soviet Math. Dokl. **2** (1961).

186. \_\_\_\_, *Divergent Fourier series*, Uspekhi Mat. Nauk **16** (1961), no. 3(99), 61–142; English transl. in Russian Math. Surveys **16** (1961).

187. \_\_\_\_, *Exact Weyl multipliers for unconditional convergence*, Dokl. Akad. Nauk SSSR **141** (1961), 1048–1049; English transl. in Soviet Math. Dokl. **2** (1961).

188. \_\_\_\_, *Weyl multipliers for unconditional convergence*, Mat. Sb. **60(102)** (1963), 39–62. (Russian)

189. \_\_\_\_, *On Haar series*, Mat. Sb. **63(105)** (1964), 356–391. (Russian)

190. \_\_\_\_, *Solved and unsolved problems in the theory of trigonometric and orthogonal series*, Uspekhi Mat. Nauk **19** (1964), no. 1 (115), 3–69; English transl. in Russian Math. Surveys **19** (1964).

191. ____, *Series with respect to a Haar system with monotone coefficients*, Izv. Akad. Nauk SSSR Ser. Mat. **28** (1964), 925–950. (Russian)

192. ____, *Some properties of series with respect to the Haar system*, Mat. Zametki **1** (1967), 17–24; English transl. in Math. Notes **1** (1970).

193. ____, *Representation of functions by series and the classes* $\varphi(L)$, Uspekhi Mat. Nauk **27** (1972), no. 2 (164), 3–52; English transl. in Russian Math. Surveys **27** (1972).

194. N. Ya. Vilenkin, *On a class of complete orthonormal systems*, Izv. Akad. Nauk SSSR Ser. Mat. **11** (1947), 363–400; English transl. in Amer. Math. Soc. Transl. (2) **28** (1963).

195. J. L. Walsh, *A closed set of normal orthogonal functions*, Amer. J. Math. **45** (1923), 5–24.

196. Chinami Watari, *Mean convergence of Walsh Fourier series*, Tôhoku Math. J. (2) **16** (1964), 183–188.

197. Ingemar Wik, *Criteria for absolute convergence of Fourier series of functions of bounded variation*, Trans. Amer. Math. Soc. **163** (1972), 1–24.

198. J. Michael Wilson, *A simple proof of the atomic decomposition for* $H^p(\mathbf{R}^n)$, $0 < p \leq 1$, Studia Math. **74** (1982), 25–33.

199. P. Wojtaszczyk, *The Franklin basis is an unconditional basis in* $H_1$, Ark. Mat. **20** (1982), 293–300.

200. Shigeki Yano, *On a lemma of Marcinkiewicz and its applications to Fourier series*, Tôhoku Math. J. (2) **11** (1959), 191–215.

201. A. Zygmund, *Trigonometric series*. 2nd rev. ed., Vols. 1, 2, Cambridge Univ. Press, 1959.

202. *Mathematical encyclopedia* (I. M. Vinogradov et al., editors). Vols. 1–5, Izdat. "Sovet. Èntsiklopediya", Moscow, 1977–1985; English transl., *Encyclopedia of mathematics*. Vols. 1–?, Reidel, 1988 ff.

# Index(1)